WHAT YOUR COLLEAGUES ARE SAYING . . .

"This text provides a clear definition of what fluency really is and provides strategies to deepen students' number sense and make them fluent mathematicians."

Meghan Schofield
Third-Grade Teacher

"I wish I'd had this book when I was in the classroom 10 years ago. The authors clearly lay out a pathway to procedural fluency, including intentional activities to understand and practice specific strategies, while also advocating for space for students to make decisions and feel empowered as mathematical thinkers and doers."

Kristine M. Gettelman
Instructional Designer
CenterPoint Education

"This book is a must-read for teachers wanting to learn more about focused math fluency instruction. The steps are clear and easy to follow. You will have all the steps to help your students become fluent math thinkers."

Carly Morales
Instructional Coach
District 93

"Are you ready to help your students connect their Number Talks and number routines to the real world? *Figuring Out Fluency* will give you the routines, games, protocols, and resources you need to help your students build their fluency in number sense (considering reasonableness, strategy selection, flexibility, and more). Our students deserve the opportunity to build a positive and confident mathematics identity. We can help support them to build this identity by providing them with access to a variety of strategies and the confidence to know when to use them."

Sarah Gat
Instructional Coach
Upper Grand District School Board

"*Figuring Out Fluency* goes beyond other resources currently on the market. It not only provides a robust collection of strategies and routines for developing fluency but also pays critical attention to the ways teachers can empower each and every student as a mathematical thinker who can make strategic decisions about their computation approaches. If you are looking for instruction and assessment approaches for fluency that move beyond getting the right answer, this is the resource for you."

Nicole Rigelman
Professor of Mathematics Education
Portland State University

"As principal of a Title I school and former mathematics specialist, I appreciate this resource for what it is; a true teacher's companion! The authors provide explicit strategy instruction in whole number multiplication and division to foster flexibility, efficiency, and true fluency in ALL students."

Allie S. Watkins
Principal
Frederick County Public Schools

"This is the book classroom teachers have been waiting for! It provides extensive support as students gain confidence with multiplying and dividing whole numbers, and it does so with a deep esteem for educators and learners. The authors, highly respected in the field of mathematics education, provide careful explanations of effective strategies that will help students build their number sense and their computational fluency. There are routines for supporting instruction in the classroom, games and centers for practicing the strategies, and prompts to encourage sense making and extend learning. Kudos to all three authors for writing such an important, no-nonsense book. I cannot wait to share it with my colleagues!"

Elisa Waingort
Grades 4 & 5 Spanish Bilingual Teacher
Calgary Board of Education

"For years research has indicated that fluency is much more than speed, yet timed assessments and traditional instruction persist for teachers without a clear vision or tools to change their practices. This series provides teachers with the explicit examples, resources, and activities needed to bring that research to life for their students and will quickly become a well-worn guidebook for every fluency-focused classroom. This is the toolkit teachers have been yearning for in their journey toward fluency with their students."

Gina Kilday
Elementary Math Interventionist and MTSS Coordinator

"Fluency isn't a dry landscape of disconnected facts—it is a rich soil for developing and connecting diverse perspectives and ideas. This book series equips you with a deep understanding of fluency and a variety of activities to engage students in co-constructing ideas about multiplication and division that will last a lifetime."

Berkeley Everett
Math Coach and Facilitator for UCLA Mathematics Project
Math Consultant for DragonBox

"This is the book we (educators) have been waiting for! *Figuring Out Fluency – Multiplication and Division With Whole Numbers* provides classroom teachers, special ed teachers, and instructional coaches with the support they need to develop true fluency for every child. This book is packed with great activities; quality practice that is not a worksheet; examples, routines, games, and centers; and a companion website with resources ready to use! Any question you may have—about how and when to teach a specific strategy, how to engage your students in meaningful practice, or assessing your students' fluency—will be answered in this book! Any educator that is serious about understanding how basic fact strategies grow into general reasoning strategies and how to advance their students' fluency will want this book! True fluency is a way of thinking, rather than a way of doing!"

Melisa Jean Hancock
Mathematics Consultant

"*Figuring Out Fluency – Multiplication and Division With Whole Numbers* is a must-have for all educators who teach multi-digit multiplication and division. With a focus on equity embedded throughout, this practical, ready-to-use resource provides everything needed to teach students to become strategic thinkers while giving *all* children access to reasoning."

Nichole DeMotte
Atkinson Academy
K–5 Mathematics Coach

"This book—indeed this *series*—is a must-read for elementary and middle level teachers, coaches, and administrators. Within this resource you will find a synthesis of important research organized to help readers develop a clear and common understanding of fluency paired with a large collection of teaching activities that provide concrete ways to support students' fluency development. *Figuring Out Fluency* provides a much-needed roadmap for teachers looking to increase computational proficiency with multiplication and division."

Delise Andrews
3–5 Mathematics Coordinator
Lincoln Public Schools

"The authors John J. SanGiovanni, Jennifer M. Bay-Williams, and Rosalba Serrano shine a bright light on how math fluency is *the* equity issue in mathematics education. How refreshing to have a book that equips math educators with the research and strategies to make a difference for *all* students! Let's implement these strategy modules in this book and help kids figure out fluency once and for all!"

Kelly DeLong
Executive Director for the Kentucky Center for Mathematics
Northern Kentucky University

"In *Figuring Out Fluency – Multiplication and Division With Whole Numbers*, John J. SanGiovanni, Jennifer M. Bay-Williams, and Rosalba Serrano have provided readers with a thorough education that guides them through the entire fluency journey. Grounded in research and packed with illustrative examples, this book is a must-have that delivers practical strategies, tools, resources, and recommendations that will immediately enhance your practice. In a sea of books on fluency, this one stands out. In a word . . . wow!"

Alison J. Mello
Assistant Superintendent Foxborough Public Schools
Author
Math Consultant

"John J. SanGiovanni, Jennifer M. Bay-Williams, and Rosalba Serrano hit the mark with *Figuring Out Fluency – Multiplication and Division With Whole Numbers*. The activities are fun and engaging and encourage students to think more flexibly and fluently as they work with multiplication and division strategies. Bravo!"

Joshua Barnes
Third-Grade Teacher

"The number one topic teachers grapple with every year is fluency. This companion book is an all-encompassing resource that takes the guess work out of multiplication and division fluency instruction and practice! With a focus on reasoning, understanding, and reasonableness, this is a practical, easy-to-follow guide for teachers. The strategy modules contain teaching prompts, routines, games, center activities, and a very critical component missing in most resources—strategy briefs for families."

Marissa Walsh
Elementary Math Instructional Coach
Blue Springs School District

Figuring Out Fluency—Multiplication and Division With Whole Numbers: A Classroom Companion
The Book at a Glance

Building off of *Figuring Out Fluency*, this classroom companion dives deep into five of the Seven Significant Strategies, plus the standard algorithms, that relate to procedural fluency when multiplying and dividing whole numbers, beyond basic facts.

FIGURE 13 ● Reasoning Strategies for Multiplying and Dividing Whole Numbers

REASONING STRATEGIES	RELEVANT OPERATIONS
1. Break Apart to Multiply (Module 1)	Multiplication
2. Halve and Double (Module 2)	Multiplication
3. Compensation (Module 3)	Multiplication
4. Partial Products and Quotients (Modules 4 and 6)	Multiplication and Division
5. Think Multiplication (Module 5)	Division

Strategy overviews and family briefs communicate how each strategy helps students develop flexibility, efficiency, accuracy, automaticity, and reasonableness.

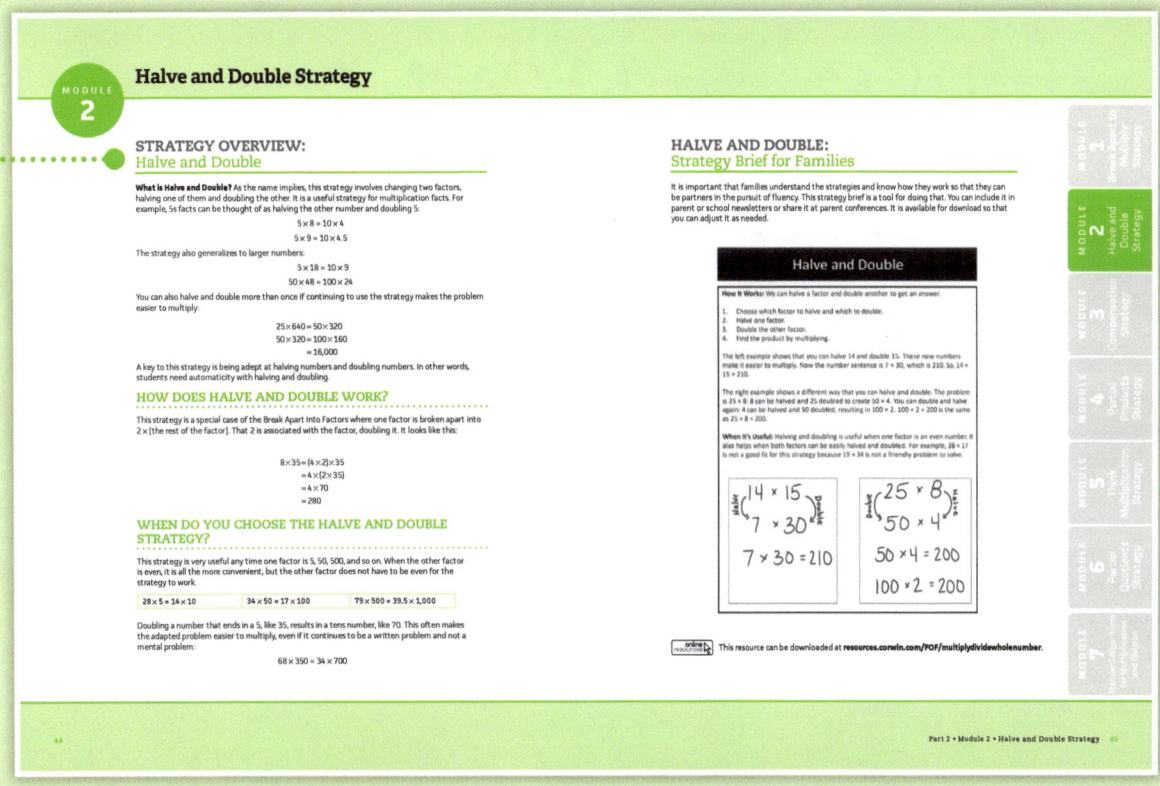

Each strategy module starts with teaching activities that help you explicitly teach the strategy.

TEACHING ACTIVITIES for Halve and Double

Before students are able to choose strategies, a key to fluency, they first must be able to understand and use relevant strategies. These activities focus specifically on Halve and Double. Halve and Double is a strategy that works well in certain cases. While students may employ other methods, which is appropriate, they also must learn this strategy to develop their flexibility and efficiency. Students must understand that halving a factor while doubling the other yields the same result. You can develop this concept by working with a variety of representations and connecting them to equations. In this section, we focus on instructional activities for helping students develop understanding of the Halve and Double strategy and for determining when it is appropriate. Fluency with this strategy will build as students develop skill with halving or doubling numbers in general.

ACTIVITY 2.1
HALVING AND REARRANGING RECTANGLES

In this activity, you pose rectangles to students for them to trace on grid paper and cut out. Rectangle dimensions to use could include 15×14, 28×5, 16×6, 25×22, and 16×16. Students record the equation for the area of one of their rectangles. Then, they cut the rectangle in half and rearrange it to make one, longer rectangle and record the equation for it. In the following figure, the original rectangle was 15×14. The student cut it in half, creating two rectangles that were each 15×7. Then, she aligned those two smaller rectangles, creating a rectangle that was 30×7. She recorded the equations for each rectangle's area.

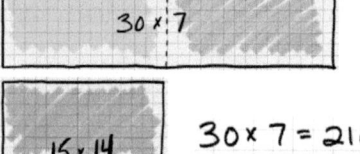

$$30 \times 7 = 210$$
$$15 \times 14 = 210$$

Students repeat halving and rearranging for the other rectangles that you pose. As students work, you want them to think about how the factors and the area of the corresponding rectangles are similar and different. After students complete the activity, a group discussion should be had about how one factor is halved and how the other is doubled yet the product (area) remains the same. Ask students to determine if this idea always works and then have them find examples to prove their case. It's possible that students will overlook that one factor is always even. If so, be sure to insert a rectangle, like 13×15 or 17×21, for them to explore.

Even though this is an introduction to the concept of halving and doubling, it isn't too soon to discuss when this technique might be a good idea. Notice that 15×14 and 28×5 from the suggested set create expressions that yield multiples of 10 (30×7 and 14×10, respectively). Because of this, the Halve and Double strategy is a good choice for these expressions.

Each strategy shares worked examples for you to work through with your students as they develop their procedural fluency.

WORKED EXAMPLES

Worked examples are problems that have been solved. Correctly worked examples can help students make sense of a strategy and incorrectly worked examples attend to common errors.

Halve and Double is useful in special cases, so a major challenge is to just remember it *is* a strategy and be on the lookout for when it will work. Other challenges include the following:

1. The student makes an error in halving an odd number.
   - 65×5: changes to 38.5×10.
2. The student has trouble choosing which number to halve and which to double.
   - 450×12: changes to 225×24, rather than 900×6.

A key idea is to look for a number that, when doubled, is more manageable to multiply. In addition, noticing a second opportunity to use Halve and Double can continue to make the problem easier to work. Worked examples can attend to these ideas. The prompts from Activity 2.5 can be used for collecting worked examples. Throughout the module are various worked examples that you can use as fictional worked examples. A sampling of additional ideas is provided in the table.

SAMPLE WORKED EXAMPLES FOR HALVE AND DOUBLE	
Correctly Worked Example (make sense of the strategy) What did _____ do? Why does it work? Is this a good method for this problem?	Trevor's work for 250×28: 250×28 500×14 $1,000 \times 7 = 7,000$
Partially Worked Example (implement the strategy accurately) Why did _____ start the problem this way? What does _____ need to do to finish the problem?	Neesha's start for 15×260: $30 \times 130 =$ Dante's start for 25×68: $50 \times 34 =$
Incorrectly Worked Example (highlight common errors) What did _____ do? What mistake does _____ make? How can this mistake be fixed?	Greta's work for 50×45: $100 \times 23.5 = 2,350$

ACTIVITY 2.6

Name: *"A String of Halves"* **Type:** *Routine*

About the Routine: *Making use of the Halve and Double strategy relies on students' ability to double and find halves. Often, students do well with doubling but finding a half proves more challenging. "A String of Halves" aims to help students improve their skill with halving. This routine uses a set of numbers to help students see how numbers are halved.*

Materials: Prepare a set of numbers that are intentionally related.

Directions: 1. Post a set of three or more related numbers as shown.

2. Ask students to mentally find the halves of as many numbers as possible.

3. If students are unable to find all of the halves, have them talk with partners about how they can use the relationships between numbers and the known halves to find the unknown half. If students find all of the halves, have them discuss with partners how the numbers and their halves are related. To extend this situation (when all halves are known), have students generate a new number and its half that is related to the set.

4.

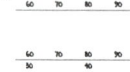

These two examples of "A String of Halves" were used in a fourth-grade classroom. The teacher started with Example A, recording 40, 16, and 56 because she wanted to develop an idea about how to halve 56. Students were asked to think of the half for each of the numbers. They could halve 40 and 16. At that point, she asked students to talk with partners about how they could figure out the half of 56 by knowing the halves of 40 and 16. The teacher then used Example B in a similar way. She posed the three related numbers, and students found the halves. She then asked students to look for patterns and relationships between the numbers and the halves to prove what half of 30 is.

Halving numbers like 76 or even 90 can be challenging for students because there is an element of regrouping within the half. This routine can be leveraged for developing skill with those halves by making use of relationships and patterns. In the following figure, four multiples of 10 (60, 70, 80, and 90) are posed. Students are tasked with halving each, but the conversation focuses first on 60 and 80 before students work to find half of 70 and 90. During these discussions, focus their attention on half of 10 being added to or taken from half of 60 or 80 and why. Of course, another strategy is to think of the half (35) between those halves (30 and 40).

C

	60	70	80	90
	30		40	

Routines, Games, and Centers for each strategy offer extensive opportunity for student practice.

ACTIVITY 2.10

Name: *Halve and Double Flips* **Type:** *Game*

About the Game: *Halve and Double Flips* practices identifying multiplication problems that can be solved with the Halve and Double strategy. Players place chips on corresponding spaces and can flip opponents' pieces that lie directly between two of their pieces (similar to *Othello*).

Materials: *Halve and Double Flips* game cards and game board, multiplication expression cards, and two-colored counters for game pieces

Directions: 1. Players take turns pulling a multiplication expression card.

2. Players consider if the expression can be solved more efficiently using the Halve and Double strategy.

3. If the expression can be solved more efficiently with the strategy, the players put their counter on any space labeled "WORKS" (i.e., the Halve and Double strategy works for this expression).

4. If the expression cannot be solved more efficiently with the strategy, the player puts their counter on any space labeled "DOESN'T WORK" (i.e., the Halve and Double strategy doesn't make the problem easier to solve).

5. When placing a game piece, any opponent pieces between a previously placed piece and the current piece are flipped (similar to *Othello*). Note that unlike *Othello*, a player does not have to place a piece at the end of a row of pieces.

6. The game ends when all of the spaces are filled.

7. The player with the most chips on the board wins the game.

RESOURCE(S) FOR THIS ACTIVITY

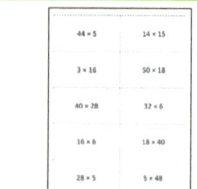

Game cards and this game board can be downloaded at **resources.corwin.com/FOF/multiplydividewholenumber.**

Download the resources you need for each activity at this book's companion website.

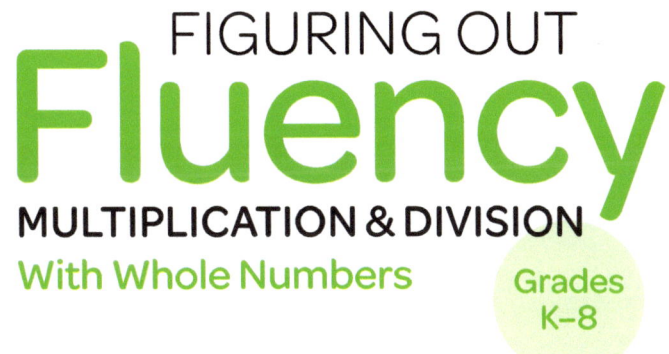

FIGURING OUT
Fluency
MULTIPLICATION & DIVISION
With Whole Numbers

Grades
K–8

FIGURING OUT
Fluency
MULTIPLICATION & DIVISION
With Whole Numbers

Grades K–8

A Classroom Companion

John J. SanGiovanni

Jennifer M. Bay-Williams

Rosalba Serrano

CORWIN **Mathematics**

For information:

Corwin
A SAGE Company
2455 Teller Road
Thousand Oaks, California 91320
(800) 233–9936
www.corwin.com

SAGE Publications Ltd.
1 Oliver's Yard
55 City Road
London, EC1Y 1SP
United Kingdom

SAGE Publications India Pvt. Ltd.
B 1/I 1 Mohan Cooperative
Industrial Area
Mathura Road, New Delhi 110 044
India

SAGE Publications
Asia-Pacific Pte. Ltd.
18 Cross Street #10–10/11/12
China Square Central
Singapore 048423

President: Mike Soulesl
Associate Vice President and Editorial
   Director: Monica Eckman
Publisher: Erin Null
Content Development Editor:
   Jessica Vidal
Senior Editorial Assistant:
   Caroline Timmings
Production Editor: Tori Mirsadjadi
Copy Editor: Christina West
Typesetter: Integra
Proofreader: Susan Schon
Indexer: Integra
Cover Designer: Rose Storey
Marketing Manager:
   Margaret O'Connor

Library of Congress Cataloging-in-Publication Data

Names: SanGiovanni, John, author. | Bay-Williams, Jennifer M, author. | Serrano, Rosalba, author.
Title: Figuring out fluency - multiplication and division with whole numbers : a classroom companion / John J. SanGiovanni, Jennifer M. Bay-Williams, Rosalba Serrano.
Description: Thousand Oaks, California : Corwin, [2022] | Series: Corwin mathematics series | Includes bibliographical references and index.
Identifiers: LCCN 2021019418 (print) | LCCN 2021019419 (ebook) | ISBN 9781071825211 (paperback) | ISBN 9781071851555 (epub) | ISBN 9781071851579 (epub)
Subjects: LCSH: Mathematical fluency. | Multiplication--Study and teaching (Elementary) | Division--Study and teaching (Elementary)
Classification: LCC QA135.6 .S2578 2022 (print) | LCC QA135.6 (ebook) | DDC 372.7/044--dc23
LC record available at https://lccn.loc.gov/2021019418
LC ebook record available at https://lccn.loc.gov/2021019419

This book is printed on acid-free paper.

25 26 27   12 11 10 9 8

Contents

Visit the companion website at
**resources.corwin.com/FOF/
multiplydividewholenumber** for
downloadable resources.

Preface

Fluency is an equity issue. In written documents and in our daily work, we (mathematics teachers and leaders) communicate that every student must be *fluent* with whole number multiplication and division, for example. But we haven't even come close to accomplishing this for each and every student. The most recent National Assessment of Educational Progress (NAEP) data, for example, finds that about two-fifths (41%) of the nation's Grade 4 students are at or above proficient and about one-third (34%) of our nation's Grade 8 students at or above proficient (NCES, 2019). We can and must do better! One major reason we haven't been able to develop fluent students is that there are misunderstandings about what fluency really means.

FIGURING OUT FLUENCY

In order to ensure every student develops fluency, we first must:

- Understand what procedural fluency is (and what it isn't),

- Respect fluency, and

- Plan to explicitly teach and assess reasoning strategies.

If you have read our anchor book *Figuring Out Fluency for Mathematics Teaching and Learning*—which we recommend in order to get the most out of *this* classroom companion—you'll remember an in-depth discussion of these topics.

WHAT PROCEDURAL FLUENCY IS AND ISN'T

Like fluency with language, wherein you decide how you want to communicate an idea, fluency in mathematics involves decision-making as you decide how to solve a problem. In our anchor book, we propose the following visual as a way to illustrate the full meaning of fluency.

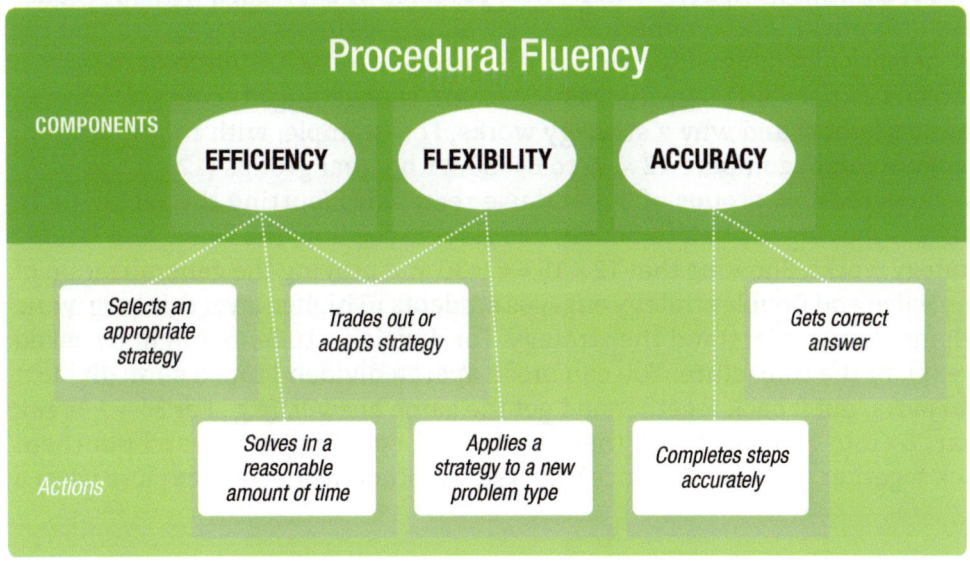

Part 1 of this book explains the elements of this procedural fluency graphic. To cut to the chase, procedural fluency is much more than knowing facts and standard algorithms. Fluency involves higher-level thinking, wherein a person analyzes a problem, considers options for how to solve it, selects an efficient strategy, and accurately enacts that strategy (trading it out for another if it doesn't go well). Decision-making is key and that means you need to have good options to choose from. This book provides instructional and practice activities so that students learn different options (Part 2) and then provides practice activities to help students learn to choose options (Part 3).

RESPECT FLUENCY

We are strong advocates for conceptual understanding. We all must be. But there is not a choice here. Fluency relies on conceptual understanding, and conceptual understanding alone cannot help students fluently navigate computational situations. They go together and must be connected. Instructional activities throughout Part 2 provide opportunities for students to discuss, critique, and justify their thinking, connecting their conceptual understanding to their procedural knowledge and vice versa.

EXPLICITLY TEACH AND ASSESS REASONING STRATEGIES

If every student is to be fluent in whole number multiplication and division, then every student needs access to the significant strategies for these operations. And there must be opportunities for students to learn how to select the best strategy for a particular problem. For example, students may learn that the Compensation strategy works well when one of the factors is close to, but less than, a ten or hundred (e.g., 49×8 or 398×7). To accomplish this, all three fluency components must have equitable attention in instruction and assessment. This is a major shift from traditional teaching and assessing, which privileges accuracy over the other two components and the standard algorithm over reasoning strategies.

Let's unpack the phrase *explicit strategy instruction*. According to the *Merriam-Webster Dictionary*, explicit means " fully revealed or expressed without vagueness" ("Explicit," 2021). In mathematics teaching, being explicit means making mathematical relationships visible. A strategy is a flexible method to solve a problem. Explicit strategy instruction, then, is engaging students in ways to clearly see how and why a strategy works. For example, with multiplication, students might compare 12×15 to 6×30 with equal groups (12 groups of 15 as compared to 6 groups of 30) and use rectangles (cutting the rectangle in half and moving the half to double one side) to see how the Halve and Double strategy works. Showing that $12 \times 15 = 6 \times 30$ and proving the generalization of this Halve and Double strategy engages students in higher-level thinking while helping them understand the strategy. For division, students might be asked to explore the conjecture: You can break apart a dividend into a sum, divide it into parts, put it back together and get the same answer. (e.g., For $84 \div 7$, break apart 84 into $70 + 14$. Divide these parts by 7, equaling $10 + 2$, and put them back together to equal 12.) Once understood, students need to explore *when* a strategy is a good option. Learning how to use and how to choose strategies *empowers* students to be able to decide how they want to solve a problem, developing a positive mathematics identity and a sense of agency.

USING THIS BOOK

This book is a classroom companion to *Figuring Out Fluency for Mathematics Teaching and Learning*. In that anchor book, we lay out what fluency is, identify the fallacies that stand in the way of a true focus on fluency, and elaborate on necessary foundations for fluency. We also propose the following:

- Seven Significant Strategies across the operations, five of which apply to whole number multiplication and division

- Eight "automaticies" *beyond* automaticity with basic facts, five of which are relevant to whole number multiplication and division

- Five ways to engage students in meaningful practice

- Four assessment options that can replace (or at least complement) tests and that focus on real fluency

- Many ways to engage families in supporting their child's fluency

In Part 1 of this book, we revisit some of these ideas in order to connect specifically to whole number multiplication and division. This section is not a substitute for the anchor book, but rather a brief revisiting of central ideas that serve as reminders of what was fully illustrated, explained, and justified in *Figuring Out Fluency in Mathematics Teaching and Learning*. Hopefully, you have had the chance to read and engage with that content with colleagues *first*, and then Part 1 will help you think about those ideas as they apply to whole number multiplication and division. Finally, Part 1 includes suggestions for how to use the strategy modules.

Part 2 is focused on explicit instruction of each significant strategy for whole number multiplication and division. Each module includes the following:

- An overview for your reference and to share with students and colleagues

- A strategy brief for families

- A series of instructional activities, with the final one offering a series of questions to promote discourse about the strategy

- A series of practice activities, including worked examples, routines, games, and center activities that engage students in meaningful and ongoing practice to develop proficiency with that strategy

As you are teaching and find your students are ready to learn a particular strategy, pull this book off the shelf, go to the related module, and access the activities and ready-to-use resources. While the modules are sequenced in a developmental order overall, the order and focus on each strategy may vary depending on your grade and your students' experiences. Additionally, teaching within or across modules does not happen all at once; rather, activities can be woven into your instruction regularly, over time.

Part 3 is about becoming truly fluent—developing flexibility and efficiency with multiplication and division of whole numbers. Filled with more routines, games, and centers, the focus here is on students *choosing* to use the strategies that make sense to them in a given situation. Part 3 also provides assessment tools to monitor students' fluency. As you are teaching and find your students are needing opportunities to choose from among the strategies they are learning, pull this book off the shelf and select an activity from Part 3.

In the Appendix, you will find lists of all the activities in order to help you easily locate what you are looking for by strategy or by type of resource.

This book can be used to complement or supplement any published mathematics program or district-created program. As we noted earlier, elementary mathematics has tended to fall short in its attention to efficiency and flexibility (and the related Fluency Actions illustrated in the earlier graphic). This book provides a large collection of activities to address these neglected components of fluency. Note that this book is part of a series that explores other operations and other numbers. You may also be interested in *Figuring Out Fluency for Whole Number Addition and Subtraction* as well as the classroom companion books for decimal and fraction operations.

WHO IS THIS BOOK FOR?

With 120 activities and a companion website with resources ready to download, this book is designed to support classroom teachers as they advance their students' fluency with whole number multiplication and division. Special education teachers will find the explicit strategy instruction, as well as the additional practice, useful in supporting their students. Mathematics coaches and specialists can use this book for professional learning and to provide instructional resources to the classroom teachers they support. Mathematics supervisors and curriculum leads can use this book to help them assess fluency aspects of their mathematics curriculum and fill potential gaps in resources and understanding. Teacher preparation programs can use this book to galvanize preservice teachers' understanding of fluency and provide teacher candidates with a wealth of classroom-ready resources to use during internships and as they begin their career.

Acknowledgments

Just as there are many components to fluency, there are certainly many components to having a book like this come to fruition. The first component is the researchers and advocates who have defined procedural fluency and effective practices that support it. Research on student learning is hard work, as is defining effective teaching practices, and so we want to begin by acknowledging this work. We have learned from these scholars, and we ground our ideas in their findings. It is on their shoulders that we stand. Second are the teachers and their students who have taken up "real" fluency practices and shared their experiences with us. We would not have taken on this book had we not seen firsthand how a focus on procedural fluency in classrooms truly transforms students' learning and shapes their mathematics identities. It is truly inspiring! Additionally, the testimonies from many teachers about their own learning experiences as students and as teachers helped crystalize for us the facts and fallacies in this book. A third component to bringing this book to fruition was the family support to allow us to actually do the work. We are all grateful to our family members—expressed in our personal statements that follow—who supported us 24/7 as we wrote during a pandemic.

From Jennifer: I am forever grateful to my husband, Mitch, who is supportive and helpful in every way. I also thank my children, MacKenna and Nicolas, who often offer reactions and also endure a lot of talk about mathematics. I also want to express my deep gratitude to my extended family—a mother who served on the school board for 13 years and helped me make it to a second year of teaching and a father who was a statistician and leader and helped me realize I could do "uncomfortable" things. My siblings—an accountant, a high school math teacher, and a university statistician—and their children have all supported my work on this book and the others in this series.

From John: I want to thank my family—especially my wife—who, as always, endure and support the ups and downs of taking on a new project. Thank you to Jenny and Rosalba for being exceptional partners. And thank you for dealing with my random thoughts, tangent conversations, and fantastic humor. As always, a heartfelt thank you to certain math friends and mentors for opportunities, faith in me, and support over the years. And thank you to my own math teachers who let me do math "my way," even if it wasn't "the way" back then.

From Rosalba: I want to thank my three boys, Daniel, Dylan, and Declan. All of the long days and hard work I do is to make the world a better place for you. Thank you for being so patient with your Madre and listening to countless math conversations. To John and Jenny, I cannot express enough thanks. Thank you for your advice, support, and encouragement throughout this process. I am forever grateful for this learning experience and the opportunity you have given me.

A fourth component is vision and writing support. We are so grateful to Corwin for recognizing the importance of defining and implementing procedural fluency in the mathematics classroom. Our editor and publisher, Erin Null, has gone above and beyond as a partner in the work, ensuring that our ideas are as well stated and useful as possible. The entire editing team at Corwin has been creative, thorough, helpful, and supportive.

As with fluency, no component is more important than another, and without any component, there is no book, so to the researchers, teachers, family, and editing team, thank you. We are so grateful.

PUBLISHER'S ACKNOWLEDGMENTS

Corwin gratefully acknowledges the contributions of the following reviewers:

Becky Evans
Grades 3–5 Math Teacher Leader
Lincoln Public Schools

Jamie Fraser
Math Consultant
Developer and Educational Distributer
Bound2Learn

Sarah Gat
Instructional Coach
Upper Grand District School Board

Kristine M. Gettelman
Instructional Designer
CenterPoint Education

Christina Hawley
Grade 1 and Grade 2 Teacher

Cathy Martin
Associate Chief of Academics
Denver Public Schools

Carly Morales
Instructional Coach
District 93

Margie Pearse
Math Coach

Nicole Rigelman
Professor of Mathematics Education
Portland State University

Meghan Schofield
Third-Grade Teacher

About the Authors

John J. SanGiovanni is a mathematics supervisor in Howard County, Maryland. There, he leads mathematics curriculum development, digital learning, assessment, and professional development. John is an adjunct professor and coordinator of the Elementary Mathematics Instructional Leadership graduate program at McDaniel College. He is an author and national mathematics curriculum and professional learning consultant. John is a frequent speaker at national conferences and institutes. He is active in state and national professional organizations, recently serving on the board of directors for the National Council of Teachers of Mathematics (NCTM) and currently as the president of the Maryland Council of Supervisors of Mathematics.

Jennifer M. Bay-Williams is a professor of mathematics education at the University of Louisville, where she teaches preservice teachers, emerging elementary mathematics specialists, and doctoral students in mathematics education. She has authored numerous books as well as many journal articles, many of which focus on procedural fluency (and other aspects of effective mathematics teaching and learning). Jennifer is a frequent presenter at national and state conferences and works with schools and districts around the world. Her national leadership includes having served as a member of the NCTM Board of Directors, on the TODOS: Mathematics for All Board of Directors, and as president and secretary of the Association of Mathematics Teacher Educators (AMTE).

Rosalba Serrano is an elementary mathematics consultant in New York and is the founder of Zenned Math, where she provides online professional development and coaching for elementary mathematics teachers. Rosalba has used her experience as a classroom teacher and mathematics coach to support teachers in deepening their understanding of mathematics and their use of effective teaching practices. A frequent speaker at both regional and national conferences, Rosalba is also active in mathematics organizations, such as the NCTM, where she contributes to multiple committees. She has also worked as a professional development facilitator for various mathematics organizations and consults as a math editor for a number of publishing companies.

PART 1

FIGURING OUT FLUENCY

Key Ideas

WHAT IS FLUENCY WITH WHOLE NUMBER MULTIPLICATION AND DIVISION?

To set the stage for figuring out fluency for whole number multiplication and division, take a moment to do some math. Find the solution to each of these.

35×19	$2{,}458 \div 8$	$756 \div 7$
320×5	$240 \div 3$	24×15

How did you find the products and quotients? Did you use the same approach or strategy for each? Did you move between different strategies? Did you change out a strategy based on the numbers within the problem? Did you start with one strategy and shift to another? You likely said "yes" to all of these questions because you fluently multiply and divide whole numbers. Yet another reader can answer "yes" to each of these questions as well but solve each problem differently. This is true because fluency is a way of thinking rather than a way of doing. Thinking is unique to each individual. Thinking is grounded in parameters but its execution is left to the understanding, preference, and creativity of each thinker.

Of course, there are strategies used most frequently for certain problems (based on the numbers in the problem), but even in those examples, efficient alternatives are likely. For example, 320×5 may seem like a problem that "fits" Partial Products (Module 4). Find the product of 300×5 (1,500) and 20×5 (100) and add them together (1,600). But a Halve and Double approach (Module 2) may "jump out" to another person who sees 320×5 as 160×10. Which method is closer to your way of thinking? Or did you think about it differently?

Real fluency is the ability to select efficient strategies; to adapt, modify, or change out strategies; and to find solutions with accuracy. Real fluency is not the act of replicating someone else's steps or procedures for doing mathematics. It is the act of thinking, reasoning, and doing mathematics on one's own. Before fluency can be taught well, you must understand what fluency is and why it matters.

Procedural fluency is an umbrella term that includes basic fact fluency and computational fluency (see Figure 1). *Basic fact fluency* attends to fluently adding,

FIGURE 1 ● The Relationship of Different Fluency Terms in Mathematics

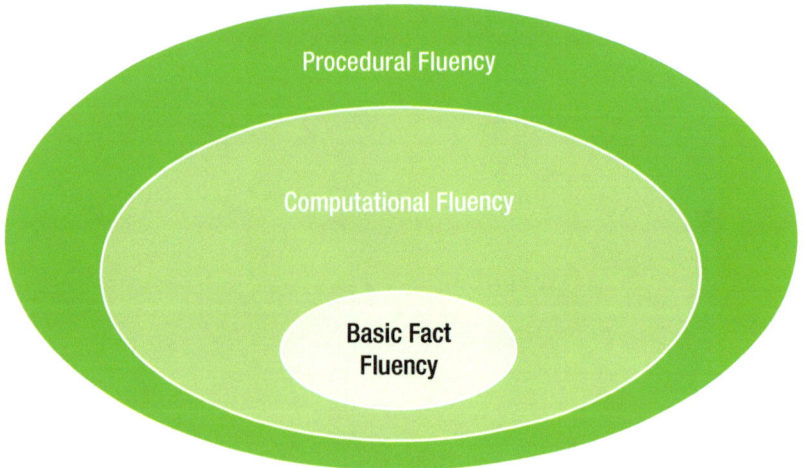

Procedural Fluency

Computational Fluency

Basic Fact Fluency

subtracting, multiplying, and dividing single-digit numbers. *Computational fluency* refers to the fluency in four operations across number types (whole numbers, fractions, etc.), regardless of the magnitude of the number. Procedural fluency encompasses both basic fact fluency and computational fluency, plus other procedures like finding equivalent fractions.

Beyond being an umbrella term that encompasses basic fact and computational fluency, procedural fluency is well defined as solving procedures efficiently, flexibly, and accurately (Kilpatrick, Swafford, & Findell, 2001; National Council of Teachers of Mathematics, 2014). These three **components** are defined as follows:

> *Efficiency:* solving a procedure in a reasonable amount of time by selecting an appropriate strategy and readily implementing that strategy
>
> *Flexibility:* knowing multiple procedures and can apply or adapt strategies to solve procedural problems (Baroody & Dowker, 2003; Star, 2005)
>
> *Accuracy:* correctly solving a procedure

Strategies are not the same as algorithms. Strategies are general methods that are flexible in design, compared to algorithms that are established steps implemented the same way across problems.

To focus on fluency, we need specific observable actions that we can look for in what students are doing in order to ensure they are developing fluency. We have identified six such actions. The three components and six Fluency Actions, and their relationships, are illustrated in Figure 2.

FIGURE 2 • Procedural Fluency Components, Actions, and Checks for Reasonableness

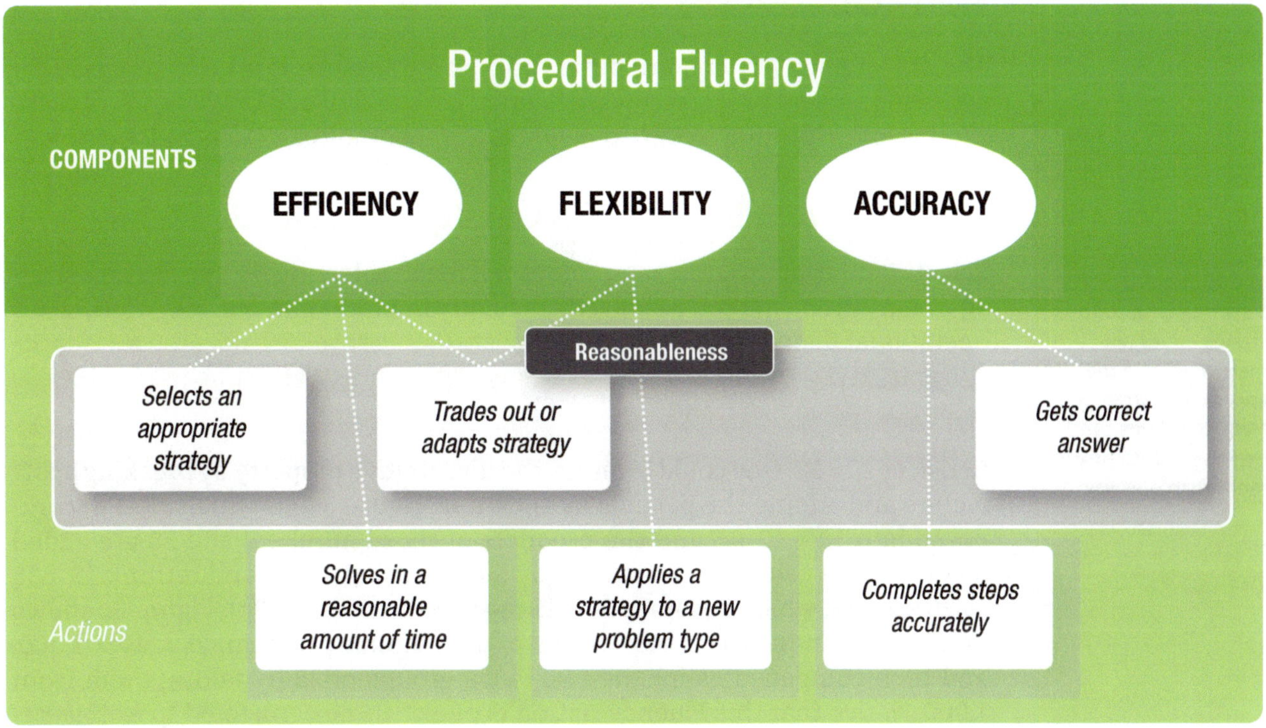

Source: Adapted with permission from D. Spangler & J. Wanko (Eds.), *Enhancing Classroom Practice with Research behind Principles to Actions*, copyright 2017, by the National Council of Teachers of Mathematics. All rights reserved.

Three of the six Fluency Actions (should) attend to reasonableness. Fluency Actions and reasonableness are described later in Part 1, but first, it is important to consider why this "bigger" (comprehensive) view of fluency matters.

WHY FOCUS ON FLUENCY FOR WHOLE NUMBER MULTIPLICATION AND DIVISION?

There are two key reasons why it's important to focus on fluency with whole number multiplication and division. First, it is a critical foundation for ensuring that students fully realize procedural fluency in general, with all kinds of numbers. Fluency with whole numbers begins with developing fluency with single-digit multiplication and division (the basic facts) and seeing how strategies such as Subtract-a-Group can be transferred to using Compensation in order to multiply 29×8 efficiently (e.g., 29×8 is adjusted to 30×8 and to compensate, 8 is subtracted from the product). Basic facts are addressed in more detail later in Part 1.

Second and most importantly, developing fluency is an equity issue. Equipping students with options for how to solve multiplication and division problems and positioning students to choose a method that works best for them develops a positive mathematics identity and sense of agency. Conversely, trying to remember algorithms and feeling anxiety about being correct or fast develops a negative mathematics identity and lack of agency.

WHAT DO FLUENCY ACTIONS LOOK LIKE FOR WHOLE NUMBER MULTIPLICATION AND DIVISION?

The six Fluency Actions are observable and therefore form a foundation for assessing student progress toward fluency. Let's take a look at what each of these actions looks like in the context of whole number multiplication and division.

FLUENCY ACTION 1: Select an Appropriate Strategy

Selecting *an* appropriate strategy does not mean selecting *the* appropriate strategy. Many problems can be solved efficiently in more than one way. Here is our operational definition: Of the available strategies, the one the student opts to use gets to a solution in about as many steps and/or about as much time as other appropriate options.

Consider 25×38. One could leverage the distributive property by breaking apart factors and adding the partials as shown in Figure 3a, in which 25 is decomposed into 10 + 10 + 5 and the products of those numbers and 38 are added together. One could break apart 38 by factors using 2×19 and rethink 25×38 as $25 \times 2 \times 19$, which becomes 50×19 (as shown in Figure 3b). Figure 3c shows Compensation in which one thinks of a friendlier problem, $25 \times 40 = 1,000$, and then compensates for the two extra groups of 25 by taking them from 1,000. Notice that this Fluency Action is one of three connected to *reasonableness*. Within this action is noticing when a strategy "fits" the numbers in the problem. This particular problem fits well with Compensation because 38 is close to (but less than) 40. The other strategies are reasonable if you are adept at multiplying by 5s. In other words, what is an "appropriate" strategy depends on the problem and on the person (their knowledge and experiences).

FIGURE 3 ● Multiplying 25 × 38

a. Break Apart to Multiply
(break apart into addends)

$$25 \times 38$$

$$10 \times 38 = 380$$
$$10 \times 38 = 380$$
$$5 \times 38 = 190$$
$$\overline{950}$$

b. Break Apart to Multiply
(break apart into factors)

$$25 \times 38$$

$$25 \times 2 \times 19$$
$$50 \times 19 = 950$$

c. Compensation
(change a factor and adjust the answer)

$$25 \times 38$$

$$25 \times 40 = 1,000$$
$$25 \times 2 = 50$$
$$1,000 - 50 = 950$$

A strategy cannot be used until it is understood. Once understood, a strategy becomes part of a student's repertoire of options and they are then able to select the strategy.

Importantly, we name strategies so that we can talk about them. But one approach may fit within various types of strategies and have different names. For example, solving 23×9 as $20 \times 9 + 3 \times 9$ may be using Break Apart thinking or Partial Products thinking. The focus must be on the ideas (not the naming of the strategy).

FLUENCY ACTION 2: Solve in a Reasonable Amount of Time

There is no set amount of time that should be expected for solving any whole number multiplication or division problem. Students should be able to work through a problem without getting stuck or lost. The amount of time is relative to the student's grade and mathematical maturity. Keep in mind that appropriate strategies can be carried out in inefficient, unreasonable ways. For example, Figure 3a shows a Break Apart approach that can be enacted in a reasonable amount of time. In contrast, Figure 4 shows this approach with many more parts, which requires significantly more time. The approach is likely the result of the student's ability to recall or find certain products (×10, ×2). While this may be an appropriate beginning strategy, it is ultimately not an approach that can be completed in a reasonable amount of time.

FIGURE 4 ● Solving 25 × 38 by Breaking Apart 38

$$25 \times 38$$

$$10 \times 25 = 250 \qquad 2 \times 25 = 50$$
$$10 \times 25 = 250 \qquad 2 \times 25 = 50$$
$$10 \times 25 = 250 \qquad 2 \times 25 = 50$$
$$\overline{750} \qquad 2 \times 25 = 50$$
$$\overline{200}$$

$$25 \times 38 = 950$$

FLUENCY ACTION 3: Trade Out or Adapt a Strategy

As students' number sense and understanding of strategies advances, they are able to adapt and trade out strategies. Let's revisit 25×38. A student might attempt 25×38 as illustrated in Figure 3a but get stuck because they don't know 5×38. Thus, they decide to *adapt* the strategy by breaking apart 38 instead:

$$25 \times 30 = 750$$
$$25 \times 8 = 200$$
$$750 + 200 = 950$$

Or a student might *trade out* the Break Apart strategy for another strategy. For example, the student might use Compensation (see Figure 3c) or Partial Products (two options are illustrated in Figure 5a). Notice that this Fluency Action is one of three connected to *reasonableness*. Within this action is noticing how the use of the strategy is going. If it isn't going well or if a student is getting bogged down, then the strategy needs to be adapted or traded for another, more efficient option.

FIGURE 5 ● Partial Products

a. Partial Products Decomposing Both Factors

25×38

$20 \times 30 = 600$
$5 \times 30 = 150$
$20 \times 8 = 160$
$5 \times 8 = 40$
$\overline{950}$

b. Partial Products Decomposing One Factor

25×38

$30 \times 25 = 750$
$8 \times 25 = 200$
$\overline{950}$

FLUENCY ACTION 4: Apply a Strategy to a New Problem Type

The Compensation work shows how, when multiplying 25×38, 38 can be thought of as 40 to find $25 \times 40 = 1,000$ and then two groups of 25 can be taken away. That individual's first experience with compensation was likely finding products of $\times 9$ basic facts that transferred to multidigit factors that were one group away from a friendly computation such as 18×9 or 35×39. In time, this strategy will be applied to multiplying fractions by thinking of $6 \times 2\frac{3}{4}$ as 6×3, taking away $6 \times \frac{1}{4}$ $(18 - 1\frac{1}{2} = 16\frac{1}{2})$ or multiplying decimals by thinking of 6×1.9 as 6×2 and taking away six tenths $(12 - 0.6 = 11.4)$.

FLUENCY ACTIONS 5 AND 6: Complete Steps Accurately and Get Correct Answers

These two Fluency Actions are about accuracy. An error at the end of a problem may be due to an error in how a strategy was enacted. For example, in using Compensation for 25×38, a student changes the problem to 25×40 to get 1,000 and then *subtracts* 2 from the answer, resulting in the answer 998. This incorrect answer is due to a misconception of the steps in implementing the Compensation strategy (specifically not recognizing that they need to compensate by subtracting two groups of 12). Conversely, a student may enact the strategy accurately but make a computational error. The student's work in Figure 6 is a good example of this. Here, the student correctly enacts the steps for Partial Products but makes an error with 20×8 that leads to the wrong answer.

FIGURE 6 ● Incorrect Partial Product Example

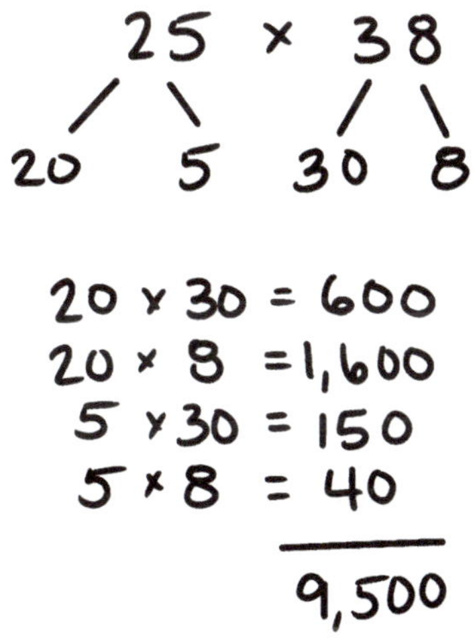

$$25 \times 38$$

$$20 \quad 5 \quad 30 \quad 8$$

$$20 \times 30 = 600$$
$$20 \times 8 = 1,600$$
$$5 \times 30 = 150$$
$$5 \times 8 = 40$$
$$\overline{\quad 9,500 \quad}$$

Fluency Action 6 is one of three connected to *reasonableness*. Within this action is noticing if your answer makes sense. While reasonableness has been woven into the discussion of Fluency Actions, it is critical to fluency and warrants more discussion.

REASONABLENESS

As described earlier, reasonableness is more than "checking your answer." Reasonableness occurs in three of the six Fluency Actions as shown in Figure 2 and described within the related Fluency Actions. Let's revisit 25×38. This example has several reasonable options, including Break Apart to Multiply, Compensation, and Halve and Double (Action 1). Skip-counting by 25s or 38s is not a reasonable strategy. Let's say a student chooses Break Apart (using addends). They break apart 25 ($20 \times 38 + 5 \times 38$) but don't know these partial products. The student recognizes that this attempt is not going well and adapts or trades out the strategy (Action 3). Finally, they notice that 950 is a reasonable answer (Action 6). Knowing that $25 \times 38 = 950$ is reasonable can be determined in (at least) three different ways:

1. You might estimate the product to be less than 1,200 because both factors round up to 30 and 40, respectively.

2. You might estimate it to be about 1,000 using friendly numbers (25×40).

3. You might think about it being in the range of 600 (20×30) and 1,200 (30×40).

It takes time to develop reasonableness. It should be practiced and discussed as often as possible. Students can develop reasonableness by practicing three moves (a match to the Fluency Actions 1, 3, and 6).

THREE 'Cs' OF REASONABLENESS

Choose: Choose a strategy that is efficient based on the numbers in the problem.
Change: Change the strategy if it is proving to be overly complex or unsuccessful.
Check: Check to make sure the result makes sense.

You can encourage and support student thinking about reasonableness by providing Choose, Change, Check reflection cards (see Figure 7). These cards can be adapted into anchor charts for students to use while working on problems or during class discussions about multiplying and dividing.

FIGURE 7 ● Choose, Change, Check Reflection Card for Students

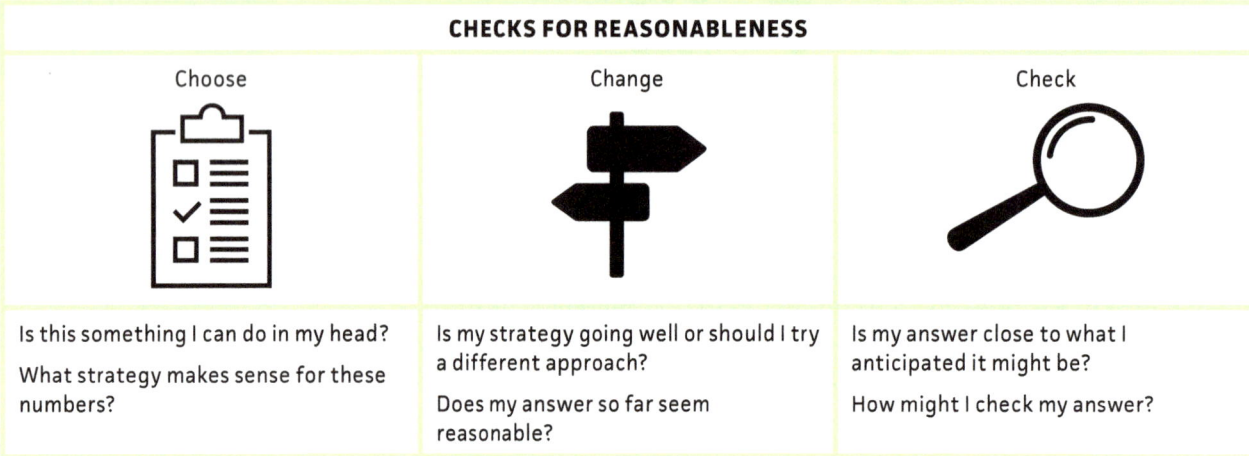

CHECKS FOR REASONABLENESS		
Choose	Change	Check
Is this something I can do in my head? / What strategy makes sense for these numbers?	Is my strategy going well or should I try a different approach? / Does my answer so far seem reasonable?	Is my answer close to what I anticipated it might be? / How might I check my answer?

Icon sources: Choose by iStock.com/Enis Aksoy; Change by iStock.com/Sigit Mulyo Utomo; Check by iStock.com/Indigo Diamond

 This resource can be downloaded at **resources.corwin.com/FOF/multiplydividewholenumber**.

WHAT FOUNDATIONS DO STUDENTS NEED TO DEVELOP FLUENCY WITH MULTIPLICATION AND DIVISION?

To develop fluency, students need a strong foundation in five domains:

- *Conceptual understanding:* knowing the meaning of the operations

- *Properties:* being able to use the operations in order to manipulate numbers and retain equivalence

- *Utilities:* small skills that make a big difference, such as knowing or easily finding half of a number

- *Computational estimation:* being able to quickly and flexibly determine a "close" answer

- *Basic facts:* single-digit addition, subtraction, multiplication, and division facts that are needed for multidigit work

Rushing students to strategy instruction before these foundations are firmly in place can be disastrous.

CONCEPTUAL UNDERSTANDING

Developing fluency from conceptual understanding begins with including concrete experiences for students so they can make sense of the quantities. Hence, developing fluency *begins* with stories. It is a mistake to save story problems as an application, as stories give students a context from which they can reason.

As students work with multiplication and division story problems, there are few important ideas to keep in mind:

1. Stories need to be *relevant* to students, meaning that students are familiar with the context and it is interesting to students.

2. Avoid key-word strategies for solving story problems so that students must make sense of the problem and how multiplication or division can be used to find a solution.

3. Balance experiences with division between both partitive and quotative (measurement) interpretations of division.

4. Vary and balance experience with the different situations for multiplication and division as shown in Figure 8.

5. Use representations that are true to the situation (e.g., equal group representations for equal group problems, area representations for area problems) until students have a deep understanding and are able to work symbolically and move between representations and tell explicitly how they are related.

Figure 8 can be used as a resource to be sure you are varying your story types (for example, you can tally which types of stories you are telling as part of action research).

TEACHING TAKEAWAY

Developing fluency *begins* with stories and contexts. It is a mistake to save story problems as an application, as stories give students a context from which they can reason.

FIGURE 8 ● Multiplication and Division Situations

MULTIPLICATION AND DIVISION SITUATIONS			
Equal Groups	**Area/Array**	**Compare**	**Combinations**
Story is about a quantity of same-sized grouped amount.	Story is about a quantity of equal-length rows.	Story compares two quantities multiplicatively.	Story is about finding how many pairings are possible (and beyond).
Ex: AJ has ____ stacks of books. Each stack has ____ books. She has a total of ____ books.	AJ has ____ books on each shelf. She has ____ shelves. She has a total of ____ books.	AJ has ____ books. Ian has ____ books. AJ has ____ times more books than Ian.	AJ has ____ books and ____ magazines. She takes one of each to school. There are ____ different combinations.

In addition to stories, visuals provide concrete, conceptual beginnings for students. For example, young children count collections of objects in order to start learning to skip count (Franke, Kazemi, & Turrou, 2018). Counting objects progresses to counting visuals and representations, which eventually leads to abstract counting strategies and representation of groups.

PROPERTIES AND UTILITIES FOR STRATEGIC COMPETENCE

In addition to conceptual foundations, fluency is grounded in using properties of the operations and a few other skills that we refer to as "utilities" because students must utilize them in their reasoning. First, fluency with multiplication relies heavily on students using the commutative, associative, and distributive properties of multiplication. Note that knowing properties does not equal using properties. It is *not* useful to have students simply name the associative property. It is absolutely necessary that students *utilize* this property in solving problems efficiently. For example, in the following problem, students break apart factors (44) by their factors (4 × 11) and rearrange these numbers (mentally or in writing) using the commutative property to create friendly computations.

TEACHING TAKEAWAY

Knowing properties does not equal using properties.

$$44 \times 25$$

$$4 \times 11 \times 25$$
$$4 \times 25 \times 11$$
$$100 \times 11$$
$$1,100$$

The distributive property is most critical for fluency with multiplication because so many strategies, including the standard algorithm, make use of it. On the left in the following example, the distributive property is applied to 8 × 398 in order to find partial products. Numerically, this is 8 × 398 = 8 (300 + 90 + 8). On the right, the distributive property is used for a compensation strategy. Numerically, this is 8 × 398 = 8(400 + –2) = 8(400 – 2).

$$8 \times 398$$

	300	90	8
8	2,400	720	64

$$3,184$$

$$8 \times 398$$

$$8 \times 400 = 3,200$$
$$-8 \times 2 = 16$$
$$8 \times 398 = 3,184$$

See Chapter 3 (pp. 47–75) of *Figuring Out Fluency* for more about foundations and good beginnings for fluency.

Beyond the properties is a short list of utilities that support fluency, which is presented in Figure 9.

FIGURE 9 ● Utilities for Strategic Competence With Multiplication and Division

UTILITY	WHAT IT IS	RELATIONSHIP TO FLUENCY
Distance From a 10	Knowing that 9 is 1 away from 10 (and 8 is 2 away and so on).	Knowing how far a number (e.g., 5, 6, 7, 8, 9) is from 10 is necessary for finding how many groups of a number are away from a friendly computation (e.g., 28 is 2 away from 30, so 28 × 6 is two groups of 6 away from 30 groups of 6).
Composing and Decomposing Numbers Flexibly	Understanding diverse, flexible ways to compose and decompose, including but not limited to place value decomposition.	Flexibly decomposing numbers supports strategy selection and facility with any of the strategies.
Skip Counting	Skip counting by multiples of tens, hundreds, and thousands as well as other useful benchmarks such as 25.	Efficiency comes from skip-counting with multiples. This applies to all multiplication and division strategies.
Multiplying and Dividing by Tens, Hundreds, and Thousands	Understanding that 6 × 40 is similar to 6 × 4, in that 6 × 40 is saying six groups of 4 tens, which is 24 tens or 240.	Flexible decomposition of factors relies on recognition and knowing of products of factors that are multiples of 10, 100, or 1,000.

COMPUTATIONAL ESTIMATION

Just like computation, there are strategies for estimation and the use of those strategies should be *flexible*. For multiplication and division of whole numbers, students might use any of these methods:

1. *Rounding*: Flexible rounding means that one or both numbers might be rounded. Students may round to the nearest number or they may round one number up and one number down to have a more accurate estimate. Rounding is a well-known strategy but is often approached in a step-by-step manner, which can interfere with the point of estimating—getting a quick idea of what the answer will be close to. Use conceptual language, such as, "Which tens/hundreds/thousands is that number close to?" Help students understand that they choose how to round. For example, for 24 × 34, rounding both to the nearest 10 will give a low estimate, whereas rounding one up and one down gives a closer estimate.

2. *Front-end estimation*: In its most basic form, students just multiply or divide using the largest place value. More flexibly, though, students may use the largest two place values or adjust their estimate because of what they notice with the rest of the numbers. Front-end estimation is *quick*. For example, 74 × 55 is about 3,500 (relying on 70 × 50). To adjust, take a quick look to the right and decide to keep estimate or adjust. With division, that front end must focus on compatibles. For example, to estimate 4,576 ÷ 7, the front end is altered to be 4,200 or 4,900 because these "fronts" are multiples of 7 (this is an overlap to the next strategy, compatible numbers!).

3. *Compatible numbers*: With flexibility in mind, students change one or both of the numbers to a nearby number so that the numbers are easy to multiply or divide. For example, estimating the quotient of 337 ÷ 72 could be to think of 337 as 350 and 72 as 70 to estimate a quotient of about 5. Compatibles are particularly useful with division. Consider estimating 345 ÷ 8. Compatible alternatives might be 320 ÷ 8, 400 ÷ 8, or 350 ÷ 7.

DEVELOPING BASIC FACT FLUENCY

The teaching of basic facts must attend to conceptual understanding and strategies for reasoning rather than rote instruction (Bay-Williams & Kling, 2019; O'Connell & SanGiovanni, 2014). Reasons to focus on *strategies* when teaching the basic facts (as opposed to memorizing) include the following:

1. It is well established across many studies that students actually learn and retain their facts better when they focus on conceptual understanding versus memorization. In fact, students don't just learn and retain their facts better, they perform better in math *in general* (e.g., Baroody, Purpura, Eiland, Reid, & Paliwal, 2016; Brendefur, Strother, Thiede, & Appleton, 2015; Jordan, Kaplan, Ramineni, & Locuniak, 2009; Locuniak & Jordan, 2008; Purpura, Baroody, Eiland, & Reid, 2016).

2. Students need to know and use these strategies to support whole number multiplication and division (as well as decimal and fraction operations). In our *Figuring Out Fluency* anchor book, we elaborate more on the key strategies and ideas for effectively developing basic fact fluency.

3. Students who learn to use and choose strategies for basic facts develop confidence. Students who memorize often develop anxiety. A student who knows how to generate an answer to a quotient such as 9 ÷ 6 (beyond counting) doesn't have to worry if they forget the fact. This sense of agency is critical to student success in mathematics!

Figure 10 lists the basic fact strategies for multiplication and division. To be clear, automaticity is the goal for learning basic facts. Students become automatic through learning the strategies and practicing them over and over again. In so doing, students develop automaticity with the facts *and with implementing the strategies*. Examples are illustrated in Figure 11. Keep in mind that there are many ways to implement a reasoning strategy, and only one way is shown for each example.

FIGURE 10 ● Reasoning Strategies for Basic Fact Multiplication and Division

STRATEGY NAME	HOW THE STRATEGY WORKS	EXAMPLE STUDENT TALK
Multiplication	**Example: 6 × 7**	
Doubling	Student sees an even factor, finds the product of half of that factor, and doubles their answer.	I got 42. I know 3 times 7 is 21 and I doubled 21.
Add-a-Group	Student thinks of a known fact where one of the factors is one less, multiplies, then adds a group back on.	When I see a 6, I use my 5s: 5 times 7 is 35 and 7 more is 42.
Subtract-a-Group	Student thinks of a known fact where one of the factors is one more, multiplies, then adds a group back on.	I know 7 groups of 7 is 49, so I subtract one group of 7 and I have 42.
Near Squares	Student uses a square fact they know and then adds or subtracts a group. *Note: This is an undertaught but useful strategy.*	Well, 6 times 6 is 36, and I add 6 more and get 42.
Division	**Example: 36 ÷ 9**	
Think Multiplication	Student thinks, *How many groups of 9 make 36?*	I know 9 times 4 is 36, so it's 4. Or I used doubling to get to 18, doubled again, and got 36, so it is 4.

Basic fact strategies evolve directly to the significant reasoning strategies for multidigit multiplication and division, which are the focus of Part 2 of this book. You can see this transformation in the examples in Figure 11.

FIGURE 11 ● How Basic Fact Strategies Grow Into General Reasoning Strategies

REASONING STRATEGY	EXAMPLES WITH MULTIDIGIT NUMBERS
Add-a-Group(s) leads to **Break Apart to Multiply**, and **Partial Products** strategies	$42 \times 6 \rightarrow (40 \times 6) + (2 \times 6)$
Subtract-a-Group(s) leads to **Compensation**	$79 \times 5 \rightarrow (80 \times 5) - (1 \times 5)$
Doubling applies to the **Halve and Double** strategy and extends to **Break Apart to Multiply**, involving breaking apart into factors	$36 \times 5 \rightarrow 18 \times 10$ $45 \times 6 \rightarrow 45 \times 2 \times 3 \rightarrow 90 \times 3$ $25 \times 68 \rightarrow 25 \times 4 \times 17 \rightarrow 100 \times 17$
Think Multiplication extends to larger numbers and to **Computational Estimation** involving division.	$49 \div 7 \rightarrow 490 \div 7$ [or $490 \div 70$] $4871 \div 85 \rightarrow$ About how many 8s are in 48? [6, so estimate is 60]

WHAT AUTOMATICITIES DO STUDENTS NEED BEYOND THEIR BASIC FACTS?

Unlike the foundation of conceptual understanding, automaticities are not prerequisites for, but coincide with, strategy instruction. For example, automaticity with basic facts (just discussed) begins with strategy instruction and leads to eventual automaticity with the facts. But there are automaticities beyond the basic facts that support student reasoning!

Automaticity is the ability to complete a task with little or no attention to process. Little thought, if any, is given to skills that are automatic (Cheind & Schneider, 2012). We consider automaticities to be those skills that a fluent person can do without much attention to process. For example, you know that four 25s are 100 and a drawing or repeated addition is not needed; it is intuitive or reflexive. Figure 12 identifies automaticities that are particularly critical for multiplying and dividing whole numbers. Of course, this is not a complete list. These automaticities are strengthened through strategy instruction (and conversely, having these automaticities strengthens students' capacities to use strategies).

See Chapter 5 (pp. 107–129) of *Figuring Out Fluency* for more about automaticities for fluency.

FIGURE 12 ● Automaticities for Multiplying and Dividing Whole Numbers

AUTOMATICITY	WHAT IT IS	HOW IT COMPLEMENTS STRATEGY INSTRUCTION
Basic facts	Quickly recognizing how a problem relates to a basic fact (e.g., 30×80 relates to 3×8).	Identifying relationships to basic facts helps students consider which numbers to decompose and how to decompose them.
Using 25s	Knowing multiples of 25 (within reason) and how they are related to multiples of 250, 2,500, and so on.	This helps students think about how to decompose factors and find partials when multiplying and dividing.
Using 15s and 30s	Knowing multiples of 15 and 30 and ways to decompose them efficiently (e.g., 75 is 60 and 15 or two 30s and 15 or five 15s).	This helps students think about how to decompose factors and find partials when multiplying and dividing.
Doubling	Doubling a given number.	Although explicitly connected to the Halve and Double strategy, Doubling is useful for Compensation and strategies with partials.
Halving	Finding a half of a given number.	Though explicitly connected to the Halve and Double strategy, Halving is useful for Compensation and strategies with partials.

WHAT ARE THE SIGNIFICANT STRATEGIES FOR MULTIPLYING AND DIVIDING WHOLE NUMBERS?

Teaching strategies beyond the common algorithms has been a challenge, as there has been pushback and criticism from families and in social media. Two questions require attention:

1. Why do students need strategies when they can use the standard algorithm?

One way to quickly respond to this question is to share an example for which the standard algorithm takes much more time than an alternative. Examples include 24×5, $125 \div 5$, or a problem that has various possibilities, like 16×35. Why learn other methods? Because many problems can be solved more efficiently another way. Fluent students look for efficient methods; if students are limited in the methods they are taught, they have little to choose from, which limits flexibility and efficiency.

2. What strategies are worthy of attention?

Let's just take some pressure off here. The list is short, and we must help students see that they are not necessarily learning a *new* strategy, but they are applying a strategy they learned with basic facts and transferring it to other numbers. In *Figuring Out Fluency* we propose Seven Significant Strategies. Of these, five relate to multiplying and/or dividing whole numbers and they are listed in Figure 13.

TEACHING TAKEAWAY

Students don't need to constantly learn dozens of new strategies, but rather connect how the key strategies they learned for basic facts are transferred to other numbers.

FIGURE 13 ● Reasoning Strategies for Multiplying and Dividing Whole Numbers

REASONING STRATEGIES	RELEVANT OPERATIONS
1. Break Apart to Multiply (Module 1)	Multiplication
2. Halve and Double (Module 2)	Multiplication
3. Compensation (Module 3)	Multiplication
4. Partial Products and Quotients (Modules 4 and 6)	Multiplication and Division
5. Think Multiplication (Module 5)	Division

Using an area model or a number line is *not* a strategy. It is a representation. If a student multiplies using an area model, they are implementing a strategy—perhaps Break Apart or Partial Products. When a student says, "I used an area model," ask *how* they used it—then you will learn what strategy they used. Teaching for fluency means that each of these strategies is explicitly taught to students. We teach students to *use* the strategy, and then we give students many opportunities to engage in *choosing* strategies (Part 3 of this book). Explicitly teaching a strategy does not mean turning the strategy into an algorithm. Strategies require flexible thinking. Each module provides instructional ideas and practice to ensure students become adept at using each strategy flexibly. There is also a module on **standard algorithms** (Module 7), so that they are integrated into the use of strategies.

HOW DO I USE THE PART 2 MODULES TO TEACH, PRACTICE, AND ASSESS STRATEGIES?

Part 2 is a set of modules, each one focused on understanding why a specific reasoning strategy works and learning how to use it well. The seven modules in Part 2 each have a consistent format. First, each module provides an overview of the strategy—unpacking what it is, how it works, and when it is useful. Then, each module provides a series of instructional activities for explicit strategy instruction, followed by a collection of practice activities, including routines, games, and centers.

EXPLICIT STRATEGY INSTRUCTION

Strategies must be explicitly taught so that students understand them and can use them. Each module provides core teaching activities for explicit instruction. The activities are designed so that you can modify and extend them as needed. Any one activity might form the focus of your instruction over the course of multiple lessons. Keep in mind that you can swap out tools and representations as well as adjust the numbers within the task.

The last teaching activity in each module is a collection of *investigation prompts* that you can use to develop reasoning and understanding of the strategy. Each investigation prompt itself can easily become a core teaching task.

We intend for students to work with instructional activities in collaborative partner or group settings. We encourage you to let students make their own meaning and to make mistakes. After students engage in the activities, a group discussion is needed to focus student thinking on the concepts within the strategy, how the strategy works, and the different ways a strategy might be carried out.

QUALITY PRACTICE

Students need access to quality practice that is not a worksheet. Quality practice is focused on a strategy, varied in type of engagement, processed by the student to make sense of what they did, and connected to what they are learning.

See Chapter 6 (pp. 130– 153) of *Figuring Out Fluency* for more about quality practice.

Each practice section begins with **worked examples**. Worked examples are opportunities for students to attend to the thinking involved with a strategy, without solving the problem themselves. We feature three types to get at all components of fluency:

1. *Correctly worked example*: efficiency (selects an appropriate strategy) and flexibility (applies strategy to a new problem type)

2. *Partially worked example*: efficiency (selects an appropriate strategy) and accuracy (completes steps accurately; gets correct answer)

3. *Incorrectly worked example*: accuracy (completes steps accurately; gets correct answer)

Also, comparing two correctly worked examples is very effective in helping students learn to choose efficient methods. Throughout the modules are dozens of examples, which can be used as worked examples (and adapted to other similar worked examples). Your worked examples can be from a fictional "student" or authentic student work. Some of the prompts from teaching the strategy section are, in fact, worked examples.

The remaining practice activities include routines, games, and centers. Each activity provides a brief "About the Activity" statement to help you quickly match what your students need with a meaningful activity. Game boards, recording sheets, problem cards, and all other activity resources, including modified versions, are available as online resources that you can download, modify, print, and copy. General resources, including number cards, mini ten-frame cards, multiplication charts, and more, are also available for download on the companion website.

ASSESSING STRATEGY USE

Each module offers a plethora of practice activities. As students are practicing, you can observe and assess the extent to which they are able to apply the selected strategy. **Observation tools** help you keep track of where each student is and monitor their progress. An observation tool can be simple, such as a class list with an extra column. Your observations can be codes:

+ Is regularly implementing the strategy adeptly

✔ Understands the strategy, takes time to think it through

– Is not implementing the strategy accurately

A note-taking observation tool provides space for you to insert notes about how a student is doing (see Figure 14). You can laminate the tool and use dry-erase markers to reuse it for different observations, use sticky notes, or just write in the boxes.

FIGURE 14 ● Example Note-Taking Observation Tool

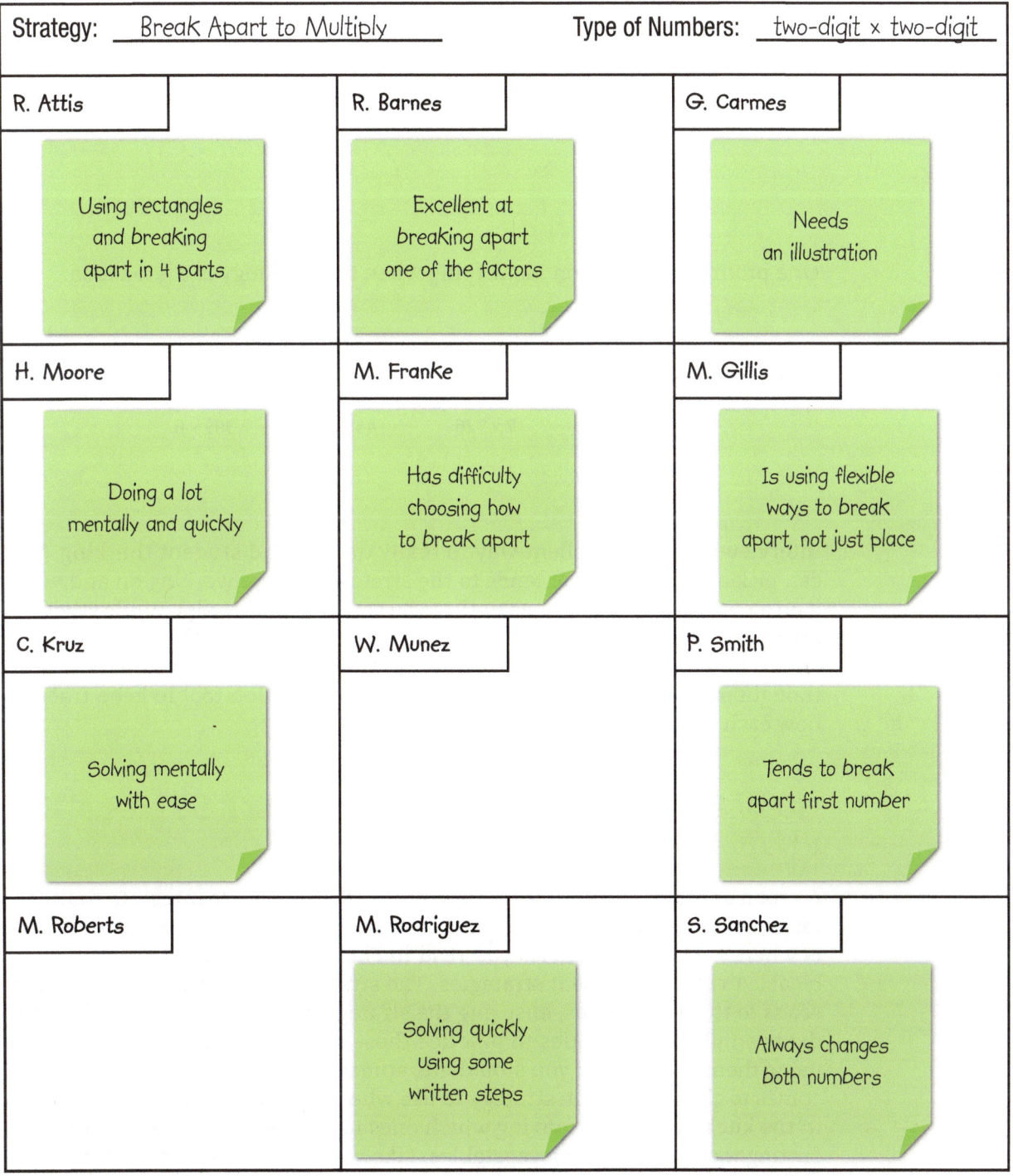

Strategy: Break Apart to Multiply		Type of Numbers: two-digit × two-digit
R. Attis	R. Barnes	G. Carmes
Using rectangles and breaking apart in 4 parts	Excellent at breaking apart one of the factors	Needs an illustration
H. Moore	M. Franke	M. Gillis
Doing a lot mentally and quickly	Has difficulty choosing how to break apart	Is using flexible ways to break apart, not just place
C. Kruz	W. Munez	P. Smith
Solving mentally with ease		Tends to break apart first number
M. Roberts	M. Rodriguez	S. Sanchez
	Solving quickly using some written steps	Always changes both numbers

online resources ☞ This resource can be downloaded at **resources.corwin.com/FOF/multiplydividewholenumber**.

Some days, you collect data on some students; other days, you collect data on other students. The data can help in classroom discussions and in planning for instructional next steps.

Journal prompts provide an opportunity for students to write about their thinking process. Each module provides a collection of prompts that you might use for journaling. You can modify those or easily craft your own. The prompt can specifically ask students to explain how they used the strategy:

> **Explain and show how you can use Compensation to solve 6 × 498.**

Or a prompt can focus on identifying when that strategy is a good idea:

> **Circle the problems that are good choices for solving with the Compensation strategy and tell why you selected them:**
>
> 495 × 3   7 × 726   440 × 2   399 × 6

Interviewing is an excellent way to really understand student thinking. You can pick any problem that lends to the strategy you are working on and write it on a note card (or record two or three on separate notecards). While students are engaged in an instructional or practice activity, roam the room, select a child, show them a card, and ask them (1) to solve it and (2) explain how they thought about it. You can pair this with an observation tool to keep track of how each student is progressing.

HOW DO I USE PART 3 TO SUPPORT STUDENTS' FLUENCY?

As soon as students know more than one way, it is time to integrate routines, tasks, centers, and games that focus on choosing when to use a strategy. That is where Part 3 comes in. As you read in Fluency Action 1, students need to be able to choose efficient strategies. The strategy modules provide students *access* to those strategies, ensuring the strategies make sense and giving students ample opportunities to practice those strategies and become adept at using them. However, if you stop there, students are left on their own when it comes to choosing which strategy to use when. It is like having a set of knives in the kitchen but not knowing which ones to use for slicing cheese or bread, cutting meat, or chopping vegetables. Like with food, some items can be cut with various knives, but other food really needs a specific knife.

Do not wait until after all strategies are learned to focus on when to use a strategy—instead weave in Part 3 activities regularly. Each time a new strategy is learned, it is time to revisit activities that engage students in making choices from among the strategies in their repertoire. Students must learn what to look for in a problem to decide which strategy they will use to solve the problem *efficiently* based on the numbers in the problem. This is *flexibility* in action, and thus leads to fluency.

TEACHING TAKEAWAY
Teaching for fluency means teaching strategies as core instruction, routinely practicing them, and offering opportunities for students to choose among strategies.

"FACTORS" IN GETTING THE BEST "PRODUCT"

Part 1 has briefly described factors that are important in developing fluency, and these ideas are important as you implement activities from the modules. We close Part 1 with five key factors to figuring out fluency.

1. Be clear on what fluency means (three components and six actions). This includes communicating it to students and their parents.

2. Attend to readiness skills: conceptual understanding, properties, utilities, computational estimation, and, of course, basic fact fluency.

3. Through activities and discussion, help students connect on the features of a problem and how that relates to good strategy options.

4. Reinforce student reasoning and choice selection, rather than focus on speed and accuracy. Getting the strategies down initially takes more time, but eventually will become more efficient.

5. Assess fluency, not just accuracy.

Time invested in strategy work has big payoffs—confident and fluent students (and that is the "best product"). That is why we have so many activities in this book. Teach the strategies as part of core instruction, *and* continue to practice throughout the year, looping back to strategies that students might be forgetting to use (with Part 2 activities) and offering ongoing opportunities to choose from among strategies (with Part 3 activities).

PART 2

STRATEGY MODULES

Break Apart to Multiply Strategy

STRATEGY OVERVIEW:
Break Apart to Multiply

What is Break Apart to Multiply? As the name implies, this strategy breaks apart (decomposes) one or both of the factors. They may be decomposed into addends (or "by addends") or broken apart into factors (or "by multiplication"). The strategy begins with multiplication facts, as illustrated here:

STRATEGY	STUDENT THINKING FOR 6×7	SYMBOLIC REPRESENTATION
Add-a-Group (Break Apart Into Addends)	"I don't know this fact, but I know 5×7 equals 35. I can break apart 6 into $5 + 1$. I know 5×7 is 35, and one more group of 7 equals 42."	6×7 $(\mathbf{5+1}) \times 7$ $5 \times 7 + 1 \times 7$ $35 + 7$ 42
Doubling (Break Apart Into Factors)	"I don't know this fact, but I know 3×7 equals 21; 6 is double 3, so I can double 21. The answer is 42."	6×7 $(\mathbf{2 \times 3}) \times 7$ $2 \times (3 \times 7)$ 2×21 42

Let's look at these two ways to Break Apart for the two-digit by two-digit problem of 25×16:

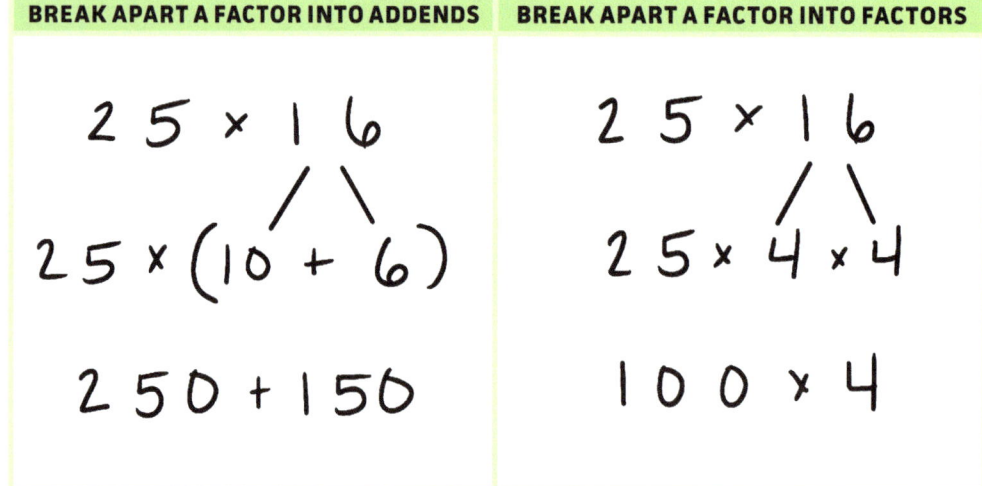

BREAK APART A FACTOR INTO ADDENDS	BREAK APART A FACTOR INTO FACTORS
25×16 $25 \times (10 + 6)$ $250 + 150$	25×16 $25 \times 4 \times 4$ 100×4

Notice that Break Apart Into Addends is an overlap to Partial Products and may be referred to either way. Halve and Double is an example of Break Apart Into Factors.

HOW DOES BREAK APART TO MULTIPLY WORK?

Breaking apart numbers into addends utilizes the distributive property. Breaking apart a factor into smaller factors and reassociating a factor utilizes the associative property, and sometimes the commutative property. The earlier examples show how one factor can be broken apart, but you can break apart both factors. By addends, this is (20 + 5)(10 + 6) and an area model can be used to show the four partial products (this is what is actually happening in the standard algorithm). For breaking apart by factors, 25×16 can be decomposed into $5 \times 5 \times 4 \times 4$ as follows:

$$25 \times 16$$
$$5 \times 5 \qquad 4 \times 4$$

$$5 \times 4 \times 5 \times 4$$

$$20 \times 20$$

WHEN DO YOU CHOOSE BREAK APART TO MULTIPLY?

This strategy is a good choice when you are able to break apart a factor and then compute the parts mentally. For two-digit by one-digit multiplication, Break Apart by Addends is almost always a good option. Break Apart Into Factors is useful when reassociating a factor such as a 2, 4, or 5 leads to a benchmark. In the earlier example, the student noticed the **25** and knew (1) that $25 \times 4 = 100$ and (2) that 16 has 4 as a factor. As problems get larger, Break Apart is not always an efficient strategy. The key is to look at each factor and consider which one can be decomposed into addends or factors and how they can be decomposed so that they can be multiplied efficiently if not mentally.

MODULE 1 Break Apart to Multiply Strategy

MODULE 2 Halve and Double Strategy

MODULE 3 Compensation Strategy

MODULE 4 Partial Products Strategy

MODULE 5 Think Multiplication Strategy

MODULE 6 Partial Quotients Strategy

MODULE 7 Standard Algorithms for Multiplication and Division

BREAK APART TO MULTIPLY:
Strategy Briefs for Families

It is important that families understand the strategies and know how they work so that they can be partners in the pursuit of fluency. These strategy briefs are a tool for doing that. You can include them in parent or school newsletters or share them at parent conferences. They are available for download so that you can adjust them as needed.

Break Apart by Factors to Multiply

How It Works: We can break apart either factor into its factors and find the product of the original problem.

1. Choose which factor to break apart by its factors.
2. Multiply the other factor by one of the parts.
3. Then, multiply that product by the other part.

The left example shows that 8 is broken apart into 2 × 4. Multiply the first part: 15 × 2 = 30. Then multiply the 30 by 4: 30 × 4 = 120. So, 15 × 8 = 120.

The right example shows that 15 is broken apart into 3 × 5. Multiply the first part: 3 × 8 = 24. Then multiply the 24 by 5: 24 × 5 = 120. So, 15 × 8 = 120.

When It's Useful: Breaking apart by factors to multiply is most useful when one factor can be broken apart easily. In the example, breaking apart 8 works well because 2 × 15 and 30 × 4 are friendly. Breaking apart 15 into 3 × 5 works, but 24 × 5 may be a difficult one to multiply.

```
Factor  Factor          Factor  Factor
  15  ×  8                15  ×  8
        /\                        /\
       2 × 4                     3 × 5

  15 × |2| = 30          5 × |8| = 40
  30 × |4| = 120         40 × |3| = 120

  15 × 8 = |120|         15 × 8 = |120|
           product                product
```

Break Apart by Addends to Multiply

How It Works: We can break apart either factor.

1. Choose which factor to break apart.
2. Break apart that factor into a sum. Multiply each addend by the other factor.
3. Add all the products together. This gives you the answer to the original problem.

The left example shows that, 15 is broken apart into 10 and 5. Multiply the first part, 10 × 8 is 80. Then multiply the second part, 5 × 8 is 40. The 80 and 40 are added. 80 + 40 = 120. The answer to the original problem is 120, 15 × 8 = 120.

The right example shows that 8 is broken apart into 4 and 4. Multiply the first part: 15 × 4 is 60. Then multiply the second part: 15 × 4 is 60. The 60 and 60 are added: 60 + 60 = 120. The answer to the original problem is 120, 15 × 8 = 120.

When It's Useful: Breaking apart by addition to multiply is often useful. Any factor can be broken apart in different ways. The goal is to break apart factors in a way that makes the most sense.

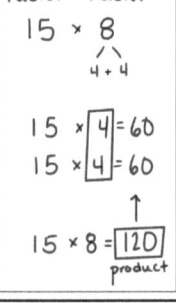

```
Factor  Factor          Factor  Factor
  15  ×  8                15  ×  8
        /\                        /\
      10 + 5                    4 + 4

  |10| × 8 = 80         15 × |4| = 60
  |5|  × 8 = 40         15 × |4| = 60
           ↑                     ↑
  15 × 8 = |120|         15 × 8 = |120|
           product                product
```

 These resources can be downloaded at **resources.corwin.com/FOF/multiplydividewholenumber**.

NOTES

_____

_____

TEACHING ACTIVITIES for Break Apart to Multiply

Break Apart to Multiply is one of the first multiplication strategies students learn. It often begins with decomposing basic facts and evolves into multidigit whole numbers. Factors can be broken apart into addends or into factors. In this section, we focus on instructional activities for helping students develop efficient ways to break apart. The goal is that students become adept at decomposition for efficiency and accuracy and also to consider when the strategy is useful.

ACTIVITY 1.1
MULTIPLES OF 10

Students will use their knowledge of basic facts and the associative property to multiply and divide by multiples of 10. Using base-10 blocks is one way to help students see the multiples of 10. First, pose a basic fact expression, such as 3×4. Briefly discuss the product before posing 3×40. Provide base-10 blocks so that students can create representations of both expressions.

Multiples of 10			
3 x 40			
Thousands	Hundreds	Tens	Ones

After they make models of these problems, record the expressions for each problem making the connections explicit. For example, 3×40 can be thought of as 3×4 tens; 40 can be thought of as 4×10, and so 3×40 is the same as $3 \times 4 \times 10$ or 12×10 (which is also shown with the base-10 blocks).

$$3 \times 40$$
$$4 \times 10$$

$$3 \times 40$$
$$3 \times 4 \times 10$$

$$3 \times 4 \times 10$$
$$12 \times 10$$
$$120$$

As the numbers get larger, base-10 blocks are not efficient. For example, you would not use base-10 blocks to solve 30×40. This is why the connections to expressions (as shown) and the deep understanding of breaking apart factors is so important. To solve 30×40, the factors can be decomposed into $3 \times 10 \times 4 \times 10$ and then rearranged to become $3 \times 4 \times 10 \times 10$ and eventually 12×100. This work is the underpinning as to why it looks like we are adding zeros when multiplying whole numbers. For instance, 3×4 is 12, which is multiplied by 10 and by 10 again.

(*Continued*)

(*Continued*)

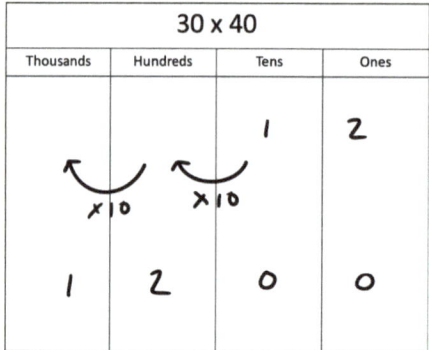

There are patterns when multiplying and dividing by multiples of 10. Those have to do with the number of zeros in the product or quotient. Students need to be aware of the patterns, but we mustn't say "just multiply and add zeros." This shortcut, without understanding, is detrimental to work with decimals and leads to many misconceptions.

 **ACTIVITY 1.2**
COLOR TILE ARRAYS

Using color tiles to build arrays and to break them apart is a good way to help students *see* that factors can be broken apart. This is a good introductory activity for students' early learning with multiplication. It is also wise to revisit an activity like this to reinforce a firm understanding of single-digit factors before extending to two- and three-digit factors. First, prompt all partner groups to build an array with 24, 36, 40, 48, or 60 color tiles. After arrays are built, ask students to decompose the arrays into two smaller rectangles and to record their findings. Notice in Deryn's array that the prompt is open so that students come up with a variety of decompositions. You might consider asking students to come up with some or all of the different ways Deryn could have done this. These different possibilities are important for establishing that a factor can be decomposed in many ways.

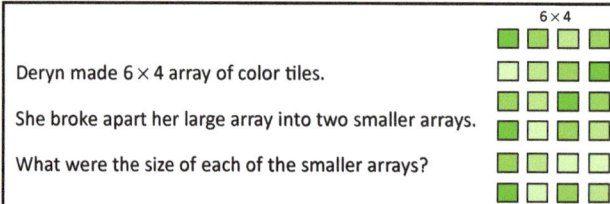

Possible solutions show some of the ways students might break apart the array. Your priority in this activity is to show how the original factors of 6 and 4 are distributed in these examples. Be sure to make explicit connections and ensure that students recognize that the total number of color tiles (24) remains unchanged. Students can then work with arrays of different sizes to see if breaking apart a factor always works.

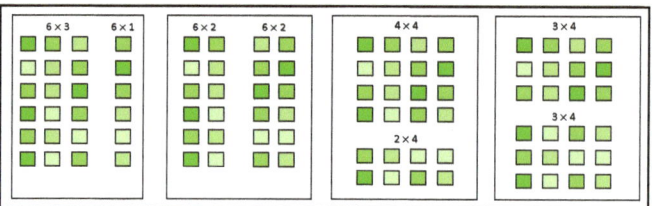

ACTIVITY 1.3
BREAKING APART BASIC FACTS

The strategies students learn for recalling basic facts are the groundworks for multiplying multidigit factors. This activity revisits how basic facts can be decomposed.

First, pose a collection of basic facts like those noted and ask students to find the products by breaking apart one or both factors. Note that some students simply know the basic facts, which is good! For this activity, you want them to show how factors can be decomposed so that the concept can be extended.

Basic Facts to Use for Breaking Apart Factors

6×3	8×3	7×6	7×4	7×8

Students should show that any of the factors in these problems might be decomposed. As you know, factors of 3 can be decomposed into $2 + 1$, factors of 6 can be decomposed into $5 + 1$ or 3×2, factors of 7 can be thought of as $5 + 2$, and factors of 8 can be thought of as $5 + 3$ or $2 \times 2 \times 2$.

After confirming that students recognize and can make use of breaking apart factors, prompt them to think about how they might break apart these problems:

16×3	8×23	7×16	7×24	15×8

Give students time to think about ways that they might decompose these factors. Look for students who continue to break apart the single-digit factor and ask them to think about how they might decompose the two-digit factor instead. It is likely that at least one group of your students will decompose the two-digit factors. However, if this doesn't happen, be sure to acknowledge their thinking, give them an example of how you decomposed a two-digit factor, and then charge them with determining if it can be done.

Note that this early work with decomposing two-digit factors is likely most successful with relatively small, two-digit factors (less than 30). As students show their understanding with these, you can begin to move to two-digit factors greater than 30.

ACTIVITY 1.4
TWO CUTS

Centimeter grid paper is a useful tool for representing multiplication with two-digit factors. In this activity, each student group has a rectangle with the same dimensions. In this example, students have rectangles that are 24×27 cm. Students are tasked with making two cuts so that they create four smaller rectangles. Students should use calculators for this activity.

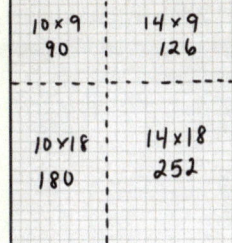

(Continued)

(Continued)

Students find the area of each smaller rectangle and record the equations. Groups of students will create a variety of examples, as shown here. Bring the class together to discuss the patterns they see within the expressions. Highlight that a factor can be broken apart in any way.

$$10 \times 9 = 90$$
$$14 \times 9 = 126$$
$$10 \times 18 = 180$$
$$14 \times 18 = 252$$

$$12 \times 14 = 168$$
$$12 \times 14 = 168$$
$$12 \times 13 = 156$$
$$12 \times 13 = 156$$

$$7 \times 8 = 56$$
$$17 \times 8 = 136$$
$$7 \times 19 = 133$$
$$17 \times 19 = 323$$

TEACHING TAKEAWAY

Help students align their equations as they record them so the patterns within the factors can be recorded easily.

After discussion, ask students if they think that breaking apart a factor will always work. Then, have them work with rectangles of different sizes to see if it does always work. Consider extending the activity so that students create six rectangles by breaking apart one of the factors into three parts.

ACTIVITY 1.5
THE BETTER BREAK

As students show understanding of the Break Apart to Multiply strategy, you want to begin to shift their attention to efficient ways to break apart a factor. After all, 105×13 could be thought of as 21 sets of 5×13, but that isn't what an efficient, fluent person would do. In this activity, you want students to decompose a problem and then consider different decompositions that you offer to determine which is the better break (apart). For example, you might pose 43×15. After students break it apart, you might share it broken apart into $(10 + 10 + 10 + 10 + 3) \times (10 + 5)$, $(40 + 3) \times (5 + 5 + 5)$, or $(20 + 20 + 3) \times (10 + 5)$. Then, as a class, discuss which seem to be the better breaks. Keep in mind that efficiency may be different for each student and that accuracy obviously plays a significant role. Because of this, students' early preferences might rely on breaking factors into groups of 10 $(10 + 10 + 10 + 10 + 3)$ because they can find products of 10. Some problems for students to work within this activity include the following:

ONE-DIGIT BY TWO-DIGIT	TWO-DIGIT BY TWO-DIGIT	ONE-DIGIT BY THREE-DIGIT	ONE-DIGIT BY FOUR-DIGIT
34×5	54×16	5×670	$2 \times 1,458$
23×8	22×63	353×8	$6 \times 7,080$
55×7	75×13	240×6	$4,125 \times 5$
6×82	29×44	515×8	$3,199 \times 4$
94×4	86×30	925×3	$3,450 \times 6$

 # ACTIVITY 1.6
BREAKING APART WITH MULTIPLICATION

Factors can be decomposed by addends or by factors (i.e., with multiplication). This instructional activity focuses on the latter approach. Pose an example of a problem with sets of equal groups. Ask students how they would find the total amount. Give students time to work and then bring the group together to discuss ideas.

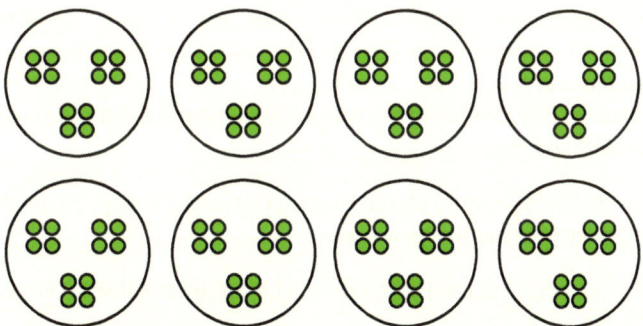

After discussion, ask to make an argument for the image showing $8 \times 3 \times 4$ or 8×12. Student thinking might look something like this:

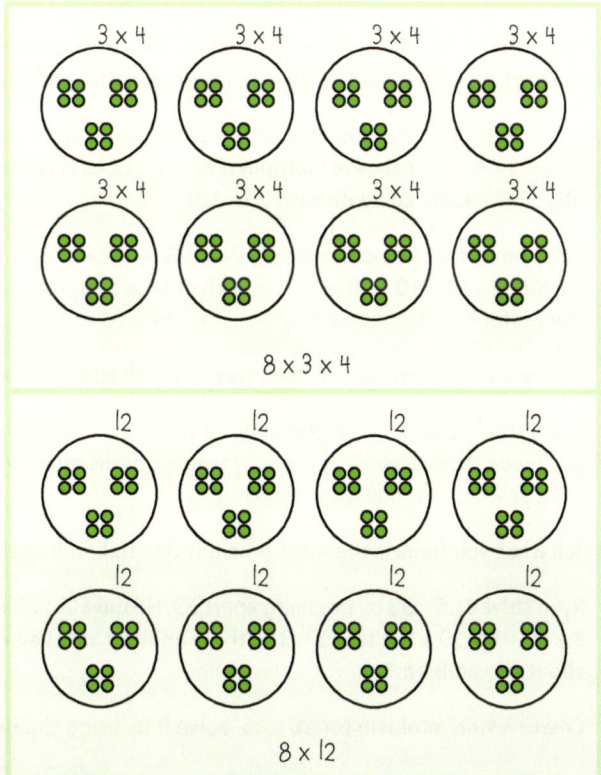

During discussion, it should become clear to students that $8 \times 3 \times 4$ is the same as 8×12. You can prompt them to think about why this is so and when it might be useful. You can have them explore the idea further by posing problems and having them find different ways to break apart factors. For example, challenge students to also explain how the visual shows 24×4 and 32×3. Also ask, "Which of these options would you choose and why?" At first, use representations like those in this example. Good problems include 15×14, 16×7, 22×8, and so on. Each of these is relatively easy to draw. In time, work with problems like 25×16, 32×14, or 48×30, in which students work solely with numeric expressions.

ACTIVITY 1.7
PROMPTS FOR TEACHING BREAK APART TO MULTIPLY

Use the following prompts as opportunities to develop understanding of and reasoning with the strategy. Have students use representations and tools to justify their thinking, including base-10 models, number lines, arrays, and so on. After students work with the prompt(s), bring the class together to exchange ideas. These could be useful for collecting evidence of student understanding. Any prompt can be easily modified to feature different numbers (e.g., three-digit or four-digit numbers) and any prompt can be offered more than once if modified.

- In the expression 15×8, does it matter which way you break apart the factors? Why or why not? Explain your answer.

- How can you use the Break Apart strategy to solve 8×424?

- Jo solved 27×8 by breaking apart 27. How do you think she broke it apart?

- Lee thinks that you can only break apart one factor when multiplying. Do you agree or disagree with Lee? Create a problem to justify your thinking.

- Tyler solved 34×13. First, he multiplied 30×10 and found 300. What do you think he did next? What else did he have to do to solve the problem?

- Jewel found 16×9 by breaking apart 16 into 4×4 and 9 into 3×3. She rearranged it to be $4 \times 3 \times 4 \times 3$. Then, she multiplied 12×12. Does this create the same product? What do you think she could do with 25×12?

- Dezi shows that you can multiply 24×40 by breaking 24 into 10 + 10 + 4 or by breaking 40 into 10 + 10 + 10 + 10. Do both of Dezi's approaches work? If they do, which do you prefer?

- Create an argument for the best way to break apart 435×7.

- Jonah thinks that 48×6 is the same as 46×8 because you can break apart the numbers and move them around. Tell why you agree or disagree with Jonah. Use pictures, numbers, or words to show your thinking.

- Tell what you think is the most efficient way to break apart 34×25.

- Ryan solved 65×33 by breaking apart 33. He did $60 \times 10 + 60 \times 10 + 60 \times 10 + 60 \times 3 + 5 \times 10 + 5 \times 10 + 5 \times 10 + 5 \times 3$. Is there another way you would suggest that he thinks about the problem?

- Create a story problem for 16×45. Solve it by using the Break Apart strategy.

NOTES

_____

_____

PRACTICE ACTIVITIES for Break Apart to Multiply

Fluency is realized through quality practice that is focused, varied, processed, and connected. The activities in this section focus students' attention on how this strategy works and when to use it. The activities are a collection of varied engagements. The discussion you facilitate after an activity or the reflection prompts you attach to it should help students think about what they did mathematically, how they reasoned about the activity, and when the math they did (namely the strategy) might be useful. Debriefing should also help students see how the practice activity connects to recent instruction or how the strategy connects to other strategies they know. Game boards, recording sheets, digit cards, and other required materials are available as online resources for you to download, possibly modify, and use. As students work with activities, look for how well they are acquiring the strategy and assimilating it into their collection of strategies.

FLUENCY COMPONENT	WHAT TO LOOK FOR AS STUDENTS PRACTICE THIS STRATEGY
Efficiency	• Are students using the Break Apart to Multiply strategy or are they reverting to previously learned and/or possibly less appropriate strategies? • Are students using the Break Apart to Multiply strategy efficiently? • Are they breaking apart factors in convenient ways? • Do they break apart factors regardless of appropriateness for the problem at hand? • Do they recognize after carrying out the strategy that there was a more efficient way to break apart the factors?
Flexibility	• Are students carrying out the strategy in flexible ways? (e.g., Do they break apart by addends only? Do they break apart factors in different ways?) • Do they change their approach to how they are breaking apart factors if the results are proving too complicated?
Accuracy	• Are students using the Break Apart to Multiply strategy accurately? (e.g., Do they multiply the correct numbers? Do they add the correct products?) • Are students finding accurate solutions? • Are they considering the reasonableness of their solutions?* • Are students estimating before finding solutions?*

*This consideration is not unique to this strategy and should be practiced throughout the pursuit of fluency with whole numbers.

WORKED EXAMPLES

Worked examples are problems that have been solved. Correctly worked examples can help students make sense of a strategy and incorrectly worked examples attend to common errors.

A strength (and a challenge) of Break Apart to Multiply is that it is flexible. Here are some challenges that students commonly encounter:

1. The quantity that is broken apart (e.g., 2) is added back on (rather than adding on two groups).
   - 32×22: changes to 32×20 to equal 640 and then incorrectly adds 2 to equal 642.

2. The student breaks apart a factor but not into parts that they can multiply mentally.
   - 25×48: breaks apart into $20 \times 48 + 5 \times 48$ and gets stuck.

3. The student thinks of breaking apart using addends (distributive property) or by factors (associative property) but forgets about the other option.

   - 25×48: breaks apart by addends, forgetting to break apart into factors ($25 \times 4 \times 12$).

4. The student becomes confused about place value when there are zeros in a problem.

   - 305×15: breaks apart into $30 \times 15 = 450$ and $5 \times 15 = 75$ and gets 525.

5. The student breaks apart as they would for adding two numbers, multiplying tens by tens and ones by ones and omitting the other parts.

   - 34×62: multiplies $30 \times 60 = 1,800$ and $4 \times 2 = 8$, then adds them to get 1,808.

The prompts from Activity 1.7 can be used for collecting worked examples. Throughout the module are various worked examples that you can use as fictional worked examples. A sampling of additional ideas is provided in the following table.

SAMPLE WORKED EXAMPLES FOR BREAK APART TO MULTIPLY

	ONE WAY	ANOTHER WAY
Correctly Worked Example (make sense of the strategy) What did _____ do? Why does it work? Is this a good method for this problem?	Patrick's work for 25×48: $25 \times 48 =$ $25 (40 + 8)$ $25 \times 40 = 1,000$ $25 \times 8 = \underline{\quad 200}$ $\qquad\qquad 1,200$	Theresa's work for 25×48: $25 \times 4 \times 12$ $100 \times 12 = 1,200$
Partially Worked Example (implement the strategy accurately) Why did _____ start the problem this way? What does _____ need to do to finish the problem?	Rosi's start for 44×26: $44 \times 26 = 11 \times 4 \times 26$	Daniel's start for 560×14: $500 \times 10 = 5,000$ $500 \times 4 = 2,000$
Incorrectly Worked Example (highlight common errors) What did ____ do? What mistake does ____ make? How can this mistake be fixed?	Denisa's work for 708×60: $70 \times 60 = 4,200$ $8 \times 60 = \underline{\quad 480}$ $\qquad\qquad 4,680$	Christopher's work for 26×46: $20 \times 40 = 860$ $6 \times 6 = 36$ $800 + 36 = 836$

ACTIVITY 1.8

Name: "Complex Number Strings"   **Type:** Routine

About the Routine: Fluency with multiplying two-, three-, and four-digit factors relies on prowess with multiplying multiples of tens, hundreds, and thousands. Students also must be capable beyond just adding zeros. Instead, they must recognize that 7×60 is 7×6 tens and so on. A complex number string helps students practice relationships between basic facts and relationships between basic facts and multiplying multiples.

Materials: This routine does not require any materials.

Directions: 1. Provide a matrix of related number strings with one known product. The first row in this routine is always left blank and found first as an anchor for understanding and conversation as needed.

2. Students use the known product to work across the rows and down the columns.

3. After students signal that they know the products of each, you hold a class discussion about how the first known relates to the others. Draw students' attention to how basic facts relate to multiplying multiples of tens, hundreds, and thousands. Keep in mind that students will recognize the pattern of zeros in the products. Be sure to reinforce why this pattern makes sense.

1×1 =	10×1 =	10×10 =	100×1 =	100×10 =
$9 \times 5 = 45$	$90 \times 5 =$	$90 \times 50 =$	$900 \times 5 =$	$900 \times 50 =$
$9 \times 6 =$	$90 \times 6 =$	$90 \times 60 =$	$900 \times 6 =$	$900 \times 60 =$
$9 \times 7 =$	$90 \times 7 =$	$90 \times 70 =$	$900 \times 7 =$	$900 \times 70 =$
$9 \times 8 =$	$90 \times 8 =$	$90 \times 80 =$	$900 \times 8 =$	$900 \times 80 =$
$9 \times 16 =$	$90 \times 16 =$	$90 \times 160 =$	$900 \times 16 =$	$900 \times 160 =$

ACTIVITY 1.9

Name: "The Breaks"　　　　　　　　　　　　　　**Type:** Routine

About the Routine: Students are most likely comfortable decomposing numbers before they work with multidigit factors. Even so, they may not think to decompose numbers in strategic ways. This routine intends to have students make connections between the different ways factors can be broken apart and then determine which may be most advantageous. This routine can use feature Break Apart by Factors or Addends, but the examples focus on Break Apart by Addends.

Materials: Prepare a collection of related expressions in one row as shown.

Directions:　1. Pose a multiplication expression to students, charging them with thinking about how they might break apart the first factor or the second factor.

2. Students share some ideas with a partner.

3. Bring the group together to discuss different decompositions. Record the decompositions in two columns, with one column for decompositions of the first factor and one column for decompositions of the second factor. After recording a handful of each, ask students to identify which may be most useful or most friendly. Then, discuss if there are any ideas about how they might break apart both factors for an even better option.

The Breaks: 32×8

BREAK APART 32	BREAK APART 8
$30 \times 8 + 2 \times 8$	$32 \times 4 + 32 \times 4$
$10 \times 8 + 10 \times 8 + 10 \times 8 + 2 \times 8$	$32 \times 5 + 32 \times 2 + 32 \times 1$
$15 \times 8 + 15 \times 8 + 2 \times 8$	$32 \times 2 + 32 \times 2 + 32 \times 2 + 32 \times 2$
$20 \times 8 + 10 \times 8 + 2 \times 8$	

In this example, the teacher posed 32×8. The teacher recorded student ideas for breaking apart 32 in the first column and ideas for breaking apart 8 in the second column. Many of the students agreed that the shaded expression was the best for them. Some students preferred the first expression ($30 \times 8 + 2 \times 8$). Then, the teacher prompted students to consider if there was a good option for breaking apart both factors. The class agreed that there wasn't for this problem.

ACTIVITY 1.10

Name: "Strategize Your Break Apart" **Type:** Routine

About the Routine: Efficiency is part of fluency. A problem can be broken apart in a variety of ways but we want students to think about the most efficient way to do so. This routine helps students consider different ways to break apart to efficiently solve a problem. Note that this routine can also focus on ways to break apart without then solving the problems. This allows for more problems to be explored, although sometimes seeking the answer provides insights into whether the break apart option was, in fact, efficient.

Materials: list of three or four problems of the same operation

Directions:
1. Show and discuss a series of problems, one at a time, like a Number Talk (Parrish, 2014).

2. Ask students to mentally break apart the first expression seen, and signal when ready.

3. Ask different students to share how they would break apart the expression to solve. Record all ideas.

4. Ask students to pick one of the Break Apart methods discussed and solve the problem.

5. Share answers and discuss whether the first Break Apart method mentioned worked out well or not.

6. Repeat with other problems (as time allows).

7. Conclude the series with a synthesizing discussion, asking questions such as these:

   • What were efficient ways to break the factors apart?

   • What other ways might you have broken apart the factors?

   • Are there any ways that wouldn't have worked? Are there any ways that wouldn't have worked well?

   • What might be a problem similar to this that wouldn't work well with the break apart strategy?

Examples of problems to use with this routine include the following:

TWO-DIGIT BY ONE-DIGIT	TWO-DIGIT BY TWO-DIGIT	THREE-DIGIT BY ONE-DIGIT	FOUR-DIGIT BY ONE-DIGIT
9×26	12×35	5×125	$7 \times 2{,}015$
8×42	72×15	8×350	$9 \times 2{,}700$
6×36	24×25	4×275	$3 \times 4{,}520$
7×54	16×45	6×120	$4 \times 1{,}625$

ACTIVITY 1.11

Name: *Factor Race*   **Type:** *Game*

About the Game: The skill of multiplying by multiples of 10 is necessary for the Break Apart strategy as well as the other multiplication strategies. *Factor Race* is a practice opportunity for multiplying by multiples of 10, 100, or 1,000.

Materials: *Factor Race* cards and game board, two counters, calculator (one per player)

Directions:
1. Stack the *Factor Race* cards face down and place their counter on Start.

2. Players take turns flipping cards and finding the missing factors.

3. Players place their counter on top of the next space with a factor that corresponds to their missing factor.

4. Opponents check for accuracy. If incorrect, the player returns to their previous space.

5. The player who reaches the Finish block first wins.

In this example, Travis first selected 240 × ___ = 2,400 and moved his counter to 10. On his second turn, he selected 2 × ___ = 2,000 and moved his counter to the next 1,000 space.

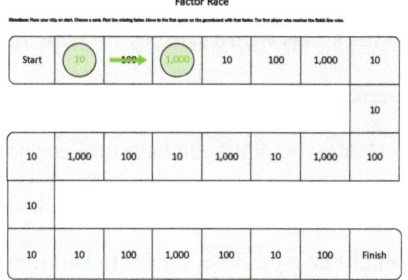

RESOURCE(S) FOR THIS ACTIVITY

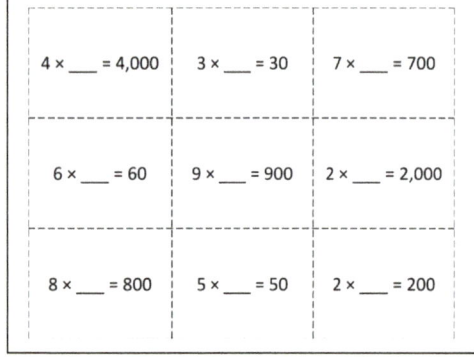

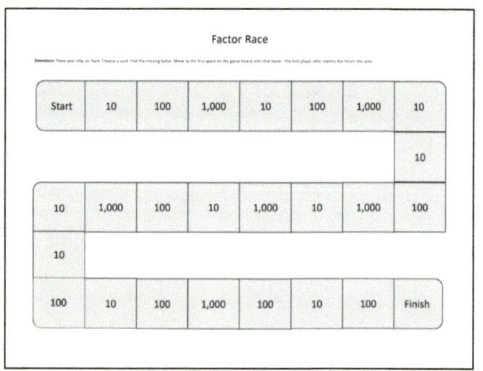

 These resources can be downloaded at **resources.corwin.com/FOF/multiplydividewholenumber**.

ACTIVITY 1.12

Name : *Mult-zee*   **Type:** *Game*

About the Game: *Mult-zee* is a take on the classic dice game, *Yahtzee*. It is an engaging way to practice multiplication using the Break Apart to Multiply strategy. *Mult-zee* can be modified in a variety of ways. As you change the game, you will need to change the conditions. For example, if you were to feature 2 two-digit factors, you would want to change "product is greater than 500" to "product is greater than 5,000" or "product is less than 500" to "product is less than 5,000."

Materials: 10-sided dice, four decks of digit cards (0–9), or playing cards (queens = 0, aces = 1, kings and jacks removed); *Mult-zee* game board

Directions: 1. Players deal themselves three digit cards to make a two-digit and a one-digit factor.

2. Players find the product of the factors.

3. Players use the product to satisfy one of the conditions on the game board by recording it in the corresponding space.

4. After each turn, the player must choose one of the places on the game board for the product they found or take a zero for that condition.

5. After 11 turns, the players add their scores and the highest score wins.

Note that there are two "chance" spaces on the game board. In these spaces, the product is the point value for the chance space.

RESOURCE(S) FOR THIS ACTIVITY

Mult-tzee

Directions: Make a multiplication problem with digit cards. Find the product. Choose a space to record your product. You only have 11 turns.

Player 1	Points	My Problem	Player 2	Points	My Problem
Even Product	100		Even Product	100	
Odd Product	100		Odd Product	100	
Product with a 5 in it	100		Product with a 5 in it	100	
Product with a 3 in it	100		Product with a 3 in it	100	
Product has two digits that are the same	500		Product has two digits that are the same	500	
Product is greater than 500	100		Product is greater than 500	100	
Product is less than 500	100		Product is less than 500	100	
Product is between 500 and 750	200		Product is between 500 and 750	200	
Chance (Product is your score)			Chance (Product is your score)		
Chance (Product is your score)			Chance (Product is your score)		
Product is 1,000	1,000		Product is 1,000	1,000	
Total score			Total score		

ACTIVITY 1.13

Name: Break Apart Bingo   **Type:** Game

About the Game: *Break Apart Bingo* helps students think about how they might break apart factors when multiplying.

Materials: *Break Apart Bingo* boards (one per student), counters, and expression cards (or list of expressions)

Directions: 1. Players take turns pulling expression cards.

2. Players solve the expression by using the Break Apart to Multiply strategy.

3. Then, players place a counter on decomposed parts they used to solve the problem.

4. Bingo is five in a row in any direction.

For example, the expression card 78 × 4 was called. Player 1 (left game board) broke apart 78 into 70 and 8. She covered the 8 because there was no 70 on her game board. Player 2 (right game board) broke it apart in the same way and was able to cover both numbers. Note that a player might have thought about 78 as 50 + 20 + 8 and could potentially cover three spaces. It is important that students first solve the problem before looking at their spaces, so they don't create inefficient decompositions to win the game.

Break Apart Bingo

Directions: Use the numbers at the top to fill in the boxes on the game board. You can write a number more than once.

2 3 4 5 6 7 8 9	20 30 40 50 60 70 80 90

2	30	20	50	4
9	(8)	60	20	5
40	20	4	40	7
60	50	5	30	4
3	80	20	9	8

Break Apart Bingo

Directions: Use the numbers at the top to fill in the boxes on the game board. You can write a number more than once.

2 3 4 5 6 7 8 9	20 30 40 50 60 70 80 90

5	60	(70)	(8)	3
90	40	6	5	20
3	80	20	9	8
5	6	50	40	3
20	7	30	8	4

You can find a sample list of expressions for playing the game on the companion website, although it will work with any examples. Expressions can also be written on index cards if you want small groups of students to play *Bingo* without you.

Blank 5 × 5 game boards can be copied or downloaded. Students can then create their own boards by writing the numbers 2–9 and multiples of 10 (20–90) anywhere on their board. Note that students can break apart factors by 1 and 10 but that might challenge the performance of the game. If a student uses 8 tens on his board and the first problem is 83 × 6, he could cover all 8 of his tens.

online resources Game cards and game boards can be downloaded at **resources.corwin.com/FOF/multiplydividewholenumber**.

ACTIVITY 1.14

Name: Break Apart to Multiply Math Libs   **Type:** Game

About the Game: Games do not always have to be ripe with strategy. Students can enjoy rather simplistic games of chance. *Break Apart to Multiply Math Libs* is a race to complete the equations. It can be adjusted for use with any strategy.

Materials: *Break Apart to Multiply Math Libs* game board; digit cards (0–9), playing cards (queens = 0, aces = 1, kings and jacks removed), or a spinner

Directions:
1. Players take turns pulling a digit card and using the number to fill a space on their board.

2. Once a player fully completes a problem, they use the Break Apart to Multiply strategy to find the unknown. Their opponent also finds the unknown to confirm they are correct (but cannot record it on their game board).

3. A player loses their turn if a digit they pull can't be used for any space on their board.

4. The first player to find all of the unknowns correctly wins.

For example, on Jackson's first turn he pulled a 5 and used it to fill the third 5 × 10 in the third row. Then, he pulled a 1 and completed the first row, finding the product. On his third turn, he pulled a 4 and completed the third row, finding the product of 170.

$21 \times 6 = ?$	$(20 \times 6) + (\underline{\textbf{1}} \times 6) = ?$	$? = \textbf{126}$
$45 \times 7 = ?$	$(20 \times \underline{}) + (20 \times \underline{}) + (5 \times 7) = ?$	$? =$
$5 \times 34 = ?$	$(5 \times 10) + (5 \times 10) + (\underline{\textbf{5}} \times 10) + (5 \times \underline{\textbf{4}}) = ?$	$? = \textbf{170}$

RESOURCE(S) FOR THIS ACTIVITY

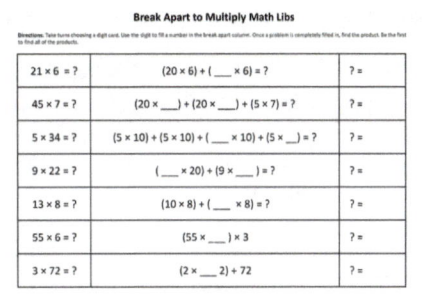

 Digit cards and this recording sheet can be downloaded at **resources.corwin.com/FOF/ multiplydividewholenumber**.

ACTIVITY 1.15

Name: Same But Different   **Type:** Center

About the Center: This center helps students recognize that numbers can be broken apart in different ways when multiplying. This strategy develops efficiency as students compare various ways to decompose a factor.

Materials: Same But Different cards, recording sheet

Directions:
1. Students select a Same But Different card.

2. Students copy the first Break Apart expression and solve.

3. Students copy the second Break Apart expression and solve.

4. After working with the center, students should be asked to explain how the problems are the same and how they are different.

In the following example, a student picks up the card 8×15. The student solves side A. Side A shows that the 8 was broken apart into $4 \times 15 + 4 \times 15$, which equals 120. Then the student solves side B. This card is missing a factor, so the student would have to think about which of the factors was broken apart. In this case, it was the 8. The students would then complete the expression and solve: $2 \times 15 + 6 \times 15 = 120$.

8 × 15	
A	B
4 × 15 4 × 15	2 × 15 __ × 15

RESOURCE(S) FOR THIS ACTIVITY

Same But Different Cards

8 × 15		5 × 44	
A	B	A	B
4 × 15 4 × 15	2 × 15 __ × 15	1 × 44 4 × 44	5 × 40 5 × 4

12 × 30		424 × 8	
A	B	A	B
2 × 30 10 × 30	6 × 30 6 × 30	2 × 424 2 × 424 4 × 424	4 × 424 4 × 424

online resources ↖ This resource can be downloaded at **resources.corwin.com/FOF/multiplydividewholenumber**.

ACTIVITY 1.16

Name: Backward Breaks   **Type:** Center

About the Center: Reversing an activity is a good way to enrich a task (SanGiovanni, Katt, & Dykema, 2020). In this center, students are given a decomposed multiplication problem and they have to find the product and recompose it.

Materials: Backward Breaks cards, recording sheet

Directions: 1. Students choose a card and record the information.

2. Students find the product and record it.

3. Then, students determine the original problem.

4. Students determine if the decomposed expression card was the most efficient way (for them) to think about the problem or they write another way that the problem might have been solved.

Decomposed expression cards, including the examples here, are available for download.

RESOURCE(S) FOR THIS ACTIVITY

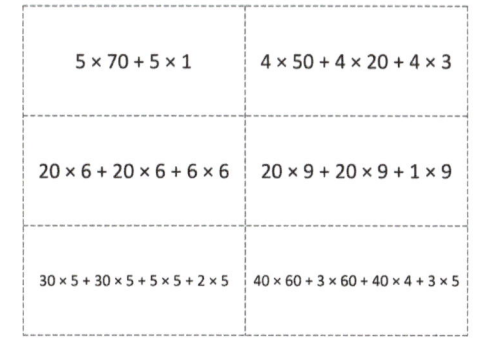

 Center cards and this recording sheet can be downloaded at **resources.corwin.com/FOF/ multiplydividewholenumber**.

ACTIVITY 1.17

Name: Break Apart With an Area Model Representation   **Type:** Center

About the Center: Early work with the Break Apart to Multiply strategy, and multidigit multiplication in general, uses precise area models. In time, instruction shifts to a more generic representation of an area model and eventually to a rather simplistic, abstract representation in which the parts are similar in size. This center is an opportunity for students to practice using this generic, area model representation and connecting it to the related equations.

Materials: digit cards (0–9) or playing cards (queens = 0, aces = 1, kings and jacks removed); Break Apart With an Area Model Representation recording sheet; generic area model recording sheet

Directions: 1. Students pull digits to create a multiplication problem relative to the size of factors they have been working with. For example, they pull three digit cards to make a two-digit by one-digit problem or four digit cards to make a two-digit by two-digit problem.

2. Students record the multiplication expression.

3. Students break apart one or both factors and represent their thinking with a generic area model.

4. Students record the equations represented by the area model.

Although this center activity seems rather simplistic, it offers a window into students' thinking. They are not directed how to break apart the factors, so you will see if they rely on groups of 10 as in the example or if they work with groups of tens. You can extend the center by asking students to create a problem and show two different ways to break apart the factors, or you can ask students to tell how they selected the break aparts that they did. This center also has potential as an assessment artifact.

RESOURCE(S) FOR THIS ACTIVITY

Breaking Apart

Directions: Use digit cards to create a multiplication problem. Break apart the factors to show how you multiplied. Show your thinking with an area model. Show your thinking with numbers.

Problem	Show It With an Area Model	Show It With Numbers
32 × 8	8 | 80 | 80 | 80 | 16	10 × 8 = 80 10 × 8 = 80 256 10 × 8 = 80 8 × 2 = 16

1	2	3	4
5	6	7	8
9	1	2	3
4	5	6	7
8	9	0	0

online resources → Digit cards and this recording sheet can be downloaded at **resources.corwin.com/FOF/ multiplydividewholenumber**.

ACTIVITY 1.18

Name: Which and How?   **Type:** Center

About the Center: As students become comfortable with the Break Apart to Multiply strategy, you want them to begin to think strategically about which factor to decompose and how to decompose it. This center offers an opportunity to practice choosing which factor to decompose and how it might be decomposed.

Materials: multiplication expression cards, math journal or recording sheet

Directions: 1. Students select a multiplication expression card and record it.

2. They circle the factor they choose to decompose.

3. Then, students record how they decomposed the factor and find the product.

In this example, the student first selected the problem 36×22. She decided to break apart 22 into $10 + 10 + 1 + 1$ and multiplied each by 36 finding a product of 792.

Problem	How
3 6 × ②②	$36 \times 10 + 36 \times 10 + 36 \times 1 + 36 \times 1$ $360 + 360 + 36 + 3$ $720 + 72$ 792
①④ × 5 3	$10 \times 50 + 10 \times 3 + 2 \times 53 + 2 \times 53$ $500 + 30 + 106 + 106$ $530 + 212$ 742

Problem	How
15 × ③②	15×32 $\diagup \diagdown$ $15 \times 4 \times 8$ 60×8 480
⑤⑤ × 20	55×20 $11 \times 5 \times 20$ 11×100 $1,100$

online resources A multiplication expression card template can be downloaded at **resources.corwin.com/FOF/ multiplydividewholenumber**.

Halve and Double Strategy

STRATEGY OVERVIEW:
Halve and Double

What is Halve and Double? As the name implies, this strategy involves changing two factors, halving one of them and doubling the other. It is a useful strategy for multiplication facts. For example, 5s facts can be thought of as halving the other number and doubling 5:

$$5 \times 8 = 10 \times 4$$
$$5 \times 9 = 10 \times 4.5$$

The strategy also generalizes to larger numbers:

$$5 \times 18 = 10 \times 9$$
$$50 \times 48 = 100 \times 24$$

You can also halve and double more than once if continuing to use the strategy makes the problem easier to multiply:

$$25 \times 640 = 50 \times 320$$
$$50 \times 320 = 100 \times 160$$
$$= 16,000$$

A key to this strategy is being adept at halving numbers and doubling numbers. In other words, students need automaticity with halving and doubling.

HOW DOES HALVE AND DOUBLE WORK?

This strategy is a special case of the Break Apart Into Factors where one factor is broken apart into $2 \times$ [the rest of the factor]. That 2 is associated with the factor, doubling it. It looks like this:

$$8 \times 35 = (4 \times 2) \times 35$$
$$= 4 \times (2 \times 35)$$
$$= 4 \times 70$$
$$= 280$$

WHEN DO YOU CHOOSE THE HALVE AND DOUBLE STRATEGY?

This strategy is very useful any time one factor is 5, 50, 500, and so on. When the other factor is even, it is all the more convenient, but the other factor does not have to be even for the strategy to work.

$28 \times 5 = 14 \times 10$	$34 \times 50 = 17 \times 100$	$79 \times 500 = 39.5 \times 1,000$

Doubling a number that ends in a 5, like 35, results in a tens number, like 70. This often makes the adapted problem easier to multiply, even if it continues to be a written problem and not a mental problem:

$$68 \times 350 = 34 \times 700$$

HALVE AND DOUBLE:
Strategy Brief for Families

It is important that families understand the strategies and know how they work so that they can be partners in the pursuit of fluency. This strategy brief is a tool for doing that. You can include it in parent or school newsletters or share it at parent conferences. It is available for download so that you can adjust it as needed.

Halve and Double

How It Works: We can halve a factor and double another to get an answer.

1. Choose which factor to halve and which to double.
2. Halve one factor.
3. Double the other factor.
4. Find the product by multiplying.

The left example shows that you can halve 14 and double 15. These new numbers make it easier to multiply. Now the number sentence is 7 × 30, which is 210. So, 14 × 15 = 210.

The right example shows a different way that you can halve and double. The problem is 25 × 8: 8 can be halved and 25 doubled to create 50 × 4. You can double and halve again: 4 can be halved and 50 doubled, resulting in 100 × 2. 100 × 2 = 200 is the same as 25 × 8 = 200.

When It's Useful: Halving and doubling is useful when one factor is an even number. It also helps when both factors can be easily halved and doubled. For example, 38 × 17 is not a good fit for this strategy because 19 × 34 is not a friendly problem to solve.

Halve ⤵ 14 × 15 ⤴ Double
7 × 30

7 × 30 = 210

Double ⤵ 25 × 8 ⤴ Halve
50 × 4

50 × 4 = 200

100 × 2 = 200

MODULE 1 — Break Apart to Multiply Strategy
MODULE 2 — Halve and Double Strategy
MODULE 3 — Compensation Strategy
MODULE 4 — Partial Products Strategy
MODULE 5 — Think Multiplication Strategy
MODULE 6 — Partial Quotients Strategy
MODULE 7 — Standard Algorithms for Multiplication and Division

TEACHING ACTIVITIES for Halve and Double

Before students are able to choose strategies, a key to fluency, they first must be able to understand and use relevant strategies. These activities focus specifically on Halve and Double. Halve and Double is a strategy that works well in certain cases. While students may employ other methods, which is appropriate, they also must learn this strategy to develop their flexibility and efficiency. Students must understand that halving a factor while doubling the other yields the same result. You can develop this concept by working with a variety of representations and connecting them to equations. In this section, we focus on instructional activities for helping students develop understanding of the Halve and Double strategy and for determining when it is appropriate. Fluency with this strategy will build as students develop skill with halving or doubling numbers in general.

ACTIVITY 2.1
HALVING AND REARRANGING RECTANGLES

In this activity, you pose rectangles to students for them to trace on grid paper and cut out. Rectangle dimensions to use could include 15×14, 28×5, 16×6, 25×22, and 16×16. Students record the equation for the area of one of their rectangles. Then, they cut the rectangle in half and rearrange it to make one, longer rectangle and record the equation for it. In the following figure, the original rectangle was 15×14. The student cut it in half, creating two rectangles that were each 15×7. Then, she aligned those two smaller rectangles, creating a rectangle that was 30×7. She recorded the equations for each rectangle's area.

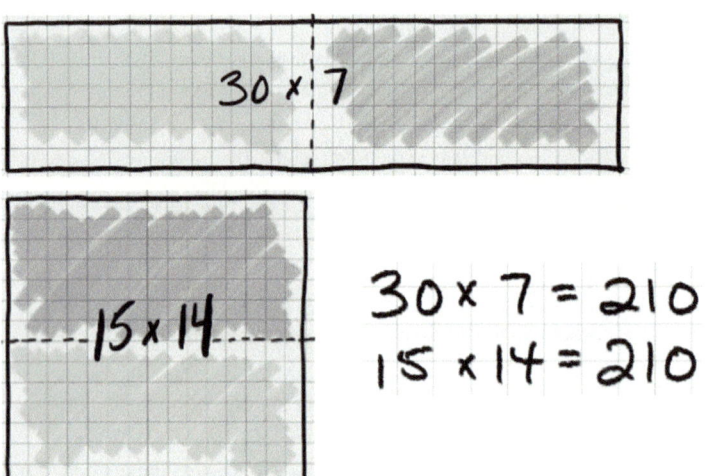

Students repeat halving and rearranging for the other rectangles that you pose. As students work, you want them to think about how the factors and the area of the corresponding rectangles are similar and different. After students complete the activity, a group discussion should be had about how one factor is halved and how the other is doubled yet the product (area) remains the same. Ask students to determine if this idea always works and then have them find examples to prove their case. It's possible that students will overlook that one factor is always even. If so, be sure to insert a rectangle, like 13×15 or 17×21, for them to explore.

Even though this is an introduction to the concept of halving and doubling, it isn't too soon to discuss when this technique might be a good idea. Notice that 15×14 and 28×5 from the suggested set create expressions that yield multiples of 10 (30×7 and 14×10, respectively). Because of this, the Halve and Double strategy is a good choice for these expressions.

 # ACTIVITY 2.2
CUISENAIRE ROD PROOFS

Cuisenaire rods are useful for developing a variety of concepts beyond fractions. In this activity, students work with orange and yellow Cuisenaire rods to investigate halving and doubling. First, be sure that students recognize that a yellow rod is exactly half of an orange rod. Then, you can create a train of oranges with an assigned value. For example, you might begin with orange having a value of 14 and ask students what the value of a yellow rod would be (7).

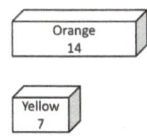

Then, ask students to build a train of five orange rods and have them determine how long the train is. They will find that a train of five orange rods is 70.

Orange 14	Orange 14	Orange 14	Orange 14	Orange 14

Then, ask students to predict and find how many yellow rods it would take to create a train of yellow rods with the same length of 70.

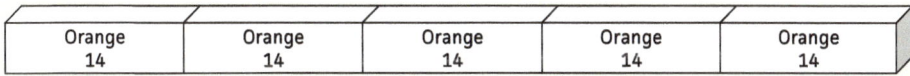

Orange 14	Orange 14	Orange 14	Orange 14	Orange 14

Yellow 7	Yellow 7	Yellow 7	Yellow 7	Yellow 7	Yellow 7	Yellow 7	Yellow 7	Yellow 7	Yellow 7

Be sure to have students record equations that describe the trains that they make. Some students may need a calculator to support their accuracy, which is fine. In this example, the two equations to record are $5 \times 14 = 70$ (orange train) and $10 \times 7 = 70$ (yellow train). You can repeat the activity endlessly by asking, "What if the orange had a value of ___?" You can also change the number of orange rods in a train. For example, you could pose that an orange rod has a value of 22 and there are 7 of them, yielding a length with the value of 154 (22×7). Then, students would work to find a yellow train with 14 rods and the same length of 154. Keep in mind that you want to look for patterns in the equations, how they represent the strategy of halving and doubling, and what values (factors) this strategy works best with.

ACTIVITY 2.3
REPRESENTING HALVE AND DOUBLE WITH GRAPHIC ORGANIZERS

Graphic organizers can be a useful tool for understanding the concept of a strategy and how it works. Graphic organizers, like the following Halve and Double organizer, work well when joined with other instructional activities such as Activities 2.1 or 2.2. This organizer has students identify which factor to halve and which to double. The middle section focuses the action on halving and the resulting products. The bottom section shows how the halved product can be repositioned to create the new problem. In the examples, the student works with 40×12. She halves 12 into 6 and 6 and eventually 80×6 or 480. The examples show the possible rich conversations that can come from these activities. In it, the student halves 40 instead of 12 and their subsequent work is shown. That rich conversation then comes from the questions you might ask: Which factor do you think was a better choice for halving? How does it help getting to a related problem that connects with basic facts? Why do some problems let us halve and double either factor?

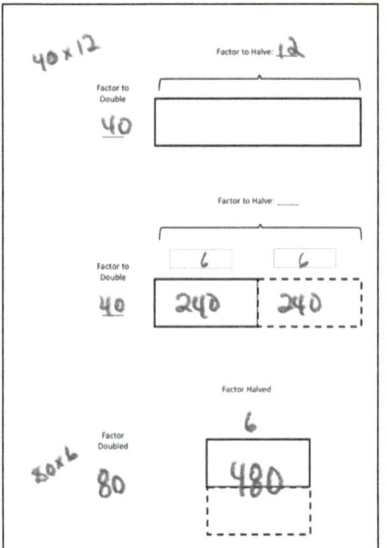

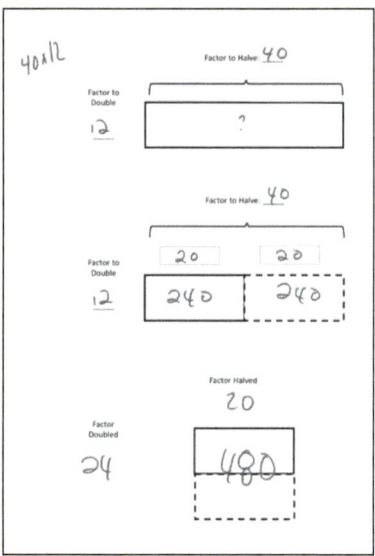

online resources — This graphic organizer can be downloaded at **resources.corwin.com/ figuringoutfluency.**

ACTIVITY 2.4
NICKELS AND DIMES

Nickels and dimes may be one of the most effective contexts for developing the halving and doubling strategy. They also represent a situation in which a problem can be thought of with multiples of 10. The activity is a rather straightforward problem. You tell students that Bag A has a certain amount of nickels and students work to find how many dimes would be in another bag (Bag B) so that the two bags have the same amount of money. Here are two problems that students solved.

Bag A has 28 nickels.

$28 \times 5 = 140 ¢$

Bag B has 14 dimes.

$14 \times 10 = 140 ¢$

28×5 is the same as 14×10

Bag A has 44 nickels.

$44 \times 5 = 220 ¢$

Bag B has 22 dimes.

$22 \times 10 = 220 ¢$

44×5 is the same as 22×10

Students created the following table to show the number of nickels and dimes and to show how they recorded the corresponding equations after each problem. The table and equations help students see how the factors are halved and doubled.

n	d	
16	8	$16 \times 5 = 8 \times 10$
22	11	$22 \times 5 = 11 \times 10$
28	14	$28 \times 5 = 14 \times 10$
32	16	$32 \times 5 = 16 \times 10$
44	22	$44 \times 5 = 22 \times 10$

This activity is also quite useful for showing how the halving and doubling strategy can work with odd numbers. For example, if you have 7 nickels you can think of this as 3 dimes and a nickel (35¢) or you can think of it as 3.5 dimes, and $3.5 \times 10 = 35$.

This activity can be adapted to exploring quarters (25¢). For example 40 quarters (40×25) is the same as 20×50 or 10×100, to equal 1,000¢ or $10.

ACTIVITY 2.5
PROMPTS FOR TEACHING HALVE AND DOUBLE

Use the following prompts as opportunities to develop understanding of and reasoning with the strategy. Have students use representations and tools to justify their thinking, including base-10 models, number lines, arrays, and so on. After students work with the prompt(s), bring the class together to exchange ideas. These could be useful for collecting evidence of student understanding. Any prompt can be easily modified to feature different numbers (e.g., three-digit or four-digit numbers) and any prompt can be offered more than once if modified.

- David stated that 18×6 is the same as 9×3. Kayce disagreed and said that 18×6 is the same as 9×12. How do we know that Kayce is correct? What was David's mistake?

- Sara said the halve and double strategy works with division too, but it is halve and halve instead (or double and double). For example, $240 \div 6$ is the same as $120 \div 3$. Justify whether or not this is always true.

- Ciera said that doubling and halving always works. Is that correct? If not, give an example of when it doesn't work.

- Is Halve and Double the most effective strategy when solving for 20×15? Why or why not?

- Does halving and doubling always work well? Create problems to justify your thinking.

- Che says you can't use Halve and Double for 15×17. Maria says she can use the strategy. She says she just has to make one of the factors even. She gives 1 from 17 to 15, creating 16×16. What do you think about Maria's strategy?

- Russ says 4×15 is perfect for using the Halve and Double strategy but that 12×15 and 9×15 aren't good examples of when to use the strategy. Explain why you agree or disagree with Russ.

- Write your own multiplication word problems that could be solved using the Halve and Double strategy.

- Michael used the Halve and Double strategy to solve 8×15. He halved 8 and doubled 15. Michael came up with 4×30 and found a product of 120. Would the strategy still be efficient if Michael halved 15 and doubled 8 instead?

- Create a story problem for 5×22. Solve using Halve and Double.

NOTES

_____

_____

_____

PRACTICE ACTIVITIES for Halve and Double

Fluency is realized through quality practice that is focused, varied, processed, and connected. The activities in this section focus students' attention on how this strategy works and when to use it. The activities are a collection of varied engagements. The discussion you facilitate after an activity or the reflection prompts you attach to it should help students think about what they did mathematically, how they reasoned about the activity, and when the math they did (namely the strategy) might be useful. Debriefing should also help students see how the practice activity connects to recent instruction or how the strategy connects to other strategies they know. Game boards, recording sheets, digit cards, and other required materials are available as online resources for you to download, possibly modify, and use. As students work with activities, look for how well they are acquiring the strategy and assimilating it into their collection of strategies.

FLUENCY COMPONENT	WHAT TO LOOK FOR AS STUDENTS PRACTICE THIS STRATEGY
Efficiency	• Are students using the Halve and Double strategy or do they rely on other strategies regardless for how appropriate they are for the problem at hand? • Are students using the Halve and Double strategy efficiently? • Do they use the strategy regardless of its appropriateness for the problem at hand?
Flexibility	• Are students carrying out the strategy in flexible ways? (e.g., With this strategy do they make certain tens well and others not as well?) • Do they change their strategy to Halve and Double as they realize it will be more useful? (e.g., When solving 35×8, a student begins to break apart and then realizes they can make the problem 70×4.) • Do they change their approach from Halve and Double as it proves too complicated for the problem at hand?
Accuracy	• Are students using the Halve and Double strategy accurately? (e.g., Do they halve and double or double without halving?) • Are students finding accurate solutions? • Are they considering the reasonableness of their solutions?* • Are students estimating before finding solutions?*

*This consideration is not unique to this strategy and should be practiced throughout the pursuit of fluency with whole numbers.

WORKED EXAMPLES

Worked examples are problems that have been solved. Correctly worked examples can help students make sense of a strategy and incorrectly worked examples attend to common errors.

Halve and Double is useful in special cases, so a major challenge is to just remember it *is* a strategy and be on the lookout for when it will work. Other challenges include the following:

1. The student makes an error in halving an odd number.

   - 65×5: changes to 38.5×10.

2. The student has trouble choosing which number to halve and which to double.

   - 450×12: changes to 225×24, rather than 900×6.

A key idea is to look for a number that, when doubled, is more manageable to multiply. In addition, noticing a second opportunity to use Halve and Double can continue to make the problem easier to work. Worked examples can attend to these ideas. The prompts from Activity 2.5 can be used for collecting worked examples. Throughout the module are various worked examples that you can use as fictional worked examples. A sampling of additional ideas is provided in the table.

SAMPLE WORKED EXAMPLES FOR HALVE AND DOUBLE	
Correctly Worked Example (make sense of the strategy) What did _____ do? Why does it work? Is this a good method for this problem?	Trevor's work for 250×28: 250×28 500×14 $1{,}000 \times 7 = 7{,}000$
Partially Worked Example (implement the strategy accurately) Why did _____ start the problem this way? What does _____ need to do to finish the problem?	Neesha's start for 15×260: $30 \times 130 =$ Dante's start for 25×68: $50 \times 34 =$
Incorrectly Worked Example (highlight common errors) What did _____ do? What mistake does _____ make? How can this mistake be fixed?	Greta's work for 50×45: $100 \times 23.5 = 2{,}350$

ACTIVITY 2.6

Name: "A String of Halves"   **Type:** Routine

About the Routine: Making use of the Halve and Double strategy relies on students' ability to double and find halves. Often, students do well with doubling but finding a half proves more challenging. "A String of Halves" aims to help students improve their skill with halving. This routine uses a set of numbers to help students see how numbers are halved.

Materials: Prepare a set of numbers that are intentionally related.

Directions: 1. Post a set of three or more related numbers as shown.

2. Ask students to mentally find the halves of as many numbers as possible.

3. If students are unable to find all of the halves, have them talk with partners about how they can use the relationships between numbers and the known halves to find the unknown half. If students find all of the halves, have them discuss with partners how the numbers and their halves are related. To extend this situation (when all halves are known), have students generate a new number and its half that is related to the set.

4.
A		
40	16	56

B		
30	40	50

40	16	56
20	8	

30	40	50
	20	25

These two examples of "A String of Halves" were used in a fourth-grade classroom. The teacher started with Example A, recording 40, 16, and 56 because she wanted to develop an idea about how to halve 56. Students were asked to think of the half for each of the numbers. They could halve 40 and 16. At that point, she asked students to talk with partners about how they could figure out the half of 56 by knowing the halves of 40 and 16. The teacher then used Example B in a similar way. She posed the three related numbers, and students found the halves. She then asked students to look for patterns and relationships between the numbers and the halves to prove what half of 30 is.

Halving numbers like 76 or even 90 can be challenging for students because there is an element of regrouping within the half. This routine can be leveraged for developing skill with those halves by making use of relationships and patterns. In the following figure, four multiples of 10 (60, 70, 80, and 90) are posed. Students are tasked with halving each, but the conversation focuses first on 60 and 80 before students work to find half of 70 and 90. During these discussions, focus their attention on half of 10 being added to or taken from half of 60 or 80 and why. Of course, another strategy is to think of the half (35) between those halves (30 and 40).

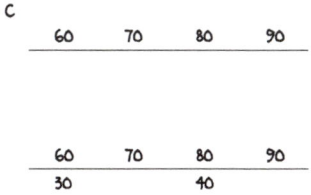

ACTIVITY 2.7

Name: "The Same As"　　　　　　　　　**Type:** Routine

About the Routine: This routine helps students think about how to use the Halve and Double strategy. As students become comfortable with the strategy, you might sprinkle in expressions that don't call for the strategy (6×50) or those that prove problematic for the strategy (7×13).

Materials: Prepare two or three expressions for students to restate.

Directions:　1. Pose a few expressions to students.

2. Students talk to each other about another way they could think about the problem by either doubling or halving. To note, students do not need to solve the expressions, just rewrite them.

3. Students discuss which of the expressions would be easier to solve.

THE SAME AS		THE SAME AS	
6×12 is the same as _____	35×8 is the same as _____	350×6 is the same as _____	16×450 is the same as _____

These are two different examples of the routine. The left example shows two-digit by one-digit problems. First, students were tasked with rethinking 6×12. Students share that it can be thought of as the same as 3×24. The second prompt, 35×8, is the same as 70×4. After students shared suggestions for these two, they then discussed which situation was a good use of the strategy (35×8 because it becomes a friendly problem). The right example shows how the routine can be used with larger factors. In this case, both problems are good for the Halve and Double strategy.

ACTIVITY 2.8

Name: The Splits   **Type:** Game

About the Game: Skill with halving is useful in many ways and necessary for leveraging the Halve and Double strategy. *The Splits* is a practice opportunity for halving. The object of the game is to generate numbers, halve them, and put the halves in order from least to greatest.

Materials: place value dice, digit cards (0–9), or playing cards (queens = 0, aces = 1, kings and jacks removed); *The Splits* game board

Directions:

1. Players take turns generating a two-digit number. Note that three- and four-digit numbers are also options.

2. Players halve the number and record the result in one of the five boxes.

3. Each subsequent half must be placed in a box so that halves are in order from least to greatest.

4. If a half cannot be placed, the player loses their turn. (Optional: If an odd number is generated, the player loses their turn also.)

5. The first player to fill their five boxes wins.

For example, students are playing *The Splits* by finding half of two-digit numbers. Toby's first number is 44. He halves it and places 22 in the second box. After his partner's turn, Toby's second number is 70. Half of 70 is 35 and he places it to the right of 22. Toby's third number is 62. He loses his turn because there is no space to place the resulting half, 31.

TEACHING TAKEAWAY

Games are great homework assignments. Families can also use math games to keep kids busy, like when waiting for dinner at a restaurant.

Player 1

	22	35		

RESOURCE(S) FOR THIS ACTIVITY

The Splits

Directions: Take turns rolling dice or pulling cards to make a number. Find half of that number. Record the half in one of the spaces. The halves you put in the boxes have to be in order from least to greatest. If a number doesn't fit because of other numbers you have already played, you lose your turn. Be the first to fill all of your boxes.

online resources

Digit cards and this game board can be downloaded at **resources.corwin.com/FOF/ multiplydividewholenumber**.

ACTIVITY 2.9

Name: *Get 10*   **Type:** *Game*

About the Game: *Get 10* is a quick-and-easy game for practicing finding halves. The goal of the game is to be the first player to correctly halve 10 numbers.

Materials: 19 bingo chips or counters, four decks of digit cards or playing cards with faces and tens removed

Directions: 1. Players take turns pulling two digit cards.

2. Players have the chance to earn two points, finding half of each two-digit number that can be formed with their cards.

3. Opponents confirm accuracy with a calculator, and the player who correctly halved their numbers claims two chips.

4. First player to get 10 chips wins. For example, Chala draws a 5 and a 2; he says half of 52 is 26 and half of 25 is 12.5 or 12 and a half. He is correct both times and gets two chips.

This game can be played with three- and four- digit numbers, but rather than have multiple options, the player organizes the three or four cards in any order and finds half of that number.

TEACHING TAKEAWAY

Calculators ensure accurate practice. For this game, you might need to train students how to find half on a calculator.

ACTIVITY 2.10

Name: Halve and Double Flips   **Type:** Game

About the Game: *Halve and Double Flips* practices identifying multiplication problems that can be solved with the Halve and Double strategy. Players place chips on corresponding spaces and can flip opponents' pieces that lie directly between two of their pieces (similar to *Othello*).

Materials: *Halve and Double Flips* game cards and game board, multiplication expression cards, and two-colored counters for game pieces

Directions:
1. Players take turns pulling a multiplication expression card.

2. Players consider if the expression can be solved more efficiently using the Halve and Double strategy.

3. If the expression can be solved more efficiently with the strategy, the players put their counter on any space labeled "WORKS" (i.e., the Halve and Double strategy works for this expression).

4. If the expression cannot be solved more efficiently with the strategy, the player puts their counter on any space labeled "DOESN'T WORK" (i.e., the Halve and Double strategy doesn't make the problem easier to solve).

5. When placing a game piece, any opponent pieces between a previously placed piece and the current piece are flipped (similar to *Othello*). Note that unlike *Othello*, a player does not have to place a piece at the end of a row of pieces.

6. The game ends when all of the spaces are filled.

7. The player with the most chips on the board wins the game.

RESOURCE(S) FOR THIS ACTIVITY

44 × 5	14 × 15
3 × 16	50 × 18
40 × 28	32 × 6
16 × 6	18 × 40
28 × 5	5 × 48

Halve and Double Flips

Directions: Pull an expression card. Choose a space that tells if you can or can't uses the Halve and Double strategy with the expression. After placing a piece, flip over any opponent pieces caught between your new piece and a piece you already had on the board. The player with the most pieces when the board is filled wins.

Halve and Double **WORKS**	Halve and Double **DOESN'T WORK**	Halve and Double **WORKS**	Halve and Double **DOESN'T WORK**	Halve and Double **WORKS**	Halve and Double **DOESN'T WORK**
Halve and Double **DOESN'T WORK**	Halve and Double **WORKS**	Halve and Double **DOESN'T WORK**	Halve and Double **WORKS**	Halve and Double **DOESN'T WORK**	Halve and Double **WORKS**
Halve and Double **WORKS**	Halve and Double **DOESN'T WORK**	Halve and Double **WORKS**	Halve and Double **DOESN'T WORK**	Halve and Double **WORKS**	Halve and Double **DOESN'T WORK**
Halve and Double **DOESN'T WORK**	Halve and Double **WORKS**	Halve and Double **DOESN'T WORK**	Halve and Double **WORKS**	Halve and Double **DOESN'T WORK**	Halve and Double **WORKS**
Halve and Double **WORKS**	Halve and Double **DOESN'T WORK**	Halve and Double **WORKS**	Halve and Double **DOESN'T WORK**	Halve and Double **WORKS**	Halve and Double **DOESN'T WORK**
Halve and Double **DOESN'T WORK**	Halve and Double **WORKS**	Halve and Double **DOESN'T WORK**	Halve and Double **WORKS**	Halve and Double **DOESN'T WORK**	Halve and Double **WORKS**

 Game cards and this game board can be downloaded at **resources.corwin.com/FOF/ multiplydividewholenumber**.

ACTIVITY 2.11

Name: *The First to Seven* **Type:** *Game*

About the Game: *The First to Seven* is a game for considering expressions that might be best solved with the Halve and Double strategy. Players attempt to get 7 points by collecting seven expression cards that can be efficiently solved with the Halve and Double strategy. After each successful flip, students must decide if they want to keep going or pass to their opponent.

Materials: four decks of digit cards (0–9) or playing cards (queens = 0, aces = 1, kings and jacks removed)

Directions:
1. Player 1 begins by using digit cards to make a multiplication problem.

2. Player 1 determines if the problem can be solved with the Halve and Double strategy and renames it accordingly and solves it mentally. If they solve it correctly, they score a point and have the choice of going again or passing.

3. If player 1 continues playing, they can earn another point by creating a second problem that can be solved with the Halve and Double strategy. If the player continues playing and cannot mentally solve the problem using Halve and Double, the player loses all of the points they have earned.

4. A player only goes back to zero if they attempt to get consecutive points.

5. The first player to 7 points wins the game.

For example, player 1 turns up 2, 4, 5, 1 and forms the expression 24×15. They say aloud, "I can halve 24 and double 15. That is 12 times 30, which is 360." Their opponent checks the answer on a calculator. Player 1 draws again. They draw 3, 6, 1, and 0. They form 30×16, and explain that they can halve and double to get 60×8, which is 480. They decide to stop, and play goes to their opponent.

TEACHING TAKEAWAY

During the game, have students record Halve and Double expressions. After the game, have students determine which expressions are truly best for the strategy.

ACTIVITY 2.12

Name: Fives and Tens   **Type:** Center

About the Center: This center is designed as an independent practice intended to focus on situations where multiples of tens are useful applications of the Halve and Double strategy. Often, that applies to leveraging multiples of 10. This center might pair well with Activity 2.4 ("Nickels and Dimes"). Fives and tens can be modified to focus students' attention on a similar relationship between 15 and 30, 25 and 50, or any other multiple of 5 and corresponding multiple of 10 as shown.

Materials: digit cards (0–9) or playing cards (queens = 0, aces = 1, kings and jacks removed), 10-sided dice, and Fives and Tens recording sheets

Directions:
1. Students generate a number by drawing two cards or rolling the dice two times..

2. Even numbers can be recorded as a factor in the ×5 column or the ×10 column.

3. Odd numbers are recorded in the ×10 column.

4. After placing the number as a factor in one of the columns, students write the related ×5 of ×10 problem that is generated by the Halve and Double strategy.

RESOURCE(S) FOR THIS ACTIVITY

5s and 10s

Directions: Generate a number. Use even numbers as a factor for the ×5 or ×10 column and use odd numbers for the ×10 column only. Write the related multiplication problem and find the product.

Number I Made	×5 Column	×10 Column	Product of Both
16	16 × 5	8 × 10	80
	____ × 5	____ × 10	

15s and 30s

Directions: Generate a number. Use even numbers as a factor for the ×15 or ×30 column and use odd numbers for the ×30 column only. Write the related multiplication problem and find the product.

Number I Made	×15 Column	×30 Column	Product of Both
16	16 × 15	8 × 30	240
	____ × 15	____ × 30	

25s and 50s

Directions: Generate a number. Use even numbers as a factor for the ×25 or ×50 column and use odd numbers for the ×50 column only. Write the related multiplication problem and find the product.

Number I Made	×25 Column	×50 Column	Product of Both
16	16 × 25	8 × 50	400
	____ × 25	____ × 50	

35s and 70s

Directions: Generate a number. Use even numbers as a factor for the ×35 or ×70 column and use odd numbers for the ×70 column only. Write the related multiplication problem and find the product.

Number I Made	×35 Column	×70 Column	Product of Both
16	16 × 35	8 × 70	560
	____ × 35	____ × 70	

 Digit cards and recording sheets can be downloaded at **resources.corwin.com/FOF/ multiplydividewholenumber**.

ACTIVITY 2.13

Name: Write It as Halve and Double   **Type:** Center

About the Center: This center prompts students to think about how they can rethink a multiplication problem using the Halve and Double strategy. It also asks them to think about if the strategy is useful for the problem. Students select an expression card, rewrite it with the strategy, and tell if using the strategy is efficient for the problem.

TEACHING TAKEAWAY

Like other centers, this one can be modified for other strategies by renaming expressions with the Compensation or Think Multiplication strategies.

Materials: Write It as Halve and Double cards and recording sheet

Directions: 1. Students pull an expression card.

2. They rewrite the expression using the Halve and Double strategy.

3. They tell if the Halve and Double strategy is useful for the problem.

RESOURCE(S) FOR THIS ACTIVITY

Write It as Halve and Double

Directions: Select a multiplication problem card. Rewrite it using the Halve and Double strategy. Tell if using the strategy is a good choice for the problem.

Problem on the multiplication card	Rewrite the problem using the Halve and Double strategy	Is the strategy a good choice for the problem? Why?

40 × 4	45 × 4
16 × 5	32 × 5
22 × 5	5 × 48
38 × 5	5 × 32
5 × 42	7 × 45

 Game cards and this recording sheet can be downloaded at **resources.corwin.com/FOF/ multiplydividewholenumber**.

NOTES

Compensation Strategy

STRATEGY OVERVIEW:
Compensation

What is Compensation? Compensation involves making a change or adjustment to one of the factors and then compensating for that change to maintain equivalence. Often, Compensation involves rounding up one of the factors, multiplying, then compensating for the extra. For example, with basic facts, the 9s facts can be developed by thinking of 9×6 as 10×6 and then subtracting one group of 6 (compensating for the extra group of 6). This also extends to larger numbers:

$$99 \times 6 = 100 \times 6 - 6$$
$$= 600 - 6$$
$$= 594$$

HOW DOES COMPENSATION WORK?

Compensation employs the distributive property:

$$12 \times 48 = 12(50 - 2)$$
$$= 12 \times 50 - 12 \times 2$$
$$= 600 - 24$$
$$= 576$$

A key is to keep track of the number of *groups* (or sets or rows or columns) added and subtract the same number. In this example, 48 was adjusted to 50 (adding two extra columns of 12). In the end, those two extra groups (24) are subtracted to compensate.

WHEN DO YOU CHOOSE COMPENSATION?

This strategy is very useful when one of the factors is close to a benchmark number like 70 or 50 or 200 and that benchmark can be mentally multiplied by the other factor.

$68 \times 7 = 70 \times 7 - 14$	$8 \times 49 = 50 \times 8 - 8$	$198 \times 15 = 200 \times 15 - 30$

Notice that in these three examples, Break Apart to Multiply is also a reasonable option. However, because the ones place is a larger digit (i.e., is close to the next benchmark), Compensation is likely easier. Compensation can be applied to other benchmark numbers, such as 25. For example, 6×24 can be solved with Compensation by thinking $6 \times 25 = 150$ and then subtracting a group of 6.

COMPENSATION:
Strategy Brief for Families

It is important that families understand the strategies and know how they work so that they can be partners in the pursuit of fluency. This strategy brief is a tool for doing that. You can include it in parent or school newsletters or share it at parent conferences. It is available for download so that you can adjust it as needed.

MODULE 1 Break Apart to Multiply Strategy

MODULE 2 Halve and Double Strategy

MODULE 3 Compensation Strategy

MODULE 4 Partial Products Strategy

MODULE 5 Think Multiplication Strategy

MODULE 6 Partial Quotients Strategy

MODULE 7 Standard Algorithms for Multiplication and Division

Compensation

How It Works: We can change a factor to find a product and then you can compensate for the change.

1. Decide on a friendlier computation.
2. Change a factor and find the product.
3. Compensate (undo) the amount that was changed.

The left example shows that there are 49 groups of 7: 49 × 7. You can change 49 to 50 because 50 is easier to multiply with. This means you are adding an extra group to make it 50 groups instead of 49. Multiply 50 groups of 7: 50 × 7 = 350. Then, you can take away the extra group of 7: 350 − 7 = 343.

The right example shows that 23 × 8 can be changed to 25 × 8, which means you added 2 groups to 23. You now have 25 groups: 25 × 8 = 200. Then, take away the two extra groups of 8 (i.e., 16): 200 − 16 = 184. So, 23 × 8 = 184.

When It's Useful: Compensation is useful any time a factor can be easily changed to a a factor that is easier to multiply. For example, 23 × 17 can be thought of as 23 × 20 − 23 × 3, but that number sentence may not be easier to multiply.

Factor   Factor
49 × 7

49 × 7
(50−1)

50 × 7 = 350
350 − 7 = 343
↗ product

Factor   Factor
23 × 8

23 × 8
(25−2)

25 × 8 = 200
200 − 16 = 184
↗ product

 This resource can be downloaded at **resources.corwin.com/FOF/ multiplydividewholenumber.**

TEACHING ACTIVITIES for Compensation

Before students are able to choose strategies, a key to fluency, they first must be able to understand and use relevant strategies. These activities focus specifically on Compensation. Compensation is a strategy that works well in certain cases. While students may employ other methods, which is appropriate, they also must learn this strategy to develop their flexibility and efficiency. Students must understand how expressions that have common factors are related and how a number of groups must be taken away or added on when expressions are adjusted. You can develop this concept by working with a variety of representations and connecting them to equations. In this section, we focus on instructional activities for helping students develop understanding of the Compensation strategy. Compensation with multiplication is different from compensation with other operations, so we also offer activities to expose this difference.

 ACTIVITY 3.1
SHOW ME

In this activity, students compare the area of related expressions to prove that compensation works. At first, you might use grid paper with relatively small factors so that students can find the exact area of each. Pose two expressions with a common factor such 12×9 and 12×10. Have students use grid paper to make two rectangles with those dimensions and find the area of both. You might consider having students cut out each rectangle to see how they are related when one is placed on top of the other. Have students discuss how the dimensions of the rectangles are similar and how they are different.

$$12 \times 9 = 108$$
$$12 \times 10 = 120$$

After discussion about the first set of rectangles, prompt students to repeat the activity with pairs of different rectangles, including 15×9 and 15×10, 20×9 and 20×19, and 25×9 and 25×10. Before students work with each set, have them discuss how the factors are related and how the expressions are similar to the first investigation. Bring the class together after they have found the area of each rectangle described above. Record their findings as shown in the following figure, so that patterns might be observed by students.

$$15 \times 10 = 150 \qquad 20 \times 10 = 200 \qquad 25 \times 10 = 250$$
$$15 \times 9 = 135 \qquad 20 \times 9 = 180 \qquad 25 \times 9 = 225$$

Again, ask students to think about how the expressions are the same and different. Extend the activity by asking them to create sets of expressions that show a similar relationship. The next day you can repeat the activity with expressions that multiply by 8 and by 10.

As you move to developing the Compensation strategy with 2 two-digit factors, you can have students use open area representations. For example, you might have them work with 12×30 and 12×28 to determine how the products are related.

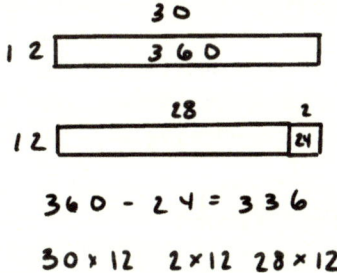

As with the earlier expressions, help students see how the products of two expressions with common factors are related. In this case, 30×12 is two groups of 12 more than 28×12.

 **ACTIVITY 3.2**
COMPENSATED TRAINS

Area models work well to show how products compare when the Compensation strategy is used. Students often benefit from a different representation to galvanize their understanding of a concept. This activity functions similarly to Activity 3.1 in that you want to have students compare the results of two related expressions. Also know that this physical model loses functionality as the size of the factors increases. Have students create a train of linking cubes of a certain size such as 13. Then, have students predict and find how many cubes would be in 9 trains of 13 cubes and how many cubes would be in 10 trains of 13 cubes. As with Activity 3.1, be sure to record the equations for each and have students examine the equations ($10 \times 13 = 130$ and $9 \times 13 = 117$).

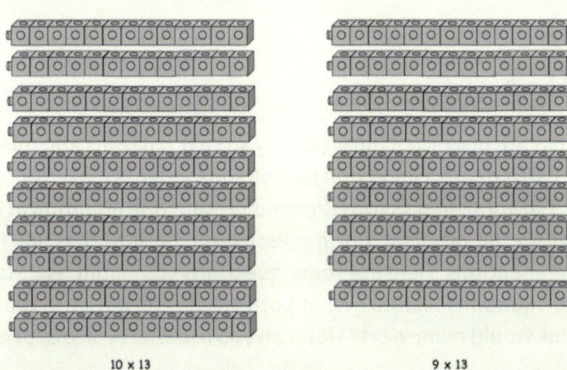

Alternatively, you can have each student in your class build a train of 13 and then have 10 students link their trains and compare its overall length with a train built by 9 students.

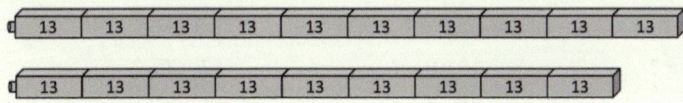

ACTIVITY 3.3
TRUE OR FALSE COMPENSATION STATEMENTS

Students may believe that Compensation for multiplication functions similarly to Compensation with addition. For example, they may believe that 38×8 is the same as $40 \times 8 - 2$ because they added 2 to 38, rather than recognize they are adding 2 groups of 8. In this activity, you provide students with a collection of statements and they have to work to prove if they are **true or false**. Students should be asked to justify their thinking. After working with the activity, the class should be brought back together to debrief and solidify that when you use Compensation for multiplication, you are adjusting with groups.

Here are some possible prompts for two-digit by one-digit multiplication:

True or False:

- To solve 9×15, I can add 1 to 9 to get 10×15 (150) and compensate by subtracting 1.

- To solve 58×5, I can add 2 to 58. Then 60×5 is 300, and I subtract 2 fives to equal 290.

- To solve 9×11, I go up 1 for one, down 1 for the other to get 10×10. My answer is 100.

Here are some examples of prompts for other multidigit multiplication situations:

True or False:

- To solve 19×25, I multiply 25 times 20, which is 500, and subtract 25 to equal 475.

- 175×8 is the same as $175 \times 10 - 2$.

- 25×38 is the same as 25×40, and then subtract 50.

- 35×99 is the same as 35×100, then subtract 1.

For more examples see Activity 3.13.

ACTIVITY 3.4
NUMBER STRINGS

Number strings are a set of related equations that are organized intentionally so that patterns and relationships can be easily observed and used. Compensation number strings are a good tool for helping students make sense of the strategy. First, prepare a string for students to work with. Gather students together and present one equation at a time. Students solve each new problem and share how they solved the problem. During discussion, be sure that you help establish or reinforce the connections between consecutive equations. Here are some questions you might ask: How are the equations similar and different? How are the first two equations related? What patterns do you notice in the equations? Why do you think that pattern is so? What do you think would come next? How can you use what you discovered to help you?

STRING A	STRING B	STRING C	STRING D
$4 \times 10 = ?$	$4 \times 10 = ?$	$6 \times 10 = ?$	$7 \times 10 = ?$
$4 \times 9 = ?$	$4 \times 9 = ?$	$6 \times 8 = ?$	$7 \times 7 = ?$
$14 \times 9 = ?$	$4 \times 19 = ?$	$16 \times 8 = ?$	$17 \times 7 = ?$
$24 \times 9 = ?$	$4 \times 29 = ?$	$26 \times 8 = ?$	$27 \times 7 = ?$
$34 \times 9 = ?$	$4 \times 49 = ?$	$36 \times 8 = ?$	$37 \times 7 = ?$
$44 \times 9 = ?$	$4 \times 69 = ?$	$46 \times 8 = ?$	$47 \times 7 = ?$

STRING E	STRING F	STRING G	STRING H
4 × 10 = ?	45 × 1 = ?	4 × 8 = ?	7 × 3 = ?
4 × 4 = ?	45 × 10 = ?	6 × 8 = ?	70 × 3 = ?
40 × 40 = ?	45 × 20 = ?	60 × 80 = ?	70 × 30 = ?
40 × 39 = ?	45 × 40 = ?	4 × 80 = ?	70 × 300 = ?
40 × 38 = ?	45 × 80 = ?	64 × 80 = ?	69 × 300 = ?
	45 × 79 = ?	64 × 79 = ?	69 × 290 = ?

Number strings can be challenging to create at first. A good tip for creating them is to first think about the problem you want to build toward. In the third example of the second row, we thought about 64 × 79. Then, we considered that we would use 64 × 80 and put it right above our end result. We then thought that 64 × 80 could be found by using 60 × 80 and 4 × 80 and positioned them before 64 × 80. Finally, 60 × 80 and 4 × 80 naturally connected to the basic facts, so we started with those problems.

ACTIVITY 3.5
PROMPTS FOR TEACHING COMPENSATION

Use the following prompts as opportunities to develop understanding of and reasoning with the strategy. Have students use representations and tools to justify their thinking, including base-10 models, number lines, arrays, and so on. After students work with the prompt(s), bring the class together to exchange ideas. These could be useful for collecting evidence of student understanding. Any prompt can be easily modified to feature different numbers (e.g., three-digit or four-digit numbers) and any prompt can be offered more than once if modified.

- How can you use Compensation to find the product of 49 × 40?

- Chloe uses Compensation for 9 × 7 and 19 × 7. How do you think she uses Compensation for these two problems?

- Create examples of multiplication problems that are good for using the Compensation strategy.

- Jimmy says that anytime a factor has a 9 in it, you should use the Compensation strategy. Do you agree or disagree with Jimmy? Use examples to justify your thinking.

- Jade says that Compensation only works with a two-digit and a one-digit factor like 39 × 8. Do you agree or disagree? Use examples to justify your thinking.

- Heidi solved 23 × 17 by thinking 20 × 20 and giving 3 from 23 to 17. Do you agree with Heidi's thinking?

- Samuel used Compensation to solve 3 × 296. He first changed the 296 to 300. Then he multiplied 3 × 300, which gave him a product of 900. Now Samuel does not know what to do. What should Samuel do? Explain your answer.

- Why are compatible numbers helpful when using the Compensation strategy?

- Nia used Compensation to solve 9 × 21. She changed the expression to 9 × 20, then added a group of 9. Tyler used Compensation to solve 9 × 21. He changed the expression to 10 × 21, then subtracted a group of 21. Both students had the same product. How are Nia and Tyler's work similar? How are they different?

- Hines wonders if he has to use Compensation for 8 × 61 by making it 10 × 61. Would you suggest he use Compensation or would you suggest something different? Explain your thinking.

- Create a story problem for 8 × 65. Solve using Compensation.

PRACTICE ACTIVITIES for Compensation

Fluency is realized through quality practice that is focused, varied, processed, and connected. The activities in this section focus students' attention on how this strategy works and when to use it. The activities are a collection of varied engagements. The discussion you facilitate after an activity or the reflection prompts you attach to it should help students think about what they did mathematically, how they reasoned about the activity, and when the math they did (namely the strategy) might be useful. Debriefing should also help students see how the practice activity connects to recent instruction or how the strategy connects to other strategies they know. Game boards, recording sheets, digit cards, and other required materials are available as online resources for you to download, possibly modify, and use. As students work with activities, look for how well they are acquiring the strategy and assimilating it into their collection of strategies.

FLUENCY COMPONENT	WHAT TO LOOK FOR AS STUDENTS PRACTICE THIS STRATEGY
Efficiency	• Are students using the Compensation strategy or are they reverting to previously learned and/or possibly less appropriate strategies? • Are students using the Compensation strategy efficiently? • Do they use the strategy regardless of its appropriateness for the problem at hand?
Flexibility	• Are students carrying out the strategy in flexible ways? (e.g., Do they use Compensation by adding on groups or taking them away? Do they compensate for one group [13 × 9] and two groups [13 × 8]?) • Do they change their approach to or from Compensation as it proves inappropriate or overly complicated for the problem?
Accuracy	• Are students using the Compensation strategy accurately? (e.g., Do they remove one or one group 13 when compensating with 13 × 9?) • Are students finding accurate solutions? • Are they considering the reasonableness of their solutions?* • Are students estimating before finding solutions?*

*This consideration is not unique to this strategy and should be practiced throughout the pursuit of fluency with whole numbers.

WORKED EXAMPLES

Worked examples are problems that have been solved. Correctly worked examples can help students make sense of a strategy and incorrectly worked examples attend to common errors.

A strength (and a challenge) of Compensation is that it is flexible. As with Break Apart to Multiply, the most common errors are as follows:

1. The student compensates by a quantity (e.g., 2) rather than groups of that quantity (e.g., two groups).
   - 14 × 18: changes to 14 × 20, multiplies to equal 280, and then subtracts 2 (rather than two groups of 14) to incorrectly get 278.
2. The student adjusts the problem but does not compensate for that change.
   - 499 × 4: changes problem to 500 × 4, multiplies to equal 2,000, and stops.

Sometimes both factors are good candidates for adjusting. Comparing both options through worked examples can help students gain insights into how to use the strategy. The prompts from Activity 3.5 can be used for collecting worked examples. Throughout the module are various worked examples that you can use as fictional worked examples. A sampling of additional ideas is provided in the following table.

SAMPLE WORKED EXAMPLES FOR COMPENSATION

Correctly Worked Example (make sense of the strategy) What did ____ do? Why does it work? Is this a good method for this problem?	James' work for 398×14: $400 \times 14 = 5,600$ $2 \times 14 = 28$ $\overline{5,572}$
Partially Worked Example (implement the strategy accurately) Why did ____ start the problem this way? What does ____ need to do to finish the problem?	Kanish's start for 49×9: 49×10 Keera's start for 49×9: 50×9
Incorrectly Worked Example (highlight common errors) What did ____ do? What mistake does ____ make? How can this mistake be fixed?	Greta's work for $2,995 \times 8$: $3,000 \times 8 = 24,000$ $24,000 - 5 = 23,995$

ACTIVITY 3.6

Name: "A Little More or A Little Less"   **Type:** Routine

About the Routine: Compensation can be applied in different ways. This routine helps students consider different ways to adjust a problem in order to solve it efficiently.

Materials: Prepare three or four expressions that are good for the Compensation strategy.

Directions:

1. Pose a few expressions to students.

2. Give students time to think and talk to each other about ways they could adjust the problem by either adding on a little more or taking a little less to create a new expression. Students do not need to solve the expressions, but instead just rethink them.

3. Then, talk with the whole group about how problems were adjusted, which approaches were most efficient, and why.

TEACHING TAKEAWAY

Trying to find a solution can distract students from making sense of the process. You can alleviate that by not asking for an answer.

A LITTLE MORE OR A LITTLE LESS		
9×21	39×9	8×28

For example, 9×21 is one of three expressions presented in the routine. One student suggests that a little less is 9×20. Another student suggests that a little more is 10×21. You record the students' suggestions. At this point, the class discusses which is the easiest to think about and why. Be careful to avoid suggesting that there is always a better or best way. Some expressions to use may include the following:

8×31	4×42	5×51	3×62	2×73	8×398
4×82	2×94	9×19	8×29	7×39	3×698
5×49	3×58	2×67	4×78	2×86	4×199
4×201	3×704	8×402	9×501	5×602	7×299

ACTIVITY 3.7

Name: *"Or You Could"*   **Type:** *Routine*

About the Routine: As students encounter multiplication problems, they often begin to work on the computation before thinking about another, more efficient approach to the problem. And, within the compensation strategy, students need experience selecting the number they will change and considering how to compensate for that change. This routine asks students to consider how they might think of the problem in a friendlier way.

Materials: Prepare two or three expressions for students to work with.

Directions: 1. Pose two or three expressions and ask students to consider how they might think of them for a friendlier computation.

2. Students think to themselves and then share with partners about an alternative.

3. Bring the class together to discuss how students thought about the problem.

For example, you might pose 48×4 to a fourth-grade class. Students might share that you could do 48×4 or you could do 50×4 and take 8 away. Another student might share that you could break apart 48×4, finding $10 \times 4 + 10 \times 4 + 10 \times 4 + 10 \times 4 + 8 \times 4$. Or you could break 48×4 into $24 \times 4 + 24 \times 4$ or even $40 \times 4 + 8 \times 4$. After these ideas are collected, the conversation then focuses on which approaches are easiest to carry out and which are most efficient. Because this routine is situated here, you may choose to limit the experience to applying the Compensation strategy to the expressions you pose.

This routine works with any problem. Some expressions to begin with include the following:

39×6	498×6	4×135	25×48
7×78	399×7	$7 \times 2,178$	39×15
5×27	219×3	$5 \times 2,527$	99×34
9×18	5×496	$9 \times 3,618$	22×88
48×4	2×799	$1,248 \times 4$	19×46
39×4	4×898	$4,639 \times 4$	98×51
8×35	297×4	$8 \times 4,035$	69×77
87×5	198×9	$7,587 \times 5$	49×81

ACTIVITY 3.8

Name: Compensation Points   **Type:** Game

About the Game: Students need ample opportunities to practice with the Compensation strategy. *Compensation Points* is an engaging way to practice Compensation and to reason about how factors influence products.

Materials: 10-sided dice, digit cards (0–9), or playing cards (queens = 0, aces = 1, kings and jacks removed); *Compensation Points* game board for each student; calculator

Directions:
1. Players take turns generating two digits to complete a line on their game board.

2. The player places the two digits on a line item and uses Compensation to find the product.

3. Players check their opponents' accuracy with a calculator. Inaccurate products cannot be used as a score.

4. After all lines are complete, players add up their products to determine their score.

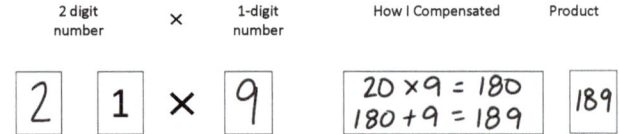

In this example, the player pulls a 2 and a 9, making 21 × 9 (note that the "1" was prefilled). The player uses Compensation accurately, finding a product of 189.

RESOURCE(S) FOR THIS ACTIVITY

Compensation Points

Directions: Roll a die two times or pick up two digit cards. Fill in the empty boxes to create a two-digit number (x) and a one-digit number. Write down how you compensated, then solve and write your product. After the final round, combine all of your products. Compare your score with your partner's. The highest score wins.

2-Digit Number		×	1-Digit Number	How I Compensated	Product
☐	1	×	☐	☐	☐
☐	9	×	☐	☐	☐
☐	2	×	☐	☐	☐
☐	8	×	☐	☐	☐
☐	7	×	☐	☐	☐

Player:	Total Score:

online resources ⬆ Digit cards and this game board can be downloaded at **resources.corwin.com/FOF/ multiplydividewholenumber**.

ACTIVITY 3.9

Name: *Give Some to Get Some*   **Type:** *Game*

About the Game: *Give Some to Get Some* focuses students' attention on an amount they would give to one factor and the related extra quantity that needs to be compensated in the end. It helps students focus on how many groups of a factor are added and then removed.

Materials: one *Give Some to Get Some* game board, one regular six-sided die per pair of students

Directions:
1. Players take turns rolling the die to generate a number.

2. The player uses that number to look for an expression in which the amount rolled could be used to adjust one of the factors to make a friendlier problem. For example, a player rolls a 2. She selects 68×30 because she could rethink the problem as 70×30 and then take two groups of 30 (60) away. She places her counter on 68×30.

3. If a player rolls a 4, 5, or 6, they lose their turn because factors adjusted by these amounts do not typically yield an efficient problem.

4. The game ends when all of the spaces are filled.

5. The player who has the most three-in-a-rows wins the game.

TEACHING TAKEAWAY

Hold students accountable by having them record their play. A player might record that 68×30 is the same as $70 \times 30 - 60$.

RESOURCE(S) FOR THIS ACTIVITY

Give Some to Get Some

Directions: Take turns rolling a number. Choose a space on the gameboard where the amount you roll could be given from one factor to the other to show the Compensation strategy. The game ends when all spaces are covered. The player with the most three-in-a-rows wins.

99 × 98	38 × 12	39 × 16	79 × 24	16 × 80
37 × 69	17 × 25	38 × 63	48 × 23	89 × 26
97 × 5	88 × 17	47 × 54	89 × 27	25 × 38
68 × 30	59 × 13	68 × 29	77 × 17	99 × 58
27 × 44	57 × 43	68 × 27	46 × 27	35 × 46

This resource can be downloaded at **resources.corwin.com/FOF/multiplydividewholenumber**.

ACTIVITY 3.10

Name: Compensation Concentration   **Type:** Game

About the Game: *Compensation Concentration* helps students think about how the Compensation strategy can be applied to multiplication problems, creating new, friendly expressions. It is played similarly to a traditional game of *Memory* or *Concentration*.

Materials: one deck of *Compensation Concentration* cards

Directions:
1. Players place cards face down in an array.

2. Players take turns flipping over cards, attempting to match an expression with a card that shows how to solve the problem with partial products.

3. The player who makes the most matches wins the game.

RESOURCE(S) FOR THIS ACTIVITY

28 × 9	28 × 10 = 280 and take 28 away
8 × 45	10 × 45 = 450 and take 90 away
8 × 41	8 × 40 = 320 and add 8
12 × 9	12 × 10 = 120 and take away 12
49 × 7	50 × 7 = 350 and take away 7

This resource can be downloaded at **resources.corwin.com/FOF/multiplydividewholenumber**.

ACTIVITY 3.11

Name: *Estimation Tic-Tac-Toe*   **Type:** *Game*

About the Game: *Estimation Tic Tac Toe* practices estimating products of problems that might be solved with the Compensation strategy.

Materials: *Estimation Tic-Tac-Toe* game board and game cards

Directions:
1. Players choose estimation cards and place their mark (X or O) on an expression that is a good estimate for their card.

2. The first player to get three in a row wins.

For example, player 1 (X) selects the card with 19×38, uses 20×40 as an expression for estimating the product, and places their X on the board. Player 2 (O) selects the card with 29×20 and uses 20×30 as their estimate.

20×20	20×30	20×40
20×50	20×60	20×70
20×80	20×90	20×100

RESOURCE(S) FOR THIS ACTIVITY

Estimation Tic-Tac-Toe

Directions: Player 1 is X and Player 2 is O. Pick up a card. Choose a space that would be a good estimation for the expression picked. Place your mark on that space. The first player with three marks in a row wins.

2×20	2×30	2×40
2×50	2×60	2×70
2×80	2×90	2×100

3×20	3×30	3×40
3×50	3×60	3×70
3×80	3×90	3×100

4×20	4×30	4×40
4×50	4×60	4×70
4×80	4×90	4×100

5×20	5×30	5×40
5×50	5×60	5×70
5×80	5×90	5×100

2×18	2×72
2×27	2×83
2×36	2×89
2×45	2×98
2×61	2×11

online resources Game cards and this game board can be downloaded at **resources.corwin.com/FOF/ multiplydividewholenumber**.

ACTIVITY 3.12

Name: Creating Compensations   **Type:** Center

About the Center: Countless problems can be adjusted in very similar ways and even by the same amount. For example, 6 × 229 and 6 × 239 are both likely to be adjusted by the same amount of 1. However, students can develop a misconception that only certain problems can be adjusted by 1, 2, or 3. This center is a high-order thinking activity for students to create problems that could be adjusted by the given amount.

Materials: six-sided die, Creating Compensations recording sheet

Directions: 1. Students generate a number by rolling the die.

2. Students use the number to generate an expression that they would solve by adjusting one of the factors by the amount they generated.

3. Students capture their thinking on the recording sheet.

4. Note that Compensation always works but it is not always efficient. When a student rolls a 4, 5, or 6, they should still create a problem that can be compensated by that amount. However, you may add a direction for students to identify if the compensation is an efficient way to think of the problem.

For example, a student rolls a 3. She creates the expression 4 × 497 because she can adjust 497 by 3, creating 4 × 500. She describes that adjustment in her writing, noting that she would have to take away 12 because she adjusted by four groups of 3.

RESOURCE(S) FOR THIS ACTIVITY

Creating Compensations

Directions: Roll a number. Create a problem that you would adjust by that amount to solve. Record your thinking.

Number I Pulled	Problem I Created	How I Would Adjust It by the Number I Pulled
3	4 × 497	I can think of 4 × 500 and then take away 12.

 This resource can be downloaded at **resources.corwin.com/FOF/multiplydividewholenumber**.

ACTIVITY 3.13

Name: Prove It   **Type:** Center

About the Center: Students can confuse how compensation works in different situations. This center helps them reinforce or refine their reasoning.

Materials: Compensation Prove It cards and recording sheet

Directions: 1. A student selects a Compensation Prove It card.

2. The student explains that the card is true or false and records their thinking on the recording sheet.

For example, a student pulls a card that says 4×47 is the same as 4×50 and take away 12. The student determines that the statement is true, since 4×3 was added to 4×47 to make it 4×50. The product for 4×50 is 200. Then we have to remove 4×3 from 200. Note that students don't have to find products necessarily to prove their thinking.

RESOURCE(S) FOR THIS ACTIVITY

A 59×3 is the same as 60×3 and take away 3	A 59×3 is the same as 60×3 and take away 3
B 6×28 is the same as 6×30 and take away 2	B 6×28 is the same as 6×30 and take away 2
C 5×68 is the same as 5×70 and take away 10	C 5×68 is the same as 5×70 and take away 10

Prove It

Directions: Pull a Prove It card. Prove if it is true or false

Card I Pulled	Proof That It Is TRUE or FALSE

Game cards and this recording sheet can be downloaded at **resources.corwin.com/FOF/multiplydividewholenumber**.

ACTIVITY 3.14

Name: Use This for Solving   **Type:** Center

About the Center: Two very different expressions can be adjusted to the same expression. This center provides a given equation and students determine what other problems it can be used to solve. The example offered is for multiplying two-digit factors but can be easily modified.

Materials: Use This for Solving center cards and recording sheet

Directions:

1. Students select a Use This for Solving equation like $40 \times 82 = 3{,}280$.

2. Students then generate two different expressions that could be adjusted to this problem to be solved. For $40 \times 82 = 3{,}280$, the student might create 39×82 and 41×82.

3. Students then use the known equation to help them find the product of the two created problems: $39 \times 82 = 3{,}198$ because it has one less group of 82 and $41 \times 82 = 3{,}362$ because it has one more group of 82.

RESOURCE(S) FOR THIS ACTIVITY

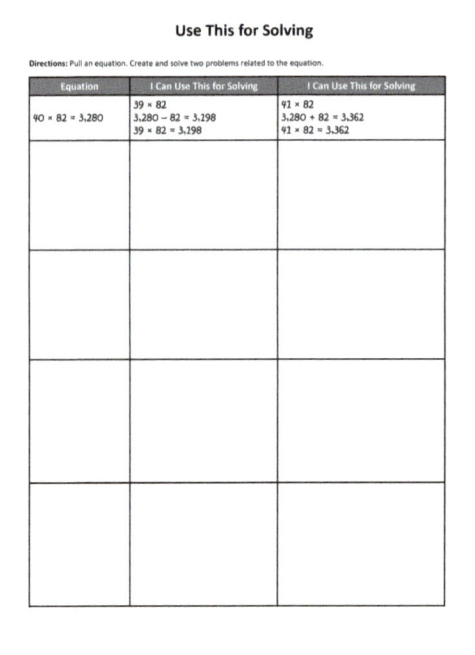

20 × 53 = 1,060	40 × 66 = 2,640
50 × 32 = 1,600	42 × 60 = 2,520
60 × 25 = 1,500	35 × 80 = 2,800
40 × 34 = 1,360	62 × 50 = 3,100
35 × 60 = 2,100	40 × 75 = 3,000

online resources ♟ Center cards and this recording sheet can be downloaded at **resources.corwin.com/FOF/ multiplydividewholenumber**.

NOTES

Partial Products Strategy

STRATEGY OVERVIEW:
Partial Products

What is Partial Products? Partial Products breaks numbers apart (usually by place value), and then makes sure each part of one factor is multiplied by each of the other parts. Break Apart to Multiply (by addends) and Partial Products are essentially two names for the same idea. The idea of Partial Products is to start with the largest place value and move to the smallest (the reverse of many standard algorithms). Here are examples for 27×4:

Partial Products by Place Value	**Break Apart Into Addends**
$27 \times 4 =$	$27 \times 4 =$
$20 \times 4 = 80$	$25 \times 4 = 100$
$7 \times 4 = 28$	$2 \times 4 = 8$
$80 + 28 = 108$	$100 + 8 = 108$

Partial Products can be used with large factors, and both factors can be broken into partials. In such cases, rectangles are often used to keep track of each partial product.

347×7

68×36

Note that the rectangles only roughly represent the size of the numbers; they are not meant to be proportional representations.

HOW DOES PARTIAL PRODUCTS WORK?

Partial Products utilizes the distributive property, breaking apart one or both factors into addends, multiplying the parts, and putting them back together. For 347 × 7, this is what happens numerically: 7(300 + 40 + 7) = 2,100 + 280 + 49. When there are 2 two-digit numbers, one or both can be broken apart. The decision relies on whether the two-digit number can be computed mentally as-is. For example, for 68 × 36, neither 68 × 3 or 6 × 36 is likely to be recalled mentally. So then, both factors are decomposed.

Now try 23 × 15.

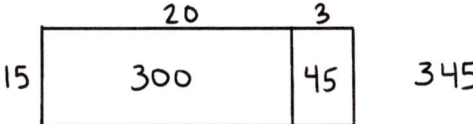

There are products you likely recognize: 20 × 15 = 300 and 3 × 15 = 45. Because of this, you might only decompose one factor (23).

WHEN DO YOU CHOOSE PARTIAL PRODUCTS?

Partial Products is very versatile and always works. However, it can require more steps and take more time than other strategies like Halve and Double or Compensation depending on the factors. For example, 99 × 5 can be solved with Partial Products (90 × 5 + 9 × 5), but Compensation is likely a bit easier (100 × 5 − 5). Partial Products would be selected after other strategies are ruled out.

The standard algorithm is also an application of partial products, but the standard algorithm begins with the ones place and works out to larger place values and combines partial products per line. The Partial Products strategy is as versatile as the standard algorithm and oftentimes more accessible to students (because they can use the area model), so it is almost always a good choice for more complicated multiplication. That said, Partial Products can become unwieldy as the number of digits in each factor increases because there are more and more partials to keep track of.

MODULE 1 Break Apart to Multiply Strategy

MODULE 2 Halve and Double Strategy

MODULE 3 Compensation Strategy

MODULE 4 Partial Products Strategy

MODULE 5 Think Multiplication Strategy

MODULE 6 Partial Quotients Strategy

MODULE 7 Standard Algorithms for Multiplication and Division

PARTIAL PRODUCTS:
Strategy Brief for Families

It is important that families understand the strategies and know how they work so that they can be partners in the pursuit of fluency. This strategy brief is a tool for doing that. You can include it in parent or school newsletters or share it at parent conferences. It is available for download so that you can adjust it as needed.

Partial Products

How It Works: We can break apart factors by place value to multiply. The parts of each factor are multiplied by each of the parts of the other factor. The partial products are added together to get your answer.

1. Decompose each factor.
2. Multiply the parts of each factor by each of the parts of the other factor.
3. Add each of the partial products together to get your final answer.

In the example, 26 × 47, each factor is decomposed by place value: 20 + 6 and 40 + 7. Then, each part is multiplied by each part of the other factor. 20 is multiplied by 40. 20 is also multiplied by 7, since those are the parts from the other factor. Also, 6 is multiplied by 40 and 6 is multiplied by 7, since those are the parts from the other factor. The partial products are then added together.

The example on the right is an area model. The rectangle represents how the area of these two factors can be decomposed.

When It's Useful: Partial Products is a useful strategy for two- and three-digit factors. It is not as useful when the numbers get larger.

Factor Factor
26 × 47
/ \
(20+6) (40+7)

20 × 40 : 800
6 × 40 : 240
20 × 7 : 140
6 × 7 : 42

↓
product : 1,222

Factor Factor
26 × 47

	40 + 7	
20 +	800	140
6	240	42

800
240
140
+ 42
1,222

This resource can be downloaded at **resources.corwin.com/FOF/multiplydividewholenumber**.

TEACHING ACTIVITIES for Partial Products

Partial Products is the same thing as the Break Apart (with addition) to Multiply strategy. In this module, we provide activities that focus solely on breaking apart by place value as an extension of the break apart activities in Module 1. This strategy bridges initial decomposition strategy with standard algorithms in the future. In this section, we focus on instructional activities for developing the Partial Products strategy. The goal is for students to become adept with the strategy and also to consider when the strategy is useful. Because of its relationship to the Break Apart strategy, many of the activities in Module 1 can be modified for use with the Partial Products strategy.

 ## ACTIVITY 4.1
MULTIPLES MATRIX

Both estimation and computation of multidigit whole numbers rely on the foundation of students' capability to multiply and divide by 10, 100, 1,000, and their multiples (e.g., 40, 400, 4,000). However, this capability is not the rote recall of moving zeros or place values. Rather, *such a* shortcut is grounded in understanding and explicit connection to and application of basic facts. In this activity, students use basic facts to make examples of related problems. Students are given a basic fact and identify what problems it can help with by thinking of it as __ groups of ___ tens, __ groups of ___ hundreds, and __ groups of __ thousands.

4 groups of 3	4 × 3	12
4 groups of 30	4 × 30	120
4 groups of 300	4 × 300	1,200
4 groups of 3,000	4 × 3,000	12,000

In this example, a group of students worked with the basic fact 4 × 3 and restated it as groups of tens, hundreds, and thousands before writing the expression and product. Note that it's entirely reasonable to provide students with place value disks or some other tool to find products as they first work with the activity.

Ideally, different groups of students work with three or four different basic facts and record their work on chart paper or something similar. Doing so allows for a gallery walk after all students have completed their work. They can look for similarities, differences, and patterns so they can establish generalizations for working with multiples of 10, 100, and 1,000 as well as how they can leverage basic facts to do so. Note that eventually, students will notice that zeros are "just added" to the product. This is true and useful. You simply want to make sure they understand *why* this is so. As you get into work with Partial Quotients, you can reuse this activity with division facts.

ACTIVITY 4.2
BASE-10 PARTIALS

Base-10 blocks help students make sense of the Partial Products strategy by providing a concrete example. Using manipulatives, properties of operations, and patterns within factors and products can help them develop a deep understanding of multiplication. In this instructional activity, students use base-10 blocks to represent partial products. Start by providing students with one-digit by two-digit expressions. Show the students how to arrange the blocks in an array form. Ensure students have a foundational understanding before moving on to multidigit numbers. The following example models how students could break apart the factors of 23 × 11 by place value. One side shows 23 split into 10 + 10 + 3. On the opposite side, students would break apart the 11 into 10 + 1. Then students would work to fill in the base-10 blocks that would form the array.

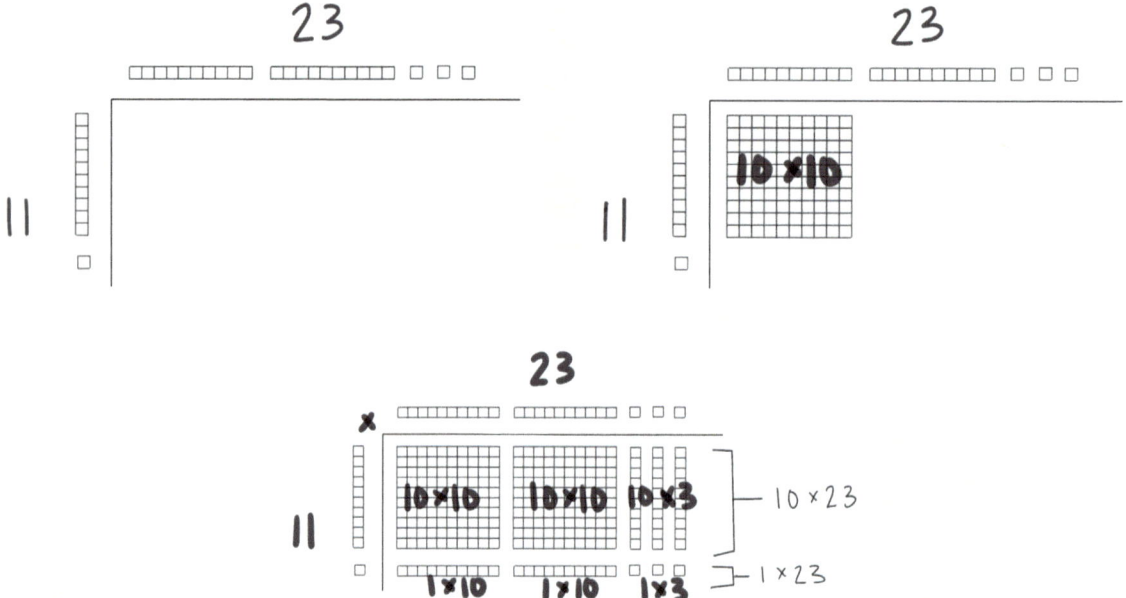

This activity transitions nicely to area models, since it is the physical representation of how students use the area model.

ACTIVITY 4.3
PARTIAL PRODUCTS WITH AREA MODELS

Area models do well to help students make sense of the Partial Products strategy. This is especially true when students have the opportunity to connect precise area models with a more general, open or abstract model. In this activity, students make precise models of multiplication problems using grid paper. Those precise models are connected to open, abstract models and the expressions that represent the partial products. Students will need a few exposures to this activity before shifting to practice with open area models alone. The following example is from a class beginning to work with 2 two-digit factors. The precise model is on the left and the open, general model is on the right. In both examples, the student has shown how the factors are broken apart by place value as well as Partial Products equations.

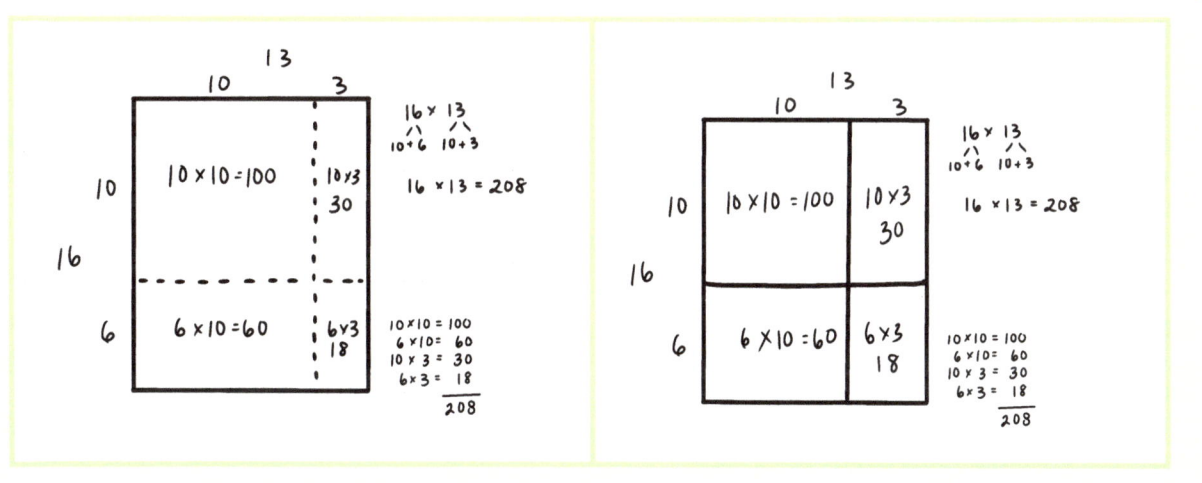

ACTIVITY 4.4
IS IT THE SAME?

Students may tend to overly rely on breaking apart factors into tens and ones because the multiplication is more comfortable for them. For example, they may decompose 26 × 3 into 10 × 3 + 10 × 3 + 2 × 3 + 2 × 3 + 2 × 3. They may even decompose the 6 into 6 ones. This can become quite problematic as problems with 2 two-digit factors (26 × 38) or three-digit by one-digit problems (263 × 8) become unwieldy. The Partial Products strategy is simply more efficient. This teaching activity focuses on establishing that the two decomposition approaches yield the same result and that the Partial Products strategy is more efficient. In this activity, you pose a problem such as 26 × 38. You discuss with students how these two numbers might be broken apart and record their thinking. Then, groups of students work to find the product by decomposing into tens and ones and groups of tens and groups of ones (place value), as shown in the following example. After students find the product, discuss how the models are the same and how they are different. Ask which approach is more efficient or at least easier to keep track of and if they think it will always work. They can then create their own problems to find out.

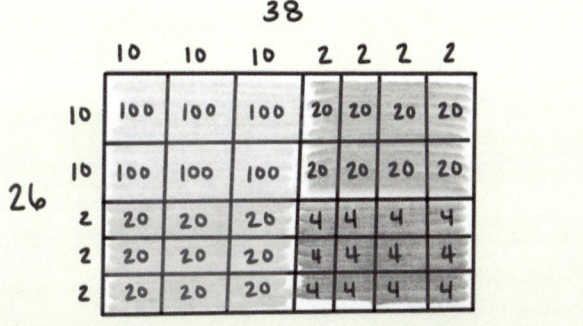

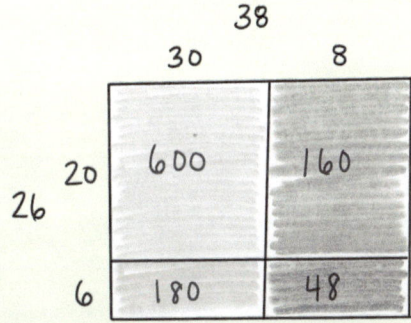

You can see in the example how students might go about this. Note that this group was permitted to use groups of 2 instead of singles. During discussion, students noted that there are 6 hundreds in the left model and there is a single 600 in the right example but that they are the same. Other students noted that nine groups of 20 is 180 and that there is an area of 180 in the right model. When this was shared, the teacher took time to confirm that nine groups of 20 was, in fact, 180. The same was done for 160 and 48. Students then went off to create new problems to see if this would remain true. This resulted in many different examples for the class to discuss, each showing that when you break apart by place value the product remains the same.

ACTIVITY 4.5
ONE OR BOTH?

The Partial Products strategy allows for one or both factors to be decomposed. Often it makes sense to decompose both. But there are situations in which it is more efficient to decompose only one factor. Those problems include factors that students can easily multiply (like 25) or problems in which no grouping will be needed (22×43). This activity establishes for students that decomposing one factor yields the same results as decomposing both.

Arrange students in groups of two or three. Have half of the groups solve a problem by breaking apart both factors. Have the other groups solve the problem by breaking apart only one of the factors. Be sure that both groups justify their solutions with drawings and equations. Once finished, bring both groups together and have them share their findings. During discussion, ask students questions like these:

- Why do you think the products were the same?

- Does this always work? Why?

- Why would this be useful?

- How did you decide which factor to break apart (for groups that break apart only one)?

- Which approach do you find better? Why?

After discussion, you can pose new problems for students to work with. During the extension, you may choose to have students solve each set of problems with both approaches. Or, you might have groups change the approach they used initially to replicate the initial experience. The activity will work with any set of problems but you might decide to use the following examples:

	DECOMPOSING BOTH FACTORS	DECOMPOSING ONE FACTOR		
25×43	$(20 \times 40) + (20 \times 3) + (5 \times 40) + (5 \times 3)$	$(25 \times 40) + (25 \times 3)$	OR	$(20 \times 43) + (5 \times 43)$
75×52	$(70 \times 50) + (70 \times 2) + (5 \times 50) + (5 \times 2)$	$(75 \times 50) + (75 \times 2)$	OR	$(70 \times 52) + (5 \times 52)$
62×31	$(60 \times 30) + (60 \times 1) + (2 \times 30) + (2 \times 1)$	$(62 \times 30) + (62 \times 1)$	OR	$(60 \times 31) + (2 \times 31)$
35×55	$(30 \times 50) + (30 \times 5) + (5 \times 50) + (5 \times 5)$	$(35 \times 50) + (35 \times 5)$	OR	$(30 \times 55) + (5 \times 55)$

ACTIVITY 4.6
SAME DIGITS, DIFFERENT PRODUCT

Students may misinterpret the commutative property when working with multidigit factors. For example, they may believe that 24×8 is the same as 28×4. This instructional activity takes aim at this misconception by having students find products and then rearranging digits before finding a new product. Using Partial Products will help them see how and why the products change and that rearranged digits do not yield the same product. For the activity, pose a prompt like this: "24×8, 28×4, 42×8, and 48×2 all have the same digits. Do they all have the same product?"

The following example shows how students might go about determining if you can rearrange the digits in factors. Here, the student uses Partial Products and area models to show that the products are not the same.

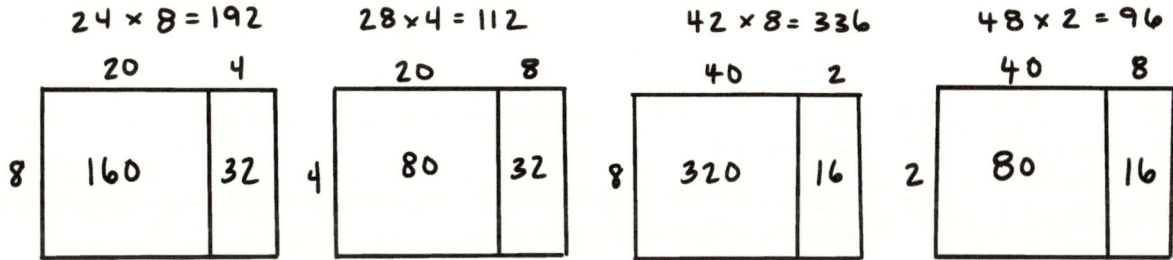

This activity can be easily modified in a variety of ways. You might ask if 24×8 and 28×4 have the same product in which only the digits in the ones place are swapped. Or you can, like the initial prompt, have students investigate all of the possible combinations for the digits like 24×83, 24×38, 23×84, and 23×48. Or you could pose one problem like 35×71 and ask students to find different ways to rearrange the digits and determine if any of the products are the same.

ACTIVITY 4.7
PROMPTS FOR TEACHING PARTIAL PRODUCTS

Use the following prompts as opportunities to develop understanding of and reasoning with the strategy. Have students use representations and tools to justify their thinking, including base-10 models, number lines, arrays, and so on. After students work with the prompt(s), bring the class together to exchange ideas. These could be useful for collecting evidence of student understanding. Any prompt can be easily modified to feature different numbers (e.g., three-digit or four-digit numbers) and any prompt can be offered more than once if modified.

- Leticia says that $8 \times 3{,}240$ has 1,600 as a partial product. Erin says it does not. Who is correct? Explain your answer.

- Why is it important to understand place value when you are multiplying multi-digit numbers?

- Marie used Partial Products to solve 32×73. She found 30×70 and 2×3. How do you know if she is finished?

- Connie solves 31×7 by thinking of $10 \times 7 + 10 \times 7 + 10 \times 7 + 1 \times 7$. How else might she have approached the problem?

- Is the product of 49×37 over or under 1,200? Explain how you know.

- Tell why 437×8 is larger than 374×8 even though the digits are the same.

- Describe how you use the Partial Products strategy. Use examples to show your thinking.

- Create examples of when the Partial Products strategy isn't efficient. Justify your thinking.

- Sorsha solved 32×45 by thinking of $30 \times 45 + 2 \times 45$. Deryn solved it by thinking $30 \times 40 + 2 \times 40 + 30 \times 5 + 2 \times 5$. Who do you agree with? Explain your thinking.

- Kaz and O'Neil disagree about the best way to solve 66×42. Kaz thinks $60 \times 42 + 6 \times 42$. O'Neil thinks $60 \times 40 + 60 \times 2 + 6 \times 40 + 6 \times 2$. Who do you agree with? Explain your thinking.

- Create a story problem for 42×13. Solve the problem using Partial Products.

NOTES

_____

_____

_____

_____

PRACTICE ACTIVITIES for Partial Products

Fluency is realized through quality practice that is focused, varied, processed, and connected. The activities in this section focus students' attention on how this strategy works and when to use it. The activities are a collection of varied engagements. The discussion you facilitate after an activity or the reflection prompts you attach to it should help students think about what they did mathematically, how they reasoned about the activity, and when the math they did (namely the strategy) might be useful. Debriefing should also help students see how the practice activity connects to recent instruction or how the strategy connects to other strategies they know. Game boards, recording sheets, digit cards, and other required materials are available as online resources for you to download, possibly modify, and use. As students work with activities, look for how well they are acquiring the strategy and assimilating it into their collection of strategies.

FLUENCY COMPONENT	WHAT TO LOOK FOR AS STUDENTS PRACTICE THIS STRATEGY
Efficiency	• Are students using the Partial Products strategy or are they reverting to previously learned and/or possibly less appropriate strategies? (e.g., Do they break apart a factor of 54 into 50 and 4 or 10, 10, 10, 10, 10, and 4?) • Are students using the Partial Products strategy efficiently? • Do they decompose one or both factors as appropriate? • Do they use Partial Products regardless of its appropriateness for the problem at hand?
Flexibility	• Are students carrying out the strategy in flexible ways? (e.g., Are they more comfortable breaking apart two-digit factors rather than three-digit factors?) • Do they change their approach to or from this strategy as it proves inappropriate or overly complicated for the problem?
Accuracy	• Are students using the Partial Products strategy accurately (e.g., decomposing factors accurately, finding all partial products)? • Are students finding accurate solutions (e.g., adding partials correctly)? • Are they considering the reasonableness of their solutions?* • Are students estimating before finding solutions?*

* This consideration is not unique to this strategy and should be practiced throughout the pursuit of fluency with whole numbers.

WORKED EXAMPLES

Worked examples are problems that have been solved. Correctly worked examples can help students make sense of a strategy and incorrectly worked examples attend to common errors.

Like Break Apart to Multiply, here are some common challenges for Partial Products:

1. The student breaks apart as they would for addition, multiplying tens by tens and ones by ones, but omits the other parts.

    • 54×67: multiplies $50 \times 60 = 3{,}000$ and $4 \times 7 = 28$, then adds them to get 3,028.

2. The student breaks apart a factor but not into parts that can be readily multiplied mentally.

    • 25×48: breaks this apart into $20 \times 48 + 5 \times 48$ and gets stuck.

3. The student becomes adept at breaking apart by all the parts and doesn't notice when a more efficient option fits the problem.

    • 350×15: breaks this apart into $(300 + 50)(10 + 5) = 3{,}000 + 1{,}500 + 500 + 250$, rather than breaking apart just one factor, such as $350 (10 + 5)$, to solve.

4. The student has place value challenges when breaking numbers apart.
   - 804×5: multiplies 80×5 (400) and adds it to 4×5 (20) to get 420.

Partial products are often determined by place value but there are other options, which can be highlighted through worked examples. Comparing both options can help students gain insights into how to use the strategy. The prompts from Activity 4.7 can be used for collecting worked examples. Throughout the module are various worked examples that you can use as fictional worked examples. A sampling of additional ideas is provided in the following table.

SAMPLE WORKED EXAMPLES FOR PARTIAL PRODUCTS

Correctly Worked Example (make sense of the strategy) What did _____ do? Why does it work? Is this a good method for this problem?	Angelica's work for 240×12: 240×12 $200 \times 10 = 2,000$ $40 \times 10 = 400$ $200 \times 2 = 400$ $40 \times 2 = \underline{80}$ $2,880$	Theresa's work for 240×12: 240×12 $240 \times 10 = 2,400$ $240 \times 2 = \underline{480}$ $2,880$
Partially Worked Example (implement the strategy accurately) Why did _____ start the problem this way? What does _____ need to do to finish the problem?	Paisley's start for $1,135 \times 4$: $1,135 \times 4$ $1,100 \times 4 = 4,400$	Leo's start for $1,525 \times 42$: $1,525 \times 42$ $1,525 \times 2 = 3,050$ $1,525 \times 4 = 6,100$
Incorrectly Worked Example (highlight common errors) What did _____ do? What mistake does _____ make? How can this mistake be fixed?	Karen's work for 78×45: 78×45 $70 \times 40 = 2,800$ $8 \times 5 = \underline{40}$ $2,840$	Lina's work for 408×90: 408×90 $400 \times 9 = 3,600$ $8 \times 9 = \underline{72}$ $3,672$

ACTIVITY 4.8

Name: "Why Not?"   **Type:** Routine

About the Routine: As students learn new strategies, they may begin to believe that every problem should be solved with the new strategy. Fluency comes about by choosing when a strategy is a good choice for a certain problem. In this routine, students encounter three or four expressions and determine why one or more of them are not good choices for using the Partial Products strategy.

Materials: Prepare three or four problems for students to consider.

Directions: 1. Pose three or four expressions for students to think about.

2. Have students examine each expression and consider if Partial Products would be a good strategy for solving the problem. Be sure to ask students to be prepared to justify their thinking.

3. Then, give students time to discuss their ideas with partners before bringing the whole class together for a group discussion.

4. Discuss as a whole group. As you discuss student thinking, be sure to recognize that the Partial Products strategy always works. Focus students' attention on why partials are unnecessary in certain situations.

WHY NOT			
8×31	4×42	5×50	6×39

For example, 25×4 could be thought of as $20 \times 4 + 5 \times 4$ but often we just know that 25×4 is 100. More prevalent examples are those that are clearly related to basic facts such as 70×8, 60×40, or 300×9. In this example, 5×50 is not suited for Partial Products; 6×39 might also be identified because it can be solved with Compensation.

This routine works with any set of problems. These next two sets focus on larger multiplication problems.

WHY NOT			
4×808	295×6	919×7	8×300

WHY NOT			
99×15	25×11	48×12	45×22

ACTIVITY 4.9

Name: "About or Between"   **Type:** Routine

About the Routine: Determining reasonableness, through estimation, plays an important role in being fluent. There are different ways to estimate and in this routine, About or Between (SanGiovanni, 2019), students have an opportunity to choose between how they would estimate products.

Materials: Prepare two or three expressions for students to work with.

Directions:
1. Pose a problem to students.

2. Have students determine how they would estimate the product and share their thinking with partners. Students can choose to estimate the products by thinking of nearby friendly numbers (about) or by finding a range (between).

3. Students share their approaches with partners.

4. Discuss ideas with the entire class. As you discuss estimates, stress that estimates give an idea of the final result and that there is no one correct way to estimate. Remind students that an efficient method is one that can be done relatively quickly.

THE PRODUCT IS ABOUT OR BETWEEN		
38×5	14×7	46×6

In this example, the class first works with 38×5. Some students suggest that it is about 200 because 38 is close to 40 and 40×5 is 200. Others say that it is about 150 because 38 is close to 30 and 30×5 is 150, but when they add it will be more than 150. After these are shared, a student suggests that the product of 38×5 will be between 150 and 200 by manipulating the first factor.

THE PRODUCT IS ABOUT OR BETWEEN	
84×77	15×37

In another example, a teacher only poses two expressions because of the size of the factors. For the first expression, students offer that it will be between 80×70 and 90×80. Other students say it will be about 80×70. In each of these cases, students offer why each option is useful for estimating the product.

ACTIVITY 4.10

Name: Make It Close   **Type:** Game

About the Game: *Make It Close* is a target-based game for practicing multiplication and developing both number and operational sense. The unique twist in this game is that the target changes from round to round. The player closest to the target wins the round.

TEACHING TAKEAWAY

When more than two players play a game, change features of the game to minimize disengagement.

Materials: four decks of digit cards or playing cards (aces equal to 1, face cards and 10s removed); *Make It Close* recording sheet

Directions:

1. Players deal four cards to make 2 two-digit factors. Both players use Partial Products to find the product, which becomes the target for the round. In the example, the target is 2,562.

2. Each player deals themselves four digit cards to make a multiplication problem with 2 two-digit factors.

3. The players arrange the four digits so that their product is close to the target. Player 1 arranges digits for a product of 2,698. Player 2 arranges digits for a product of 2,058.

4. The player who has a product closest to the target for round 1 gets a point. This is player 1 in the example.

5. The first player to 5 points wins the game.

Target	4 2	×	6 1	2,562
Player 1	3 8	×	7 1	2,698
Player 2	4 9	×	4 2	2,058

RESOURCE(S) FOR THIS ACTIVITY

Make It Close

Directions: Use digit cards to make a multiplication problem. The product of the problem is the target. Deal new digit cards to make a multiplication problem that is close to the target. The player closest to the target gets a point.

Target Problem	My Problem (give yourself a ✓ if you are closest to the target)

1	2	3	4
5	6	7	8
9	1	2	3
4	5	6	7
8	9	0	0

online resources This resource can be downloaded at **resources.corwin.com/FOF/multiplydividewholenumber**.

ACTIVITY 4.11

Name: Connect Four Partials   **Type:** Game

About the Game: *Connect Four Partials* is an opportunity for students to apply problem-solving to Partial Products as they generate factors in order to find the partial products.

Materials: *Connect Four Partials* game board and recording sheet, a 10-sided die, and two-color counters

Directions:

1. Player 1 places counters on the game board, making a four-digit number in expanded form.

2. Player 1 rolls a number.

3. Player 1 then multiplies the partials they chose on the game board by the number rolled.

4. Player 1 writes the expression they created and shows their work on the recording sheet.

5. Player 2 checks their partner's work and then uses a calculator to confirm. If player 1's answer is incorrect, player 1 removes all of their chips from the board. If correct, the counters stay on the board and it is player 2's turn.

6. Player 2 repeats the steps above.

7. The game ends when all spaces are covered. The winner is the player with the most rows of four.

For example, player 1 placed counters on 3,000, 400, 50, and 8. They rolled a 5 and wrote the expression $3,458 \times 5$ and solved it on the recording sheet. Player 2 placed their counters on 6,000, 500, 40, and 1 and rolled a 3. Once all spaces are filled, players count the number of rows of four they have.

RESOURCE(S) FOR THIS ACTIVITY

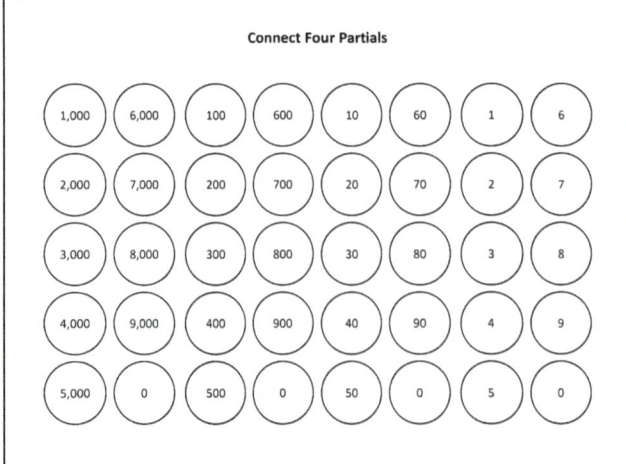

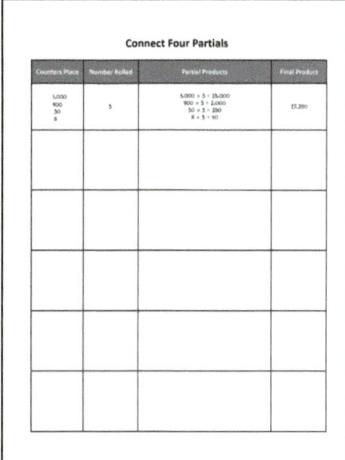

online resources — Game boards and this recording sheet can be downloaded at **resources.corwin.com/FOF/ multiplydividewholenumber.**

ACTIVITY 4.12

Name: *Target 2,500*

Type: *Game*

About the Game: This target-based game is an opportunity to estimate and find products of two-digit numbers. *Target 2,500* also offers the opportunity to observe how changing factors influence products. The target of 2,500 is set because 50×50 is half of 100×100. Players get points for getting close to the target each round.

Materials: digit cards (0–9) or playing cards (queens = 0, aces = 1, kings and jacks removed) and *Target 2,500* recording sheet

Directions:

1. Each player pulls four digit cards to make 2 two-digit factors.

2. Each player arranges the digit cards to create a product as close to 2,500 as possible.

3. The player closest to 2,500 gets a point.

4. Then, each player can rearrange their digit cards to get closer to 2,500.

5. The player closest to 2,500 after rearranging earns another point. Note that each player must rearrange regardless of how close they are on the first try.

6. The first player to 10 points wins the game.

TEACHING TIP

Many recording sheets are optional. You can have students use a piece of paper in their Math Journals to keep track of play.

RESOURCE(S) FOR THIS ACTIVITY

Target 2,500

Directions: Use four digit cards to make a multiplication problem with a product as close to 2,500 as possible. Then, rearrange the digits to make another problem with a product as close to 2,500 as possible.

First Problem	Second Problem (after rearranging digits)

online resources → Digit cards and this recording sheet can be downloaded at **resources.corwin.com/FOF/ multiplydividewholenumber**.

ACTIVITY 4.13

Name: What's the Problem?   **Type:** Center

About the Center: This center is an opportunity for students to apply problem-solving as they generate parts of a factor in order to find the partial products and eventually solve the original problem.

Materials: tens cards, digit cards (0–9) or playing cards (queens = 0, aces = 1, kings and jacks removed), What's the Problem? recording sheet

Directions:
1. Students pull tens cards to generate a number for the ?? spaces on the area model.

2. Students pull a digit card to generate a number for the ? on the area model.

3. Students find the partial products and record the original multiplication problem.

RESOURCE(S) FOR THIS ACTIVITY

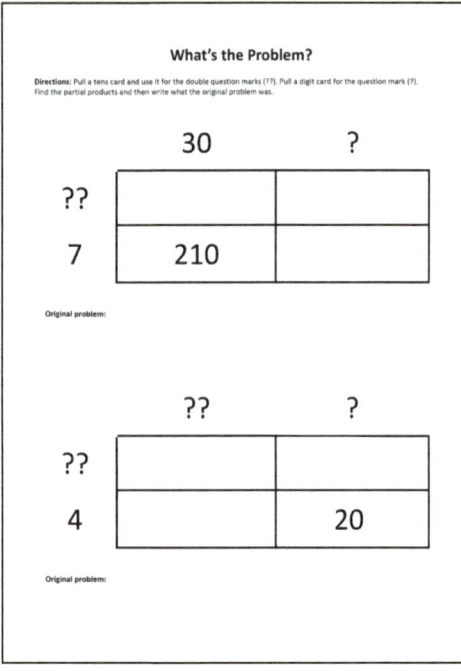

Tens cards, digit cards, and this recording sheet can be downloaded at **resources.corwin.com/FOF/ multiplydividewholenumber**.

ACTIVITY 4.14

Name: Pick Your Partials   **Type:** Center

About the Center: This center helps students consider different partial products to solve a multiplication problem.

Materials: Pick Your Partials recording sheet

Directions: 1. Provide students with the Pick Your Partials recording sheet.

2. Students circle the partial products used to solve each expression.

For example, a student has the expression 3 × 625. The student has several partials to choose from. The student multiplies 3 × 600, which equals 1,800. The student would circle the 1,800 on their recording sheet. Then the student multiplies 3 × 20, which equals 60. The student would circle the 60 on their recording sheet. Lastly, the student multiplies 3 × 5, which equals 15. The student would circle the 15 on their recording sheet. The student finished the problem. However, there were other choices the student could have made. The student could have picked 3 × 600 and 3 × 25. This would be fewer partials but would still result in the same product.

3 x 625		
180 (1800) 6 (60) 150 (15) 75 600 25 300		

$$\begin{array}{r} 625 \\ \times\ 3 \\ \hline 1800 \quad (3\times600) \\ 60 \quad (3\times20) \\ +\ \ 15 \quad (3\times5) \\ \hline 1875 \end{array}$$

3 x 625		
180 (1800) 6 60 150 15 (75) 600 25 300		

$$\begin{array}{r} 625 \\ \times\ 3 \\ \hline 1800 \quad (3\times600) \\ +\ \ \ 75 \quad (3\times25) \\ \hline 1875 \end{array}$$

RESOURCE(S) FOR THIS ACTIVITY

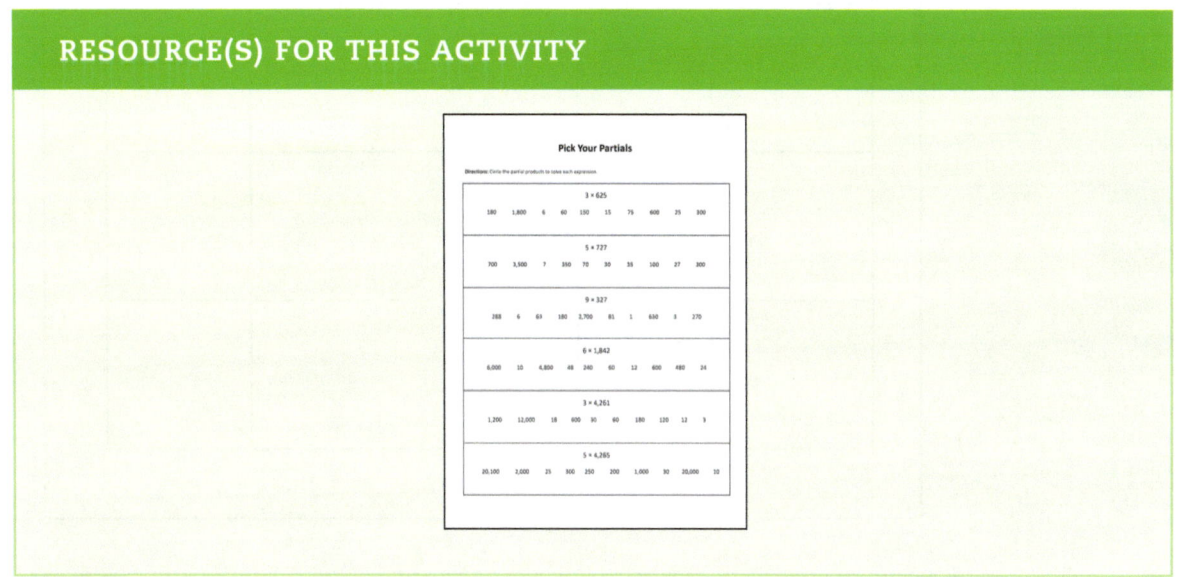

online resources This resource can be downloaded at **resources.corwin.com/FOF/multiplydividewholenumber**.

ACTIVITY 4.15

Name: Guess and Check—The Largest Product

Type: Center

About the Center: Estimating products is a way to help determine the reasonableness of a solution. In this center, students estimate and compare results and then check their thinking using partial products.

Materials: two sets of digit cards (0–9), playing cards (queens = 0, aces = 1, kings and jacks removed), or 10-sided dice; Guess and Check—The Largest Product recording sheet

Directions:
1. Students select digit cards to make multiplication expressions. Students pull three cards to make two-digit by one-digit problems, four cards for two-digit by two-digit problems, and so on.

2. Students estimate which expression will have the largest product and which will have the smallest product.

3. Students use Partial Products to find the actual products and to see if their estimates were correct.

For example, Stephon creates these problems: 37×84, 15×46, and 33×23. He thinks 37×84 will have the greatest product and that 15×46 will have the smallest product.

RESOURCE(S) FOR THIS ACTIVITY

Guess and Check—The Largest Product

Directions: Use digit cards to make three multiplication problems. Decide which problem will have the largest product and which will have the smallest product. Use Partial Products to find the exact products and to see if your guesses were correct.

Problem	Largest Product or Smallest Product	How I Found the Exact Product With Partial Products	Was My Guess Correct?

 This resource can be downloaded at **resources.corwin.com/FOF/multiplydividewholenumber**.

ACTIVITY 4.16

Name: Bag of Blocks

Type: Center

About the Center: Base-10 blocks can be a good tool for helping students see equal groups. This center presents students with a bag of base-10 blocks that shows one group of a certain amount. They then find how many would be in ___ bags (groups) of that amount.

Materials: bags of base-10 blocks, Bag of Blocks recording sheet, 10-sided die

Directions:
1. Students empty a bag and count the amount of blocks in the bag.

2. They record the amount.

3. Students then roll a number to find how many blocks would be in (the number rolled) bags.

4. Students use the Partial Products strategy and record their thinking.

TEACHING TAKEAWAY

Many centers can be used first as teaching activities. This provides an additional activity and it helps ensure students know how the center works.

RESOURCE(S) FOR THIS ACTIVITY

Bag of Blocks

Directions: Find the number of blocks in the bag. Roll a number to find how many bags of blocks you could have. Find the total number of blocks in that many bags using the Partial Products strategy.

Bag	Number of Blocks in the Bag	Number of Bags	Product (showing Partial Products)
A	234	4	200 × 4 = 800 30 × 4 = 120 4 × 4 = 16 800 + 120 + 16 = 936

online resources ⟲ This resource can be downloaded at **resources.corwin.com/FOF/multiplydividewholenumber**.

ACTIVITY 4.17

Name: *One or Both?*   **Type:** *Center*

About the Center: We break numbers apart by their place value to use the Partial Products strategy. One or both factors can be decomposed. Occasionally, students overly generalize this strategy and assume that each factor has to be broken apart. This center helps students see that either or both factors can be decomposed. One or Both is a follow-up center for Activity 4.5 ("One or Both?").

Materials: One or Both multiplication expression cards and recording sheet

Directions: 1. Students select a multiplication expression card.

2. Students find the product by breaking apart both factors and show their thinking with partial products.

3. Students find the product of the same problem by breaking apart one of the factors and show their thinking with partial products.

4. Students determine which approach was easiest for them to complete and circle that solution.

Possible expression cards are available to download.

RESOURCE(S) FOR THIS ACTIVITY

One or Both

Directions: Select a problem. Solve it by breaking apart both factors. Then, solve it by breaking apart one factor. Use drawings and numbers to show your thinking.

Problem	Break Apart Both Factors	Break Apart One Factor

32 × 46	22 × 84
19 × 46	32 × 63
25 × 66	23 × 65

online resources Expression cards and this recording sheet can be downloaded at **resources.corwin.com/FOF/ multiplydividewholenumber**.

NOTES

Think Multiplication Strategy

STRATEGY OVERVIEW:
Think Multiplication

What is Think Multiplication? This is a division strategy wherein, as the title suggests, the problem is reframed as a multiplication problem with a missing factor. For example, 240 ÷ 3 = ? is thought of as ? × 3 = 240. Think Multiplication is first introduced as a strategy for recalling basic division facts. For 24 ÷ 3, a student would ask themself, "What times 3 equals 24?" Knowing that the missing factor is 8, they then translate the missing factor to be the quotient of the original problem. The same thinking applies as numbers get larger, as illustrated in the following examples.

DIVISION PROBLEM	STUDENT THINKING	THINK MULTIPLICATION PROBLEM
24 ÷ 3 = ?	How many groups of 3 in 24? or What times 3 equals 24?	? × 3 = 24
249 ÷ 3 = ?	How many groups of 3 in 249?	? × 3 = 249
475 ÷ 25 = ?	How many 25s in 475?	? × 25 = 475

HOW DOES THINK MULTIPLICATION WORK?

Think Multiplication is an example of using an inverse relationship. Symbolically, this means that if a × b = c, then c ÷ b = a and c ÷ a = b. For 249 ÷ 3 = ?, it can be interpreted as "How many groups of 3 in 249?" or "How many rows of 3 to have 249?" If the missing factor is known, then that is the end. If it is not known, then Partial Quotients can be employed. Here are some examples:

78 ÷ 6 → How many 6s in 78? I don't know, so I will break apart the dividend into 60 + 18. How many 6s in 60? 10 How many 6s in 18? 3 The answer is 13.	330 ÷ 15 → How many 15s in 330? I don't know. I can break apart 330 into 150 + 150 + 30. How many 15s in 150? 10 How many 15s in 30? 2 The answer is 22.

WHEN DO YOU CHOOSE THINK MULTIPLICATION?

Anytime there is a division situation, Think Multiplication is one of only three options. It is a particularly good option when there is an overt relationship to a basic fact. The following examples of division problems clearly relate to 6 × 6 = 36 and can be found using that known fact.

36 ÷ 6	360 ÷ 6	360 ÷ 60	3,600 ÷ 6	3,600 ÷ 60	3,600 ÷ 600

Another good opportunity for Think Multiplication is when the divisor is a well-known multiple or automaticity (i.e., 15s, 25s, 30s), such as 45 ÷ 15 or 375 ÷ 25, or is another divisor for which the multiples are well known, like 250 (1,800 ÷ 250).

THINK MULTIPLICATION:
Strategy Briefs for Families

It is important that families understand the strategies and know how they work so that they can be partners in the pursuit of fluency. These strategy briefs are a tool for doing that. You can include them in parent or school newsletters or share them at parent conferences. They are available for download so that you can adjust them as needed.

Think Multiplication

How It Works: We can use the relationship between multiplication and division to solve division problems. Division is also a missing factor multiplication problem.

1. Rethink the division problem as a missing factor multiplication problem.
2. Use multiplication to find the unknown factor.
3. Once known, the factor becomes the quotient of the original problem.

The left example shows that the problem is thought as 6 × ? = 366. The relationships between 366 and 6 are recognized. 6 × 60 = 360. 6 × 1 = 6. 6 × 61 = 366, so 366 ÷ 6 = 61.

The right example is similar. 154 is recognized as 140 and 14, which are both products of 7.

When It's Useful: Think Multiplication is useful when division problems have clear relationships to multiplication problems.

Dividend Divisor	
$366 \div 6 = ?$	

Factor
$6 \times \boxed{?} = 366$

$6 \times \boxed{60} = 360$
$6 \times 1 = 6$

$366 \div 6 = 61$ Quotient

Factor
$7 \times \boxed{?} = 154$

$7 \times \boxed{20} = 140$
$7 \times \boxed{2} = 14$

$154 \div 7 = 22$ Quotient

Dividend Divisor
$154 \div 7 = ?$

Think Multiplication

How It Works: We can use the relationship between multiplication and division to solve division problems. Division is also a missing factor multiplication problem.

1. Rethink the division problem as a missing factor multiplication problem.
2. Use multiplication to find the unknown factor.
3. Once known, the factor becomes the quotient of the original problem.

The left example shows that the problem is thought as 30 × ? = 240. The person then uses a known basic fact (3 × 8) to find the related problem (30 × 8). The unknown factor, 30, is the quotient of the original problem.

The right example does not rely on a basic fact but a familiar problem of 12 × 2. Several equations were used to find the missing factor.

When It's Useful: Think Multiplication is useful when division problems have clear relationships to multiplication problems.

Dividend Divisor
$240 \div 30 = ?$

Factor
$30 \times \boxed{?} = 240$

$3 \times 8 = 24$
$30 \times \boxed{8} = 240$

$240 \div 30 = 8$ Quotient

Dividend Divisor
$240 \div 12 = ?$

Factor
$12 \times \boxed{?} = 240$

$12 \times 2 = 24$
$12 \times \boxed{20} = 240$

$240 \div 12 = 20$ Quotient

 These resources can be downloaded at **resources.corwin.com/FOF/
multiplydividewholenumber**.

NOTES

_____

_____

_____

_____

MODULE 1 — Break Apart to Multiply Strategy
MODULE 2 — Halve and Double Strategy
MODULE 3 — Compensation Strategy
MODULE 4 — Partial Products Strategy
MODULE 5 — Think Multiplication Strategy
MODULE 6 — Partial Quotients Strategy
MODULE 7 — Standard Algorithms for Multiplication and Division

TEACHING ACTIVITIES for Think Multiplication

Before students are able to choose strategies, a key to fluency, they first must be able to understand and use relevant strategies. These activities focus specifically on Think Multiplication. Think Multiplication relies on understanding how multiplication and division are related and recognizing basic facts within dividends and divisors. This strategy builds toward standard algorithms in the future. While students may employ other methods, which is appropriate, they also must learn this important strategy to support their flexibility and efficiency. In this section, we focus on instructional activities for developing understanding of Think Multiplication and consideration for when to use it.

 ## ACTIVITY 5.1
IS IT STILL TRUE?

Students are introduced to the inverse relationship between multiplication and division when they first learn about the operations and as they learn strategies for recalling basic facts. Although students may show solid understanding of this concept with basic facts, they may be challenged to transfer it to multidigit situations. This activity is an opportunity for students to revisit this inverse relationship to prove that it is still true. Here is an example of the prompts you might post.

IT IS TRUE THAT . . .	BUT IS IT TRUE THAT . . . ?
$5 \times 4 = 20$ is the inverse of $20 \div 4 = 5$	$50 \times 4 = 200$ is the inverse of $200 \div 4 = 50$
$5 \times 4 = 20$ is the inverse of $20 \div 4 = 5$	$5 \times 40 = 200$ is the inverse of $200 \div 40 = 5$
$8 \times 3 = 24$ is the inverse of $24 \div 3 = 8$	$800 \times 3 = 2,400$ is the inverse of $2,400 \div 3 = 800$
$8 \times 3 = 24$ is the inverse of $24 \div 3 = 8$	$8 \times 300 = 2,400$ is the inverse of $2,400 \div 300 = 8$

For these examples, students work collaboratively to determine if the related statement is true. However, at this time in their learning, they cannot focus on the number of zeros in the dividends, divisors, and quotients. Instead, they should use place value models or drawings to prove their thinking. Students might draw 5 tens in four groups along with the multiplication and division equations that their drawing represents.

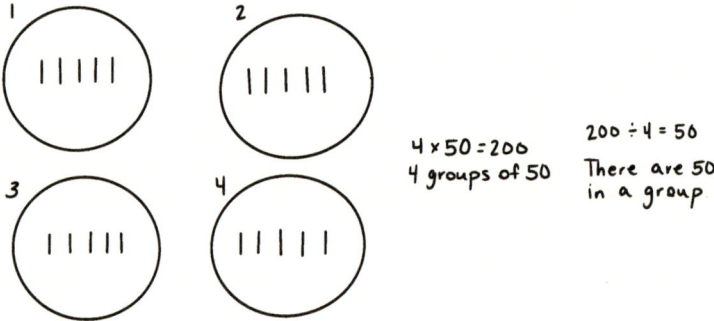

After students work to prove the relationships, you should lead a discussion with the group about how the equations in the left column are related to those in the right and how that can be useful when multiplying and dividing.

ACTIVITY 5.2
THE DIRECT CONNECT: THREE STRINGS

This activity is another experience for students to make connections between basic facts, multiplication, and division. It helps them see patterns as you eliminate everything but the related problems as you would in a number string. The twist with this activity is that it presents three number strings that are connected through the inverse relationship of multiplication and division. As this is an early experience, do not expect students to quickly solve the problems. Instead, ask groups to solve each problem within the table using tools, drawings, and numbers or equations. Then, convene the groups to share their findings for each equation. During discussion, first confirm the missing values. But then, focus students' attention on the relationship between the multiplication and division problems.

$3 \times 8 =$	$3 \times 80 =$	$3 \times 800 =$	$3 \times 8,000 =$	$30 \times 80 =$	$30 \times 800 =$
$24 \div 3 =$	$240 \div 3 =$	$2,400 \div 3 =$	$24,000 \div 3 =$	$2,400 \div 30 =$	$24,000 \div 30 =$
$24 \div ? = 3$	$240 \div ? = 3$	$2,400 \div ? = 3$	$24,000 \div ? = 3$	$2,400 \div ? = 30$	$24,000 \div ? = 30$

In this example, you might ask debriefing questions like these:

- How did knowing 3×8 help you solve all of the problems, including the division problems?

- What patterns did you notice in the problems?

- Why does it make sense that $3 \times 8 = 24$ and so 3×8 tens equals 24 tens or 240? Where do you see this thinking in other problems?

- Tell how you used any multiplication problem to help you solve the division problem in that column.

- Which of the problems were most challenging to think about? Why?

- Now that you have done this, what multiplication and division problems could you make if I gave you the basic fact $6 \times 3 = 18$?

The last bulleted question is an example of how you can extend the activity for other lessons. You can also modify it a bit to become a more problem-based variation that students would work with after they have worked with the first version a few times.

$7 \times 5 =$	$? \times ? = 350$	$? \times ? = 3,500$	$? \times ? = 3,500$	$? \times ? = 35,000$	$? \times ? = 35,000$
$35 \div 7 = ?$	$350 \div 7 = ?$	$3,500 \div 7 = ?$	$3,500 \div 70 = ?$	$35,000 \div 7 = ?$	$35,000 \div 70 = ?$
$35 \div ? = 7$	$350 \div ? = 7$	$3,500 \div ? = 7$	$3,500 \div ? = 70$	$35,000 \div ? = 7$	$35,000 \div ? = 70$

This table shows what that variation might look like. Notice that the top row has the multiplication fact but the other problems have unknown factors. Students use the information in the related division facts to determine the unknowns in each column.

ACTIVITY 5.3
RECOGNIZING PRODUCTS
AND DECOMPOSING DIVIDENDS

Using the inverse is helpful in division situations beyond known multiplication problems. In some problems, such as 154 ÷ 7, you can decompose the dividend into known multiples of the divisor. For example, 14 is a familiar multiple of 7; 154 can be decomposed into 140 (10 groups of 14) and 14. Then, it can be reasoned that 154 ÷ 7 is 20 groups of 7 and two more groups. This use of Think Multiplication is very handy but it takes time to develop. It will need to be taught and practiced. You may have to be more direct when teaching it by giving a direct prompt for students to explore. For example, you might pose this example:

Kristen thinks she can find 72 ÷ 6 by thinking of 60 ÷ 6 + 12 ÷ 6. Prove why Kristen is correct or incorrect. Create other examples to show how her thinking will or won't work.

As you might imagine, students will embark on proving this in a variety of ways. Look for students who rely simply on expressions and equations to prove their thinking as in the example. This may signal those who are advancing to a point that they can make arguments symbolically.

$$72 \div 6$$

$$60 \div 6 = 10$$
$$12 \div 6 = 2$$
$$\overline{\quad\quad} \quad \overline{\quad\quad}$$
$$72 \quad\quad 12$$

So Kristen is right.

$$72 \div 6 = 12$$

Also look for students who stray from the original prompt and attempt to decompose the divisor. As you know, the divisor cannot be decomposed. This must be explored and discussed if it comes up in student work. If it does not, you will need to provoke it so that students are aware that it doesn't work. You can use a prompt like this:

Brian thinks he can find 72 ÷ 6 by thinking 72 ÷ 2 + 72 ÷ 3. Prove why Brian is correct or incorrect. Create other examples to show if his thinking will or will not work.

The example proves what you know to be true. This student also uses equations alone to show his thinking, which is fine. In both cases, the student used a calculator to prove the accuracy of his work.

$$72 \div 3 = 24$$
$$72 \div 2 = 36$$
$$\overline{\quad} \quad \overline{\quad}$$
$$6 \quad 60$$

72 ÷ 6 does not equal 60

The additional examples that students create in both situations will enrich the discussion and provide further evidence that the dividend can be decomposed. This leads to a very important prompt for you to pose: *Why* is breaking apart the dividend helpful? Students will have all sorts of ideas. Help them "discover" the idea by providing the next few true/false problems to explore, as each helps develop understanding of how to break apart to support division.

65 ÷ 5 is the same as 50 ÷ 5 + 15 ÷ 5

48 ÷ 12 is the same as 48 ÷ 3 + 48 ÷ 4

42 ÷ 3 is the same as 30 ÷ 3 + 12 ÷ 3

60 ÷ 15 is the same as 60 ÷ 5 and 60 ÷ 3

84 ÷ 4 is the same as 40 ÷ 4 + 40 ÷ 4 + 4 ÷ 4

Activity 5.8 ("Breaking Dividends") is a routine for providing more practice with this idea.

ACTIVITY 5.4
WHAT'S MISSING?

The triangle cards used in this activity vary slightly from triangle fact cards that students are familiar with, in that one of the numbers is unknown. In this activity, students determine the missing value on a triangle card using multiplication and division. Students should justify their thinking with pictures and numbers. After finding solutions, students should discuss how they found their solutions and should share the resulting multiplication and division equations that the cards represent. During this discussion, reinforce that think multiplication can be an effective strategy for finding quotients.

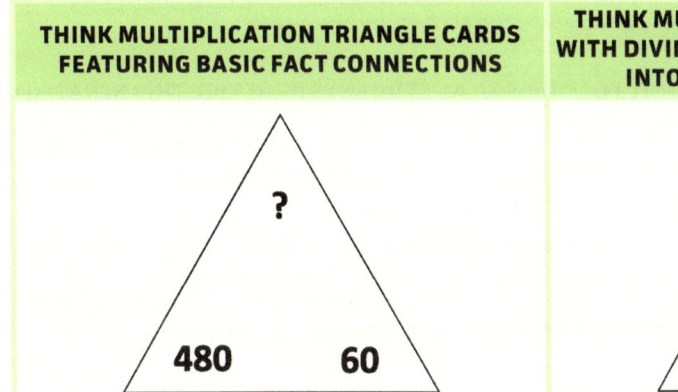

| THINK MULTIPLICATION TRIANGLE CARDS FEATURING BASIC FACT CONNECTIONS | THINK MULTIPLICATION TRIANGLE CARDS WITH DIVIDENDS THAT CAN BE DECOMPOSED INTO MULTIPLES OF THE DIVISOR |

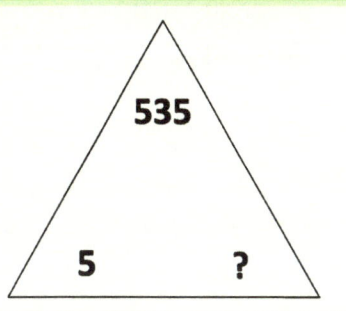

Note that early work with this strategy makes use of division problems that are clearly connected to basic facts. The card on the left is an example. As students advance, this activity can be used with dividends that are composed of recognizable multiples of the divisor. The card on the right shows this as 535 ÷ 5, which can be thought of as 500 ÷ 5 + 35 ÷ 5.

ACTIVITY 5.5
PROMPTS FOR TEACHING THINK MULTIPLICATION

Use the following prompts as opportunities to develop understanding of and reasoning with the strategy. Have students use representations and tools to justify their thinking, including base-10 models, number lines, arrays, and so on. After students work with the prompt(s), bring the class together to exchange ideas. These could be useful for collecting evidence of student understanding. Any prompt can be easily modified to feature different numbers (e.g., three-digit or four-digit numbers) and any prompt can be offered more than once if modified.

- Does the relationship between multiplication and division change when you are no longer working with basic facts? Use examples to justify your thinking.

- Tell how $5 \times 7 = 35$ can help you with 350 ÷ 5, 350 ÷ 7, 350 ÷ 50, and 350 ÷ 70.

- Ellio solves 105 ÷ 7 by thinking of 70 + 35. How does this help Ellio find the quotient?

- Create a story problem for 213 ÷ 3.

- How does knowing 4×8 help you solve 3,232 ÷ 4?

- Which basic facts does not help you solve 4,218 ÷ 6 (6×7, 6×5, or 6×3)? Explain your answer.

- Which basic facts help you solve 155 ÷ 5 (5×3, 5×9, or 5×1)? Explain your answer.

- What is the missing factor in $70 \times \_\_ = 630$ that will help you solve 630 ÷ 70?

- How could you break apart 372 to help you solve 372 ÷ 6?

PRACTICE ACTIVITIES for Think Multiplication

Fluency is realized through quality practice that is focused, varied, processed, and connected. The activities in this section focus students' attention on how this strategy works and when to use it. The activities are a collection of varied engagements. The discussion you facilitate after an activity or the reflection prompts you attach to it should help students think about what they did mathematically, how they reasoned about the activity, and when the math they did (namely the strategy) might be useful. Debriefing should also help students see how the practice activity connects to recent instruction or how the strategy connects to other strategies they know. Game boards, recording sheets, digit cards, and other required materials are available as online resources for you to download, possibly modify, and use. As students work with activities, look for how well they are acquiring the strategy and assimilating it into their collection of strategies.

FLUENCY COMPONENT	WHAT TO LOOK FOR AS STUDENTS PRACTICE THIS STRATEGY
Efficiency	• Are students using the Think Multiplication strategy or are they reverting to previously learned and/or possibly less appropriate strategies like repeated subtraction?
	• Are students using the Think Multiplication strategy efficiently? (i.e., Do they linger with a dividend trying to determine if it can be decomposed into multiples of the divisor? For example, in 616 ÷ 7, 616 is composed of 560 and 56 but that may be a challenging composition for a student to find.)
	• Do they use the strategy regardless of its appropriateness for the problem at hand?
Flexibility	• Are students carrying out Think Multiplication in flexible ways? (i.e., Do they use it for problems with clear connections to basic facts, such as 350 ÷ 7, and problems with reasonable composite dividends, such as 84 ÷ 7?)
	• Do they change their approach to or from this strategy as it proves inappropriate, inaccurate, or overly complicated for the problem?
Accuracy	• Are students using the Think Multiplication strategy accurately? (e.g., They understand 75 ÷ 25 = ? as ? × 25 = 75, not 25 × 75 = ?)
	• Are students finding accurate solutions?
	• Are they recalling and applying basic facts accurately?
	• Are they considering the reasonableness of their solutions?*
	• Are students estimating before finding solutions?*

*This consideration is not unique to this strategy and should be practiced throughout the pursuit of fluency with whole numbers.

WORKED EXAMPLES

Worked examples are problems that have been solved. Correctly worked examples can help students make sense of a strategy and incorrectly worked examples attend to common errors.

Worked examples can help students make connections between multiplication and division. The Think Multiplication strategy is commonly used to help students make sense of division. The following, however, are some challenges students may have with the strategy:

1. After the student restates the expression as a missing factor equation, they don't just know the answer and are "stuck."

   • 104 ÷ 8: interprets this as "How many 8s in 104?" Since that is not a basic fact, the strategy (or problem) is abandoned.

2. The student breaks apart the dividend, but not into parts that are multiples of the divisor:
   - 104 ÷ 8: breaks this apart into 100 ÷ 8 and 4 ÷ 8, rather than 80 ÷ 8 and 24 ÷ 8.

The prompts from Activity 5.5 can be used for collecting worked examples. Throughout the module are various examples that you can use as worked examples. A sampling of additional worked examples is provided in the following table.

SAMPLE WORKED EXAMPLES FOR BREAK APART TO MULTIPLY

Correctly Worked Example (make sense of the strategy) What did _____ do? Why does it work? Is this a good method for this problem?	Angelica's work: $495 \div 16 \rightarrow$ How many 16's in 495? I don't know products I know $320 \div 16 \rightarrow$ How many 16's in 320? 20 $160 \div 16 \rightarrow$ How many 16's in 160? 10 $320 + 160 = 480$ That leaves 15. Answer: 30 r 15
Partially Worked Example (implement the strategy accurately) Why did _____ start the problem this way? What does _____ need to do to finish the problem?	Which way to start: Tensly's way or Daryl's way? $72 \div 6 \rightarrow$ How many 6's in 72? Tensly's Way $(60 \div 6) + (12 \div 6)$ Daryl's Way $(36 \div 6) + (36 \div 6)$
Incorrectly Worked Example (highlight common errors) What did _____ do? What mistake does _____ make? How can this mistake be fixed?	Sarah's work for 3,400 ÷ 250: My problem: How many 250's in 3,400? I'll find how many 100's to start: $3,400 \div 100 = 34$ $3,400 \div 100 = 34$ $3,400 \div 50 = 68$ Answer 136

ACTIVITY 5.6

Name: "If I Know . . ."   **Type:** Routine

About the Routine: Think Multiplication relies on noticing and using multiplication facts to find quotients. In the "If I Know . . ." routine (SanGiovanni, 2019), students are provided a known multiplication fact that is related to a series of division problems. Students use the known to find all of the quotients. Then, the class talks about how each problem relates back to the known.

Materials: Prepare a series of related problems.

Directions: 1. Pose a multiplication fact with the product.

2. Then, unveil a list of division problems that are related to the multiplication problem.

3. Have partners work together to determine the unknown in each of the division problems.

4. After a few moments, bring the class together to discuss what each unknown is and how the known multiplication fact is related to the division problem.

IF I KNOW 3 × 7 = 21		
21 ÷ 3 =	21 ÷ 7 =	21 ÷ 3 =
210 ÷ 3 =	210 ÷ 7 =	42 ÷ 3 =
210 ÷ 30 =	210 ÷ 70 =	84 ÷ 3 =
2,100 ÷ 3 =	2,100 ÷ 7 =	231 ÷ 3 =
2,100 ÷ 30 =	2,100 ÷ 70 =	2,121 ÷ 3 =
2,100 ÷ 300 =	2,100 ÷ 700 =	

In this example, the known multiplication fact is 3 × 7. The first two columns show division problems that have a clear relationship to the known fact. Typically, the routine would feature these two columns. The third column might be reserved for advanced practice but is offered here to show other possibilities for the routine. The last four dividends connect with 21. The first (42) is twice as much; next is 84, which is twice as much as 42 or 4 times as much as 21; 231 is 210 and 21; while 2,121 is 2,100 and 21 more.

ACTIVITY 5.7

Name: "The Count Up"   **Type:** Routine

About the Routine: This routine has students estimate and skip-count. It is a good opportunity for practicing multiples while supporting the Think Multiplication strategy.

Materials: This routine does not require any materials.

Directions: 1. Set a clear counting path so that students know how they will count in the room. Having them gather in a large circle is one option.

2. Pose a division problem for students to consider.

3. Then, ask students how many counts of the divisor will be needed to get to the dividend. Other questions to ask before counting include the following:

   ● What are some numbers we will say as we count?

   ● What number do you think you will say?

   ● Who will be the last person to count?

4. Identify the first student to count and then students skip count up by the divisor around the room. As students count, record the numbers they say on the board, as this will help determine how many groups of the divisor are in the dividend and support the debriefing discussion.

5. Discuss how student predictions about the count compared to the actual count. Then, discuss how the count up connects with the division problem. Be sure to make explicit connections to multiplication.

For example, you pose 85 ÷ 5 = ? Students think about what the quotient might be. Then, they discuss the precount questions you ask, such as the bulleted examples in the third step in the directions above. Then, they count up to 85; while they do so, you record the numbers they say. After the count is complete, you return to the problem and ask students what they thought the quotient was. Pointing at the multiples recorded, count the number of multiples (groups), helping students see that each skip count is one more group. Ask students if they could have thought about counting up differently. In this example, 10 counts of 5 was 50, so there would be 35 more to go; 10 counts of 5 (50) plus 7 counts of 5 (35) equals 17 counts of 5 (85).

In another example, you might pose 420 ÷ 70 = ? Ask students to predict who will be the last person to count and what numbers will be said. During discussion, listen for students who are sharing how the division problem is related to another multiplication problem.

ACTIVITY 5.8

Name: *"Breaking Dividends"*   **Type:** *Routine*

About the Routine: Using the inverse is efficient for problems that are composites of multiples. That is, 72 ÷ 6 can be solved by thinking of 72 as a composite of multiples of 6 such as 60 + 12 or 30 + 42. Students need lots of practice with this idea. This routine provides an opportunity to do just that. This is an overlap with the Partial Quotients strategy (see Module 6).

Materials: Prepare a few problems for students to consider.

Directions: This routine explicitly focuses on decomposing the dividend to find friendly computations that make use of the inverse relationship.

1. Post or state a divisor to students (e.g., 6).

2. Have students identify multiples of the divisor.

3. Pose the division problem and have partners discuss how they might decompose the dividend into multiples of the divisor to take advantage of the Think Multiplication strategy.

For example, a teacher first asks students to think about multiples of 6 and record their ideas. Then, she poses 72 ÷ 6, asking students how they might break apart 72 into multiples of 6. Some may offer that 72 can be thought of as 60 and 12, so they can think of it as 60 ÷ 6 + 12 ÷ 6. Others might say that they see 72 as 36 and 36, so they can think of it as 36 ÷ 6 + 36 ÷ 6. Then, the teacher asks students how they might use their ideas about breaking apart for 84 ÷ 6 and then 96 ÷ 6. Here are some other examples to use in this routine:

- 84 ÷ 7 followed by problems like 91 ÷ 7, 98 ÷ 7, 105 ÷ 7, or 112 ÷ 7

- 48 ÷ 3 followed by problems like 54 ÷ 3, 57 ÷ 3, 66 ÷ 3, or 72 ÷ 3

- 70 ÷ 5 followed by 75 ÷ 5, 80 ÷ 5, 95 ÷ 5, 110 ÷ 5, or 115 ÷ 5

- 90 ÷ 6 followed by 96 ÷ 6, 102 ÷ 6, 126 ÷ 6, or 132 ÷ 6

Creating expressions for students to work with may seem challenging. Here is one way to generate problems: First, think of a divisor (like 6 in the last example). Then, think of a dividend that is composed of 10 times the divisor (60) and a multiple of the divisor more (36). As students progress, begin to use dividends that use multiples of 10 times the divisor (120).

ACTIVITY 5.9

Name: The Take   **Type:** Game

About the Game: *The Take* reinforces division problems that have clear connections to multiplication and basic facts.

Materials: *The Take* division cards, 1 set per group; 1 game board per group; and 10 counters per player

Directions:

1. Players place 10 counters on their game board. Players can place multiple counters or no counters on any number.

2. Players take turns choosing a division card.

3. The player reads the problem and tells the quotient.

4. The player removes a chip from the space that matches the quotient. The player loses their turn if there is no counter on that space.

5. The first player to remove all of their counters wins.

For example, a player places their 10 chips on the board as shown. The first card they choose is 320 ÷ 4 and they would remove one counter from 80. On their next turn, their card is 350 ÷ 7. They don't have a counter on 50, so they lose their turn.

TEACHING TAKEAWAY

To keep both students engaged, have both students work the problem even though it may be their opponent's turn.

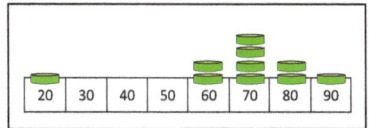

| 20 | 30 | 40 | 50 | 60 | 70 | 80 | 90 |

RESOURCE(S) FOR THIS ACTIVITY

160 ÷ 2	480 ÷ 6
240 ÷ 3	560 ÷ 7
320 ÷ 4	640 ÷ 8
400 ÷ 5	720 ÷ 9

The Take

Directions: Place 10 counters on the numbered spaces. You can put more than one counter on a number. Pull a division card. Remove a counter if it is on a number that is a quotient of the problem. Be the first to remove all 10 counters.

| 2 | 3 | 4 | 5 | 6 | 7 | 8 | 9 |

The Take TOO

Directions: Place 10 counters on the numbered spaces. You can put more than one counter on a number. Pull a *The Take TOO* division card. Remove a counter if it is on a number that is a quotient of the problem. Be the first to remove all 10 counters.

| 20 | 30 | 40 | 50 | 60 | 70 | 80 | 90 |

online resources Game cards and game boards can be downloaded at **resources.corwin.com/FOF/ multiplydividewholenumber**.

ACTIVITY 5.10

Name: *Paper Clip Connections*　　　　**Type:** *Game*

About the Game: *Paper Clip Connections* is a take on the classic Product Game in which students try to find products by moving their paper clips strategically. The game reinforces the relationship between multiplication and division with larger numbers. As students find products, they must share the related division problem in order to place their game piece.

Materials: *Paper Clip Connections* game board, two paper clips, about 40 two-color counters, and calculators

Directions:　　1. Player 1 places a paper clip on one of the factors in the left column and one of the factors in the bottom row. Factors are not within the grid.

　　2. The player finds the product of these two numbers and then tells their opponent the related division problem.

　　3. If accurate, player 1 places their counter on the space of the corresponding product of the two factors.

　　4. Player 2 then chooses to move each paper clip one space in either direction and repeats finding the product and telling the related division problem. Or player 2 can move one paper clip two spaces in either direction.

　　5. Play continues with each player choosing to move two paper clips one direction or moving one paper clip two spaces in either direction.

　　6. The first player to get five counters in a row (horizontally, vertically, or diagonally) wins the game.

For example, player 1 places paper clips on 40 and 30. They say that 40 × 30 = 1,200 and so 1,200 ÷ 40 = 30. Their opponent confirms and they place their counter on 1,200. Player 2 moves the paper clip on 40 to 30 and the paper clip on 30 to 20. Player 2 says 30 × 20 = 600 and so 600 ÷ 30 = 20. Player 2 is correct and places their counter on 600. Player 1 then moves the paper clip on 30 two spaces to 50 and cannot move the other paper clip. They share 50 × 20 = 1,000 and that 1,000 ÷ 50 = 20 and place their counter.

RESOURCE(S) FOR THIS ACTIVITY

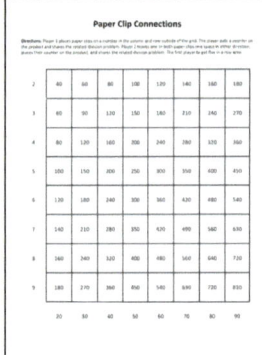 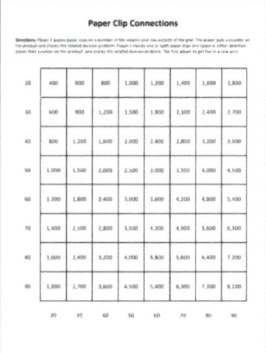

online resources · Game boards can be downloaded at **resources.corwin.com/FOF/multiplydividewholenumber**.

ACTIVITY 5.11

Name: Covering the Breaks   **Type:** Game

About the Game: When using the inverse to divide, it is useful to decompose the dividend into multiples of the divisor. Students need practice with this approach. *Covering the Breaks* is an opportunity for practice with this idea.

Materials: *Covering the Breaks* game board, 10-sided dice, about 30 two-color counters, math journal or scrap paper, and a calculator

Directions:
1. Players take turns rolling a digit using a 10-sided die. If the player rolls a 1 or 2, they roll again.

2. When a player rolls any other number (3–9), the player identifies a number on the game board that can be decomposed into multiples of the number rolled.

3. The player shares how they would decompose the number and record their thinking.

4. The player then works the problem to find the quotient of the space selected and the number rolled.

5. If accurate, the player places their counter on the space and it's their opponents turn.

6. The first player to get three in a row wins.

For example, player 1 rolls an 8. They choose to cover 96, stating that "96 can be decomposed into 80 and 16." The player records 80 ÷ 8 = 10 and 16 ÷ 8 = 2, so 96 ÷ 8 = 12 (10 + 2). Player 2 then rolls a 4. Player 2 decides to cover 52 because 52 can be thought of as 40 and 12. Player 2 records 40 ÷ 4 = 10 and 12 ÷ 4 = 3, so 52 ÷ 4 = 13. Player 1 then rolls a 4. Player 1 chooses 88, noting that it can be decomposed into 40, 40, and 4, and records their work as described.

TEACHING TIP

To keep students engaged, have both players take their turn at the same time. Here, both players would roll, compute, and record at the same time.

RESOURCE(S) FOR THIS ACTIVITY

Covering the Breaks

Directions: Take turns rolling a number. Choose a number on the game board to divide by the number you roll. Tell how the number on the space can be broken up to make division efficient. Record your thinking on a piece of paper and show your opponent. If you are correct, place your counter on the space. Get three spaces in a row to win.

42	52	63	65	66
66	72	75	77	84
84	85	88	91	92
95	96	98	102	104
105	108	110	126	126
132	135	147	154	168

online resources ↖ This resource can be downloaded at **resources.corwin.com/FOF/multiplydividewholenumber**.

ACTIVITY 5.12

Name: *Area Model Duel*　　　　　　　**Type:** *Game*

About the Game: *Area Model Duel* helps students think about ways to efficiently break apart a dividend into convenient numbers. Students show their thinking with an area model.

Materials: *Area Model Duel* cards and expression cards for each group, whiteboards and dry-erase markers for each player

Directions:
1. Players place cards face down in two piles. One pile is the expression cards (worth 1 point). The other deck are the duel cards (worth 2 points).

2. Players take turns choosing a card from the expression deck.

3. Players solve the expression using an area model.

4. Players explain how they solved the problem. The player who solved the expression with the least amount of partitioning keeps the expression card.

5. If both players solved the problem with the same area model, they declare a duel. The players flip over one card from the duel deck. The player who first states an accurate division equation that matches the area model keeps the card.

6. The player with the most cards/points at the end wins.

In this example, players solve 92 ÷ 4. Player 2 would keep the card, since they used two steps to solve the problem.

Area Model Duel can be modified in a variety of ways. For example, the expression cards can have two-, three-, or four-digit dividends. Create duel cards that have easier expressions if the goal is for students to practice basic facts. Create duel cards that have more complex expressions if the goal is efficiency.

RESOURCE(S) FOR THIS ACTIVITY

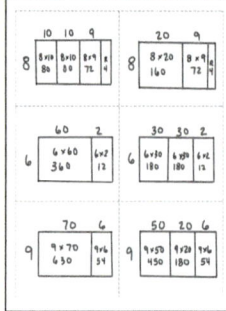
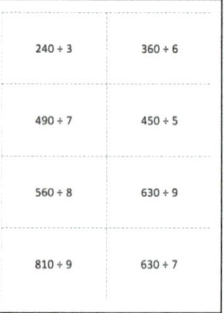

online resources　Expression cards (on the right) and duel cards (on the left) can be downloaded at **resources.corwin.com/FOF/multiplydividewholenumber**.

ACTIVITY 5.13

Name: Make the Quotient   **Type:** Center

About the Center: This center is a different take on a target-based activity. Here, students are presented with a target quotient and must make a division problem that has a quotient as close to the target as possible. You can make the center more challenging by having students deal themselves six digit cards to select from in order to make the quotient.

Materials: digit cards (0–9) or playing cards (queens = 0, aces = 1, kings and jacks removed), Make the Quotient cards and recording sheet

Directions: 1. Students spread out all of the digit cards face up.

2. Students choose a Make the Quotient card.

3. Students select from the digit cards to make two problems that satisfy the prompt and write related multiplication problems that they might have used to help them create the problem.

4. Students record their work and choose a new card.

For example, Marianna pulls "Make the Quotient 12. Use 3 digit cards." She selects 3, 6, and 3 to make the expression 36 ÷ 3, and then she selects 2, 4, and 2 to make the expression 24 ÷ 2. She records 36 ÷ 3 = 12 and 24 ÷ 2 = 12 on her recording page.

RESOURCE(S) FOR THIS ACTIVITY

Make the Quotient

Directions: Choose a Make the Quotient card. Use digit cards to make two different problems with that quotient. Write a multiplication problem related to the division problem.

Card	Division Problem 1	Related Multiplication Problem	Division Problem 2	Related Multiplication Problem

A. Make the Quotient 12 Use 3 digit cards __ __ ÷ __	B. Make the Quotient 15 Use 3 digit cards __ __ ÷ __
C. Make the Quotient 20 Use 3 digit cards __ __ ÷ __	D. Make the Quotient 24 Use 3 digit cards __ __ ÷ __
E. Make the Quotient 16 Use 3 digit cards __ __ ÷ __	F. Make the Quotient 30 Use 3 digit cards __ __ ÷ __
G. Make the Quotient 11 Use 3 digit cards __ __ ÷ __	H. Make the Quotient 13 Use 3 digit cards __ __ ÷ __

online resources Center cards and this recording sheet can be downloaded at **resources.corwin.com/FOF/ multiplydividewholenumber**.

ACTIVITY 5.14

Name: Rewrite It and Solve It   **Type:** Center

About the Center: Rewrite It and Solve It is an opportunity for students to practice rewriting division problems using the inverse.

Materials: Rewrite It and Solve It center cards and recording sheet

Directions:

1. Students select a card at random and record it on their recording sheet.

TEACHING TAKEAWAY

To save paper, print the recording sheet and put it in a plastic sleeve or laminate for students to write on with dry-erase markers.

2. Students rethink the division expression as a missing value multiplication equation and record it in column 2.

3. Students use Think Multiplication to solve and record the quotient in column 3.

Note: This could also be done in a journal rather than on a recording sheet.

RESOURCE(S) FOR THIS ACTIVITY

Rewrite It and Solve It

Directions: Pull a problem card. Rewrite it as a multiplication problem and find the product. Then, record the division equation.

Division Problem	Related Multiplication Problem	So...
120 ÷ 6 = ?	6 × ? = 120 6 × 20 = 120	120 ÷ 6 = 20

240 ÷ 3	360 ÷ 6
490 ÷ 7	450 ÷ 5
560 ÷ 8	630 ÷ 9
810 ÷ 9	630 ÷ 7

360 ÷ 12	480 ÷ 24
450 ÷ 5	600 ÷ 30
640 ÷ 32	320 ÷ 16
750 ÷ 25	396 ÷ 36

176 ÷ 8	372 ÷ 6
497 ÷ 7	455 ÷ 5
576 ÷ 8	648 ÷ 9
837 ÷ 9	644 ÷ 7

online resources ▸ Center cards and this recording sheet can be downloaded at **resources.corwin.com/FOF/ multiplydividewholenumber**.

ACTIVITY 5.15

Name: Triangle Cards   **Type:** Center

About the Center: Triangle fact cards help students develop multiplication and division basic fact recall. These cards focus students on the relationship between the three numbers through either of the operations. This center builds on that by using larger numbers. Any activity that makes use of a basic fact triangle card could be modified for use with these cards.

Materials: triangle cards and recording sheet

Directions: 1. Students select a triangle card.

2. The student writes the two multiplication equations the card represents, the two division equations the card represents, and one of the basic facts that all of these connect with.

For example, a student picks the card on the left. They record $9 \times 90 = 810$, $90 \times 9 = 810$, $810 \div 9 = 90$, $810 \div 90 = 9$, and the basic fact $9 \times 9 = 81$.

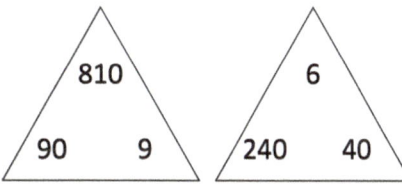

A modified version of this center presents two of the three numbers within the equation. Students must determine what the missing number is and record all of the equations. The product is the circled number, which can also be the missing number.

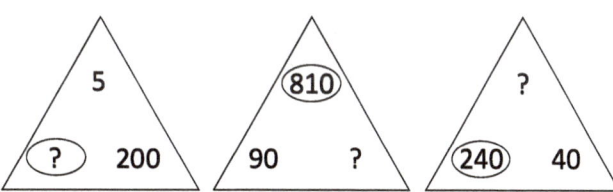

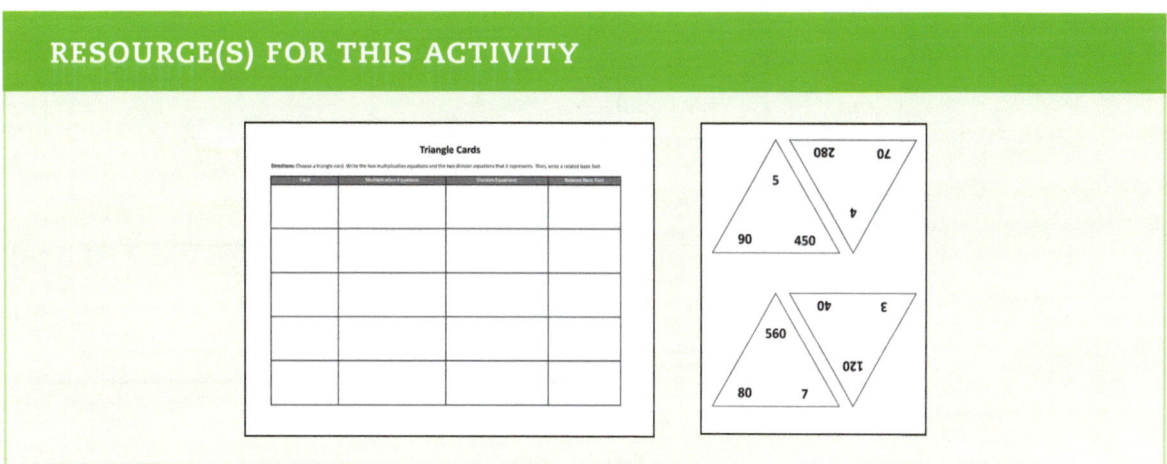

RESOURCE(S) FOR THIS ACTIVITY

Triangle cards and this recording sheet can be downloaded at **resources.corwin.com/FOF/ multiplydividewholenumber.**

ACTIVITY 5.16

Name: Finding the Breaks   **Type:** Center

About the Center: Breaking apart dividends to Think Multiplication relies on exposure, experience, and practice. Finding the Breaks provides opportunity for each of those. With this center, students choose a division problem and tell how they can decompose the dividend into multiples of the divisor. This center is intended to complement the routine in Activity 5.8 ("Breaking Dividends"). Division problems used with that routine can be applied to this center or you can use the problems from this center for that routine. As with any quality practice opportunity, students should have an opportunity to reflect on the experience and tell how using multiplication can help them solve more complex division problems.

Materials: division expression cards, Finding the Breaks recording sheet

Directions: 1. The student pulls a division card and writes it on the recording sheet.

           2. The student identifies how to break apart the dividend into multiples of the divisor so that they can use Think Multiplication to find the quotient.

For example, a student pulls 288 ÷ 8. The student breaks 288 into 240 and 48 and uses the inverse to find 240 ÷ 8 (30) and 48 ÷ 8 (6). She then adds the quotients of 30 and 6 to find the quotient of the original problem of 288 ÷ 8.

RESOURCE(S) FOR THIS ACTIVITY

Finding the Breaks

Directions: Select a division problem. Break apart the dividend into multiples of the divisor. Use multiplication to divide the breaks.

Problem	Break Apart the Dividend	Divide the Breaks Using Multiplication to Help	So...
288 ÷ 8	288 ⟋ ⟍ 240 48	240 ÷ 8 = 30 because 8 × 30 = 240 48 ÷ 8 = 6 because 8 × 6 = 48	288 ÷ 8 = 36 because I put 30 and 6 together.

 Division expression cards and this recording sheet can be downloaded at **resources.corwin.com/ FOF/multiplydividewholenumber.**

Partial Quotients Strategy

STRATEGY OVERVIEW:
Partial Quotients

What is Partial Quotients? It is a division strategy in which the dividend is broken apart into convenient parts. Those parts are divided separately, then these partial quotients are added together. This strategy is foundational for learning and using the standard algorithm for division. Unlike the standard algorithm, students can find any partial quotient, rather than seek the exact largest partial quotient for a place value.

HOW DOES PARTIAL QUOTIENTS WORK?

Partial Quotients is the distributive property in action. The way the dividend is partitioned is flexible. Let's look at three ways to use Partial Quotients to solve 216 ÷ 3.

On the left, Partial Quotients is used by first taking out 10 groups of 3 (30). This is repeated again. Then, the student recognizes they could remove 50 groups of 3 (150), which leaves 2 groups of 3. The partials add to 72. In the middle column, the first group size used is 50. The right column shows the largest partial by place value, 70. This is more efficient, and it is the same as the standard algorithm.

WHEN DO YOU CHOOSE PARTIAL QUOTIENTS?

Partial Quotients always works and is a good choice when Think Multiplication isn't a viable strategy and/or when students are struggling with the standard algorithm. A key to Partial Quotients being a good option is being able to choose partial quotients well, as each of the three earlier examples illustrate. The work shown here, however, is not as efficient. This may be a good beginning, but with experience, the partial quotients need to be grouped more efficiently.

PARTIAL QUOTIENTS:
Strategy Briefs for Families

It is important that families understand the strategies and know how they work so that they can be partners in the pursuit of fluency. These strategy briefs are a tool for doing that. You can include them in parent or school newsletters or share them at parent conferences. They are available for download so that you can adjust them as needed.

Partial Quotients

How It Works: We can remove easy-to-find groups from the dividend until there are no more groups to remove.

1. Find a friendly amount to take from the dividend.
2. Find a friendly amount to take from what is left.
3. Continue removing friendly amounts until nothing is left to remove.
4. Find the total number of groups taken away.

The left example shows that 100 is as a friendly amount to remove: 20 groups of 5 are in 100. 100 is removed and 265 is left. Another 100 can be removed, leaving 165. This is continued until nothing is left to remove. Then you add up the number of groups removed: 20 + 20 + 30 + 5 = 73.

The same process is applied to the right example. However, 350 was a friendly amount removed. Again, there are 73 groups of 5 in 365, so 365 ÷ 5 = 73.

When It's Useful: Partial Quotients is almost always useful. It can become difficult to use as the number of digits increases in the divisor or dividend.

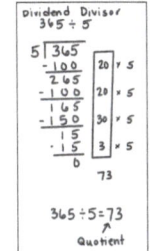 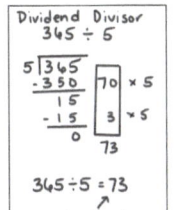

Partial Quotients

How It Works: We can remove easy-to-find groups from the dividend until there are no more groups to remove.

1. Find a friendly amount to take from the dividend.
2. Find a friendly amount to take from what is left.
3. Continue removing friendly amounts until nothing is left to remove.
4. Find the total number of groups taken away.

In the example, 20 groups of 52 (1,040) were removed from the dividend (3,684). This is continued until 44 was left to remove. A group of 52 couldn't be removed from 44. So, there are 70 (20 + 20 + 20 + 10) groups of 52 in 3,684 with a remainder of 44.

When It's Useful: Partial Quotients is almost always useful. It can become unwieldy as the number of digits increases in the divisor or dividend.

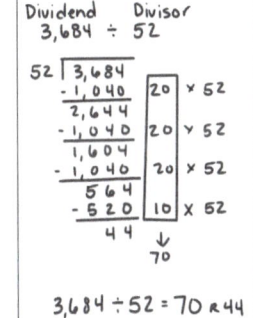

 These resources can be downloaded at **resources.corwin.com/FOF/ multiplydividewholenumber.**

NOTES

_____

_____

_____

_____

MODULE 1 Break Apart to Multiply Strategy

MODULE 2 Halve and Double Strategy

MODULE 3 Compensation Strategy

MODULE 4 Partial Products Strategy

MODULE 5 Think Multiplication Strategy

MODULE 6 Partial Quotients Strategy

MODULE 7 Standard Algorithms for Multiplication and Division

TEACHING ACTIVITIES for Partial Quotients

Partial Quotients builds on students' understanding of breaking numbers apart and finding how many groups of a certain size are in a given amount. Partial Quotients lays the foundation for understanding how the standard algorithm works in the future. In this section, we focus on instructional activities for Partial Quotients. The goal is for students to become adept with the strategy and also to consider when the strategy is useful.

ACTIVITY 6.1
BASE-10 QUOTIENTS

When students first work with division they use manipulatives to find the quotient of basic facts. We want to extend this work as students divide larger numbers. Base-10 blocks are concrete tools to help students understand division. Provide students with expressions, such as 369 ÷ 3, and have them represent the dividend using base-10 blocks. Show students how to partition the 369 into 3 equal groups and connect the process with the physical tools to the numerical method.

Start off by partitioning the 3 hundreds into three equal groups. Each group has 1 hundred. When you remove that amount from the dividend, 69 remains.

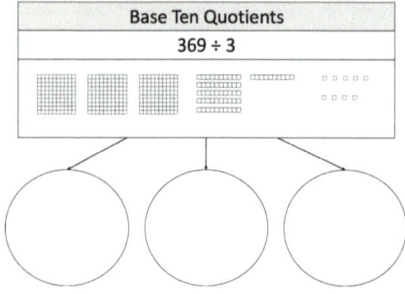
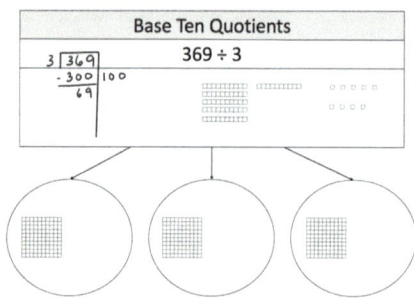

Then partition the tens into three groups. There are 2 tens in each group; 9 units remain.

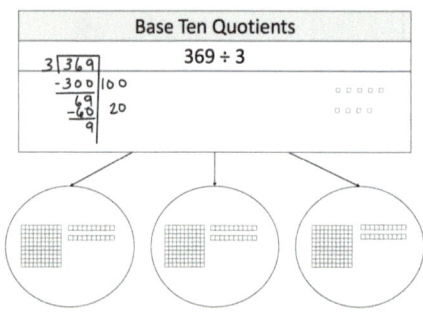
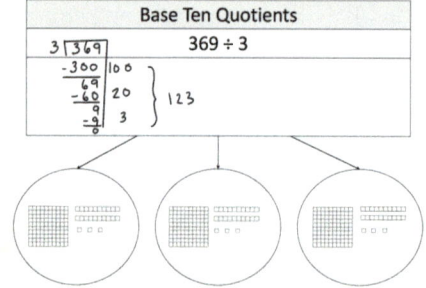

Students keep partitioning until nothing remains. By adding the partial quotients together, students then find the final quotient. For this example, there are 123 blocks in each group. It's important to provide students with problems that partition equally but also situations with remainders.

ACTIVITY 6.2
IS IT THE SAME?

Students may tend to overly rely on breaking apart a dividend into tens and ones because those multiplication factors are more comfortable for them. For example, they may solve $66 \div 3$ using $3 \times 10 + 3 \times 10 + 3 \times 2$ to find the quotient rather than consider multiples of 10, such as 3×20.

This can become quite problematic for problems with larger dividends, such as $549 \div 3$. This teaching activity focuses on establishing that the two approaches yield the same result and that one Partial Quotients strategy may be more efficient than the other. In this activity, you pose a problem such as $66 \div 3$. You discuss with students how it might be broken apart and record their thinking. Then, groups of students work to find the quotient both by using multiples of tens and also with groups of tens (as illustrated above). Explore larger dividends like $549 \div 3$. After students find the quotient, discuss how the options are the same and how they are different. Ask students to determine which approach is more efficient or at least easier to keep track of. Extend the activity by asking students if they think it will always work. Students can create their own problems to find out.

ACTIVITY 6.3
ANOTHER WAY

This activity is an opportunity for students to practice using the area model representation and connecting it to Partial Quotients. For example, a school placed a book order for 989 books. The books were to be evenly distributed to eight teachers. How many books does each teacher receive? Students solve using one strategy. Invite students to review their solutions. They may check their work by comparing answers with a partner who used the same strategy. Students then move on to show the division in another way, now using the alternate strategy. For example, if a student used an area model as their first strategy, then they would solve the problem in another way using the Partial Quotients strategy. At this point, since both strategies have been explored, debrief with the class. Ask students how the area models and partial quotients connected.

Students can discuss how they decided on the parts or factors they would need when solving. As a note, when working with story problems it's important to make sure students are interpreting the remainder correctly. When students are given expressions to solve not in context, check for students' notation of the remainder.

ACTIVITY 6.4
COMPARING PARTIALS

This activity is an opportunity for students to make connections between partial products and partial quotients. Ask the students to fold a paper in half and label the left side "Partial Products" and the right side "Partial Quotients." Provide students with a multiplication expression and a related division expression. Students then work to solve the problem on the other side of the paper. Then, students come together to share their findings. During the discussion, point out the relationship between the multiplication and division problems.

In this example, students are given the expression 6×58. Students solve using partial products. Then students move to the other side of the paper and create a division expression $348 \div 6$ based on the equation $6 \times 58 = 348$ on the left side.

ACTIVITY 6.5
PROMPTS FOR TEACHING PARTIAL QUOTIENTS

Use the following prompts as opportunities to develop understanding of and reasoning with the strategy. Have students use representations and tools to justify their thinking. After students work with the prompt(s), bring the class together to exchange ideas. Any prompt can be easily modified to feature different numbers (e.g., three-digit or four-digit numbers) and any prompt can be offered more than once with different numbers. Student work with these prompts could be useful for collecting evidence of understanding.

- How does the basic fact $21 \div 7$ help you solve $2,100 \div 7$?

- How can you use estimation to decide the first partial quotient in $496 \div 8$?

- Samuel says that $3,240 \div 6$ has a partial quotient of 1,600. Natasha disagrees with Samuel. Who is correct? Explain your reasoning.

- Create a story problem for $249 \div 3$.

- Describe an efficient way to solve $1,248 \div 6$ using Partial Quotients.

- Is the quotient of 2,495 ÷ 5 over or under 500? Explain your reasoning.

- Create an example of when the Partial Quotients strategy isn't efficient. Justify your thinking.

- Describe to a partner how you use the Partial Quotients strategy. Show an example.

- Chase solved 84 ÷ 6 by partitioning the dividend into 60 and 24, then he divided both by 6. Use Partial Quotients to prove if Chase's strategy works in finding a quotient.

- How is using an area model different from the numeric method with the long division symbol (division bracket)? How are they similar?

- Use the digits 1, 3, 5, 7, 9 to create a division problem that divides a four-digit number by a one-digit number with the greatest quotient.

- Use the digits 0, 2, 4, 6, 8, to create a division problem that divides a four-digit number by a one-digit number with the smallest quotient.

NOTES

_____

_____

_____

_____

_____

_____

_____

_____

_____

_____

PRACTICE ACTIVITIES for Partial Quotients

Fluency is realized through quality practice that is focused, varied, processed, and connected. The activities in this section focus students' attention on how this strategy works and when to use it. The activities are a collection of varied engagements. The discussion you facilitate after an activity or the reflection prompts you attach to it should help students think about what they did mathematically, how they reasoned about the activity, and when the math they did (namely the strategy) might be useful. Debriefing should also help students see how the practice activity connects to recent instruction or how the strategy connects to other strategies they know. Game boards, recording sheets, digit cards, and other required materials are available as online resources for you to download, possibly modify, and use. As students work with activities, look for how well they are acquiring the strategy and assimilating it into their collection of strategies.

FLUENCY COMPONENT	WHAT TO LOOK FOR AS STUDENTS PRACTICE THIS STRATEGY
Efficiency	• Are students using the Partial Quotients strategy or are they reverting to previously learned and/or possibly less appropriate strategies such as repeated subtraction? • Are students using the Partial Quotients strategy efficiently? (e.g., Do they work with suitable groups? Or do they take away one group or only 10 groups at a time?) • Do they use the strategy regardless of its appropriateness for the problem at hand? (e.g., 320 ÷ 8 is solved best with Think Multiplication.)
Flexibility	• Are students carrying out Partial Quotients in flexible ways? (e.g., Are they able to take away groups of different sizes based on the problem they are working?) • Do they change their approach to or from this strategy as it proves inappropriate or overly complicated for the problem?
Accuracy	• Are students using the Partial Quotients strategy accurately? (e.g., Do they subtract the accurate amount each time?) • Does student recordings of Partial Quotients match what they describe that they are doing? • Are students finding accurate solutions? (e.g., Do they multiply and/or subtract correctly?) • Are they considering the reasonableness of their solutions?* • Are students estimating before finding solutions?*

*This consideration is not unique to this strategy and should be practiced throughout the pursuit of fluency with whole numbers.

WORKED EXAMPLES

Worked examples are problems that have been solved. Correctly worked examples can help students make sense of a strategy and incorrectly worked examples attend to common errors.

Partial Quotients, like Think Multiplication, can lead to the following challenges or errors:

1. The student records partial products to start, but the partials go beyond what is needed for the problem at hand.

   • 1,834 ÷ 22: notes 22 × 10 = 220, 22 × 100 = 2,200, and 22 × 1,000 = 22,000.

2. The student continues to use ×10 facts, without looking for other possible partials.

   • 1,834 ÷ 22: 22 × 10 (220) is subtracted repeatedly, rather than noticing and using 22 × 20 (440) or 22 × 30 (660).

3. There is confusion about what "work" is collected to determine the answer.
   - Records a list of partials down the right side of the problem but does not add them together.
4. The student makes subtraction errors, in particular not regrouping when it is necessary.
   - Subtracts 2 – 5 and records 3.

Partial quotients are often obtained by considering multiples of the divisor (×10, ×100, etc.), but there are other options, which can be highlighted through worked examples. Comparing different worked examples can help students gain insights into efficient use of the Partial Quotients strategy. The prompts from Activity 6.5 can be used for collecting worked examples. Throughout the module are various worked examples that you can use as fictional worked examples. A sampling of additional ideas is provided in the following table.

SAMPLE WORKED EXAMPLES FOR PARTIAL QUOTIENTS

Correctly Worked Example (make sense of the strategy) What did _____ do? Why does it work? Is this a good method for this problem?	Thomas's work for 1,876 ÷ 14: $14 \times 10 = 140$ $14 \times 100 = 1,400$ $$14\overline{)1,876}$$ $-1,400$ 14×100 476 -140 14×10 336 -280 14×20 56 -28 14×2 28 -28 14×2 0 (134)	Alaina's work for 675 ÷ 7: $7 \times 10 = 70$ $7 \times 50 = 350$ $7 \times 90 = 630$ $$7\overline{)675}$$ -630 7×90 45 -42 7×6 3 (96 R3)
Partially Worked Example (implement the strategy accurately) Why did _____ start the problem this way? What does _____ need to do to finish the problem?	Autumn's start for 462 ÷ 6: $462 \div 6$ $6 \times 10 = 60$ $6 \times 60 = 360$	Mika's start for 625 ÷ 75: $625 \div 75$ $75 \times 2 = 150$ $75 \times \_\_ = \_\_$
Incorrectly Worked Example (highlight common errors) What did _____ do? What mistake does _____ make? How can this mistake be fixed?	Matt's work for 7,008 ÷ 15: $15 \times 10 = 150$ $15 \times 100 = 1,500$ $$15\overline{)7,008}$$ $-1,500$ 15×100 $6,508$ $-3,000$ 15×200 508 -300 15×20 208 -150 15×10 158 -150 15×10 8	

ACTIVITY 6.6

Name: "How Many Are In?" **Type:** Routine

About the Routine: When working with multidigit dividends, students are often challenged to determine how many groups of the divisor are in the dividend. Some students will simply start with 10 groups or 100 groups. This approach is acceptable but can create many numbers to keep track of during the process, which increases the likelihood of an error. This routine is a practice opportunity for determining how many groups to take from a dividend. It helps students by providing repetitions and exposes them to others' thinking.

Materials: Identify different group sizes and dividends as shown.

Directions:
1. First reveal the "are in" number, which is 537 in this example.

2. Give students a moment to think about the multiples of numbers that are close to it.

3. Then, pose a "how many" number. In this example, students think how many fives are in 537.

4. Students share their solution and how they arrived at it with a partner and then share out with the group. Some will share there are 100 fives in 537. Some might share 105 fives and some others might find 107 fives. At this time, you don't need to identify which solution is best, but rather state that each of these is a good idea about how many fives are in 537. Also note that this discussion should not take much longer than a minute or so for the first group. The intention is to get through at least four prompts within the routine. After the first number is discussed, repeat the process with each of the next numbers. If time allows, the routine can be continued with the next set of numbers as shown in the following table.

HOW MANY ___ ARE IN?				
How many	How many	How many	How many	Are in
5s	10s	50s	100s	537

HOW MANY ___ ARE IN?				
How many	How many	How many	How many	Are in
6s	7s	8s	9s	537

Sets 3 and 4 show more advanced approaches to this routine. These would be used with students who are working with multidigit divisors. In these examples, you may choose to provide all four numbers in the set at once, as it may help students see connections and move among them. For example, a student might start with the last prompt (50s) and find there are ten 50s in 537, leading them to know the solution to the first prompt is that there are 50 tens in 537.

HOW MANY ___ ARE IN?				
How many	How many	How many	How many	Are in
10s	20s	25s	50s	537

HOW MANY ___ ARE IN?				
How many	How many	How many	How many	Are in
12s	15s	18s	30s	537

ACTIVITY 6.7

Name: "Could It Be?"   **Type:** Routine

About the Routine: Students often complete a computation error and assume that their solution is correct because they have completed the steps. Asking themselves if their solution makes sense or if the solution could be ___ is not always part of their process. Determining if a solution is reasonable and/or accurate plays a significant role in students' fluency. This skill may be most elusive when dividing multidigit numbers. A routine like this provides a practice opportunity and, more importantly, access to others' ideas about how they determine reasonableness and/or accuracy. The unique feature of this routine is that students' don't have to find a quotient. Instead, a problem has been completed and students have to determine if the solution is reasonable. This approach simulates what we want students to do as they work through problems.

Materials: Provide two or three completed division problems.

Directions: 1. Pose a completed problem.

2. Have students think about the reasonableness of the solution included with the problem without solving it.

3. Ask students to share their thinking with a partner.

4. After partners talk, the whole group comes together to share their ideas about the reasonableness of the solution.

5. Pose a second problem and repeat the process. As time allows, present a third (or fourth) problem.

COULD IT BE?		
676 ÷ 6 = 121	785 ÷ 9 = 105	348 ÷ 6 = 58

In this example, students first talk about 676 ÷ 6 = 121. The quotient is not reasonable. Students might say so because 100 × 6 = 600 and 20 × 6 = 120, so 120 would be at least 720 and the dividend is 676. Another student might connect 66 ÷ 6 to 660 ÷ 6 because both are friendly and the latter, 660 ÷ 6, is close to 676 ÷ 6. The second prompt, 785 ÷ 9 = 105, cannot be reasonable because 9 × 100 is 900 and 785 is much less than 900. The last prompt is correct but students don't have to find the exact solution to prove that it is reasonable. Instead, they might suggest that 6 × 50 = 300, that 6 × 58 will be some more than 300, and that 348 is some more than 300.

Include incorrect answers that are still reasonable estimates (e.g., 512 ÷ 12 = 42). Students can follow up the discussion with finding an exact answer. This makes the point that looking for reasonableness does catch many mistakes, but not always.

ACTIVITY 6.8

Name: *Quotient Hunt*   **Type:** *Game*

About the Game: This is a game with division problems that can be solved with the Partial Quotients strategy. Players choose which numbers to work with, divide, and then "hunt" for a description that matches their quotient.

Materials: *Quotient Hunt* board, one for each group

Directions: 1. Each player chooses a divisor and a dividend. Players can only use the numbers once.

2. The player uses Partial Quotients to find the quotient.

3. Once the player has the quotient, they hunt on the game board to find the descriptions that match their quotient. The player writes the quotient in that box. If more than one box is a true statement for their quotient, they can pick which box to use (but can only take one box on a turn).

4. Player 2 repeats the same steps.

5. The player who gets four spaces arranged in a square wins the game.

For example, player 1 chooses the largest number from the dividend row and the smallest number from the divisor row. Player 1 created the largest quotient. Player 1 then places their quotient (1,800) in the space on the board that is labeled "Largest Quotient." However, the player could have placed their quotient in the space that is labeled "4-Digit Quotient With No Remainder."

Quotient Hunt

Directions: Choose a dividend and a divisor. Solve using Partial Quotients. Write your quotient in the box it describes. If your quotient describes more than one box you can choose a box but you can only select one box per turn. You can write a quotient in more than one box. Be the first player to make a square with four spaces.

Dividend								Divisor					
60	200	240	300	~~324~~	450	900	~~3,600~~		~~2~~	3	4	5	~~6~~	7	8

Odd Quotient	Quotient With No Remainder	Even Quotient	Quotient That Is a Multiple of 10
Smallest Quotient	2-Digit Quotient **54**	3-Digit Quotient With No Remainder	Largest Quotient **1800**
4-Digit Quotient With No Remainder	Quotient With an Even Remainder	3-Digit Quotient	Quotient With an Odd Remainder
2-Digit Quotient With No Remainder	Quotient That Is a Multiple of 3	Quotient That Is a Multiple of 2	4-Digit Quotient

Player 1
3600 ÷ 2

$$
\begin{array}{r}
2\overline{|3600} \\
-2000 \\
\hline
1600 \\
-1600 \\
\hline
0
\end{array}
\begin{array}{r}
1000 \\
800
\end{array}
$$

1800

Player 2
324 ÷ 6

$$
\begin{array}{r}
6\overline{|324} \\
-300 \\
\hline
24 \\
-24 \\
\hline
0
\end{array}
\begin{array}{r}
50 \\
4
\end{array}
$$

54

Quotient Hunt

Directions: Choose a dividend and a divisor. Solve using Partial Quotients. Write your quotient in the box it describes. If your quotient describes more than one box you can choose a box, but you can only select one box per turn. You can write a quotient in more than one box. Be the first player to make a square with four spaces.

Dividend								Divisor					
60	200	240	300	324	450	900	3,600	2	3	4	5	6	7	8

Odd Quotient	Quotient With No Remainder	Even Quotient	Quotient That Is a Multiple of 10
Smallest Quotient	2-Digit Quotient	3-Digit Quotient With No Remainder	Largest Quotient
4-Digit Quotient With No Remainder	Quotient With an Even Remainder	3-Digit Quotient	Quotient With an Odd Remainder
2-Digit Quotient With No Remainder	Quotient That Is a Multiple of 3	Quotient That Is a Multiple of 2	4-Digit Quotient

online resources This resource can be downloaded at **resources.corwin.com/FOF/multiplydividewholenumber**.

ACTIVITY 6.9

Name: *End of the Lane*   **Type:** *Game*

About the Game: *End of the Lane* is an opportunity for students to practice the Partial Quotients strategy. Students make division problems and move forward or backward depending on the accuracy of their solution and if it has a remainder.

Materials: digit cards (0–10) or a 10-sided die, *End of the Lane* game board, calculator, scratch paper

Directions:

1. Both players begin at the starting line of their own lane.

2. Both players generate four digits to make a division problem.

3. Each player solves their problem using Partial Quotients.

4. If the quotient has a remainder, the player moves two spaces. If the problem has no remainder, the player moves forward one space.

5. Both players check their opponent's work with a calculator. If the calculator proves the player's quotient is incorrect, the player moves back one space.

6. Players make a new problem and continue.

7. The first player to reach the end of their lane wins.

RESOURCE(S) FOR THIS ACTIVITY

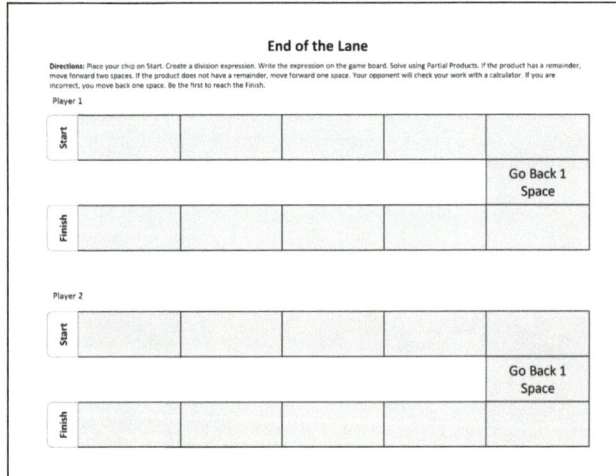

online resources ⟋ Digit cards and this game board can be downloaded at **resources.corwin.com/FOF/ multiplydividewholenumber**.

ACTIVITY 6.10

Name: *Quotient Connect*   **Type:** *Game*

About the Game: This game is an engaging way to practice division using partial quotients. It is played similarly to the classic board game, *Boggle*, as students find quotients through connected digits.

Materials: *Quotient Connect* game board; digit cards (0–9), playing cards (queens = 0, aces = 1, kings and jacks removed), or division cards; paper for recording work

Directions:

1. Players make a game board by randomly placing digits in each grid as shown in the example.

2. Players use digit cards to make a division problem. As an alternative, you can provide cards with division problems on them.

3. Both players solve the division problem.

4. Players look for the quotient on the game board they made by finding adjacent digits (similar to the classic game *Boggle*).

5. Players get 1 point for each example they find. The first player to reach a set goal (e.g., 10 points) or the most points after five problems wins.

For example, two players made the problem 189 ÷ 9. They both found the quotient to be 21. Player 1 found 21 two times, earning 2 points as shown. Player 2 found 21 three times, earning 3 points.

Player 1

Quotient Connect

7	5	4	9	8
4	2	3	8	3
1	2	9	6	4
5	5	9	3	2
6	1	1	3	7

Player 2

Quotient Connect

1	3	6	4	8
9	1	2	3	8
5	6	4	4	1
3	8	7	2	1
3	7	9	4	8

online resources   Digit cards and this game board can be downloaded at **resources.corwin.com/FOF/ multiplydividewholenumber.**

ACTIVITY 6.11

Name: *Remainder Tally*   **Type:** *Game*

About the Game: *Remainder Tally* is a quick-and-easy game for practicing Partial Quotients division. Players earn points based on the remainders in the quotients they find. You can change the dividend for endless game possibilities. Note that you can also have students use two digit cards to create two-digit divisors (as appropriate).

Materials: two decks of digit cards (2–9), one *Remainder Tally* game board per player

Directions: The game begins with setting a dividend. In the example, the dividend is 500.

1. Each round, both players choose a digit card from the deck.

2. Players use their digit cards as divisors of the set dividend.

3. Each player finds the quotient of the dividend and their divisor using Partial Quotients.

4. Players earn points equivalent to the remainder of their quotient.

5. Players shuffle digit cards and repeat the actions for round 2.

6. After five rounds, each player adds their points.

7. The player with the most points wins.

For example, Jaeona draws a 7 from the deck and uses partials to determine the answer to 500 ÷ 7:
7 x 70 = 490
7 x 1 = 7
500 ÷ 7 = 71, r 3
Her remainder is 3. She records 3 as her score.

RESOURCE(S) FOR THIS ACTIVITY

Remainder Tally

Directions: Pull a number. Divide the dividend by the number you roll. Solve the problem with Partial Quotients and show your thinking. Record your remainder as a score for the round. Add your score for five rounds. The player with the lowest score wins.

Dividend for This Game:			
	My Number (Divisor)	Solved With Partial Quotients	My Score (Remainder)
1			

online resources 🖰 Digit cards and this game board can be downloaded at **resources.corwin.com/FOF/multiplydividewholenumber**.

ACTIVITY 6.12

Name: *Division Bump*　　　　　　　　　　　　　　　　**Type:** *Game*

About the Game: *Division Bump* reinforces the skill of dividing by 10, 100, and 1,000. The goal of the game is to cover the most spaces on the game board.

Materials: *Division Bump* cards and game board, about 30 counters each (in two different colors)

Directions:
1. Each player has a set of color counters.

2. Player 1 picks a card from the pile and divides the expression.

3. Player 1 finds the quotient on the *Division Bump* game board and places their counter on that space.

4. Player 2 repeats the steps above.

5. If the opponent already has a counter on the space the player may bump the counter.

6. If a player already has their own counter on the space, the player may stack their counter on top. This space is now locked and can no longer be bumped.

7. The game ends when all the numbers are covered. The winner is the player with the most counters on the game board.

For example, player 1 picks up the card 90 ÷ 10. The player places their counter on the 9. Player 2 picks up the card 600 ÷ 60 and places their counter on the 10. Player 1 gets 8,000 ÷ 1,000 and places their counter on the 8. Next, player 2 picks up 7,000 ÷ 700. Player 2 now has the opportunity to lock their space by stacking their counter on their previous counter.

1	10	1000	100	9	0
1000	2	10	8	1	10
10	1000	3	100	1000	100
100	7	10	4	1	10
6	100	1000	0	5	100

RESOURCE(S) FOR THIS ACTIVITY

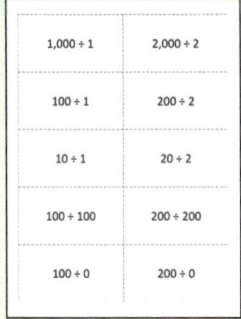

1,000 ÷ 1	2,000 ÷ 2
100 ÷ 1	200 ÷ 2
10 ÷ 1	20 ÷ 2
100 ÷ 100	200 ÷ 200
100 ÷ 0	200 ÷ 0

Division Bump

1	10	1000	100	9	0
1000	2	10	8	1	10
10	1000	3	100	1000	100
100	7	10	4	1	10
6	100	1000	0	5	100

online resources Division cards and this game board can be downloaded at **resources.corwin.com/FOF/multiplydividewholenumber**.

ACTIVITY 6.13

Name: *Tic-Tac-Take*

Type: *Game*

About the Game: This game is a take on the classic *Tic-Tac-Toe* in which players try to get three markings in a row. The twist in this version is that players can take another player's spot. The game is an opportunity to practice dividing using partial quotients.

Materials: *Tic-Tac-Take* game board, calculator

Directions:

1. Player 1 (X) chooses an expression from the game board.

2. Both players work to solve that expression using partial quotients.

3. If both players have the same answer, player 1 marks (X) that spot.

4. If players disagree about the quotient, the division is checked with a calculator.

5. The player with the correct answer takes the spot. If neither player is correct, the space is crossed off and cannot be played in the game.

6. Player 2 (O) repeats the steps above.

7. The winner is the first player with three in a row.

TEACHING TIP

Like most activities, you can swap out the division expressions on the downloadable, offering a different practice experience each time.

RESOURCE(S) FOR THIS ACTIVITY

Tic-Tac-Take

Directions: Players take turns selecting an expression on the game board. Both players solve the problem using Partial Quotients. If both players' answers are the same, the player who selected the expression places their mark on that spot. If players disagree, they check the problem with a calculator. The player who is correct takes the spot. The first player to get three in a row wins.

728 ÷ 8	592 ÷ 6	684 ÷ 6
373 ÷ 3	490 ÷ 5	963 ÷ 7
129 ÷ 4	288 ÷ 9	824 ÷ 2

 This resource can be downloaded at **resources.corwin.com/FOF/multiplydividewholenumber**.

ACTIVITY 6.14

Name: How Many Are In?   **Type:** Center

About the Center: This center is an independent practice opportunity that reinforces Activity 6.6 ("How Many Are In?"). This center affords students practice dividing with different friendly divisors. It is used here to give you an example of how any routine might be used as a center.

Materials: digit cards (0–9) or playing cards (queens = 0, aces = 1, kings and jacks removed), How Many Are In? recording sheet

Directions:
1. Students use digit cards to make a dividend. The number of digit cards used can be adjusted so that students work with appropriate division problems.

2. Students determine how many of a given group size are in that dividend. You can direct students to use the same "how many" number for each new dividend they create (left example).

3. Students explain how they know.

For example, a student creates 394 with digit cards. The student determines that there are seven groups of 50 in 394 and explains that 7×50 is 350. This justification makes sense but he could have also said that seven groups of 5 tens is 350 or 7×5 is 35 so 7×50 is 350. You can ask students to find the same "how many" number for each new dividend they create as shown in the left example. The middle example shows how you can ask students to find how many of different amounts are in the same divisor. The right example shows different divisors and four-digit dividends.

How Many	Are In	How Do You Know?
50s	362	7 because 7 × 50 = 350
50s	394	
50s	816	
50s	213	
50s	990	
50s	504	

How Many	Are In	How Do You Know?
50s	378	7 because 7 × 50 = 350
25s	378	
10s	378	
100s	378	
25s	378	
10s	378	

How Many	Are In	How Do You Know?
13s	2,435	100 because 100 × 13 = 1,400
12s	2,405	
15s	2,816	
30s	6,590	
60s	3,437	
80s	7,598	

RESOURCE(S) FOR THIS ACTIVITY

How Many Are In?

Directions: Use digit cards to make a divisor and record it in the "are in" column. Determine how many of a given amount are in the divisor and tell how you know.

How Many	Are In	How Do You Know?	How Many	Are In	How Do You Know?
50s	362	7 because 7 × 50 = 350			

 Digit cards and this recording sheet can be downloaded at **resources.corwin.com/FOF/ multiplydividewholenumber**.

ACTIVITY 6.15

Name: Estimate and Divide   **Type:** Center

About the Center: Estimating a solution before finding it is a strategy for determining reasonableness. When used with division, it helps students determine if the partial quotients they are finding make sense and if they are "on the right track." In this center, students create a division problem with digit cards, estimate the quotient, find the quotient, and determine if they made a good estimate.

Materials: digit cards (0–9) or playing cards (queens = 0, aces = 1, kings and jacks removed), Estimate and Divide recording sheet, calculator (optional)

Directions: Each student:

1. makes a division problem with digit cards.

2. estimates the quotient.

3. finds the quotient using Partial Quotients.

4. confirms that their quotient is correct with a calculator (optional).

5. determines if they made a good estimate.

TEACHING TAKEAWAY

Be sure that students understand there is no one correct estimate or one correct way to estimate.

RESOURCE(S) FOR THIS ACTIVITY

Estimate and Divide

Directions: Make a division problem. Estimate the quotient. Find the quotient. Tell if you made a good estimate.

Division Problem	Estimate	Actual Quotient	Was It a Good Estimate?

 Digit cards and this recording sheet can be downloaded at **resources.corwin.com/ FOF/multiplydividewholenumber**.

ACTIVITY 6.16

Name: The Largest Quotient, The Smallest Quotient

Type: Center

About the Center: This center reinforces partial quotients while also focusing on number sense and reasoning. It can be used as a game where students earn points for finding the largest or smallest quotient within a group. You can adjust this center by having students use the same digit cards to create four different problems, two with the largest quotient and two with the smallest quotient. Doing so gives students better insight into how quotients change as problems change. It also helps them in case they don't successfully make the largest or smallest quotient on their first try.

Materials: digit cards (0–9) or playing cards (queens = 0, aces = 1, kings and jacks removed); Largest Quotient, Smallest Quotient recording sheet

Directions:

1. Students pull four digit cards to make a three-digit divisor and a one-digit dividend. Note that this can be adjusted to make division problems aligned to what students are working with.

2. Students arrange the digit cards to make the largest quotient and find the quotient using partial quotients.

3. Students then rearrange the same cards to make a division problem with the smallest quotient, again finding the actual quotient with the Partial Quotients strategy.

RESOURCE(S) FOR THIS ACTIVITY

The Largest Quotient, The Smallest Quotient

Directions: Use digit cards to make a division problem. Arrange the digits to make the largest quotient. Show how you divided. Then, rearrange the digits to make the smallest quotient. Show how you divided again.

Largest Quotient	Smallest Quotient
Division Problem:	Division Problem:
Show how you divided:	Show how you divided:
Quotient:	Quotient:

1	2	3	4
5	6	7	8
9	1	2	3
4	5	6	7
8	9	0	0

online resources — Digit cards and this recording sheet can be downloaded at **resources.corwin.com/ FOF/multiplydividewholenumber**.

ACTIVITY 6.17

Name: Changing Dividends Make for Changing Quotients   **Type:** Center

About the Center: Quotients change when the dividend changes and the divisor remains the same. With whole numbers, the quotient increases as the dividend increases. This activity is an opportunity for students to experience and observe how quotients change. When discussed, it can help students reinforce this idea and advance both number sense and sense of reasonableness.

Materials: digit cards (0–9) or playing cards (queens = 0, aces = 1, kings and jacks removed), Changing Dividends Make for Changing Quotients recording sheet

Directions:
1. Students select digit cards to make a division problem. For example, students working with three-digit dividends divided by one-digit divisors would select four cards.

2. Students arrange the cards to make a division problem and find the quotient.

3. Students rearrange the digits in the dividend while the divisor remains unchanged. The student then predicts if the new quotient will be greater or less than the first quotient.

4. Students find the quotient of the new problem.

5. This can be repeated for a third quotient or a new problem can be created and the steps can be repeated.

In this example (recording sheet not used), the student creates 214 ÷ 8 and then makes a new problem 142 ÷ 8, predicting that the quotient will be less.

$$214 \div 8 \qquad\qquad 142 \div 8$$

The quotient will be less

$$
\begin{array}{r}
8\overline{\smash{)}214} \\
-160 \quad (8 \times 20) \\
\hline
54 \\
-48 \quad (8 \times 6) \\
\hline
6
\end{array}
\qquad
\begin{array}{r}
8\overline{\smash{)}142} \\
-80 \quad (8 \times 10) \\
\hline
62 \\
-56 \quad (8 \times 7) \\
\hline
6
\end{array}
$$

$$214 \div 8 = 26\ R6 \qquad 142 \div 8 = 17\ R6$$

RESOURCE(S) FOR THIS ACTIVITY

Changing Dividends Make for Changing Quotients	
Directions: Use digit cards to make a division problem. Find the quotient. Rearrange the digits in the dividend to make a new problem, keeping the divisor the same. Predict if the second quotient will be greater or less than the first. Show how you divided in each problem.	
First Division Problem	Second Division Problem
Show how you divided	Predict if the new quotient will be greater or less than the quotient from the first problem. Show how you divided
Quotient	Quotient

1	2	3	4
5	6	7	8
9	1	2	3
4	5	6	7
8	9	0	0

 Digit cards and this recording sheet can be downloaded at **resources.corwin.com/FOF/ multiplydividewholenumber**.

NOTES

Standard Algorithms for Multiplication and Division

STRATEGY OVERVIEW:
Standard Algorithms

What is a Standard Algorithm? Algorithms vary from country to country, and even region to region. In the United States, the standard algorithm for whole number multiplication involves using Partial Products, beginning with the ones place and working to the left (larger) place values. Notations do not define the algorithm (Fuson & Beckmann, 2012–2013). So, both of the following examples illustrate the standard algorithm, one showing each partial product and the other using shorthand notations.

RECORDING EACH PARTIAL PRODUCT IN ITS OWN ROW	USING SHORTHAND NOTATION
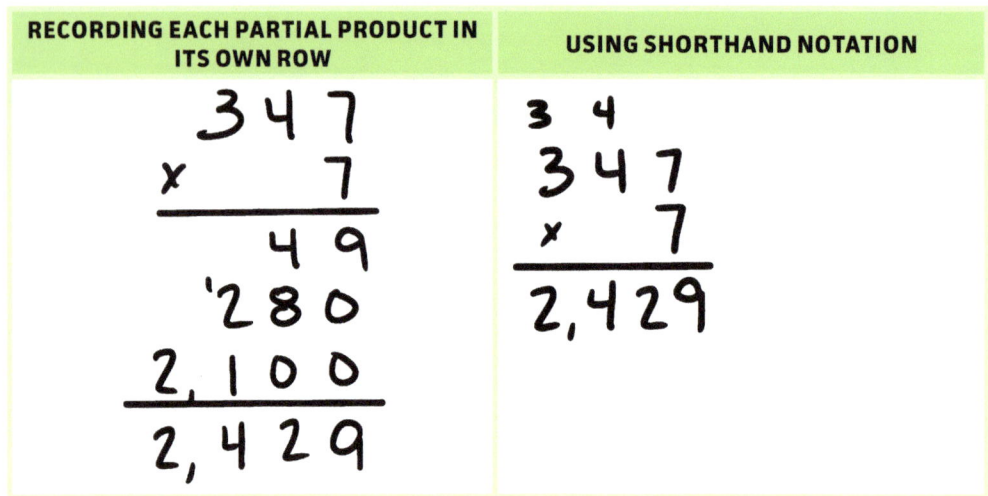	

The U.S. standard algorithm for division is a Partial Quotients approach but rather than use any partial quotient, the largest possible partial is expected. It is an iterative process of seeking the largest possible partial quotient for each place value, subtracting, and repeating for the next smaller place value.

HOW DO U.S. STANDARD ALGORITHMS WORK?

Standard algorithms are based on place value and the properties. This fact is more visible in the longhand version rather than the shorthand, which is why Partial Products and Partial Quotients are important processes to connect to the standard algorithms, as illustrated in the 347×7 example. Division is perhaps the hardest algorithm to understand, possibly because you must consider more than one place value as the dividend relates to the divisor. In the following example of 7,420 ÷ 35, you are looking not just at the thousands place but also at the hundreds place. The language associated with the algorithm is digit based (the italicized words are typically not stated). Using the italicized words has the benefits of connecting and strengthening number sense (e.g., Ask, "How many 35s are in 7,400?" rather than "How many 35s are in 74?"). Because many students have a fragile understanding of multiples, it is easier for them to answer the digit-based version.

EXAMPLE	WHAT IS SAID
$$35 \overline{\smash{\big)}\, 7{,}420} \atop \begin{array}{r} 212 \\ -7{,}000 \\ \hline 420 \\ -350 \\ \hline 70 \\ -70 \\ \hline 0 \end{array}$$	How many 35s in 7 *thousand*? (0) How many 35s in 74 *hundreds*? (2) Record **2 hundreds**. Multiply 35×200. (7,000) Subtract that quantity from the total. *Repeat:* How many 35s in 42 *tens*? (1) Record **1 ten**. Multiply 35×10. (350) Subtract that from the remaining total. (70) *Repeat:* How many 35s in 70? (2) Record **2**. Multiply 35×2. (70) Subtract that from the remaining total. (0)

WHEN DO YOU CHOOSE STANDARD ALGORITHMS?

While standard algorithms always work, they are not always the best choice. A standard algorithm is needed in the following situations:

- Numbers in the problem do not lend themselves to a mental method.
- Numbers in the problem do not lend themselves to a convenient written option (e.g., compensation).
- You don't know an alternate method.
- You want to check an answer having used a different method.

Perhaps because standard algorithms are in state standards, a major part of our own learning, and applicable to all problems, there can be a hurry to get to them and significantly more time devoted to learning them than their reasoning alternatives. This is a mistake. A rush to the standard algorithm and memorizing procedures undermines students' confidence and may cause math anxiety, negatively impacting achievement (Boaler, 2015a, b; Jameson, 2013; Ramirez, Shaw, & Maloney, 2018). Note that "fluently multiply" does not mean "use the standard algorithm adeptly." It means you use the standard algorithm adeptly *when it is the most efficient option*. Such flexibility is at the heart of ensuring students develop fluency.

MODULE 1 Break Apart to Multiply Strategy

MODULE 2 Halve and Double Strategy

MODULE 3 Compensation Strategy

MODULE 4 Partial Products Strategy

MODULE 5 Think Multiplication Strategy

MODULE 6 Partial Quotients Strategy

MODULE 7 Standard Algorithms for Multiplication and Division

STANDARD ALGORITHMS:
Strategy Briefs for Families

It is important that families understand the strategies and know how they work so that they can be partners in the pursuit of fluency. These strategy briefs are a tool for doing that. You can include them in parent or school newsletters or share them at parent conferences. They are available for download so that you can adjust them as needed.

Standard Algorithm for Multiplication

How It Works: The algorithm is a digit-based procedure for multiplying multidigit factors.

1. Multiply the digit in the ones place of the bottom factor by each digit in the top factor.
2. Continue to multiply digits by place value from right to left. Record products of each new place value below the products of the previous place value.
3. Add the products of each place value.

The left example shows a problem with no regrouping. The products of 1 and each digit in the top factor are recorded. The products of the second digit, which represents tens, are recorded on the second line. The two products are added for a sum of 9,982. So, 322 × 31 = 9,982.

The right example shows a problem with regrouping: 2 × 6 = 12. The ones are recorded and the 10 is grouped to be added to the product of the tens place. 2 × 4 tens is 8 tens and 1 more ten is 9 tens recorded as shown. Finding the products of 7 tens requires additional regrouping. The two products are added for a sum of 24,912. So, 346 × 72 = 24,912.

When It's Useful: The algorithm becomes more useful as the number of digits increases in each factor.

```
    3 2 2
  ×   3 1
  ───────
    3 2 2
  . 9, 6 6 0
  ─────────
    9, 9 8 2
```

```
    ³ ⁴
    3 4 6
  ×   7 2
  ───────
    6 9 2
  .2 4, 2 2 0
  ───────────
    2 4, 9 1 2
```

Standard Algorithm for Division

How It Works: The algorithm is a digit-based procedure for dividing.

1. Determine how many groups of the divisor are in an amount of the dividend. In the following example, there are no (whole number) groups of 35 in 3. So, it is expanded to 36. There is 1 group of 35 in 36.
2. 1 × 35 is 35. Note that the 1 is in the hundreds place so it is 1 hundred times 35, creating 3,500.
3. 3,500 is subtracted from 3,687 with a difference of 187.
4. There are no (whole number) groups of 18, so a 0 is recorded and 187 is considered.
5. There are 5 groups of 35 in 187. 5 × 35 = 175.
6. In this example, 3,687 divided by 35 is 105 with a remainder of 12.

When It's Useful: The algorithm becomes more useful as the number of digits increases in the dividend or divisor.

```
  Dividend   Divisor
   3,687  ÷   35

            1 0 5 R12
  ²35 ) 3, 6 8 7
       -3, 5 0 0
       ─────────
            1 8 7
          - 1 7 5
          ───────
              1 2
```

 These resources can be downloaded at **resources.corwin.com/FOF/multiplydividewholenumber**.

NOTES

_____

_____

TEACHING ACTIVITIES for Standard Algorithms

Explicit strategy instruction for standard algorithms has been a pillar of elementary school. The use of base-10 blocks and other tools has helped students to see how each functions conceptually. You likely have many resources for teaching the steps of each algorithm. The teaching activities offered here do not focus on that task but do not suggest that it isn't important. Instead, these tasks focus on making connections between strategies and the algorithm, uncovering errors within an algorithm, and exposing students to thinking about when an algorithm is or isn't useful. Keep in mind that determining reasonableness plays a role in fluency and it is critically important as students carry out algorithms because procedural and computational errors are quite possible.

ACTIVITY 7.1
CONNECTING PARTIAL PRODUCTS AND THE ALGORITHM

This instructional activity is an opportunity for observing the relationship between Partial Products and the standard algorithm for multiplication. It is somewhat unique in that it does not require students to find a product or to complete the steps for finding a product. Instead, students complete a problem using both approaches and are told that both are completed correctly. Then, students are charged with observing the two approaches and finding the many ways that they are the same and different.

$$
\begin{array}{r} 45 \\ \times\, 27 \\ \hline 800 \\ 280 \\ 100 \\ 35 \\ \hline 1{,}215 \end{array}
\qquad
\begin{array}{r} 45 \\ \times\, 27 \\ \hline 35 \\ 280 \\ 100 \\ 800 \\ \hline 1{,}215 \end{array}
\qquad
\begin{array}{r} 45 \\ \times\, 27 \\ \hline 315 \\ 900 \\ \hline 1{,}215 \end{array}
$$

This example shows how the activity can be used with 2 two-digit factors. The example on the left shows partial products carried out beginning with the tens place of the factors and the middle example shows partials beginning with the ones place of the factors. The right example shows the standard algorithm. Students are likely to notice the relationship between the two Partial Product approaches because the products are in reverse order. The relationship between the partials and the standard algorithm may be less obvious. Students will eventually see that 315 is the sum of 35 and 280 and that 900 is the sum of 800 and 100. This can buoy a discussion as to why this may be (because the algorithm combines steps or partials) and if it will always happen (yes!).

To continue the activity, you can pose additional completed problems for students to observe and comment on. Then, you can extend the activity by charging students to solve a problem on their own with both processes.

ACTIVITY 7.2
CONNECTING PARTIAL QUOTIENTS AND THE ALGORITHM

Activity 7.2 is directly related to the previous activity. Here, students are presented with a division problem correctly completed with Partial Quotients and the standard algorithm. Students are prompted to identify how the work is similar and different. It too should be complemented with multiple examples to reinforce the relationship. This instructional activity may be even more powerful with division because it shows the utility and friendliness of partial quotients and the streamlined work of the standard algorithm. In the example, the standard algorithm is not accompanied by all of the "scratch work" to determine products, differences, and other computations. This is the case because the divisor is somewhat friendly. That would not be the case if the divisor was something like 29 or 37.

$$
\begin{array}{r}
28\ \text{R}15 \\
24\overline{)687} \\
-240 \quad 24\times 10 \\
\overline{447} \\
-240 \quad 24\times 10 \\
\overline{207} \\
-120 \quad 24\times 5 \\
\overline{87} \\
-48 \quad 24\times 2 \\
\overline{39} \\
-24 \quad 24\times 1 \\
\overline{15} \\
\\
28\ \text{R}15
\end{array}
\qquad
\begin{array}{r}
28\ \text{R}15 \\
24\overline{)687} \\
-480 \\
\overline{207} \\
-192 \\
\overline{15}
\end{array}
$$

This example shows a three-digit dividend divided by a two-digit divisor. Naturally, students would start with more simplistic problems for working with division but the activity would be carried out the same way. This activity and Activity 7.1 are effective ways to make sense of the shorthand used in the standard algorithms for multiplication and division, as compared to solely connecting standard algorithms to base-10 blocks or other visuals.

Also note that with this activity, you can pose one problem at a time for students to observe and discuss as described in Activity 7.1. Yet another approach would be to provide different completed problems as different "stations" around the room and have partners move from station to station observing the problems and recording their thinking to be discussed during a whole group conversation after the rotations are completed.

ACTIVITY 7.3
FIND THE ERROR

Error analysis is when students analyze an example of completed work, looking specifically for the error made. Error analysis is a higher-level thinking skill because it requires a deeper understanding of the skill or concept to be able to identify (and correct) errors. Errors could include computation miscalculation or missing step(s) misconceptions.

This activity requires students to analyze work examples, find errors, and explain how to rectify the mistakes. "Find the Error" can be done as a whole class discussion. This allows students to collaborate with one another and listen to different explanations from their peers. Display one problem at a time for students to find errors and explain the mistake. Provide students a mix of multiplication or division problems. The following examples show different types of errors that students would work through.

COMPUTATION ERROR	GREATEST FACTOR ERROR

An alternative to this activity is to provide students with problems that contain errors and some that don't, so students work to distinguish between the two. "Find the Error" can transition from a whole class activity to independent practice or centers. See also the section on Worked Examples, where a series of examples are provided.

ACTIVITY 7.4
TO ALGORITHM OR NOT TO ALGORITHM

Fluency with standard algorithms includes not only how to use an algorithm but *when* to use an algorithm. It is critical that we empower students to think critically about when they need this tool. This activity aims to do just that. Pose a collection of multiplication or division problems to small groups of students. Have the students think about which problems call for an algorithm and which don't. Students must show their strategies for problems that don't call for an algorithm. Conversely, students must tell why a certain problem is best suited for the algorithm.

use algorithm	Don't use algorithm
340 ÷ 39	486 ÷ 12
782 ÷ 16	800 ÷ 40
516 ÷ 55	650 ÷ 25

In this example, the teacher posed a list of six different problems. Students worked with partners to sort and record those problems. They determined that the three problems on the right didn't require an algorithm. For 486 ÷ 12, they recognized 48 ÷ 12 and were able to break 486 into 480 + 6. They explained (*not shown*) that to solve 650 ÷ 25, you could break apart 650 into 100 + 100 + 100 + 100 + 100 + 100 + 50 and figure out how many 25s were in each. For 800 ÷ 40, a student shared that you could use Think Multiplication using the basic fact 4 × ? = 8.

ACTIVITY 7.5
PROMPTS FOR TEACHING STANDARD ALGORITHMS

Use the following prompts as opportunities to develop understanding of and reasoning with the strategy. Have students use representations and tools to justify their thinking, including base-10 models, number lines, arrays, and so on. After students work with the prompt(s), bring the class together to exchange ideas. These could be useful for collecting evidence of student understanding. Any prompt can be easily modified to feature different numbers (e.g., three-digit or four-digit numbers) and any prompt can be offered more than once if modified.

- Find the product of $8 \times 2{,}021$ with a strategy and the algorithm. Tell how the two are similar and different.

- Create a problem that would be best solved with the standard algorithm. Tell why this problem is a good choice for solving with the standard algorithm.

- Tasha and Michael solved 24×47 but came up with different products. Who is correct? Explain how you know.

Tasha	Michael
24 × 47	24 × 47
168	168
960	96
1,128	264

- Devon said when you multiply a two-digit number by a one-digit number, you get a two-digit product. Natasha disagreed with Devon and said you get a three-digit product. Who is correct? How do you know?

- Create a story problem for $249 \div 3$. Solve using the standard algorithm, explaining how the steps connect to the story.

- Cierra says the product of 296×126 will be less than 35,000. Without solving the problem, explain how you know if Cierra is correct or not.

- Dylan divides two numbers and the quotient is 150. What could be the two numbers Dylan used?

- How is the Partial Products strategy for multiplication similar to and different from the standard algorithm for multiplication? Use a problem to show your thinking.

- Create and solve a division problem with the standard algorithm. Describe what is happening in each part of the problem.

- Create a division problem that would be best solved with the standard algorithm and a division problem that doesn't need the standard algorithm for solving. Tell how you know when you will not use the standard algorithm.

- Think about the standard algorithm for multiplication (or division). What do you think is a common error that people make when using it?

PRACTICE ACTIVITIES for Standard Algorithms

Fluency is realized through quality practice that is focused, varied, processed, and connected. The activities in this section focus students' attention on how this strategy works and when to use it. The activities are a collection of varied engagements. The discussion you facilitate after an activity or the reflection prompts you attach to it should help students think about what they did mathematically, how they reasoned about the activity, and when the math they did (namely the strategy) might be useful. Debriefing should also help students see how the practice activity connects to recent instruction or how the strategy connects to other strategies they know. Game boards, recording sheets, digit cards, and other required materials are available as online resources for you to download, possibly modify, and use. As students work with activities, look for how well they are acquiring the strategy and assimilating it into their collection of strategies.

FLUENCY COMPONENT	WHAT TO LOOK FOR AS STUDENTS PRACTICE THIS STRATEGY
Efficiency	• Are students using the algorithm when another strategy is clearly more efficient? (e.g., 45×12 can be solved with the algorithm but also aligns with Halve and Double as well as Partial Products.) • Are students using the algorithm efficiently? (i.e., Are they choosing problems that are best for solving with the algorithm?) • Do they use the algorithm regardless of its appropriateness for the problem at hand? (e.g., Do they use it for 400×6 or $250 \div 50$?)
Flexibility	• Are students using the algorithm when it is unnecessary? (e.g., $900 \div 300$) • Do students change their approach to or from the algorithm as it proves inappropriate or overly complicated for the problem? (e.g., A student abandons $455 \div 5$ after finding 9 tens, recognizing that the dividend is $450 + 5$ and that they can use Think Multiplication more efficiently.)
Accuracy	• Are students using the algorithm accurately? (e.g., Are they accurately recording or keeping track of place values?) • Are students finding accurate solutions when using the algorithm? • Are they considering the reasonableness of their solutions?* • Are students estimating before they use the algorithm?*

* This consideration is not unique to this strategy and should be practiced throughout the pursuit of fluency with whole numbers.

WORKED EXAMPLES

Worked examples are opportunities for students to analyze the thinking of another student. Worked examples include correct and incorrect examples. While algorithms always work, the enactment of established steps and notations can lead to mistakes along the way. Worked examples focused on common errors can help students develop an understanding of why that step does not work.

The prompts from Activity 7.5 can be used for collecting examples. Throughout the module are various examples that you can use as worked examples. The following table shows common errors and a sample worked example with that error. Because standard algorithms are error prone, estimating and checking for reasonableness are very important. When you analyze worked examples, you can begin by asking, "Is this answer reasonable?" If the answer is "no," then ask, "What is a reasonable answer?" Of course, there are wrong answers that turn out to be reasonable, which is why it is also good to check for accuracy, either by rechecking the steps or using the inverse operation with the answer.

COMMON ERROR	WORKED EXAMPLE
Ignores internal zeros (multiplication)	$\begin{array}{r} {}^{5}907 \\ \times\ 8 \\ \hline 776 \end{array}$ $\qquad$ $\begin{array}{r} 6{,}003 \\ \times\ 8 \\ \hline 4{,}824 \end{array}$
Has difficulty with zeros (division)	$\begin{array}{r} 35 \\ 3\,\overline{)915} \\ -900 \\ \hline 15 \end{array}$ $\qquad$ $\begin{array}{r} 20\ \text{R}5 \\ 12\,\overline{)2{,}405} \\ 24 \\ \hline 005 \end{array}$
Adds regrouped number before multiplying a place value (multiplication)	$\begin{array}{r} {}^{6}\ {}^{4} \\ 257 \\ \times\ 7 \\ \hline 5{,}639 \end{array}$
Loses track of place value (multiplication)	$\begin{array}{r} 316 \\ \times\ 23 \\ \hline 948 \\ +\ 632 \\ \hline 1{,}580 \end{array}$
Loses track of place value, especially when zeros are involved (division)	$\begin{array}{r} 15\ \text{R}16 \\ {}^{4}28\,\overline{)2{,}956} \\ -2{,}800 \\ \hline 156 \\ -140 \\ \hline 16 \end{array}$

ACTIVITY 7.6

Name: "Over/Under" **Type:** Routine

About the Routine: Determining reasonableness when using standard algorithms is critically important. In this routine, students determine if a product or quotient will be over or under (more or less) than a given number.

Materials: Identify a target number for comparison and prepare a few multiplication or division expressions that would likely be solved with the standard algorithm.

Directions:

1. Share the over/under number.

2. Pose the first expression.

3. Have students estimate the product or quotient of each problem.

4. Students share their ideas with partners.

5. The class then discusses their approaches and reasoning. After ideas are shared, the exact solution can be found and compared to the over/under number.

6. Continue the routine with the next expression.

Over/Under 50		
355 ÷ 7	248 ÷ 33	2445 ÷ 59

In this example, students might argue that 7 × 50 = 350, so the first quotient must be more than 50. The middle expression might be reasoned that 30 × 8 is 240 so the quotient is about 8, which is much less than 50. The last example shows how the routine might evolve to less friendly divisors. Here, students might suggest that 2400 ÷ 60 is about 40 and because they increased the divisor, the answer may be even smaller than 40, but certainly under 50.

Over/Under 25		
287 ÷ 7	306 ÷ 3	84 ÷ 2

The first two dividends in this example are less obviously connected to basic facts. However, students can still use multiplication to reason about quotients. In the first expression, one might think that 40 × 7 = 280, so then the quotient is clearly more than 25. Similarly, in the middle expression, 3 × 100 = 300 so that quotient must be more than 25 as well. The third may be thought of differently as students might contend that half of 84 is 40 rather than thinking about multiplication. It is a good reminder that although Think Multiplication is the focus of the routine in these examples, students can (and will) offer other, viable thoughts and reasons.

ACTIVITY 7.7

Name: "That One"   **Type:** Routine

TEACHING TAKEAWAY

With some guidance, students can help create problems for games and routines by writing them on sticky notes or index cards.

About the Routine: This routine helps students consider when to use an algorithm. It is intended as a practice opportunity after instruction of the concept (see Activity 7.4 "To Algorithm or Not to Algorithm"). Problems from the other routines as well as those throughout this book work great for this routine.

Materials: Prepare three or four expressions for students to consider.

Directions:

1. Pose a few problems to students.

2. Have students discuss with a partner which problems are good candidates for solving with an algorithm and which are not.

3. Have partners explain their decisions in a class discussion. Be sure to avoid implying that any problem should or must be solved with an algorithm.

That One		
600×3	599×6	789×4

In this example, students might identify 789×4 as best solved with an algorithm, as the first could be solved by thinking about the basic fact 6×3 and the second could be solved with compensation by thinking 600×6 and then taking a group of 6 away. Students could also argue that 789×4 might be solved with a compensation strategy ($800 \times 4 - 44$), in which case they would respond that "none" of these are best solved with an algorithm.

ACTIVITY 7.8

Name: Call Your Shot   **Type:** Game

About the Game: *Call Your Shot* offers an opportunity to practice with multiplication or division in an engaging way. In this game, players create situations and then try to make a problem that matches the condition. Players earn a point if they are unable to do so, with an object of the game being not to earn points.

Materials: two sets of digit cards (0–9) or playing cards (queens = 0, aces = 1, kings and jacks removed), list of conditions, math journal (optional)

Directions: 1. Player 1 identifies a condition (calls their shot).

2. Both players pull three or more digit cards to make a problem.

3. A player gets a point if they can't make a problem to fit the condition.

4. Player 2 then identifies a condition and both players attempt to make a problem to meet the condition.

5. The first player to get 5 points loses.

For example, player 1 says make a quotient between 0 and 50. Both players pull digit cards. Player 1 makes 391 ÷ 7 and player 2 makes 332 ÷ 8. Player 1 would get a point because their quotient is 55 r 6. Player 2 would not get a point because they met the condition.

Here are some examples for three-digit divided by one-digit problems. Players can also create their own conditions.

EVEN QUOTIENT	ODD QUOTIENT	QUOTIENT BETWEEN 0 AND 50	QUOTIENT BETWEEN 60 AND 90
Quotient with a 0 in it	Quotient with a 5 in it	Quotient greater than 70	Quotient less than 25
Quotient with a remainder	Quotient without a remainder	Two-digit quotient	Three-digit quotient
Quotient where each digit is an even number	Quotient where each digit is an odd number	Quotient with an odd remainder	Quotient with an even remainder

ACTIVITY 7.9

Name: A List of Ten   **Type:** Game

About the Game: In terms of procedural fluency, knowing when to use an algorithm is just as important as knowing how to use an algorithm. Yet students don't have frequent opportunities to consider when to use an algorithm. *A List of Ten* is a game for considering when an algorithm is useful and when it isn't. It is the perfect complement to instructional Activity 7.4 ("To Algorithm or Not to Algorithm").

Materials: *A List of Ten* game cards and game board

Directions:
1. Expression cards are shuffled and placed face down.

2. Players take turns selecting a card and determining if the expression is best solved with or without the algorithm. Note that if a player can tell their opponent why an expression is better solved without an algorithm, they can steal the card from their opponent.

3. The player records the equation in the appropriate column.

4. The first player to get a list of 10 wins.

5. After the game, both players solve each problem on their game board.

The following example shows the first three problems that player 1 flipped over. Player 2 says that 132 ÷ 6 could be thought of as 120 ÷ 6 + 12 ÷ 2 and steals the problem. Player 1 will mark it off of their page and player 2 will add it to the left column on their recording sheet.

Problems solved WITHOUT the STANDARD ALGORITHM	Problems solved WITH the STANDARD ALGORITHM
1. 640 ÷ 80 =	1. 388 ÷ 18 =
2.	2. 132 ÷ 6 =

RESOURCE(S) FOR THIS ACTIVITY

A List of Ten

Directions: Players take turns flipping over cards and deciding if the problem is best solved with or without the standard algorithm. Players can steal problem cards put in the algorithm list if they can explain how a different strategy could be used to solve the problem.

Problems to solve WITHOUT the STANDARD ALGORITHM	Problems to solve WITH the STANDARD ALGORITHM
1.	1.
2.	2.

132 ÷ 2	258 ÷ 3
640 ÷ 80	78 ÷ 3
160 ÷ 4	344 ÷ 4
118 ÷ 2	388 ÷ 18
171 ÷ xxx	132 ÷ 6

Game cards and this recording sheet can be downloaded at **resources.corwin.com/FOF/ multiplydividewholenumber**.

ACTIVITY 7.10

Name: *For Keeps*   **Type:** *Game*

About the Game: *For Keeps* is an opportunity to practice multiplication in an engaging way that also helps students develop reasoning about numbers. Players find products and determine if they want to keep the product to count toward their final score. The game can be adjusted easily so that students practice division by choosing to keep or discount quotients.

Materials: *For Keeps* game board; 10-sided dice or playing cards (queens = 0, aces = 1, kings and jacks removed), digit cards (0–9)

Directions: 1. Each player generates their own four numbers to make 2 two-digit factors.

2. Players find the product of the two factors and decide if they want to keep the product as the score for the round or if they don't want to keep it for a score. Once they decide (before the next problem), the decision is final and the product can't be moved later.

3. Players play a total of four rounds yet can only keep two of those rounds. At the end of the fourth round, each player adds the two scores that they kept.

4. The player with the higher score wins.

In the example, the player first made 22 × 31 for a product of 682 and didn't keep it. The second product of 1,156 was kept. The player thinks they can do better than their third-round score of 700 in the fourth round. No matter what product is made in the fourth round, the player will have to keep it. The sum of the two kept scores (1,156 and the fourth round) will be the player's score.

Round	Numbers Created		Product For Keeps	Product NOT Kept
1	22	31		682
2	68	17	1,156	
3	14	50		700
4				
	Sum of Keeps			

Show your thinking

RESOURCE(S) FOR THIS ACTIVITY

For Keeps

Directions: Make 2 two-digit numbers. Find the product and decide if you want to keep the product as a score or not keep it. You can only keep two of the four rounds. After the fourth round, add the two scores you kept. The player with the higher score wins.

Round	Numbers Created		Products for Keeps	Products NOT Kept
1				
2				
3				
4				
	Sum of Keeps			

online resources ⬑   Digit cards and this game board can be downloaded at **resources.corwin.com/FOF/ multiplydividewholenumber**.

ACTIVITY 7.11

Name: The V Cover Up   **Type:** Game

About the Game: *The V Cover Up* is an opportunity to practice multiplication and division in an engaging way. Players solve problems in order to cover spaces on their game board so they make a three-space V. You can have students practice both operations, as the spaces represent products and quotients of the cards, or you can remove one operation for isolated play.

Materials: *The V Cover Up* game board, counters, expression cards, calculators

Directions: 1. Players take turns pulling multiplication and division cards and finding the product or quotient.

2. The player covers the product or quotient they find on the game board. Their opponent checks their accuracy with a calculator.

3. If the product or quotient is already taken or the player computes incorrectly, the player loses their turn.

4. The first player to make a three-space V wins. (Or keep playing, and whoever gets the most Vs when the board is filled or time is up is the winner.)

The following image shows how to make a three-space V. The V can be oriented in any way.

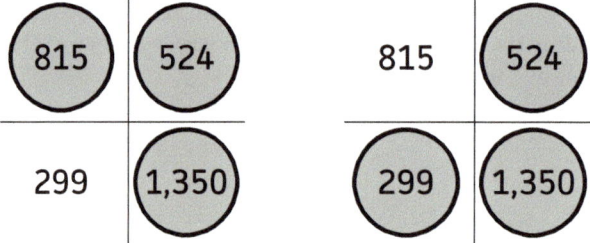

RESOURCE(S) FOR THIS ACTIVITY

9 × 64	19 × 60
6 × 72	4 × 232
54 × 25	3 × 682
18 × 25	2 × 512
42 × 24	5 × 376

The V Cover Up

Directions: Take turns choosing a card. Find the product or the quotient and cover the space on the board. Your partner will check your work with a calculator. Be the first to make a V.

576	784	1,008	928	1,024
207	1,360	307	1,082	82
432	486	1,140	2,046	1,880
49	1,072	815	524	967
516	807	299	1,350	176
1,199	119	814	754	306

 Expression cards and this game board can be downloaded at **resources.corwin.com/FOF/ multiplydividewholenumber**.

ACTIVITY 7.12

Name: Missing Numbers With Multiplication Algorithms   **Type:** Center

About the Center: Missing Numbers With Multiplication Algorithms helps students practice finding and correcting multiplication errors. This center can be complemented with a writing component. For example, after students complete a Missing Numbers card, they can be expected to write about how they found the missing digit or to justify how they know their solution is correct.

Materials: prepared Missing Number problem cards (directions follow)

Directions: 1. Students select a Missing Number problem card.

2. Students work to find the Missing Number(s) in the algorithm.

To make Missing Numbers problem cards, create a problem and then simply change out one or more digits with a question mark. The template provided is for two-digit by one-digit multiplication. Additional templates featuring two-, three-, and four-digit factors are available for download.

```
    [?][?]           [?][?]         [?][?][?][?]
       [?]           [5][8]               [?]
  X _____      X _____       X _____
    [4][8]        [1,][2][7][6]    [1][7,][0][3][2]
```

RESOURCE(S) FOR THIS ACTIVITY

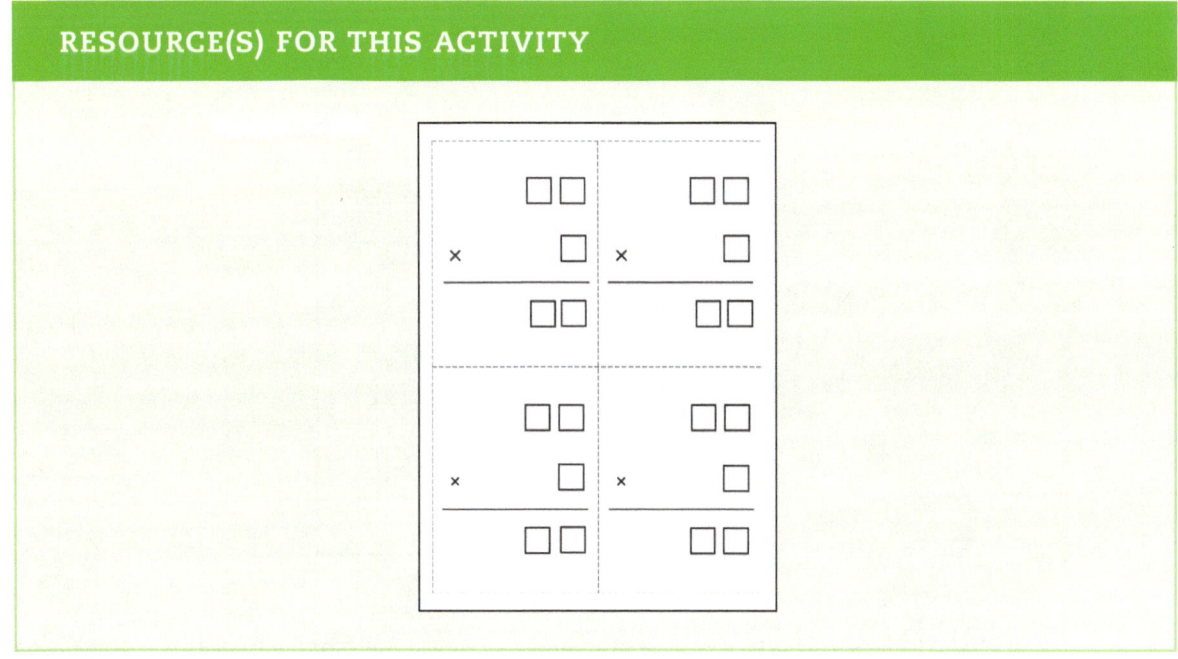

online resources ☞ This resource can be downloaded at **resources.corwin.com/FOF/multiplydividewholenumber**.

ACTIVITY 7.13

Name: Missing Digits With Division Algorithms **Type:** Center

About the Center: This is a problem-based activity for practicing the standard algorithm for division. Students work to find the missing components (digits) in problems that have already been solved.

Materials: Prepare Missing Digit problem cards.

Directions:
1. Students select a Missing Digit problem card.

2. Students work to find the Missing Digit(s) in the algorithm.

To make Missing Digit problem cards, create a problem and then simply change out one or more digits with a question mark. A template is provided for three-digit by one-digit division. In this example, the teacher made the problem 142 ÷ 2 = 71 and removed a few of the digits. The student then worked to find the missing digits to be 4, 0, 2, then 0 (from top to bottom).

```
            7  1
      2 | 1  ?  2
        - 1  4  0
            0  2
            0  ?
               ?
```

ACTIVITY 7.14

Name: Partials to Algorithms   **Type:** Center

About the Center: Connecting algorithms to other strategies, especially Partial Products and Partial Quotients, helps students make sense of the procedure. In this center, students are given a problem completed with Partial Products. Students then have to rewrite the problem using the standard algorithm.

Materials: Partials to Algorithms cards and recording sheet

Directions: 1. Students select a card that shows a problem completed with partial products.

2. Then, students solve the problem again with the standard algorithm.

3. Students then tell which approach they thought was the better choice for the problem (and why).

RESOURCE(S) FOR THIS ACTIVITY

```
46 × 27

40 × 20 = 800
40 × 7 = 280
 6 × 20 = 120
 6 × 7 =  42
        1,242
```

```
382 × 9

300 × 9 = 2,700
 80 × 9 =   720
  2 × 9 =    18
          3,438
```

```
91 × 19

90 × 10 = 900
90 × 9 = 810
 1 × 10 =  10
 1 × 9 =   9
        1,729
```

```
16 × 13

10 × 10 = 100
10 × 3 =  30
 6 × 10 =  60
 6 × 3 =  18
         208
```

Partials to Algorithms

Directions: Select a card that shows a problem solved with partials. Solve the problem with the standard algorithm. Tell which approach was the better choice for the problem.

Problem	Solved With the Standard Algorithm	Which Was the Better Choice? Why?

online resources 🔗 Cards and this recording sheet can be downloaded at **resources.corwin.com/FOF/multiplydividewholenumber**.

ACTIVITY 7.15

Name: Making Connections   **Type:** Center

About the Center: A Frayer model is a graphic organizer that helps students make connections between concepts and ideas. This center uses an adaptation of the classic Frayer model by focusing on strategies and algorithms. In this center, students are given a problem. They solve the problem with different strategies or approaches and then comment on which was most useful for the problem. You can use this center with multiplication or change out the prompts asking students to represent the problem with a diagram or write a word problem that goes with the expression in the center.

Materials: Multiple Methods graphic organizer

Directions: 1. Students are given or select a division problem (recorded in the center of the organizer).

2. Students apply three different strategies to the problem and then explain which method they thought was the best fit for that problem.

The example shows a student who completed 154 ÷ 7.

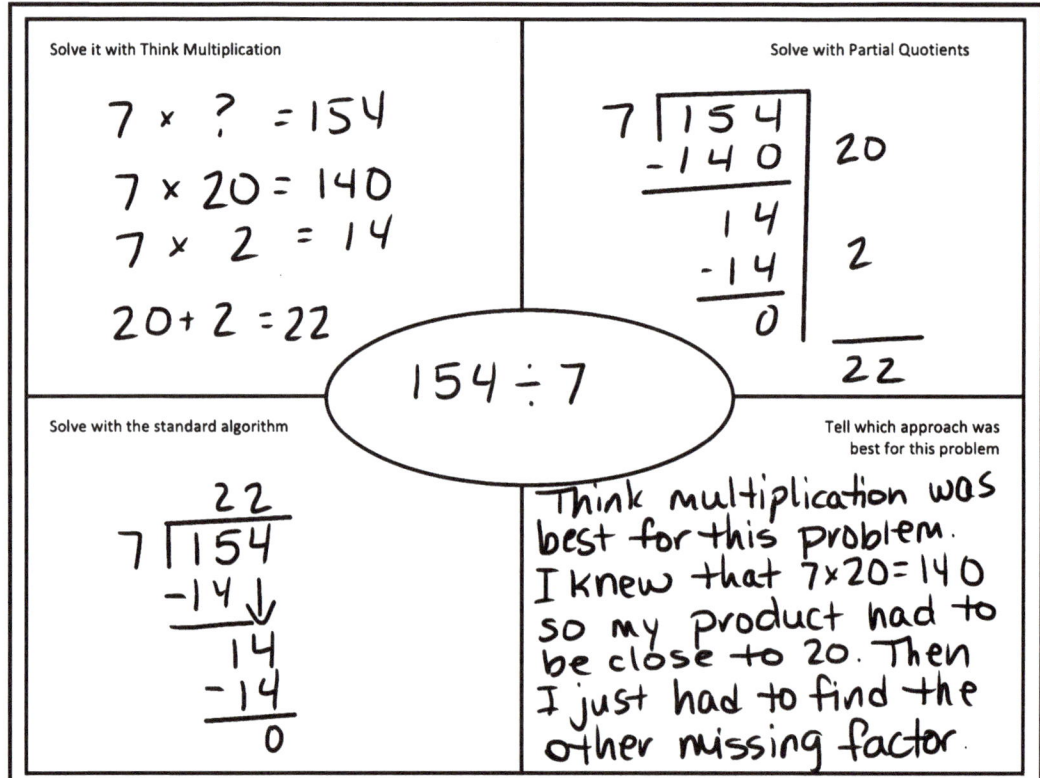

ACTIVITY 7.16

Name: Correct the Error　　　　**Type:** Center

About the Center: As students complete the steps in an algorithm, they make errors. Often students continue on without noticing. In other instances, they don't look back to see where an error occurred if their answer doesn't align with what they anticipated. This center is an opportunity to practice looking for errors. It can help students develop a habit of looking over their work when completing procedures.

Materials: Correct the Error recording sheet

Directions:　1. Students review the problem in the first column of the recording sheet.

2. Students find the error and tell what it was in the second column of the recording sheet.

3. Students work to solve the problem correctly on the last column.

The problem	What is the error?	Solve
31 R2 4⟌136 −120 6 −4 2	The error was 136 − 120 does not equal 6. 136 − 120 = 16.	34 4⟌136 −120 16 −16 0

RESOURCE(S) FOR THIS ACTIVITY

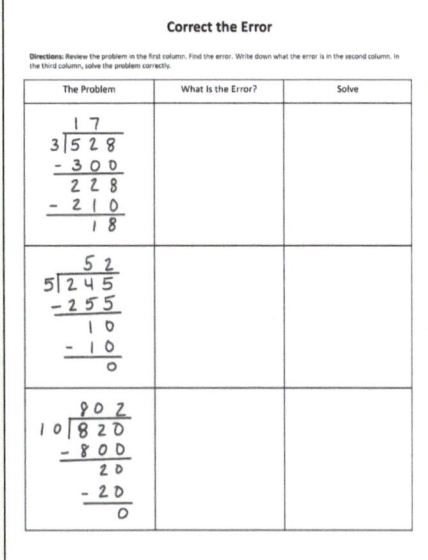

This resource can be downloaded at **resources.corwin.com/FOF/multiplydividewholenumber**.

PUTTING IT ALL TOGETHER

Developing Fluency

FLUENCY IS . . .

How might you finish a sentence that begins "Fluency is . . ."? One way is "using procedures efficiently, flexibly, and accurately." Another option is "important." Or maybe fluency is "an equity issue." All of these are true statements. This section is the capstone of this book on fluency with multiplication and division of whole numbers. If students learn a strategy in isolation and never get to practice choosing when to use it, they will not become truly fluent. Part 3 focuses on learning to *choose* strategies. The following lists reflect subsets of the Seven Significant Strategies.

MULTIPLICATION "MUST KNOW" STRATEGIES	DIVISION "MUST KNOW" STRATEGIES
• Break Apart to Multiply	• Think Multiplication
• Halve and Double	• Partial Quotients
• Compensation	
• Partial Products	

Once students know more than one strategy, they need opportunities to practice choosing strategies. In other words, you do not want to wait until they have learned all strategies and the standard algorithms before they begin choosing from among the strategies they know. For example, if you are teaching Compensation, you may have likely already taught Break Apart to Multiply. So, it is time to use a Part 3 activity to help students decide *when* they will use each strategy. Then, add a new strategy to students' repertoire and return again to activities that focus on choosing a strategy. This iterative process continues through the teaching of standard algorithms, as students continue to accumulate methods for multiplying and dividing whole numbers. Along the way, it is important that students also learn that sometimes there is more than one good way to solve a problem, and other times, one way really stands out from the others.

CHOICE

Choosing strategies is at the heart of fluency. After a strategy is learned, it should always be an option for consideration. Too often students feel like when they have moved on to a new strategy, they are supposed to use only the new one, as though it is more sophisticated or preferred by their teacher. But, the strategies are additive—they form a collection from which students can select in order to solve the problem at hand.

Once students know how and why a strategy works, they need to figure out when it makes sense to use the strategy. That is when you need questions such as these:

• When do you/might you use a Break Apart strategy for multiplication?

• When do you/might you use a Compensation strategy for multiplication?

• When do you/might you use Partial Products? Partial Quotients?

• When do you/might you use Think Multiplication?

• When do you/might you use a standard algorithm?

- When do you/might you not use _____ strategy/algorithm?
- Which strategies do you think of first when you see a problem?
- Which strategies do you choose only after other options don't work?
- Which strategies do you avoid?

The first six prompts can be mapped to the Part 2 modules (and asked during instruction on those strategies, as well as during mixed practice). As students are exploring the Compensation strategy, for example, include regular questioning about when it is a good fit and when it is not. The full set of questions is essential to the development of fluency. Activities are woven throughout the modules to help identify when a strategy works best and when it doesn't. However, those activities don't necessarily bring other strategies into consideration. Part 3 focuses on this notion of putting it all together, which includes attending to metacognition.

METACOGNITION

As students become more proficient with the strategies they have learned and when to use them, they make decisions as to what they will do for a given problem. Taking time up front to make a good choice can save time in the enactment of a not-efficient strategy. We can help students with this reasoning by sharing a metacognitive process. This could be a bulletin board or a card taped to students' desks. We can remind students that as we work, we make good choices about the methods we use. This is *flexibility* in action.

METACOGNITIVE PROCESS FOR SELECTING A STRATEGY

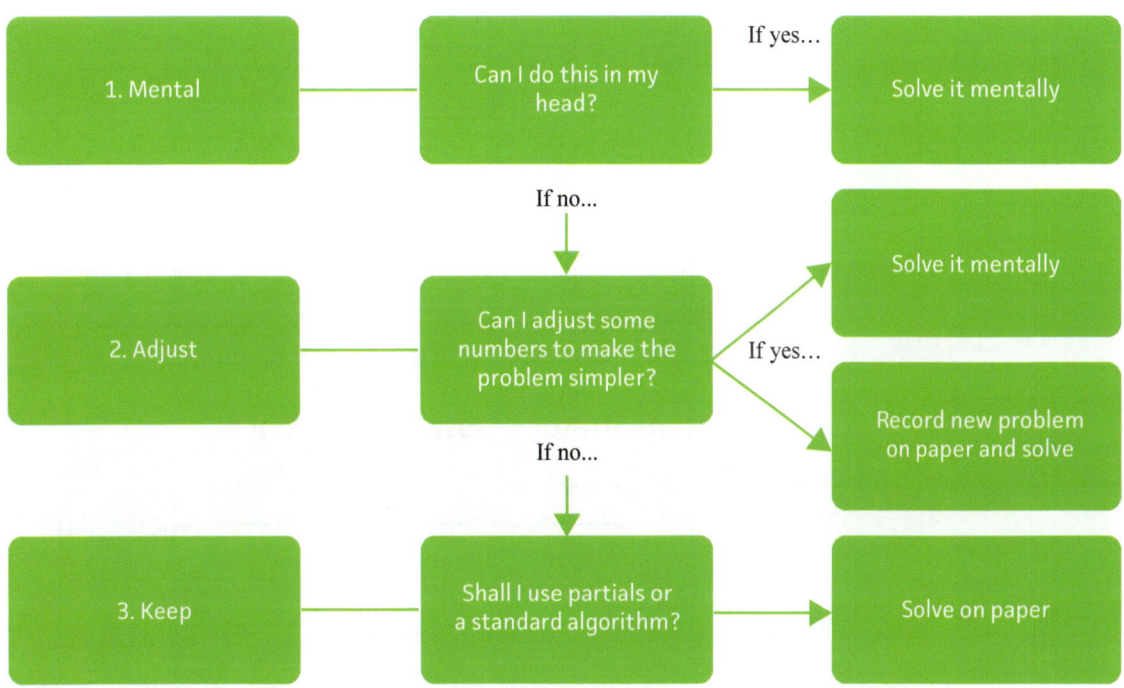

ASSESSING FLUENCY

Traditionally, fluency assessment has focused on speed and accuracy. While accuracy does matter, it is only one-third of what it means to be fluent. As you assess fluency, you want to intentionally look for each of the three fluency components and the six Fluency Actions. Flexibility and efficiency can be observed as students engage in meaningful activities. There is no need to stop and give a test.

OBSERVATION TOOLS

See Chapter 7 (pp. 154–175) of *Figuring Out Fluency* for more information about assessing fluency.

Observation tools can help us focus on the "neglected" components of fluency and serve as a way to communicate with students and their parents about what real fluency looks like. Three examples are shared here.

Student: _____   Date: _____

Problem(s): _____

Fluency Actions Checklist

Procedural Fluency Actions	Evident?			Comments
1. Selects appropriate strategy	Yes	No	Not Observed	
2. Solves in a reasonable amount of time	Yes	No	Not Observed	
3. Trades out or adapts a strategy	Yes	No	Not Observed	
4. Applies strategy to a new problem type	Yes	No	Not Observed	
5. Completes steps correctly	Yes	No	Not Observed	
6. Gets correct answer	Yes	No	Not Observed	

Comments:

Multiplication Strategy Observation Tool

Student Name	Multiplication Strategies				Algorithm
	Break Apart to Multiply	Halve and Double	Compensation	Partial Products	

Division Strategy Observation Tool

Student Name	Division Strategies		Algorithm
	Think Multiplication	Partial Quotients	

online resources 🔍 These resources can be downloaded at **resources.corwin.com/FOF/ multiplydividewholenumber**.

GRADING ASSIGNMENTS AND ASSESSMENTS

All of your efforts to teach strategies and to teach students to choose efficient strategies must be continued through assessments. Too often tests are graded for accuracy only. Instead, provide a fluency score using a rubric, such as the one that follows.

FOUR-POINT FLUENCY RUBRIC

BEGINNING 1	DEVELOPING 2	EMERGING 3	ACCOMPLISHED 4
Knows one algorithm or strategy but continues to get stuck or make errors.	Demonstrates efficiency and accuracy with at least one strategy/ algorithm, but does not stop to think if there is a more efficient possibility.	Demonstrates efficiency and accuracy with several strategies/ algorithms, and sometimes selects an efficient strategy, although still figuring out when to use and not use a strategy.	Demonstrates efficiency and accuracy with several strategies/ algorithms and is adept at matching problems with efficient strategies (knowing when to use each and when not to).

With an eye on fluency and how to assess it, select fluency activities from Part 3 based on your students' needs and what type of activity you are seeking.

FLUENCY ACTIVITIES

The nine activities in this section focus on all components of fluency, providing students with the opportunity to practice choosing an appropriate strategy, enact that strategy, and reflect on the efficiency of the strategies selected.

ACTIVITY F.1

Name: *"Strategize First Steps"*   **Type:** Routine

About the Routine: This routine involves sharing how to *start* a problem. It is at this first step that students are selecting a strategy. In its quickest version, that is the only step that is ever done. Simply ask, "Which step first and why?" A second option begins the same way but after students have shared ideas for first steps, they get to choose which first step they like and finish the problem. Then, answers can be shared and students can reflect on which first step turned out to be a good one (and why).

Materials: list of three or four problems on the same topic, but that lend to different reasoning strategies

Directions: This routine involves showing a series of problems, one at a time, like a Number Talk. But it differs in the fact that the tasks you selected *do not* lend to the same strategy. Here is the routine process:

1. Ask students to mentally determine their first step (only) and signal when they're ready (since it is only the first step, they only need a few seconds).

2. Record first-step ideas by creating a list on the board.

(Continued)

3. Discuss which first steps seem reasonable (or not).

4. Repeat with two to four more problems, referring to the list created from the first problem and adding to the list if/when new strategies are shared.

5. Conclude the series with a discussion: When will you use _____ strategy?

Possible problem sets for this routine include the following:

TWO-DIGIT BY ONE-DIGIT MULTIPLICATION	TWO-DIGIT BY ONE-DIGIT DIVISION
$59 \times 8 =$	$55 \div 5 =$
$35 \times 6 =$	$65 \div 3 =$
$52 \times 4 =$	$59 \div 7 =$
$27 \times 9 =$	$99 \div 4 =$
$67 \times 3 =$	$72 \div 6 =$
THREE-DIGIT BY ONE-DIGIT MULTIPLICATION	**THREE-DIGIT BY ONE-DIGIT DIVISION**
$185 \times 4 =$	$185 \div 6 =$
$490 \times 5 =$	$490 \div 2 =$
$258 \times 7 =$	$258 \div 4 =$
$315 \times 6 =$	$315 \div 7 =$
$207 \times 8 =$	$207 \div 5 =$
TWO-DIGIT BY TWO-DIGIT MULTIPLICATION	**FOUR-DIGIT BY TWO-DIGIT DIVISION**
$28 \times 13 =$	$6{,}420 \div 12 =$
$68 \times 59 =$	$8{,}650 \div 25 =$
$71 \times 35 =$	$1{,}360 \div 40 =$
$86 \times 57 =$	$9{,}810 \div 15 =$

Let's say you were using the list for two-digit multiplication. Here are some examples of the first steps you might hear for the first problem, 28×13:

"Change 28 to 30"

"Break apart both factors"

"Think 28×10 is 280"

You can layer in strategy names at this point. For example, label the first idea as Compensation, the second as Break Apart (or Partial Products), and so on. Then, by consistent use of the language, students should begin to say, "I used Compensation and . . ." For each problem, some of the same strategies will be named again, and new ideas might be added. If an idea doesn't come up, you can ask about it directly or just move to the next problem. Be mindful not to imply that it is the preferred approach. As noted earlier, if students are going to solve the problem, allow them to choose from the ideas that were first stated. They can later share if they stayed with their original first step or switched out to one of the ideas from their classmates.

ACTIVITY F.2

Name: "M-A-K-E a Decision"   **Type:** Routine

About the Routine: Mental-Adjust-Keep-Expressions is a routine that, like other routines, can function for assessment purposes. Project for students an illustration of the metacognitive process, like the one discussed earlier in the Part 3 overview. If a problem can be solved mentally, there is no need to use a written method such as a standard algorithm. Students tend to dive in without stopping to think if they can solve a problem mentally. There are two ways to use the tasks in this routine. One is to show one problem at a time, each time asking, "Which way will you solve it?" Or you can show the full set and ask, "Which ones might you solve mentally? Adjust and solve? Keep as-is and solve?" The former is used below.

Materials: Prepare a set of three to six expressions.

Directions:
1. Explain that you are going to display a problem. Students are not to solve it but simply "M-A-K-E a Decision: Mental, Adjust and solve, or Keep and solve on paper?"

2. Display the first problem and give students about 10 seconds to decide.

3. Use a cue to have students share their choice (one finger = mental, two fingers = adjust, three fingers = keep and solve on paper).

4. Ask a student who picked each decision to share why (and how).

5. Repeat step 3, giving students a chance to change their minds.

6. Repeat for the next problem.

7. At the end of the set, ask, "What do you 'see' in a problem that leads to you doing it mentally? Adjusting? Keep and use paper?"

Possible problem sets to use with this routine include the following:

TWO-DIGIT BY ONE-DIGIT MULTIPLICATION		TWO-DIGIT BY ONE-DIGIT DIVISION	
$31 \times 8 =$	$17 \times 9 =$	$59 \div 5 =$	$65 \div 4 =$
$68 \times 6 =$	$46 \times 3 =$	$34 \div 3 =$	$72 \div 6 =$
$55 \times 4 =$		$21 \div 7 =$	
THREE-DIGIT BY ONE-DIGIT MULTIPLICATION		**THREE-DIGIT BY ONE-DIGIT DIVISION**	
$301 \times 4 =$	$369 \times 6 =$	$357 \div 6 =$	$497 \div 7 =$
$648 \times 5 =$	$465 \times 8 =$	$508 \div 2 =$	$214 \div 5 =$
$135 \times 7 =$		$522 \div 4 =$	
TWO-DIGIT BY TWO-DIGIT MULTIPLICATION		**FOUR-DIGIT BY TWO-DIGIT DIVISION**	
$57 \times 17 =$	$32 \times 19 =$	$8,064 \div 12 =$	$1,425 \div 15 =$
$68 \times 52 =$	$83 \times 49 =$	$1,825 \div 25 =$	$2,280 \div 30 =$
$94 \times 76 =$		$3,350 \div 50 =$	

ACTIVITY F.3

Name: "Share–Share–Compare"   **Type:** Routine

About the Routine: This routine can also be a longer classroom activity, depending on how many problems students are asked to solve. Each person first solves problems independently, then has the chance to have a one-on-one with a peer to compare their thinking on the same problem.

Materials: Prepare a list of three to five problems that lend to being solved different ways.

Directions:
1. Students work independently to solve the full set of problems.

2. Students write if they solved it mentally by naming the strategy they used. If they solved it by writing, they do not need to name a strategy.

3. Once complete, everyone stands up with their page of worked problems.

4. Students find a partner who is *not* at their table. When they find a partner, they high-five each other and begin Share–Share–Compare for the first problem:

   ● Share: Partner 1 **shares** their method.

   ● Share: Partner 2 **shares** their method.

   ● Compare: Partners discuss how their methods **compared**:

If their methods are different, they compare the two, discussing which one worked the best or if both worked well.

If their methods are the same, they think of an alternative method and again discuss which method(s) worked well.

After the exchange, partners thank each other, raise their hands to indicate they are in search of a new partner, find another partner, and repeat the process of Share–Share–Compare for problem 2 (or any problem they haven't yet discussed).

Possible problem sets for this routine include the following:

TWO-DIGIT BY ONE-DIGIT MULTIPLICATION		TWO-DIGIT BY ONE-DIGIT DIVISION	
$78 \times 8 =$	$65 \times 9 =$	$26 \div 5 =$	$67 \div 4 =$
$59 \times 6 =$	$47 \times 3 =$	$34 \div 3 =$	$48 \div 6 =$
$37 \times 4 =$		$98 \div 7 =$	
THREE-DIGIT BY ONE-DIGIT MULTIPLICATION		**THREE-DIGIT BY ONE-DIGIT DIVISION**	
$482 \times 4 =$	$835 \times 7 =$	$287 \div 6 =$	$523 \div 4 =$
$292 \times 5 =$	$244 \times 6 =$	$327 \div 2 =$	$497 \div 7 =$
TWO-DIGIT BY TWO-DIGIT MULTIPLICATION		**FOUR-DIGIT BY TWO-DIGIT DIVISION**	
$61 \times 52 =$	$37 \times 19 =$	$4,450 \div 50 =$	$7,248 \div 12 =$
$19 \times 76 =$	$83 \times 49 =$	$1,125 \div 25 =$	$8,520 \div 40 =$

ACTIVITY F.4

Name: Strategy Spin   **Type:** Game

About the Game: This game can be played in many ways. You can have spinners that focus on metacognition (mental, adjust, keep; see metacognition visual), you can pick the strategies students know, or you can have the options "Standard Algorithm" and "Not Standard Algorithm." There are many online spinner tools, such as Wheel Decide (https://wheeldecide.com/), which allow you to enter the categories you want and actually spin virtually. This game can be played with two to four players.

Materials: strategy spinner (one per group) and expression cards

Directions: 1. Players take turns spinning the strategy spinner.

2. Once the spinner lands on a strategy, the player looks through the expression bank to find a problem they want to solve using that strategy.

3. The player tells how to solve the selected problem using that strategy. Opponents check solutions using a Hundred Chart, number line, or calculator. If correct, the player claims that expression card.

4. Repeat steps 2 and 3 three times for four players, four times for three players, and five times for two players.

5. Together the group looks at the remaining expressions and labels which strategy they think is a good fit for each.

RESOURCE(S) FOR THIS ACTIVITY

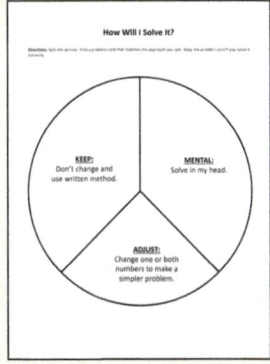

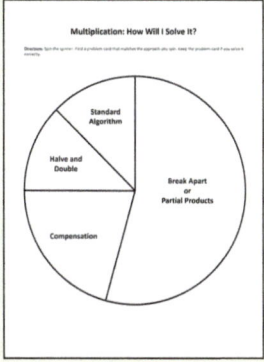

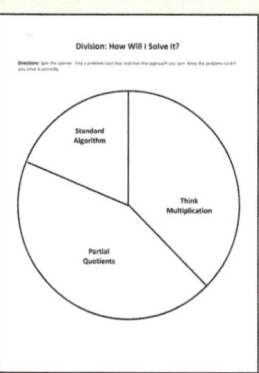

online resources — Game cards and spinners can be downloaded at **resources.corwin.com/FOF/ multiplydividewholenumber**.

ACTIVITY F.5

Name: Just Right   **Type:** Game

About the Game: This game board can be adapted to incorporate the strategies students have learned. Earlier in the year, it may have only two options; later in the year, it may have five options. Students may be tempted to use inefficient strategies in order to get four in a row. To counter this, require students to record the equations and the strategy they used.

Materials: expression cards (mixed strategies), *Just Right* game board, and two-sided counters

Directions: 1. Players take turns flipping cards.

2. Player selects a "just right" strategy, and solves it thinking aloud or on paper.

3. Opponents confirm accuracy with calculators.

4. Having correctly solved the problem, the player places a marker on the strategy they used on the *Just Right* game board.

5. The first player to place four markers in a row on the game board wins.

For example, Sydney draws 55×12. She says, "My 'just right' strategy is Halve and Double. I think double 55 is 110 and half of 12 is 6. 110×6 is 660." Her opponent, Annie, confirms that she is correct. Sydney covers her choice of a Halve and Double square.

RESOURCE(S) FOR THIS ACTIVITY

online resources These resources can be downloaded at **resources.corwin.com/FOF/multiplydividewholenumber**.

ACTIVITY F.6

Name: Strategories   **Type:** Game

About the Game: This game is an excellent opportunity for practice or assessment once a variety of strategies are learned. Students generate three to six examples of problems that lend to using that strategy, such as 29 × 7 (Compensation), 58 × 5 (Halve and Double), or 38 × 48 (Partial Products), relative to the size of factors to which they have been exposed. The directions describe this as a small group activity, but it can be modified to a center. An alternative to this game is to just focus on one strategy (e.g., Partial Products).

Materials: *Strategies* game card (one per student)

Directions:
1. Each player works independently to generate a problem for which they would use that strategy (alternatively, students can work with a partner to discuss problems that would fit in each strategy).

2. After players have completed their *Strategies* game card (i.e., has one example expression in each strategy), place students in groups of three.

3. On a player's turn, they ask one of the other group members to share the problem on their *Strategies* card for _____ strategy. If the player explains the problem using that strategy, they score 5 points. If not, the third player gets a chance to "steal" by explaining the problem using that strategy. If the third player cannot, the author of the problem must explain using that strategy. If they cannot, they lose 10 points.

4. Play continues until all strategies have been solved, or students have played three rounds. The high score wins!

5. To facilitate discussion after play, ask questions such as these: What do you notice about the problems in _____ strategy? When is this strategy useful?

RESOURCE(S) FOR THIS ACTIVITY

Strategies: Division

Strategy	My Problem
Think Multiplication	
Think Multiplication	
Partial Quotients	
Partial Quotients	

Strategies: Multiplication

Strategy	My Problem
Break Apart	
Halve and Double	
Compensation	
Partial Products	

online resources   These resources can be downloaded at **resources.corwin.com/FOF/ multiplydividewholenumber**.

ACTIVITY F.7

Name: Make the Product   **Type:** Game

About the Game: As students encounter different two-digit factors, they will find expressions that lend to different strategies. That makes this a good game for observing what strategies students are commonly choosing to use (and whether they are using a variety of strategies). The game can be played as a center in which students play to get their personal best score.

Materials: two sets of digit cards or half a deck of playing cards (with face cards removed), *Make the Product* recording sheet

Directions:

1. Establish a target product. Note that the range for 2 two-digit factors is between 100 (10×10) and 9,801 (99×99). In this case, the product to make is 2,500.

2. Each player is dealt six cards.

3. Each player uses four of the six cards to make 2 two-digit factors with a product as close to 2,500 as possible.

4. The player records their equation and shows how they found the product.

5. Players record how far away they are from the target product (in this case, 2,500).

6. After five rounds, each player adds the distance from 100 from each round for a score.

7. The player with the lower score wins.

For a target of 2,500, Oscar pulls 5, 3, 4, 7, 2, and 1. He makes 43×57 and finds the product to be 2,451. He is 49 from the target.

RESOURCE(S) FOR THIS ACTIVITY

Make the Product _____

Directions: Deal six cards. Use four cards to make two factors with a product as close to the target as possible. Find the product and show how you multiplied. Record how far from the target you are for each round.

Round	My Cards	My Multiplication Problem	How Far From _____?
1			
2			

 Digit cards and this recording sheet can be downloaded at **resources.corwin.com/FOF/ multiplydividewholenumber**.

ACTIVITY F.8

Name: A-MAZE-ing Race   **Type:** Game

About the Game: This game is not about speed! Putting time pressure on students works against fluency, as the stress can block good reasoning. The purpose of this game is to practice selecting and using strategies that are a good fit for the numbers in the problem. It can be adapted for any multiplication or division problem sets. You can also use it as a center in which students highlight their path, record the problems, find the solutions, and show their strategy.

Materials: *A-MAZE-ing Race* game board (one per pair), two-sided counters, calculators

Directions: 1. Players both put their marker on Start.

2. Players take turns selecting an unoccupied square that shares a border with their current position.

3. Players talk aloud to say the answer and explain how they solved the equation. An optional sentence frame to use is, "The answer is __. I used __ strategy to solve it. I __, then I __."

4. Opponents confirm accuracy with calculators. If correct, the player moves to the new square.

5. Note that the dark lines cannot be crossed.

6. The winner is the first to reach the finish.

RESOURCE(S) FOR THIS ACTIVITY

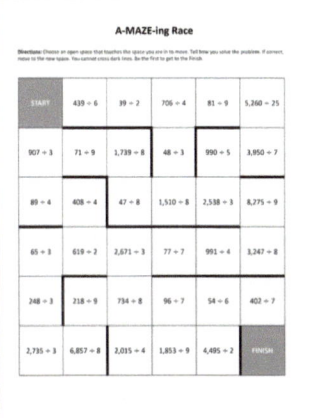

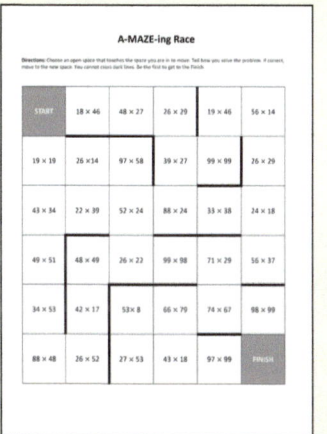

online resources ↖ These resources can be downloaded at **resources.corwin.com/FOF/ multiplydividewholenumber**.

ACTIVITY F.9

Name: Strategy Problem Sort   **Type:** Center

About the Center: Just because a problem can be solved with a given strategy does not mean it is a good fit. It is a good fit when it is the most appropriate option from a student's set of known strategies. (See the Part 1 discussion of appropriate strategies.) You can extend the activity by having students go through the "Does Not Fit the Strategy" category and identify which strategy would work well for that problem.

Materials: Strategy Problem Sort cards or a set of 12–20 mixed problems (some that fit the intended strategy and some that do not), Strategy Problem Sort placemat (optional)

Directions: 1. Students flip over a problem card.

2. Students determine if the problem fits the strategy or doesn't fit the strategy.

3. Students then solve the problem.

To create assessment artifacts, you can take a picture of a student's completed sort. Or, you can ask students to provide written responses to questions like these:

I placed _____ in the "Fits the Strategy" side because . . .

I placed _____ in the "Does Not Fit the Strategy" side because . . .

RESOURCE(S) FOR THIS ACTIVITY

Strategy Sort Placemat	
Strategy: _____	
Fits the Strategy | Doesn't Fit the Strategy	

39×7	44×74
51×62	899×3
298×4	$9 \times 4,177$
$3,299 \times 2$	95×8
61×9	374×6

 These resources can be downloaded at **resources.corwin.com/FOF/ multiplydividewholenumber**.

Appendix
Tables of Activities

Figuring out fluency is a journey. Fluency with whole number multiplication and division is absolutely essential for life and for future mathematics. Each and every child must have access and ample opportunities to develop their understanding and use of reasoning strategies. It is critical to remember these points:

- Fluency needs to be a daily part of mathematics instruction.

- There are no shortcuts or quick fixes to developing fluency.

- Fluency requires instruction *and* ongoing reinforcement.

- Different students require different types and quantities of experiences to develop fluency.

- While strategy choice is about the individual student, every student must learn and practice significant strategies so that they *can* choose to use them.

- Fluency practice must not be stressful. Stress complicates thinking.

- Fluency is more than accuracy; you must assess the other components.

This book is packed with activities for instruction, practice, and assessment to support the work that you do and to supplement the resources you use. The following pages provide a listing of all of the activities in this book. These tables can help you achieve the following:

- Jump between strategies, as you may not teach them sequentially.

- Locate prompts for teaching each strategy.

- Identify a specific type of activity to incorporate into your fluency instruction.

- Identify activities for specific strategies that you need to reteach, reinforce, or assess.

- Take notes about revisiting an activity later in the year.

- Take notes about modifying an activity for use with another strategy.

- Take notes about how you might leverage the activity in future years.

- Identify an activity that is particularly useful for assessment.

MODULE 1: BREAK APART TO MULTIPLY ACTIVITIES				
NO.	**PAGE**	**TYPE**	**NAME**	**NOTES**
1.1	25	T	Multiples of 10	
1.2	26	T	Color Tile Arrays	
1.3	27	T	Breaking Apart Basic Facts	
1.4	27	T	Two Cuts	
1.5	28	T	The Better Break	
1.6	29	T	Breaking Apart With Multiplication	
1.7	30	T	Prompts for Teaching Break Apart to Multiply	
1.8	33	R	"Complex Number Strings"	
1.9	34	R	"The Breaks" (breaking apart factors)	
1.10	35	R	"Strategize Your Break Apart"	
1.11	36	G	*Factor Race*	
1.12	37	G	*Mult-zee*	
1.13	38	G	*Break Apart Bingo*	
1.14	39	G	*Break Apart to Multiply Math Libs*	
1.15	40	C	Same But Different	
1.16	41	C	Backward Breaks	
1.17	42	C	Break Apart With an Area Model Representation	
1.18	43	C	Which and How?	

T (Teaching) • R (Routine) • G (Game) • C (Center/Independent)

MODULE 2: HALVE AND DOUBLE ACTIVITIES				
NO.	**PAGE**	**TYPE**	**NAME**	**NOTES**
2.1	46	T	Halving and Rearranging Rectangles	
2.2	47	T	Cuisenaire Rod Proofs	
2.3	48	T	Representing Halve and Double With Graphic Organizers	
2.4	49	T	Nickels and Dimes	
2.5	50	T	Prompts for Teaching Halve and Double	
2.6	53	R	"A String of Halves" (practice halving)	
2.7	54	R	"The Same As" (restating a problem with Halve and Double)	
2.8	55	G	*The Splits*	
2.9	56	G	*Get 10* (when to use the strategy)	
2.10	57	G	*Halve and Double Flips*	
2.11	58	G	*The First to Seven*	
2.12	59	C	Fives and Tens	
2.13	60	C	Write It as Halve and Double	

T (Teaching) • R (Routine) • G (Game) • C (Center/Independent)

MODULE 3: COMPENSATION ACTIVITIES				
NO.	PAGE	TYPE	NAME	NOTES
3.1	64	T	Show Me	
3.2	65	T	Compensated Trains (with linking cubes)	
3.3	66	T	True or False Compensation Statements	
3.4	66	T	Number Strings (Compensation focus)	
3.5	67	T	Prompts for Teaching Compensation	
3.6	70	R	"A Little More or A Little Less"	
3.7	71	R	"Or You Could" (restating with Compensation)	
3.8	72	G	*Compensation Points*	
3.9	73	G	*Give Some to Get Some*	
3.10	74	G	*Compensation Concentration*	
3.11	75	G	*Estimation Tic-Tac-Toe*	
3.12	76	C	Creating Compensations	
3.13	77	C	Prove It	
3.14	78	C	Use This for Solving	

T (Teaching) • R (Routine) • G (Game) • C (Center/Independent)

MODULE 4: PARTIAL PRODUCTS ACTIVITIES				
NO.	PAGE	TYPE	NAME	NOTES
4.1	83	T	Multiples Matrix	
4.2	84	T	Base-10 Partials	
4.3	84	T	Partial Products With Area Models	
4.4	85	T	Is It the Same? (comparing different ways to break apart)	
4.5	86	T	One or Both? (which factors to decompose)	
4.6	87	T	Same Digits, Different Product	
4.7	88	T	Prompts for Teaching Partial Products	
4.8	91	R	"Why Not?"	
4.9	92	R	"About or Between" (estimation)	
4.10	93	G	*Make It Close*	
4.11	94	G	*Connect Four Partials*	
4.12	95	G	*Target 2,500*	
4.13	96	C	What's the Problem	
4.14	97	C	Pick Your Partials	
4.15	98	C	Guess and Check—The Largest Product	
4.16	99	C	Bag of Blocks (partials with base-10 blocks)	
4.17	100	C	One or Both?	

T (Teaching) • R (Routine) • G (Game) • C (Center/Independent)

MODULE 5: THINK MULTIPLICATION ACTIVITIES				
NO.	PAGE	TYPE	NAME	NOTES
5.1	104	T	Is It Still True?	
5.2	105	T	The Direct Connect: Three Strings	
5.3	106	T	Recognizing Products and Decomposing Dividends	
5.4	107	T	What's Missing? (with triangle cards)	
5.5	107	T	Prompts for Teaching Think Multiplication	
5.6	110	R	"If I Know. . ."	
5.7	111	R	"The Count Up" (counting by multiples to find dividends)	
5.8	112	R	"Breaking Dividends"	
5.9	113	G	*The Take*	
5.10	114	G	*Paper Clip Connections*	
5.11	115	G	*Covering the Breaks*	
5.12	116	G	*Area Model Duel*	
5.13	117	C	Make the Quotient	
5.14	118	C	Rewrite It and Solve It	
5.15	119	C	Triangle Cards	
5.16	120	C	Finding the Breaks	

T (Teaching) • R (Routine) • G (Game) • C (Center/Independent)

MODULE 6: PARTIAL QUOTIENTS ACTIVITIES				
NO.	**PAGE**	**TYPE**	**NAME**	**NOTES**
6.1	124	T	Base-10 Quotients	
6.2	125	T	Is It the Same? (comparing different partial quotients)	
6.3	125	T	Another Way (representing partial quotients)	
6.4	126	T	Comparing Partials	
6.5	126	T	Prompts for Teaching Partial Quotients	
6.6	130	R	"How Many Are In?"	
6.7	131	R	*"Could It Be?"*	
6.8	132	G	*Quotient Hunt*	
6.9	134	G	*End of the Lane*	
6.10	135	G	*Quotient Connect*	
6.11	136	G	*Remainder Tally*	
6.12	137	G	*Division Bump*	
6.13	138	G	*Tic-Tac-Take*	
6.14	139	C	How Many Are In?	
6.15	140	C	Estimate and Divide	
6.16	141	C	The Largest Quotient, The Smallest Quotient	
6.17	142	C	Changing Dividends Make for Changing Quotients	

T (Teaching) • R (Routine) • G (Game) • C (Center/Independent)

MODULE 7: STANDARD ALGORITHM ACTIVITIES				
NO.	PAGE	TYPE	NAME	NOTES
7.1	147	T	Connecting Partial Products and the Algorithm	
7.2	148	T	Connecting Partial Quotients and the Algorithm	
7.3	148	T	Find the Error	
7.4	149	T	To Algorithm or Not to Algorithm	
7.5	150	T	Prompts for Teaching Standard Algorithms	
7.6	153	R	"Over/Under" (estimation)	
7.7	154	R	"That One" (when to use the algorithm)	
7.8	155	G	*Call Your Shot*	
7.9	156	G	*A List of Ten*	
7.10	157	G	*For Keeps*	
7.11	158	G	*The V Cover Up*	
7.12	159	C	Missing Numbers With Multiplication Algorithms	
7.13	160	C	Missing Digits With Division Algorithms	
7.14	161	C	Partials to Algorithms	
7.15	162	C	Making Connections	
7.16	163	C	Correct the Error	

T (Teaching) • R (Routine) • G (Game) • C (Center/Independent)

NO.	PAGE	TYPE	NAME	NOTES
F.1	169	R	"Strategize First Steps"	
F.2	171	R	"M-A-K-E a Decision"	
F.3	172	R	"Share–Share–Compare"	
F.4	173	G	*Strategy Spin*	
F.5	174	G	*Just Right*	
F.6	175	G	*Strategories*	
F.7	176	G/C	*Make the Product*	
F.8	177	G/C	*A-MAZE-ing Race*	
F.9	178	C	Strategy Problem Sort	

T (Teaching) • R (Routine) • G (Game) • C (Center/Independent)

References

Baroody, A. J., & Dowker, A. (Eds.). (2003). *Studies in mathematical thinking and learning. The development of arithmetic concepts and skills: Constructing adaptive expertise.* Lawrence Erlbaum Associates.

Baroody, A. J., Purpura, D. J., Eiland, M. D., Reid, E. E., & Paliwal, V. (2016). Does fostering reasoning strategies for relatively difficult basic combinations promote transfer by K–3 students? *Journal of Educational Psychology,* 108(4), 576–591. https://psycnet.apa.org/doi/10.1037/edu0000067

Bay-Williams, J., & Kling, G. (2019). *Math fact fluency: 60+ games and assessment tools to support learning and retention.* ASCD.

Boaler, J. (2015a). *What's math got to do with it? How teachers and parents can transform mathematics learning and inspire success.* Penguin Books.

Boaler, J. (2015b). *Memorizers are the lowest achievers and other Common Core math surprises.* https://hechingerreport.org/memorizers-are-the-lowest-achievers-and-other-common-core-math-surprises/

Brendefur, J., Strother, S., Thiede, K., & Appleton, S. (2015). Developing multiplication fact fluency. *Advances in Social Sciences Research Journal,* 2(8), 142–154. https://doi.org/10.14738/assrj.28.1396

Cheind, J., & Schneider, W. (2012). The brain's learning and control architecture. *Current Directions in Psychological Science,* 21(2), 78–84. https://www.jstor.org/stable/23213097?seq=1

"Explicit." (2021). *Merriam-Webster.com.* https://www.merriam-webster.com/dictionary/explicit

Fuson, K. C., & Beckmann, S. (2012–2013, Fall/Winter). Standard algorithms in the Common Core State Standards. *NCSM Journal,* 14–30.

Franke, M. L., Kazemi, E. & Turrou, A. C., (2018). *Choral counting and counting collections: Transforming the PreK–5 math classroom.* Stenhouse Publishers.

Jameson, M. M. (2013). Contextual factors related to math anxiety in second-grade children. *Journal of Experimental Education,* 82(4), 518–536. https://doi.org/10.1080/00220973.2013.813367

Jordan, N. C., Kaplan, D., Ramineni, C., & Locuniak, M. N. (2009). Early math matters: Kindergarten number competence and later mathematics outcomes. *Developmental Psychology,* 45(3), 850–867. https://doi.org/10.1037/a0014939

Kilpatrick, J., Swafford, J., & Findell, B. (2001). *Adding it up: Helping children learn mathematics.* National Academy Press. https://doi.org/10.17226/9822

Locuniak, M. N., & Jordan, N. C. (2008). Using kindergarten number sense to predict calculation fluency in second grade. *Journal of Learning Disabilities,* 41(5), 451–459. https://doi.org/10.1177/0022219408321126

National Center for Education Statistics (NCES). NAEP Report Card: 2019 NAEP Mathematics Asssessment. https://www.nationsreportcard.gov/mathematics/nation/achievement/?grade=4

National Council of Teachers of Mathematics (NCTM). (2014). *Principles to actions: Ensuring mathematical success for all.* NCTM.

O'Connell, S., & SanGiovanni, J. (2014). *Mastering the basic math facts in multiplication and division: Strategies, activities, and interventions to move students beyond memorization.* Heinemann.

Parrish, S. (2014). *Number Talks: Helping children build mental math and computation strategies, Grades K–5.* MathSolutions.

Purpura, D. J., Baroody, A. J., Eiland, M. D., & Reid, E. E. (2016). Fostering first graders' reasoning strategies with basic sums: The value of guided instruction. *The Elementary School Journal,* 117(1), 72–100. https://doi.org/10.1086/687809

Ramirez, G., Shaw, S. T., & Maloney, E. A. (2018). Math anxiety: Past research, promising interventions, and a new interpretation framework. *Educational Psychologist, 53*(3), 145–164. https://doi.org/10.1080/00461520.2018.1447384

SanGiovanni, J. J. (2019). *Daily routines to jump-start math class: Elementary school.* Corwin.

SanGiovanni, J., Katt, S., & Dykema, K. (2020). *Productive math struggle: A six-point action plan for fostering perseverance.* Corwin.

Star, J. R. (2005). Reconceptualizing conceptual knowledge. *Journal for Research in Mathematics Education, 36*(5), 404–411. https://doi.org/10.2307/30034943

Index

A SAGE Publishing Company

Helping educators make the greatest impact

CORWIN HAS ONE MISSION: to enhance education through intentional professional learning.

We build long-term relationships with our authors, educators, clients, and associations who partner with us to develop and continuously improve the best evidence-based practices that establish and support lifelong learning.

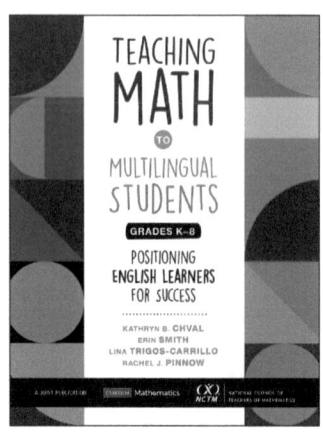

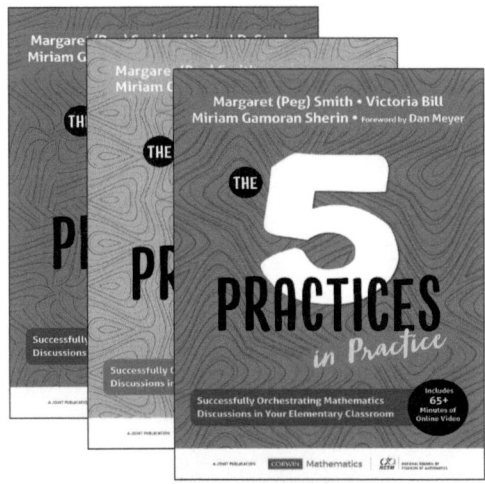
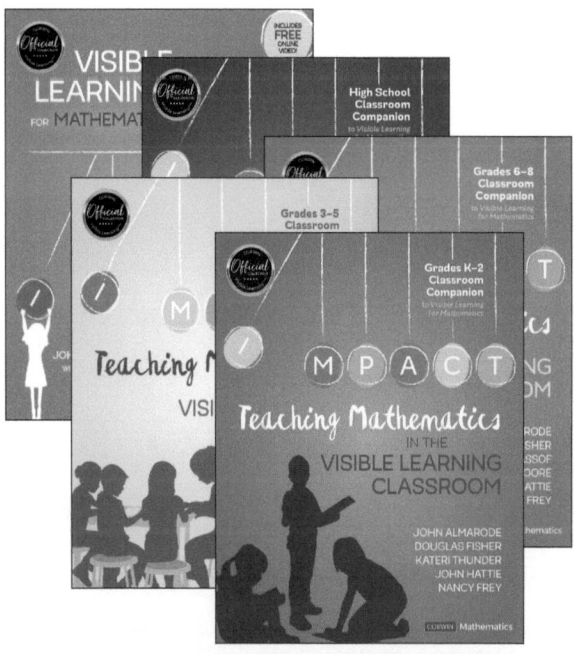

What Your Colleagues Are Saying . . .

Conway et al. have done it again! In this volume, not only do the authors share rich, standards-based, grade-level appropriate mathematical tasks with social justice contexts, they also share specific strategies and tools to engage the middle school mathematics student.

—**Kyndall Brown**
Executive Director
California Mathematics Project, Los Angeles, CA

Middle School Mathematics Lessons to Explore, Understand, and Respond to Social Injustice is a must-have resource for any middle school educator seeking to incorporate teaching mathematics for social justice in their educational environment. This book provides practical lessons that will truly enhance a child's educational experience through engaging them in relevant applications of mathematics.

—**Kristopher J. Childs**
Mathematics Consultant
Winter Garden, FL

A wonderful collection of lessons, submitted by teachers, to help students of all ages see topics they care about, and use mathematics as a tool for progress in the world.

—**Jo Boaler**
The Nominelli-Olivier Professor of Education (Mathematics)
Stanford University, Stanford, CA

If you teach middle school mathematics and have wanted to promote social justice but haven't been sure how to get started, you need to check out this book. It incorporates lessons you can use immediately and offers guidance on fostering a classroom community where students will thrive. It's the kind of book you'll want to have alongside you to support you throughout your journey.

—**Robert Kaplinsky**
Author and Consultant
Long Beach, CA

Middle School Mathematics Lessons to Explore, Understand, and Respond to Social Injustice is an outstanding addition to the growing number of texts and projects that weave the teaching of mathematics and social justice together. The authors go deep and broad to show how, why, and when this combination of curricular topics improves our students' mathematical understandings while honing their abilities and dispositions to promote social and environmental justice in their own lives and communities.

—**Bob Peterson**
Editor of *Rethinking Schools*
Editor of *Rethinking Mathematics: Teaching Social Justice by the Numbers*
Milwaukee, WI

A very compelling set of fresh ideas are offered that prepare educators to turn the corner on advocating for social justice in the mathematics classroom. Each book is full of engaging activities, frameworks, and standards that center instruction on community, worldview, and the developmental needs of all students—a much-needed resource to reboot our commitment to the next generation.

—Linda M. Fulmore
TODOS: Mathematics for ALL
Cave Creek, AZ

I imagine many people will purchase this book for the sample lesson plans. And you should; they're fabulous. But just as fabulous, and equally important, is the framework the authors lay out for a comprehensive, holistic, transformative approach to mathematics teaching, with social justice at its core.

—Paul C. Gorski
Lead Equity Specialist
Equity Literacy Institute, Columbia, SC

As a teacher educator for social justice, I am familiar with the near-constant refrain of "this isn't something you can do in math!" This book illustrates just the opposite. Indeed, not only is it possible to engage in social justice mathematics, but it is an educational imperative to do so. This much-needed and valuable collection provides practitioners with clear and compelling lessons that are grounded in theories of justice and equity. Especially timely in this text is the clear evidence that not only can middle school young adolescents engage in critical conversations, problem solving, and sociocultural analysis in their mathematics classes, but they must. The editors and contributors to this volume have curated a powerful resource that is a must-read for all mathematics educators and those who care about social justice teaching and learning.

—Alyssa Hadley Dunn
Associate Professor of Teacher Education
Michigan State University, East Lansing, MI

Middle School Mathematics Lessons to Explore, Understand, and Respond to Social Injustice is a book written for the teacher of mathematics. This diverse group of authors and educators provides everything a teacher needs to transform their practice from critical theory to teachable lesson plans! A must-read for educators who seek to transform mathematics back to its humanistic roots.

—Shraddha Shirude
Mathematics Teacher
Seattle Public Schools, Seattle, WA
Mathematics Specialist
WA Ethnic Studies Now, Seattle, WA

In addition to pedagogical tools, additional resources, and voices from the field, this book delivers over 20 lessons with extensive additional resources.

Notes tying each lesson back to Social Justice Outcomes, Mathematics Essential Concepts, and Mathematical Practices.

General overview of the lesson describing the background, learning goals, and needed materials and resources to complete the lesson.

LESSON 6.3 BILLIONAIRE POWER

Natalie Odom Pough and Y. Rhoda Latimer

ECONOMIC INEQUALITY

This lesson focuses on economic inequality, particularly a living wage and consumerism. Students are seeking ways to be successful throughout school and life, and within their community. As they obtain success, students must understand the adage, "To whom much is given, much is required." It is posited that many of the world's issues could be eradicated by some of our most wealthy individuals. It is important for younger generations to explore and troubleshoot this option as well as use mathematics to support their justifications for moving in that direction. Students need to think critically about the system that has been created and why/how it is maintained through capitalistic beliefs and practices.

DEEP AND RICH MATHEMATICS

Students will develop a deep understanding, model multiple representations, and communicate relationships of large numbers and number systems. Students will also access deep and rich meanings by selecting appropriate methods and tools for reasonable estimation, proportional reasoning, and computational fluency while solving real-world problems.

ABOUT THE LESSON

This lesson uses a launch–explore–summarize instructional model and is intended to take approximately 225 minutes to complete across three class periods.

Lesson 1: Students investigate the difference between 1 million and 1 billion as they deepen their understanding of the value of economic wealth statuses.

Lessons 2 and 3: Students are introduced to the income inequalities that have been heightened due to the COVID-19 pandemic and billionaires' economic responsibilities.

SOCIAL JUSTICE OUTCOMES

- I know that all people (including myself) have certain advantages and disadvantages in society based on who they are and where they were born. (Justice 14)
- I will work with friends, family and community members to make our world fairer for everyone, and we will plan and coordinate our actions in order to achieve our goals. (Action 20)

ESSENTIAL MIDDLE GRADES CONCEPTS

- **Number**– Grade 7: Solve real-world and mathematical problems involving the four operations with rational numbers.

MATHEMATICS PRACTICES

- Make sense of problems and persevere in solving them.
- Reason abstractly and quantitatively.
- Construct viable arguments and critique the reasoning of others.

109

LESSON 1 FACILITATION

Majority in the Supreme Court

Launch (20 minutes)

- Organize students into groups of four. Take a moment to revisit your co-constructed classroom norms for interaction and agreed-upon ways of listening and responding to one another (e.g., sentence stems for rough-draft math talk; Jansen, 2020).

- Share the following prompt and allow 3 minutes for individual work:

  + *In your own words, provide a definition for the concept of "majority." Feel free to use example(s) to describe your definition.*

- Have students discuss within their group with a shoulder partner for 5 minutes. Use a Learning to Listen, Listening to Learn Protocol and rough-draft math talk sentence stems (Jansen, 2020) to discuss definitions.

LEARNING TO LISTEN, LISTENING TO LEARN PROTOCOL

- Each person is given equal time to talk, even if they don't use it all.

- Each person has 1 minute to share their definition and ideas. The listener does not interpret, paraphrase, analyze, give advice, or break in with their ideas. The listener must remain quiet, attentively listening to their partner, until the timer goes off.

- After the talker has shared, the listener will

  + thank their peer for sharing their rough-draft definition of *majority.*

  + use the following sentence stems:

    - *What I heard you share was . . .*

    - *When you shared . . . it made me think about . . .*

    - *I used to think . . . but now I'm wondering . . .*

- Some anticipated responses include these: *more than $\frac{1}{2}$, one more than half, depends on the number we are talking about, easy to notice but tough to describe,* and so on.

- Facilitate a class discussion about the contexts, nature, and mathematical aspects of definitions brought up. Look for any connections to students, schools, and communities. Record these definitions, ideas, and connections on a shared Google slide (online) or whiteboard/poster paper (face-to-face in class) so that the learning community can revisit them throughout the lesson.

Extensive facilitation notes help educators run through the lesson with their class in a thoughtful manner.

Taking Action notes to help you and your students extend the learning outside of the classroom.

TAKING ACTION

Consider having students take action as follows:

1. Read an article and watch a video about a community working to turn corner stores into markets. Here are some videos and articles to choose from:

   + Article: "Why Do Corner Stores Struggle to Sell Fresh Produce?" by Sam Bloch, *The Counter*, February 21, 2019 (https://bit.ly/3xOJNTJ)

   + Article and video: "West Oakland Food Desert Blooms With a Single Produce Market," from KTVU (https://bit.ly/3dcOldl)

   + Video: "Grown in Oakland," from *The Atlantic* (https://bit.ly/3xL5nIY)

2. Create a social media (or poster) campaign explaining what they know about food apartheids and their effects on communities.

3. Write a reflection around how Akeelah might feel about her access to healthy foods and her journey, and how that might affect her view of the community. Consider what actions can be taken to support Akeelah's community (this is at the end of Lesson 1 as a scaffold).

We have decided to add the social media campaign piece to allow students to create advocacy on their own. Students will present their work in a gallery review so they can honor each other's creativity around the mathematics learning.

ONLINE RESOURCES

online RESOURCES Available for download at **resources.corwin/TMSJ-MiddleSchool**

▼ *Worksheet 1: Cor(o)ner Store: Where Healthy Food Comes to Die*

▼ *Worksheet 2: Grocery Trek*

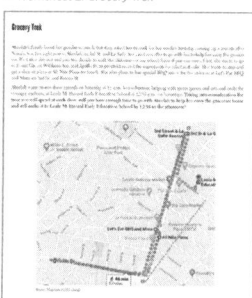

All worksheets and online resources provided in the Resources and Materials of each lesson are available for download or viewing on the companion website.

Middle School
MATHEMATICS LESSONS
TO EXPLORE, UNDERSTAND, AND RESPOND TO
Social Injustice

Middle School
MATHEMATICS LESSONS
TO EXPLORE, UNDERSTAND, AND RESPOND TO
Social Injustice

Basil M. Conway IV · Lateefah Id-Deen · Mary Candace Raygoza
Amanda Ruiz · John W. Staley · Eva Thanheiser · and Colleagues

Brian R. Lawler, *Series Editor*

Foreword by Julia M. Aguirre

A JOINT PUBLICATION

NATIONAL COUNCIL OF
TEACHERS OF MATHEMATICS

For information:

Corwin
A SAGE Company
2455 Teller Road
Thousand Oaks, California 91320
(800) 233–9936
www.corwin.com

SAGE Publications Ltd.
1 Oliver's Yard
55 City Road
London, EC1Y 1SP
United Kingdom

SAGE Publications India Pvt. Ltd.
B 1/I 1 Mohan Cooperative Industrial Area
Mathura Road, New Delhi 110 044
India

SAGE Publications Asia-Pacific Pte. Ltd.
18 Cross Street #10–10/11/12
China Square Central
Singapore 048423

ISBN 9781071845523

President: Mike Soules
Associate Vice President and Editorial Director:
  Monica Eckman
Publisher: Erin Null
Content Development Editor: Jessica Vidal
Editorial Assistant: Nyle De Leon
Production Editor: Tori Mirsadjadi
Copy Editor: Christina West
Typesetter: Integra
Proofreader: Liann Lech
Indexer: Integra
Cover Designer: Scott Van Atta
Marketing Manager: Margaret O'Connor

This book is printed on acid-free paper.

22 23 24 25 26 10 9 8 7 6 5 4 3 2 1

CONTENTS

 Visit the companion website at
resources.corwin/TMSJ-MiddleSchool
for downloadable resources.

FOREWORD

Welcome to a new generation of social justice mathematics advocates.

I am very excited and humbled to be writing the Foreword to this series of books about teaching mathematics for social justice. Over the past 30 years, there have been very few teaching resources available in one place that support teachers to embrace children as mathematical problem posers, sensemakers, and community change agents. The volumes in this series provide teachers with pedagogical and curricular tools to create mathematical learning environments that invite curiosity, social consciousness and critique, and mathematical analysis and innovation—multiple paths that lead to challenging societal inequities and making mathematics a more humanizing and just experience.

Why have such resources taken so long? This is an excellent and complex question. In my over 25 years of experience as a mathematics educator and scholar with an explicit equity ethic (McGee, 2020), the mathematics education community has been slow to embrace an equity and justice-centered approach to mathematics education. There has been a tremendous amount of emphasis on reform mathematics, mathematical thinking, and mathematical discourse. Yet that same approach has reinforced beliefs and structures about mathematics being universal and culture free. It was about making dominant mathematics—which emphasizes cisgendered male-centric and euro-centric values—accessible to more people, while failing to acknowledge that mathematics has been created and communicated by cultures and communities across the globe since time immemorial. Efforts to introduce mathematical investigations that center on community, family, and ancestral knowledge and uses of mathematics into schools—especially activities that mathematized fairness, representation, and power relationships—brought consistent criticism from mathematicians and mathematics educators, who asked, "Where's the math?" Ironically, it is the powerful mathematics inherent within these investigations that brought to light injustice, inequity, and a demand for change to make things right. Mathematics has always been created by us, with us, and for us. It is time that we embrace this idea fully.

The link between social justice and mathematics has strong roots in liberatory education as practiced by Brazilian educator and philosopher Paolo Freire, and connected to mathematics by American mathematics education scholars Marilyn Frankenstein, Arthur Powell, Eric Gutstein, Rochelle Gutiérrez, and Danny Martin. However, for me, another crucial link made between mathematics and social justice was the work of civil rights leader Bob Moses and the work of the Algebra Project. Bob Moses passed away in the summer of 2021 at the age of 86. He was the first mathematics educator to help me see mathematics as a civil rights issue. In his book *Radical Equations* (Moses & Cobb, 2001), he argued that "full citizenship" in the 21st century—including economic and political access as well as informed decision making and civic engagement—in our society is inextricably linked to mathematics literacy. We must remove the systemic barriers, beliefs, and structures that deny children the right to have a high-quality, nonviolent, and meaningful mathematics education. The guidance and resources found in this series support teachers to do just that.

I would be remiss if I did not acknowledge that this series was developed during the twin pandemics: COVID-19 and systemic racism. COVID only exacerbated systemic racism inherent in the education, economic, political, health care, and legal systems of the United States. Disproportionate deaths due to COVID in communities of Color; the deaths of Breonna Taylor and George Floyd among countless other Black and Brown people at the hands of the state; the rise of hate crimes, domestic violence, and sexual assault; increased gun violence and homelessness; the opioid crisis; the separation of families at the border; the rise of suicide among young people and veterans; anti-Asian and anti-immigrant hate; and missing and murdered Indigenous women are some examples of the pain and violence we have endured. Our planet is on fire, and many of us lack access to clean water, air, and earth. We must change and we must look to the next generation of young people to lead this effort.

This series continues the work of social justice mathematics advocacy by providing classroom-based mathematics lessons that build children's empathy and analysis skills to connect mathematics to their own lives, their communities, and the complex world around them. Relationships among people, animals, and the planet are mathematized in various ways. The investigations are grounded in social justice and mathematics standards so educators can be confident that the work meets multiple teaching goals. But what I am really excited about is the amplifying of youth voice and activism through these mathematics activities. Bob Moses said,

> We don't listen to kids enough. Really listen. It is a difficult thing for grown-ups to do—listen and actually pay attention to what young people are saying. In the Algebra Project we are still learning how to do this also. It is the voices of young people I hear every day, more than anything, that gives me hope. (Moses & Cobb, 2001, p. 191)

We must listen and learn from the voices of young people. Children can grapple with hard topics because they understand ideas of fairness, sharing, love, and friendship. They are curious about the world and they reflect the world. They are complex human beings. They are our future. They are our hope. Welcome to the new generation of social justice mathematics advocates.

—Julia M. Aguirre, PhD
December 3, 2021

ACKNOWLEDGMENTS

First, we—Basil, Lateefah, Mary, Amanda, John, and Eva—would like to thank all of the lesson authors for your willingness to write and share lessons so that this book could become a reality. We knew from the start the importance of having a diverse set of voices from the field—educators who could share lessons and experiences and were willing to trust us with their work.

We also appreciate that the lesson authors often submitted much more robust lessons than we had space to include. We have made all attempts to maintain the intent of the lesson and some of the original activities. Note that all of the lessons in this book were modified and edited to create a consistent format for presentation and ease of use for the reader. In most cases, the initial context, mathematical and social justice goals, and sequence of activities have remained the same. We trust that we have done due diligence (with many hours of editing) to share your work with others. For those whose work might not be included, we thank you for your willingness to share but due to space limitations or overlap of content, we were not able to include them in this version of the book.

Second, thank you to all of the educators who field tested a social justice mathematics lesson (SJML). Your feedback and suggestions helped us further develop some of the key features for the lessons. Your willingness to share your experiences with these SJMLs and other social justice lessons also helped enhance our thinking as we developed the chapters in this book.

Next, we would like to thank the reviewers. Your thoughtful and thorough feedback on the chapters and lessons provided us with a roadmap to make the much-needed revisions. When you read the final version of this book, our hope is that you vaguely remember the rough draft as a skeleton of ideas and see that we have attempted to address all concerns and include each suggestion.

We extend tremendous gratitude to the book series editor, Brian R. Lawler, for your leadership in birthing three additional social justice mathematics books. Lateefah, Mary, Amanda, and Eva would also like to give a special thanks to Robert, Basil, Brian, and John for the high school social justice mathematics book that has been so impactful for so many, including us, and for their continued guidance with the development of this book.

Thank you to the youth artist who designed the cover art for this book, Eda Varela Ruiz. We appreciate how you captured the essence of the social justice issues this book explores so well.

Thank you to NCTM and Corwin for your willingness to publish this book.

Last but not least, we would like to thank Erin Null, Jessica Vidal, and the Corwin team. From the inception of this book, Erin provided us with the encouragement and guidance needed to focus our work (and writing) so that we would complete this book. Your feedback on the manuscript and throughout the revision process was priceless. This book is in print today because you truly showed what it means to problem solve and persevere. Jessica, thank you for your support in all of the little details that we sometimes overlooked or remembered at the last minute. And to the copyeditors, artists, and other members of the Corwin team, thank you for your willingness to publish a book that seeks to equip mathematics teachers so that they can empower their students with the mathematical tools to explore, understand, and respond to social injustices as they become powerful learners and doers of mathematics.

Basil's Special Acknowledgments

To my wife and two boys, who persevere through my focus on the windshield (teaching and theology) rather than the majesty of what it makes visible (empowerment and God).

Lateefah's Special Acknowledgments

I am thankful for my phenomenal family, mentors, and friends who I owe more than I can give. I will always remember your unwavering and consistent prayers and support. I am grateful to every mathematics teacher and administrator who is dedicated to improving the mathematics learning experiences of historically marginalized students. More importantly, to the students who are roses that continue to grow from concrete . . . I will not stop until you get a mathematics learning experience that you deserve!

Mary's Special Acknowledgments

I have tremendous gratitude for my social justice educator partner-in-love, Juan, and my beautiful, curious, and kind 4-year-old, Rose, who daily ignites my wonder about the world and passion to support children to empower themselves as change makers. My former high school students at Esteban Torres High School and my graduate teacher education students at Saint Mary's College inspire me to never stop dreaming and building alongside youth and teachers toward a more socially just world. It is difficult to put into words how appreciative I am of my fellow book and lesson authors and all of the justice-driven mathematics teachers and scholars who have come before me, taught me so much, and paved the way for this book. Thank you!

Amanda's Special Acknowledgments

I want to thank my husband, Walter, and my children, Eda and Armando, for being my biggest motivation for making the world better. I also want to thank my parents and my aunts and uncles, who as educators demonstrated by example the value of education and the impact of being an educator.

I would also like to thank my broader community for holding me up, for being my village, and for continuing to learn with me.

Thank you also to Bob P. Moses, whose book *Radical Equations* came into my life at just the right time and taught me that *mathematics is a social justice issue*.

Lastly, without ego, I would like to thank me, and every woman like me, in solidarity as we juggle our careers, motherhood, and our responsibility to our family and communities during a global health crisis.

John's Special Acknowledgments

I want to thank my wife, Karen, for walking beside me and continuing to encourage me to follow God's plan. To our grown folks, Jonathan, Alexis, and Mariah, continue to chase your dreams and change the world. Thank you to the many educators who work to make mathematics more meaningful, relevant, and accessible for each and every student. You are *Changing the Narrative*!

Eva's Special Acknowledgments

I want to thank my three curiously brilliant daughters, who always knew how to motivate me and how to keep me in check, and my partner in life for their support throughout. I also want to thank the brilliant students I get to work with: you are the joy of mathematics education in my life. And finally, I want to thank all my amazing colleagues who allowed me to grow into the mathematics educator I am now and continue to push me to become a better human.

PUBLISHER'S ACKNOWLEDGMENTS

Corwin gratefully acknowledges the contributions of the following reviewers:

Rebecca Brooks
Associate Professor
California State University San Marcos, Spring Valley, CA

Zandra de Araujo
Associate Professor, Mathematics Education
Columbia, MO

Kelsey Macias Ingham
Math Teacher, San Juan High School & Adjunct Professor,
Saint Mary's College, CA
Sacramento, CA

ABOUT THE AUTHORS

Basil M. Conway IV is an associate professor of mathematics education in the College of Education and Health Professions at Columbus State University and serves as the mathematics education graduate programs director. He serves on numerous doctoral committees as both a chair and methodologist. He earned his BS, MS, and PhD in mathematics education from Auburn University in 2005, 2012, and 2015, respectively. He also completed his MS in statistical science at Colorado State University in 2010.

Basil previously spent 10 years teaching in public middle and high schools before he became a teacher educator. During this time, he also worked as an instructor at a local junior college. Over the past 17 years of service in teaching mathematics and future teachers of mathematics, he has served in various local mathematics education leadership positions and organizations including Transforming East Alabama Mathematics (TEAM-Math), Auburn University's Teacher Leader Academy, East Alabama Council for Teachers of Mathematics, Woodrow Wilson Fellow, National Mathematics and Science Initiative, and A+ College Ready. He has published works related to teaching mathematics for social justice in numerous books and journals and has a special interest in statistics education.

Basil's lens for teaching and student learning draws heavily from Vygotsky's theory of social constructivism in which language and culture play essential roles in human intellectual development. Thus, he believes the co-construction of knowledge is paramount in the development of students' social, religious, and mathematical identities. He believes teachers, parents, other students, cultural norms, and other cultural communicative devices play a critical role in shaping students' knowledge of themselves, faith, and mathematics.

Lateefah Id-Deen is an assistant professor for mathematics education in the Bagwell College of Education at Kennesaw State University. She earned her bachelor's degree from the University of Arkansas Pine Bluff; a master's degree from Iowa State University; and a doctorate in curriculum, instruction, and teacher education from Michigan State University, with foci in mathematics education and urban education.

She has written several articles and book chapters that examine historically marginalized students' perspectives on their experiences in mathematics classrooms and ways to support educators in hearing and developing practice in relation to students' expressed interests. She investigates social justice pedagogies and culturally responsive instructional practices that promote student–teacher relationships, affirm mathematics identities, and cultivate belongingness to support students' learning experiences in mathematics classrooms. She has engaged in projects that support Black girls' schooling experiences. She also works with curriculum developers, schools, and districts that want to incorporate culturally relevant and anti-racist mathematics instructional strategies in mathematics classrooms. Her work reflects her passion for creating equitable learning environments for historically marginalized students in mathematics classrooms. Connect with her on Twitter @Prof_IdDeenL.

Mary Candace Raygoza (she/her/hers) is a STEMinist (STEM and feminist!) teacher educator. She is an associate professor of teacher education at Saint Mary's College of California and teaches courses including Humanizing Education Methods, Teaching for Social Justice, and Praxis Seminar. She is the lead investigator on a National Science Foundation grant to design the STEM Teachers for Justice, Community, and Leadership teaching pathway at Saint Mary's. Mary earned a bachelor's degree in sociology with an education minor at University of California, Berkeley and a full mathematics teaching credential and MEd, followed by a PhD in urban schooling, at University of California, Los Angeles. Her scholarship explores teaching mathematics for social justice and critical, justice-oriented, anti-racist teacher education.

Mary is a former high school mathematics teacher in East Los Angeles, where she taught algebra and geometry, with a commitment to teaching about the social and political world through mathematics and supporting students to develop as researchers and change agents through youth participatory action research. Mary believes in fostering teachers to develop as transformative leaders who will create a more just world in solidarity with youth and communities.

Amanda Ruiz was born in Long Beach and raised in Huntington Beach, California. She left Southern California for the Bay Area to attend University of California, Berkeley, where she created her own major focused on social movements. After some experience working in secondary education and a realization that mathematics is a social justice issue, Amanda went back to school to pursue a degree in mathematics. She received a master's degree in mathematics from San Francisco State University and then her PhD in mathematics from Binghamton University in 2013. After a year as a teaching and research postdoctoral fellow in the Mathematics Department at Harvey Mudd College, Amanda joined the University of San Diego, where she is now an associate professor of mathematics.

Amanda's PhD thesis was on realization spaces of phased matroids. While her prior research is predominantly in combinatorics and matroid theory, her research has more recently expanded to include pedagogical work. She is particularly interested in using mathematics to study issues of social justice and investigating pedagogies that make mathematical spaces more inclusive, where those traditionally underrepresented in mathematics can thrive.

John W. Staley has been involved in mathematics education for over 30 years as a secondary mathematics teacher, leader, adjunct professor, and consultant. In his current role as coordinator of special projects in Baltimore County Public Schools, his primary work involves supporting schools in the continuous improvement process. He earned his bachelor of science in mathematics from the University of Maryland, College Park; a master's in secondary education from Temple University; and a PhD in mathematics education leadership from George Mason University.

John has presented at state, national, and international conferences; served on many committees and task forces; facilitated workshops and professional development sessions on a variety of topics; received the Presidential Award for Excellence in Teaching Mathematics and Science; and served in leadership roles as president for NCSM, the mathematics education leadership organization, and chair of the U.S. National Commission on Mathematics Instruction. As part of the writing teams for *Catalyzing Change in High School Mathematics: Initiating Critical Conversations* (NCTM, 2018), *Framework for Leadership in Mathematics Education* (NCSM, 2020), and *High School Mathematics Lessons to Explore, Understand, and Respond to Social Injustice* (Berry et al., 2020), John continues his life work and passion to *Change the Narrative* about who is seen as being doers and learners of mathematics.

Eva Thanheiser is a mathematics teacher educator. She is a professor of mathematics education at Portland State University in Portland, Oregon. Eva teaches mathematics content courses for elementary and middle school teachers. She contextualizes much of the mathematics content in social and political contexts. She is the lead investigator on two National Science Foundation grants, one to connect elementary mathematics to the world and another focused on anti-bias mathematics education at the K–12 level. Eva started her studies in Germany and finished a master's in mathematics in 1998 at Kansas University and a PhD in mathematics education in 2005 at the joint doctoral program between the University of California San Diego and San Diego State University. Eva's scholarship explores teaching mathematics for social justice and anti-bias mathematics education.

Eva has received the Early Career Award from the Association of Mathematics Teacher Educators (AMTE) as well as the Sigma Xi Outstanding Researcher Award. She has served in leadership roles at AMTE and the Psychology of Mathematics Education–North America as well as on editorial boards of the *Mathematics Teacher Educator* and the *Journal of Research in Mathematics Education*.

INTRODUCTION

Teachers often feel that the mathematics courses and standards mandated by their state restrict them or force them to compartmentalize the mathematics content, disconnecting topics from real and personally meaningful contexts. This makes it difficult to connect with students' questions of "When will I ever use this?" or "Why do I need to know this?" beyond the obligatory, "You need this for the next mathematics course." We meet many teachers who report feeling stuck in a rut, moving from one topic to the next, and losing opportunities for meaningful connections between mathematics topics as well as between mathematics and the real world by following curriculum pathways outlined by textbook chapters or course progressions that are mandated by district policy. Does this sound familiar? Has this ever been you? Unfortunately, students feel the results of this as well, often trudging through mathematics with this same sense of disconnection and tedium. These are missed opportunities for students to embrace how mathematics can be used for self- and community empowerment.

The exclusive focus on standards and progressions often makes teachers begin to feel discouraged about their actual impact on students. Furthermore, teachers' professionalism is gradually being diminished by policies that force them to feel like factory workers on an assembly line, teaching (or schooling) sets of students each year without regard to who they are and being unable to embrace students' identity, culture, interests, or experiences. The latest educational program turns into the next year's car model. In essence, teachers often question what students are actually learning in their classroom.

Unfortunately, the response is focused only on what mathematics must be learned for next year's course. While there is value in grounding the work in mathematics standards, how might standards be leveraged to help you create a more meaningful experience for students? How might the standards be leveraged to empower your students? By the time you finish reading and implementing ideas from this book, it is our hope that you will experience several benefits:

- Your students will see how mathematics applies to their lives, and they will become empowered to use it to change the issues that affect them the most.

- You will enhance your ability to facilitate discourse around difficult topics by incorporating them into your classroom.

- Important issues for you and your community will become better integrated with your teaching life each day, and in turn, you will feel a deeper sense of urgency—and a deeper sense of satisfaction—for effective teaching of mathematics and empowering students.

- Social disparities in your school, community, city, region, and state will begin to see improvement through grassroots efforts led by your own students. Your students will become more engaged in city, state, regional, and community grassroots efforts to address social disparities.

This book offers a collection of mathematics lessons—tied to core middle grades mathematics domains—and is grounded in issues of social importance to both you and your students. These lessons are bookended by lots of practical advice. In the opening chapters, we will discuss our ideas of what it means to teach mathematics for social justice and strategies to effectively do so. We will close by offering some ideas for how to create your own social justice mathematics lessons as well as some wisdom and advice from other teachers who have embarked on this journey.

WHY IS TEACHING MATHEMATICS FOR SOCIAL JUSTICE CRITICAL?

Whether we talk about it or not, our students regularly experience the impacts of social privilege, power, and activism. Each day, students in our schools and communities are faced with disparities in opportunity, inflating the so-called achievement gap. They listen to Hollywood actors, football players, public officials, bloggers, media outlets, and the list could go on. They use outlets like social media to express themselves, share their perspectives, and highlight their social positions—a modern format of young adolescents' participatory politics (Kahne et al., 2015).

In 2020, the unjust killing of George Floyd and several other high-profile police shootings highlighted the distance we have yet to traverse for racial equity. The lived experiences of many Black,[1] Indigenous, Latina/o,[2] Asian, and other people of Color highlight the impact that racial prejudice, health disparities, and other systematic problems are having. These experiences have also strengthened movements that are seeking to eliminate these injustices, like Black Lives Matter. Our students have concerns about their world, their community, and their family. Should we allow or encourage our students to just "let it go"?

We see and experience racial inequities in our schools. Black and Brown students are overrepresented in remedial classes and underrepresented in advanced placement classes (Seda & Brown, 2021; see Lesson 5.2 in *High School Mathematics Lessons to Explore, Understand, and Respond to Social Injustice*) and are disciplined at disproportionate rates. Kyndall Brown, for example, tells the story of his son, who did not get a recommendation for a calculus course placement from his

---

[1] Through this book and the lessons included, we have intentionally capitalized terms used for people of Color, such as Black, while leaving white written in lowercase. We follow Frances Harper (2019), one of the contributing lesson authors to this book, in her rationale: "I chose to capitalize Color but not white to challenge the ways that these standard grammar conventions reinforce systems of privilege and oppression" (p. 268).

[2] We have chosen to use the term Latina/o leaning on the leadership, expertise, and community of TODOS mathematics education scholars (TODOS Mission and Goals as of February 21, 2022).

teacher even though he fulfilled all the qualifications. Racial injustices are plaguing society and our students' lives and demand our attention.

An important aspect of our responsibility as educators is to help empower our students to be agents of change and liberation in their communities, states, nations, and world. We would like to go further than simply stating the importance of connecting mathematics teaching and learning to teachers' and students' lived experiences and interests; we argue that teaching mathematics *for* social justice (TMSJ) is critical for four reasons:

- It builds an informed society.

- It connects mathematics with students' cultural and community histories.

- It empowers students to confront and solve real-world challenges they face.

- It helps students learn to use mathematics as a tool for social change.

TMSJ can and should extend mathematics beyond the classroom. It can and should encourage students to

- learn important mathematics,

- build positive mathematical identities, and

- develop concern for the happiness of other human beings and life forms.

TMSJ is an instrument for empowering students to become agents of change.

> An important aspect of our responsibility as educators is to help empower our students to be agents of change and liberation in their communities, states, nations, and world.

THIS BOOK'S AUTHORSHIP

After the successful release of *High School Mathematics Lessons to Explore, Understand, and Respond to Social Injustice* (Berry et al., 2020), mathematics educators took to social media to express an interest in developing books for other grade levels (early elementary, upper elementary, and middle school). The group contacted the authors of the earlier book and began talking about what these books might look like at the different grade levels. The group then split into subgroups to work on each book based on their interest and expertise in the level of mathematics education. Two of the original authors of the high school book joined the middle school group to begin working on the book.

Inspired by the structure of the high school book, we wanted to invite the broader mathematics education community to submit lessons that highlighted social injustices of their own community and concerns or issues raised by their own students. And in sharing the lessons they provided, we strove to retain their and their students' authentic voices. Therefore, most of the lessons in this book come from this diverse group of lesson authors. In addition, we sent the lessons out to reviewers, many of whom implemented the lessons in their middle school mathematics or preservice mathematics teacher classrooms. These reviewers gave extensive feedback on the book and the lessons; their thoughtful insight challenges you to think even further outside of your experiences, biases, and assumptions, and to consider a broader representation of student and teacher experiences.

We value the voice of each educator who contributed a lesson, and we have made all attempts to share their work and their voice with you. Outside the major required elements for the lesson submissions, we asked lesson authors to format and submit lessons based on how they had implemented them in their classrooms. We then edited lessons for clarity, mathematical rigor, and cohesion in order to highlight the voice and authenticity of work in the field. Their lessons have been tested and refined, in their own classrooms and in others. We are grateful to them and all those who helped develop this book.

As a team, we each have our own motivations and understandings of teaching mathematics for social justice. Here, each lead author briefly describes what brought them to focus their career on equity and justice issues in mathematics education.

Basil M. Conway IV

Teaching mathematics to empower students has always been a professional passion of mine, particularly as a classroom teacher for 10 years and currently as a mathematics teacher educator. My journey toward teaching mathematics for social justice began when, while completing my mathematics education graduate studies, I read a series of Bible passages that intersected with what I was learning and inspired me. Proverbs 31:8–9 resounded in my head as I learned about mathematics research related to access, equity, and empowerment: "Speak up for those who cannot speak for themselves, for the rights of all who are destitute. Speak up and judge fairly; defend the rights of the poor and needy." I began to question what I was doing from a place of power in my own mathematics classroom. How was I empowering students to be agents of change in their and others' lives? I received Jesus as my Lord, which resulted in mimicking and modeling His behavior. He repeatedly stood against the powerful, uplifting the oppressed. I have decided to focus on Jesus' love, a love so great and so selfless that He didn't just seek justice for others; rather, He sought injustice for Himself for the sake of others. As I pray for His Beloved Community seeking His kingdom on earth, I trust He will render all things complete and settled in the end with complete justice.

Lateefah Id-Deen

I am a Black woman who taught and worked with middle and high school students in urban and suburban contexts. As a student and teacher, I experienced and witnessed the inequalities that exist in the education system. I helped increase middle and high school students' critical awareness through culturally relevant pedagogies through my mathematics teaching. I support prospective teachers in discovering how teaching matters not only for students' development of mathematical knowledge but also for their sense of self. Further, my teaching style aligns with what bell hooks described as "engaged pedagogy," whereby one teaches "in a manner that respects and cares for the souls of our students." Relatedly, my research centers on the importance of helping students recognize and analyze systems of inequality in ways that empower them to take action, capturing students' perspectives on their experiences, cultivating belongingness and student–teacher relationships, and amplifying the importance of sharpening and enhancing students' mathematics identities.

Mary Candace Raygoza (She/Her/Hers)

As a middle school student, I attended a Title I school in my neighborhood and then transferred to a well-resourced high school in an affluent neighborhood. Because I got As in algebra in middle school, I was placed in geometry at my high school. The geometry teacher gave my class a test to see what we remembered from algebra, and I scored an 11%. At the time, I was devastated and felt stupid and behind. Looking back, this was a defining moment for me in my trajectory as a social justice mathematics educator because it is when I began to sense (1) how drastically different opportunities in (math) classrooms can be and (2) how procedural tests and approaches in mathematics do not capture what young people are capable of nor the mathematics that matters to explore, understand, and respond to our complex and unjust world.

When I attended college at the University of California, Berkeley (UC Berkeley), alongside taking courses about social injustices and social movements, I found myself drawn to every quantitative methods course offered by social science departments (e.g., sociology, psychology, political science). I loved how mathematics and statistics could be used to understand the world and inform our action for a more just world. I also knew I wanted to become a public school teacher-activist and embark on a lifelong journey of interrogating and challenging the many forms of privilege I experience, notably white privilege. I pursued mathematics teaching not because I loved the way mathematics was traditionally taught, but because I knew it could be more interesting and empowering, and that we could reject mathematics as a field that is objective or neutral. That's when I came across the book Rethinking Mathematics *by Gutstein and Peterson and the growing critical mass of social justice mathematics educators and was energized to think I could join in solidarity to change mathematics education.*

As a high school teacher in East Los Angeles, home of the Chicano Power Movement, I sought to learn from both historical and present-day organizing for justice and the funds of knowledge my students brought to the classroom. I engaged students in quantitative action research, supporting them to study and make change on an issue they cared about. I then pursued researching the journeys of mathematics educators striving to teach mathematics for social justice and teaching future teachers.

I am thrilled today to be learning from movements of critical mathematics educators seeking to make mathematics more humanizing, relevant, and liberatory for and with young people.

Amanda Ruiz

My mathematics journey began at UC Berkeley, where I created my own major focused on social movements. As a student activist, learning about social justice issues nourished my education and clarified my life goals. I was inspired while learning about the impact that student activists had on the Civil Rights Movement, such as the Student Nonviolent Coordinating Committee (SNCC). While finishing my undergraduate degree, I tutored mathematics through an education nonprofit whose mission aspired to make college accessible to underserved students in

Oakland, California. At that time, I came across Robert Moses's Radical Equations: Civil Rights From Mississippi to the Algebra Project. *I recognized the author as the same Bob Moses who was a leader in SNCC. Moses taught me that mathematics is inherently a social justice issue. This book helped me realize I could pursue my interest in mathematics without abandoning my commitment to social justice. In fact, I realized as a Latina woman that my mathematics pursuit and presence was in and of itself a radical statement, and just by existing in a predominately white male space, I impact a student's access to mathematics.*

Throughout my career as a mathematics educator, social justice has guided my efforts to make mathematics accessible for students from all backgrounds. Mathematics traditionally serves as a gatekeeper to academic success; the status quo of mathematics alongside social and cultural inequities makes it difficult for marginalized students to engage authentically with mathematics. I want to broaden who thinks of themselves as mathematicians and, in doing so, create a space for marginalized practitioners like myself to bridge our social/cultural identities into the milieu of mathematics. In the classroom, my objectives always include improving students' attitudes about mathematics and building their mathematical identity. Giving students the opportunity to use mathematics as a tool to understand and respond to social injustice helps them realize mathematics as more than a gatekeeper.

John W. Staley

My career as a mathematics teacher and leader has been driven by issues related to equity, social justice, and the need to help students value and use mathematics as a tool to address social justice issues in their lives. My teaching career began in 1987 at a juvenile correctional facility for young men between the ages of 12 and 18, where I realized that teaching mathematics meant more than teaching students the steps to find the "correct" answer(s) to problems. I quickly recognized that my calling to teach students extended beyond mathematics concepts and skills to lessons that they could use in this journey called life. From the beginning, my goal as an educator has been to develop my students' self-confidence and belief in their ability to try, do, and reflect on things as they lived their lives, inside and outside of the mathematics classroom. Thus, lessons and learning opportunities were designed to encourage and empower students to become thinkers and doers of mathematics, connect to students' lives and hopes for the future, help students see relevance in the mathematics they are learning, and model for students the concept and value of respect—respect for self and others.

I also realized that as a Black man from the city of Philadelphia, Pennsylvania, I had the opportunity and responsibility to be a positive role model and advocate for my students. This opportunity eventually extended to work with other educators and adults. I thank my mother and grandparents for teaching me at an early age that things might not always be fair, but I should always act and live life as if someone was watching and treat others as I would want to be treated.

Eva Thanheiser

I grew up in Germany with a Jewish immigrant single mother from Eastern Europe. As far back as I can remember, mathematics made sense to me. However, I struggled with all other subjects to the point that teachers told my mother that I might not be able to graduate from high school. Thanks to the forceful nature of my mother, who supported me in switching schools in the eighth grade, I ended up finishing high school and went on to university to study mathematics and English to become a teacher. During an exchange year in the United States, I learned about the field of mathematics education and dove right in, changing my career path from becoming a teacher in Germany to becoming a mathematics teacher educator. The focus on social justice mathematics education came later in my career but is connected to my earliest experiences as the only Jewish kid and the only kid whose mom had an accent in my early schooling. I actually don't remember knowing any other Jewish families or families with foreign parents as a young child.

As a mathematics student, I always assisted my classmates and imagined that everyone approached mathematics from a sensemaking perspective. Later in my studies, I learned that that is not the case. A big part of what drives me is to work on allowing all students to see mathematics as (a) something that makes sense, (b) something that can be used to make sense of the world, and (c) something that can be used to influence the world.

Each lead author's introduction presents a different and personal motivation and passion for TMSJ. In part, these stories describe mathematics as a tool to empower students and future citizens to become active agents of change. In today's society, using mathematics as a tool for empowerment has become paramount to ensure that truth prevails over conviction. Reflect for a moment on what caused you to pick up this book. What story defines your personal motivations for wanting to infuse social justice into your mathematics curriculum?

Extend the array of motivations, passions, and experiences of the six lead authors to the diverse contributors to this book; this book reflects a remarkable breadth of experiences and worldviews. And in this tremendous diversity, through our collaboration, we found many common concerns about our families, neighbors, community, and country. We suspect that the variety of contexts offered in this collection of lessons will offer all teachers the opportunity to draw upon themes generated from their own context that will engage students.

THE LESSON AUTHORS

Lynette Guzmán is a mathematics education scholar who focuses on interrogating limiting discourses about people and their complexity. As a millennial who grew up with the internet, Lynette spends her time thinking about the ways digital platforms lend themselves to content creation, consumption, and remixing to promote particular kinds of discourses.

Kendrick Savage began his career as a mathematics educator 15 years ago, teaching mathematics at the university, community college, and public high school levels in Mississippi. His passion is to increase student confidence and motivation in mathematics by connecting the mathematical content to students' culture, experiences, and personal interest and helping students see the power in mathematics as a mechanism to solve social injustices.

Liza (Cope) Bondurant taught students in Grades 6–12 in urban, suburban, and rural settings at private and public schools. As a female mathematics educator of 17 years, she continues to work to help each and every learner appreciate and understand mathematics.

Becky Evans has worked as an elementary teacher and district mathematics teacher leader for 10 years in Lincoln, Nebraska. Her classroom experiences revealed the importance of utilizing teaching practices that develop each student's mathematical agency. As a white, cisgender female she supports classroom teachers in creating equitable learning spaces for all students.

Perla Lahana Myers is passionate about achieving equity in education and diversifying the STEM fields, and feels that one step toward these goals is changing the reaction people have when they hear the word *mathematics* to a smile. She enjoys collaborating with colleagues, students, and community partners to create experiences in mathematics/STEAM locally and abroad.

Travis Weiland taught high school mathematics. He works to reimagine how statistics is taught in schools, interrogating and leveraging his privilege as a white, middle-class, cisgender male to create spaces for all students to have transformative experiences with statistical practices by interrogating their world through multiple ways of knowing.

Bethany Chan is a student in the UOTeach program where she is working toward her preliminary license in teaching. Additionally, she is a student-teacher at Roosevelt Middle School in Eugene, Oregon. As a second-generation immigrant, a first-generation college student, and a female Hong Kong American, she strives to make mathematics intriguing and relatable to students while encouraging Black, Indigenous, and other people of Color and females to pursue STEM in higher education.

Jennifer Dao is a Vietnamese American and the first generation in her family to attend college. She teaches eighth-grade Algebra for All in Evanston, Illinois. She enjoys sharing her passion for mathematics and art with students, helping students see how they are connected. Currently, she is on the leadership team of Nepantla Teachers Community, a mathematics teacher group focused on social justice, and is the Grade 5–8 Director for the Illinois Council of Teachers of Mathematics.

Debasmita Basu teaches quantitative reasoning and mathematics at Eugene Lang College of Liberal Arts, The New School, in New York City. Before pursuing her doctoral studies, Debasmita was a high school mathematics teacher in India for 4 years. As a cisgender woman of Color, she aims to design mathematical activities that cultivate students' critical consciousness toward various social and environmental justice issues and help them realize the power of mathematics.

Sara Rezvi taught middle and high school mathematics with students in New York City, Chicago, and Mexico for 9 years and is currently a doctoral candidate at University of Illinois at Chicago. As a queer, South Asian, culturally Muslim woman, she sees the work of mathematics as a human endeavor. She strives to center joy, student agency, creativity, and thoughtfulness in her work with students and teachers alike as the Program Director of the Math Circles of Chicago.

Julia Novosad graduated from the University of Nebraska-Lincoln in December 2021 with a degree in mathematics and mathematics education and a minor in French. She strives to teach mathematics with equal opportunity and continue to grow in her understanding of how our backgrounds and identities shape who we are as learners.

Peggy Nayar taught Emergent Bilingual learners for over 20 years and is currently a Teacher on Special Assignment Mathematics Coach. She works to examine how race and economics impact all participants in the classroom. She is dedicated to applying a mathematical lens to issues of cultural relevance and social justice.

Maggie Lee McHugh works as an innovation specialist for a project-based learning school in La Crosse, Wisconsin. In her role, she focuses on engaging students in connecting mathematics to their lived experiences. Maggie's continued dedication to authentic learning has led her to create sustainable conditions within her school in order for students to independently apply social justice principles to mathematics concepts in order to become change agents within their community.

Lee Inmon Dean began her career in research and development for an insurance company doing statistical analysis and found that the most exciting part of her job was training other analysts. This led her to teaching college mathematics back home in the Mississippi Delta, where she's been for the last 18 years. She strives to inspire her students to appreciate the cultural richness, diversity, and beauty that the Delta area has to offer.

Courtney Koestler is a proud former public school teacher and mathematics coach in elementary and middle schools and currently serves as the Director of the OHIO Center for Equity in Mathematics and Science in the Patton College of Education at Ohio University. As a white person in Appalachia, they work to understand issues centered on diversity, equity, and justice using critical literacy and critical pedagogies in early childhood and elementary education.

Kristin Komatsubara began her work as a mathematics and science teacher 15 years ago in Los Angeles. As an Asian American, cisgender woman, she is passionate about designing STEAM experiences where students feel known, connected, and celebrated as brilliant. In her role as a course instructor at the High Tech High Graduate School of Education, she works with preservice teachers to educate for social justice and deeper learning.

(Continued)

(*Continued*)

Frances K. Harper taught mathematics, reading, and English for 9 years across Grades PreK–12 in diverse urban settings in Tennessee, Massachusetts, and Kanagawa, Japan. As a white, cis woman and first-generation college graduate, she strives to use her privilege to lift up the voices of students and families who have been systematically marginalized in mathematics by trying to understand mathematics education from their perspectives.

Jennifer Ruef identifies (and is identified) as a white, middle-class, cisgendered, dancing, diving mom and educator. She taught middle and high school mathematics for 20 years before earning her doctorate in mathematics education. Her research centers on social justice and equity concerns, with a focus on group and individual mathematical identities.

Candace Joswick began a career in education in an Upward Bound program and has taught middle school and high school mathematics in urban and suburban schools in culturally and linguistically diverse communities. Candace is dedicated to contributing to efforts that increase students' access to high-quality, inclusive, and equitable mathematics teaching and learning.

Nichole Campbell has taught elementary and middle school in both urban and suburban settings. She writes a culturally responsive mathematics curriculum for St. Paul Public Schools, where she is currently a district mathematics coach. She is also a consultant with Student Achievement Partners. Through this work, she is committed to raising awareness of social justice issues through mathematics and empowering students to be change agents.

Kari Kokka is a fourth-generation Japanese American cis nonqueer womxn who began her career in education as a public high school mathematics teacher in 1999. She worked as a mathematics teacher, instructional coach, and mathematics teacher-activist in New York City for 10 years prior to becoming a mathematics teacher educator at the University of Pittsburgh. She continues her mathematics teacher activism toward collective justice in various spaces, such as Creating Balance in an Unjust World Conference and Radical STEMM Educators.

Elizabeth O. Ayisi taught mathematics, physics, and computer science at private high schools and college applied calculus at a private university in Northeastern Ohio. As a first-generation Ghanaian American, she wants to provide better mathematics transitioning experience from high school to college for all students. She supports educators who strive to achieve social justice and equity to develop and expand students' positive mathematics identities toward programs that introduce STEM fields, professionals, and K–12 role models.

 **Odesma Dalrymple** is an associate professor in industrial and systems engineering and director of the Engineering Exchange for Social Justice at the University of San Diego. Her scholarship is in the area of engineering education, with a focus on education equity, predominantly in middle-high school settings; and equitable, reciprocal exchanges with community organizations that serve and represent persons directly impacted by marginalizing practices, systems, and policies. She identifies as an Afro-Caribbean, immigrant, and mother of a first-generation American daughter.

 **Colleen Carman** has taught middle school, high school, and university students. Colleen asks her students and herself to explore their own identities and how mathematics can play a role in their lives. Her greatest joy in teaching is seeing a student apply content to make a positive change in their school and community.

 **Oluwaseun Kudaisi** started his career as an educator in 2017 and has worked in both public and private schools. He has taught sixth-grade math, sixth-grade science, and seventh-grade science. His passion is in helping students to achieve their goals. His life motto is, "If I can help somebody, as I pass along, Then my living shall not be in vain."

 **Queshonda Kudaisi,** an African American woman, started her career as a mathematics educator in 2014 working in both public and private schools. Her experience includes being a middle and high school mathematics teacher, high school instructional mathematics coach, mathematics content writer, mathematics teacher educator, and mathematics education researcher. Her passion in mathematics is in integrating social justice issues into the mathematics classroom to empower students to enact a social transformation in the world.

 **Michelle Cody** is the mother of a vibrant young Black boy (Matthew) and the daughter of educational enthusiasts (Brian and Julia). Black womyn. University of San Francisco grad. Public school sixth-grade mathematics teacher. Cisgendered. Howard University graduate. Social Justice Warrior. Creative. A proud product of the San Francisco Unified School District. She works to use numbers to tell stories of the community. She creates space for her brilliant mathematical babies to think big. As a youngin' in this work, she knows that it is both a lifestyle and a movement.

 **Chuck Munter** taught secondary mathematics for 8 years, first in rural Missouri and then in Memphis City Schools. Currently, as a teacher educator at the University of Missouri, he is interested in using mathematics to develop more honest accounts of U.S. events past and present and to do democracy better.

(Continued)

(*Continued*)

Y. Rhoda Latimer began her career as a middle school mathematics teacher 15 years ago. She is currently a PhD candidate and serves as a middle and secondary mathematics coordinator. Her research centers on culturally responsive and equitable mathematics teaching. As a Black female mathematician, she is passionate about enhancing environments where students' talents and cultural contributions are welcomed and embraced in mathematics classrooms.

Allyson Lam is a Chinese American middle school teacher, born and raised in San Francisco. She hopes to use her mathematics classroom to equip students to think critically about inequities in their community. Her students inspire her to continually work toward dismantling systems of oppression. Allyson's mathematics class consists of 50% mathematics tasks and groupwork, 40% games, and 10% chaos. The chaos is 100% necessary.

Tashana Howse has taught high school mathematics, college-level mathematics, and mathematics education courses for over 20 years. She strives to ensure that students experience mathematics in a way that will help them develop a deep understanding of the mathematics and require them to reason about and with mathematical concepts. However, in order to accomplish this, Tashana believes it is essential to build relationships and connect with students so they know they are welcomed to the learning process first.

Andrew Reardon taught students from Grades 7–12 in districts ranging from downtown Dallas, Texas, to a vocational high school in Delaware, Ohio. He works to ensure students have the necessary mathematics skills to be agents of social change and strives to complete that work alongside his students. His goal is to help students realize that a sound knowledge of mathematics is essential to effect change.

Cara Haines began her work as a mathematics educator in 2011 at a public school district in Pittsburgh, Pennsylvania. Later, while pursuing a doctorate at the University of Missouri, Cara grew invested in interrogating racism and issues of inequity in mathematics education, and sought to support prospective teachers in doing the same. Now, a postdoctoral researcher, Cara continues her inquiry into mathematics education in hopes of learning more about what it might take to improve students' experiences in schools.

Natalie Odom Pough serves as a middle school mathematics teacher and adjunct professor of mathematics education. Her research focus is on equitable mathematics teaching practices that connect students with the content in more positive ways. As a Black female mathematician, Dr. Pough works to instill a love for the subject and a drive for mathematics achievement in all of her students.

Jennifer A. Wolfe is a biracial Asian American cisgender woman, daughter of a Thai immigrant, and has been in mathematics teacher education for over 20 years. She supports prospective and in-service secondary mathematics teachers in learning to co-create identity-affirming spaces that center student voice, equity, radical love, and joy.

Robin Keturah Anderson is a former middle and high school mathematics and physics teacher in Southern California. She currently is an assistant professor of mathematics education working with preservice and in-service teachers to develop justice-oriented pedagogies to rehumanize mathematics teaching and learning.

Lisa Skultety taught middle and high school mathematics in Houston before becoming a teacher educator. As a white woman teaching predominately white future educators, she strives to support preservice teachers in understanding their privilege and move them toward interrogating classroom practices and creating equitable and just spaces for students.

Jennifer Kosiak has been teaching mathematics for over 25 years at the middle school, high school, and university levels. She currently is a professor of mathematics education at the University of Wisconsin-La Crosse, where she strives to support teacher candidates in integrating social inequities into the PreK–12 mathematics classroom. During her time as President of the Wisconsin Mathematics Council, Jennifer focused on engaging teachers across the state in empowering all students to see themselves as mathematicians.

Rebecca Hudson taught students in Grades 9–12 in a suburban public school with a diverse population of race and culture. She has a passion for showing students how mathematics can be used in the real world and how it can be applied to explain and change social differences to improve the world around us.

Melissa Troudt is a former high school teacher and assistant professor of mathematics education. As an educator of future teachers, she aims to work alongside her students to develop equity-based practices for teaching mathematics and to learn ways to use mathematics to interrogate and challenge inequities in their worlds.

Since childhood, **Rebecca Ellis** has been involved with Social Action Tikkun Olam, the Jewish concept of repairing the world. She aims to include social justice in all her teaching, curriculum design, and assessment development. As a postdoctoral student, she researched and developed free and interactive science education materials with the Connected Biology Project. She is excited to be starting a new position as a curriculum developer at The Concord Consortium, where she can continue to integrate culturally relevant pedagogies into technology-enhanced learning environments.

Joi Spencer comes from a family and community that believes in the power of education to liberate. She was introduced to the history of Black people and their freedom struggle as a young child. Teaching mathematics with an eye toward justice is a natural extension of this upbringing. Her hope is that the youth and teachers that she works with will use mathematics to understand, uproot, and change our current arrangements toward the goal of a more just society.

(Continued)

(*Continued*)

Mathew D. Felton-Koestler teaches mathematics methods to future elementary and middle school teachers in the Department of Teacher Education at Ohio University. As a white man, Matt works to recognize and understand his privilege and to support future teachers in learning to use mathematics as a tool for understanding our world and, in particular, for revealing and countering social injustice.

Celina Gonzalez has worked alongside her mentors and colleagues learning and creating more equitable educational experiences. Originally from a small rural community and public schools, Celina's first mathematics class at University of San Diego with Dr. Myers opened her world to the beauty of mathematics. She questioned the disparities in quality and opportunity children have in their educational experiences. As a Chicana educator and leader, she continues to question and challenge injustices to understand and create equitable and transformative learning spaces.

Farshid Safi identifies strongly, fully, and proudly as Iranian American, connecting with both cultures/nationalities. His reality connects with being an immigrant but he is not considered a first-generation immigrant, being Muslim but he is not considered Muslim enough, and being an Emergent Bilingual learner but he is not considered as such. When you are not accepted by the majority and yet othered within and beyond communities that are themselves marginalized—historically, systemically, and systematically—there is a need to shift from awareness toward actions personally and professionally.

Melissa A. Gallagher was an elementary school teacher and mathematics coach both in the United States and abroad. She grew up overseas as a language learner, and this experience has influenced her research on supporting multilingual students in the mathematics classroom. She works to interrogate the privilege of her white, upper-middle-class upbringing and how that privilege afforded her different experiences than many multilinguals in the United States today.

Jeff Craig is committed to contemplating ethical questions in education. In his teaching, he reconciles ethical questioning against a backdrop of so-called "wicked problems" education, which prioritizes depth in education as it relates to students as members of communities and societies. Jeff is driven to engage students as both global and local thinkers who use mathematical and statistical techniques to understand their worlds and their positions within.

Tyrone Martinez-Black (ty, he, they) taught middle grades math, science, and reading as well as coached fellow educators in those subjects. Ty has lived an entire life "in between"—cultures, geographies, genders, and so on—working toward a just society that celebrates these pluralities and organizes with these fluidities in mind. Mathematics is a way of knowing and being that helps us pursue and create this future.

 **Emmalee Bielenberg** teaches seventh-grade mathematics in Lincoln Public Schools in Nebraska. She strives to help her students develop positive mathematical identities at the intersection of their diverse cultural and religious identities. She hopes that by believing in each and every one of her students' mathematical abilities, she can empower them to believe in themselves too.

 **Cassie Ruettiger** graduated from the University of Nebraska-Lincoln with a major in mathematics education. Her hope is for every student to feel as though they are capable of great things, both mathematically and personally.

We hope you find that the variety of contexts offered by the contributing authors in this collection of lessons will offer you lessons that you can use right away or provide the framework for developing a personalized lesson drawing upon student-generated themes, questions, or concerns that emerge in your own context—enhancing the opportunity to engage all students.

WHO IS THIS BOOK FOR?

Middle school students are broadening their worldview, and you have the opportunity to help them navigate it by collaborating with them and their communities to bring their expanded world into the mathematics classroom. This book is for middle school teachers and mathematics teacher educators who work with preservice and in-service teachers. We also believe this book can be used by mathematics coaches, center directors, and school and district leaders who want to empower their middle school students by analyzing and critiquing the world around them.

During your reading, we hope that you will grow in your understanding that mathematics may be a privileged space through which both you and your students can be empowered. Many students are not allowed the opportunity to connect mathematics with their culture and realized lives; thus your interest in TMSJ, reading of this text, and implementation in the classroom present an opportunity to shape students' lives and actions. When children learn that mathematics can be used as a tool to help them understand, explore, and investigate social situations, they are empowered to see themselves as active agents in a world of change. We hope that the lessons and the critical call for action contained in this book highlight how each and every student is capable of mathematical learning and can be empowered to use mathematics for change in their own and others' lives.

Although this book offers a number of suggestions for how to incorporate social justice mathematics lessons throughout the middle school curriculum, it is not intended as an end-all and be-all. We hope that you use the lessons as models and starters on your journey toward creating and implementing your own social justice lessons, targeting social concerns of your own students, and helping students view themselves as mathematically empowered agents of social change. Thus, we also offer suggestions for you to create your own social justice mathematics lessons for middle school students in the concluding chapter.

THE BOOK'S ORGANIZATION

This book is organized into three parts. In Parts I and III, you will find an opportunity to self-reflect and write in the book about your own identity and practice. Part I consists of Chapters 1–5, which support middle school teachers who want to implement teaching mathematics for social justice. In Chapter 1, we discuss what TMSJ means for middle school teachers. We also discuss key elements that contribute to TMSJ. In Chapter 2, we discuss how teachers can build and sustain beloved classroom communities that honor middle school students' needs. In Chapter 3, we share ways middle school teachers can foster a classroom to teach mathematics for social justice before they begin the planning process. We pose reflective questions for teachers to ask why it is important to consider content and context, and when and how teachers can plan to teach mathematics for social justice. We also provide suggestions of ways teachers can work with their interdisciplinary team to collaborate with colleagues. In Chapter 4, we look at three areas—equitable teaching practices, discourse, and assessment—and provide suggestions for you to consider as you plan for the social justice component of the lesson. In Chapter 5, we provide details about the design framework and the structure of the social justice mathematics lessons (SJMLs) in the book. It depicts a continuous cycle in which students actively investigate, understand, and reflect on challenging mathematical and social questions to empower themselves into action. This will help you understand elements of the structure that will not only help you think through the lessons as you plan to use them but also support you to develop your own SJMLs.

Part II contains SJMLs organized by mathematics content domains: the Number System (Chapter 6), Ratio and Proportional Relationships (Chapter 7), Algebra: Expressions, Equations, and Functions (Chapter 8), Statistics and Probability (Chapter 9), and Geometry (Chapter 10). An important characteristic of Part II of this book and its uniqueness to the field is the mathematical depth of the lessons. Teachers who use lessons from Part II of this text attend to the mathematical rigor required in state standards while also attending to the Social Justice Standards (Learning for Justice, 2016). We chose to organize lessons based on conceptual categories in order for teachers to easily locate lesson ideas that may be infused with mathematical course progressions from their state, district, or school.

We believe that attention to Learning for Justice's Social Justice Standards is critical in the development of middle school students; thus, a cross-reference of lessons to these Social Justice Standards may be found in Appendix E for teachers who are hoping to attend to all objectives identified in these Social Justice Standards across one academic year. In addition, each chapter in Part II is introduced with a table that highlights the lesson titles, authors, and a topic of social injustice. Teachers may use these pages to find lessons that tackle certain social injustices that are relative to their demographic, environmental, or social contexts.

Part III consists of two concluding chapters. In Chapter 11, we share reflections from the lesson plan authors and reviewers, who share their experiences with the hopes of providing inspiration and insight as you set forth to implement the

SJMLs in this book. In Chapter 12, we provide insight on creating your own and help you find your identity in teaching mathematics for social justice.

This is probably not a book to be read cover to cover, straight through. We hope that the reader gives thoughtful attention to Chapters 1–5. But next, we expect the reader to skim through the lessons of Chapters 6–10, reading those that are of most interest. Some people will be very interested to consider the thoughts and experiences from fellow teachers that are found in Chapter 11, and some will want to get into the recommendations to write their own lessons in Chapter 12. However, we imagine that both of those chapters will be read at a variety of different times—some people will want to get right to them, and others will find them most valuable after trying a few of the lessons in their classrooms.

As you consider implementing the lessons in this book, we imagine three approaches to selecting a lesson to use. First, we expect most teachers will identify lessons that align with the content standards that are assigned to the class they are teaching. Appendix E will be helpful for this approach. However, the social justice topics, as aligned to lessons at the end of Appendix E, may be of more interest to teachers who wish to respond to an important issue that is very visible in their school community. Finally, we imagine the teacher who is interested in developing students' knowledge, skills, and disposition toward social justice more generally, as opposed to topic or mathematics centered. The second chart in Appendix E aligns the lessons to the Social Justice Standards developed by Learning for Justice (2016).

We expect that the resources in this book will help you create and focus energy on authentic experiences for students while also generating mathematical analysis or modeling to probe issues of injustice relative to students' lives. These chapters may also encourage seasoned and veteran teachers using TMSJ to compare some of your guiding principles to ours, and possibly gain ideas on how to enhance the work you are currently doing. We encourage you to read each of Chapters 1–5 to become familiar with the aims of the book, recommendations for TMSJ, and the framework to understand how the lessons are organized. Next, consider selecting one or two lessons to implement that align with the content standards of the course(s) you teach and personalize them to your context if able. We hope you will come back to the book often, each new semester, to consider additional lessons. Finally, read Chapter 11 as you are ready to begin to modify the lessons provided or begin creating your own. Chapter 12 can be read at any time, and may be most insightful after implementing one or more lessons grounded in social injustices so that you can reflect on your own experiences through the wisdom of others.

We commend you for bringing your middle school students' curiosities and concerns about their lives into your mathematics classroom. We hope that the lessons in this book help you to foster student-to-student interactions that move beyond the mathematics to be learned and into actionable change in middle school students' lives and society.

PART

I

TEACHING MATHEMATICS
FOR SOCIAL JUSTICE

WHAT IS SOCIAL JUSTICE AND WHY DOES IT MATTER IN TEACHING MATHEMATICS?

Imagine a room full of 11- and 12-year-olds on their first day of middle school. This is the first big educational transition they have encountered since starting kindergarten. Their elementary school may have consolidated with many other elementary schools into one middle school, leaving them looking around the room at a bunch of people who do not look familiar. Or maybe they are in a K–8 school and have the same classmates, but for the first time they have multiple teachers. Either way, their lives are changing in and out of school. They are exploring their identities in a way that centers more around their community and their social life than their home and family life. They are navigating their preteen years and puberty, and they are starting to step into responsibility in a way that gives windows into possibilities of their future.

One day, a group of sixth graders comes into class excitedly talking about their plan to eat at different fast-food restaurants. These students often walk home together, exploring this newfound independence. Now able to explore their community with their friends, they have noticed, with awe, how many fast-food restaurants are on their route. They start listing all the places they have seen, with some disagreements about whether a certain place is considered fast food or whether one is on their route home.

A lesson starts to develop in your mind to mathematize what students are noticing about their community. Students will map out the locations of fast-food restaurants on an x-y axis with their school at the origin. They will look at the distribution of the fast-food restaurants. You will partner with the science and social studies teachers to provide lessons on nutrition and help students navigate the impact that the quality of food availability has on the health of their community. By starting with your students' expertise—what they've noticed about their community—you are helping them develop tools to think critically about the world around them. They can still enjoy comparing french fries from different fast-food restaurants, but they will also have a broader perspective of the communities in which they live. In the future, you might use Lesson 6.2 (Cor(o)ner Stores in Food Apartheid[1]) to explore access to

---

[1] Dr. Shakiyya Bland, author of a lesson in the high school book, suggested referring to "food deserts" as "food apartheid" because apartheid indicates a larger system that considers how race, geography, faith, and economics impact access to healthy foods, while deserts often bring about images of desolate places. The rephrasing shifts the focus to flaws of the food system without diminishing the vibrancy and potential of the affected communities.

fresh food in their neighborhood and how it impacts the health of their community. Or you might use Lesson 9.4 (How Many Meals Can Minimum Wage Buy?) to compare minimum wages over time from the perspective of how many burgers a worker making minimum wage could buy. As students learn to use mathematics to better understand their communities, they can highlight what they love and appreciate, even if it is Del Taco, and advocate to make improvements where needed. If the absence of a produce market is a problem, they might present their findings to the city council and write letters advocating for a new farmers market or more restaurants that provide affordable, healthy options. Students can commit to making the choice to include the produce market and healthier options in their food explorations, or they could lobby the local health food restaurant to offer affordable kid-friendly items on the menu, or discounts for students with their student IDs, as an alternative to the McDonalds across the street.

As students begin to broaden what they consider their community, it is important for us to understand how students see their role in the community, embrace the students' experiences and expertise, and help them navigate through this time of growth. This broadening of their community is an opportunity to connect their learning with, about, and for their community. In order to accomplish this, we need to invest attention into understanding the communities students are coming from and their views about their communities.

As middle school students begin to expand their worldview, you have the opportunity to help them navigate it by collaborating with them and their communities to bring their greater world into the mathematics classroom. You can show students how mathematics can be used as a tool to help them interpret issues they care about and empower them with evidence that can be leveraged to create change. Collaborating with your middle school students on issues they care about empowers them to understand and contribute to their communities, widen their view of what mathematics is, and reimagine what mathematics can be used for.

> **Collaborating with your middle school students on issues they care about empowers them to understand and contribute to their communities, widen their view of what mathematics is, and reimagine what mathematics can be used for.**

Until recently, embedding mathematics pedagogy within social and political contexts was not a serious consideration in mathematics education. The act of counting was viewed as a neutral exercise, unconnected to politics or society. Yet when do you ever count just for the sake of counting? Only in and for school are young adolescents required to count without a social purpose of some kind. Outside of school, mathematics is used to advance or block a particular agenda (Tate, 2013, p. 48).

Students can become empowered when they have access to deep, rigorous mathematics that offers opportunities to understand and use mathematics in their world. Empowering students to use mathematics to critique and understand the world requires us to take a social justice position. Larnell et al. (2016) remind us that "whether inside or outside of school, mathematics is political" (p. 26). That is, choosing not to incorporate tasks that require students to critique and understand the world is itself a political position, one of political passivity. Conversely, incorporating tasks requiring students to critique and understand the world is an act of empowerment and develops their critical consciousness, their ability to identify, understand, and take action against oppression. Development of a critical consciousness has beneficial effects on academic achievement, especially for students of Color (Seider et al., 2020). The social justice math lessons (SJMLs) in this book seek to (1) have students engage in the sociopolitical reality that is meaningful and important to them and (2) use mathematical tools to answer questions emerging in these sociopolitical contexts that relate to their own, their classmates', and others' lives. When students have opportunities to use mathematics to solve problems in their everyday lives, they are empowered to engage in political and social acts as a way to seek justice.

WHAT DOES SOCIAL JUSTICE MEAN TO YOU?

Before we tell you our definition of social justice, we invite you to reflect on what the term *social justice* means to you and how it relates to you as a mathematics educator.

PAUSE AND REFLECT

What do you think about when you hear the term *social justice*? What role does social justice play in your practice as an educator? We have left some space here for you to reflect on this.

If you could see everyone's response to the question above, you would see a wide variety of ideas about and associations with what social justice means to each person. For us, as authors of this book, social justice means considering the contributions, rights, and access to opportunity for each and every person in society. We draw on the Social Justice Standards (formerly known as the Teaching Tolerance Standards from the Southern Poverty Law Center; Learning for Justice, 2016) to inform our understanding of social justice and its implementation in the mathematics classroom. The lesson authors also use the standards to guide the social justice aspects of their lessons. The Social Justice Standards provide a common language to guide teaching; increase understanding of difference; and actively challenge bias, stereotyping, and all forms of discrimination in schools and communities. The Social Justice Standards are composed of anchor standards and learning outcomes grouped by grade band (K–2, 3–5, 6–8, and 9–12) and divided into the four domains of Identity, Diversity, Justice, and Action, summarized in Figure 1.1. For example, the following standards could be applied to the lesson about fast-food restaurants described in the vignette at the beginning of this chapter:

- Justice 12 (JU.6–8.12): I can recognize and describe unfairness and injustice in many forms including attitudes, speech, behaviors, practices, and laws.

- Action 19 (AC.6–8.19): I will speak up or take action when I see unfairness, even if those around me do not, and I will not let others convince me to go along with injustice.

The emphasis on Social Justice Standards resonates with the joint position paper by NCSM and TODOS: Mathematics for ALL (2016), as well as the position paper of the Benjamin Banneker Association (2017), both of which argue that embracing social justice moves us beyond noticing issues and concerns about societal inequities and requires actions that confront oppression and/or marginalization. And this is why we chose to include *to explore, understand, and respond to social injustice* in the title of this book. Figure 1.1 gives a broad explanation of each domain in the Social Justice Standards as well as some reasons why they are relevant in the middle school mathematics classroom.

> **Embracing social justice moves us beyond noticing issues and concerns about societal inequities and requires actions that confront oppression and/or marginalization.**

Figure 1.1. Social Justice Standards and How They Are Relevant to a Middle School Mathematics Classroom

Social Justice Domain	Broad Explanation	Why It Is Important in Middle School Mathematics Class
Identity	Understand their identity in the context of their family and how it relates to emerging individual and group identities. They feel good about their own identities, and understand that their identities don't make them better than people with different identities.	Students' mathematical identities often decline at this age.
Diversity	Can interact with, values, and is curious about people with different identities. Can be respectful even when they disagree. Can articulate how the identity and culture of different groups are shaped by how they are treated in society.	Students can appreciate the value that their classmates' diverse perspectives bring to the mathematics classroom. They are also able to recognize that people bring different expertise to the mathematical conversation.
Justice	Relates to people as individuals, not stereotypes, and can recognize injustice. Understands that privilege is a factor in how people navigate the world.	Students will begin to recognize aspects of gatekeeping that affect them and their peers. An orientation that allows them to critique such barriers can protect them from the barriers.

(Continued)

Social Justice Domain	Broad Explanation	Why It Is Important in Middle School Mathematics Class
Action	Is concerned about how people are treated and can stand up for themselves and others when faced with injustice. Will work with friends and family to take action to make the world better.	Students are able to collaborate and invest in sharing their understandings with peers.

WHAT IS TEACHING MATHEMATICS FOR SOCIAL JUSTICE?

Let's revisit the earlier vignette about fast-food restaurants. As the class went through a process and as the classroom walls broadened, the students started to notice something about their community. The teacher mathematized it in a way that gave students another viewpoint of what they were noticing—a mathematical point of view. Students moved forward using mathematics to understand their community better. By integrating mathematics with nutrition and social studies, they were able to identify how their community's well-being is affected by the availability of healthy food options.

Building on our ideas of social justice, we see that teaching mathematics for social justice (TMSJ) is about us emphasizing equitable opportunities for each and every student as well as developing an orientation toward using mathematics to read and write their world. For us, this view of TMSJ builds on Laurie Rubel's (2017) examination of the relationship between four other equity-driven instructional practices in mathematics: Standards-Based Mathematics Instruction (SBMI), Complex Instruction (CI), Culturally Relevant Pedagogy (CRP), and Critical Mathematics Education (CME) (Figure 1.2).

Each of these bodies of work emphasizes important qualities of an equitable mathematics classroom, building on the previous and adding attention to the new in a sort of nested relationship. SBMI grounds our instruction in meaningful engagement with mathematics and classmates. CI can help us ensure equitable engagement among students in the discourse-rich SBMI classroom. And CRP reminds us to draw upon cultural practices, experiences, and assets, and it challenges us to awaken our students' critical consciousness. CME provides teachers a framework to design instruction to build critical consciousness, and TMSJ provides the how. Let's discuss some key ideas from each of these five nested frameworks.

Standards-Based Mathematics Instruction

TMSJ must be grounded in pedagogical principles widely recognized by the profession, referred to as SBMI. These principles emphasize learning for understanding over attending only to fluency with algorithms and facts. A second point of emphasis is the recognition that understanding develops in a discourse-rich learning environment marked by conjecture, reasoning, and justification (Rubel, 2017). Paired with these ideas about what to learn and how to learn is an emphasis on the responsibility to ensure *all* students learn meaningful mathematics (NCTM, 1989).

Complex Instruction

Sociologists (Cohen & Lotan, 2014) recognized that the inequities of the larger society are replicated in small groups, systematically ensuring some students have less access to the discourse-rich SBMI classroom and thus fewer opportunities to learn. CI is a pedagogical theory to counteract this trend. A key feature of CI is valuing many different ways of being mathematically "smart" (Featherstone et al., 2011). In this sort of "multidimensional" mathematics classroom (Boaler, 2006), the teacher has more opportunity to raise the lowered expectations students may have for members of their group, inviting greater access to interactions for the otherwise excluded student.

Culturally Relevant Pedagogy

Both SBMI and CI attend to shifts in curriculum and teaching that ensure all students have the opportunity to learn mathematics. By studying effective teachers, Ladson-Billings (1994/2009) recognized they built upon each child's unique strengths, including mathematical as well as social and cultural assets. CRP (Ladson-Billings, 1995) reminds us to ensure that equitable instruction draws upon students' cultural practices, experiences, and assets. As these cultural experiences and artifacts become central to the process of learning, students see themselves and their interests in the curriculum. Thus, not only is academic achievement valued, but so is the growth of students' cultural competence and critical consciousness.

Critical Mathematics Education

CME builds on the idea of critical consciousness identified by CRP. This body of work is about teaching mathematics in a way that attends to the concerns of fairness and social justice in the relations between the individual and the society. It centers learning as identity work (Rubel, 2017) focused on who the student is becoming. These perspectives often build upon the pedagogical practices developed by Paulo Freire (1970/2000) in the Brazilian context.

Freire describes critical consciousness (his term was *conscientização*, or conscientization) as "a capacity to confront reality as transformable, and to intervene subsequently in it to effect that transformation" (Garcia, 1974, p. 15). Freire's critical education puts an emphasis on a shift in the power dynamic between student and teacher, shifting the authority for knowledge to the social context of the classroom community rather than held by the teacher. Students are positioned as doers, or authors (Lawler, 2012), of mathematics. As a result, students view themselves as both "an actor and author of history" (Garcia, 1974, p. 16). Figure 1.2 captures highlights of four equity-driven teaching frameworks that underscore TMSJ.

Figure 1.2. Key Elements of Equity-Driven Mathematics Teaching Frameworks

Equity-Driven Mathematics Teaching Frameworks	Key Elements Contributing to TMSJ
Standards-Based Mathematics Instruction	• Learning for understanding over fluency with algorithms and facts • Discourse-rich learning environment marked by conjecture, reasoning, and justification • Responsibility for *all* students to learn meaningful mathematics • Additional resources: NCTM (2014)
Complex Instruction	• Inequities of the larger society are replicated in small-group work, creating status differences. • Status differences ensure some students have less access to interaction, and thus fewer opportunities to learn (Expectations States Theory). • The teacher can impact this by creating a multidimensional classroom, raising classmates' expectations for contributions from each and every student. • Additional resources: Featherstone et al. (2011) and Horn (2012)
Culturally Relevant Pedagogy	• Curriculum and instruction must draw upon students' own cultural practices, experiences, and assets. • Three aims: academic achievement, cultural competence, and critical consciousness • Additional resources: Emdin (2016) and Ladson-Billings (1995)
Critical Mathematics Education	• The common teacher–student relationship reflects and reinforces inequitable power dynamics of the broader culture. • Banking model of education: students are containers to receive knowledge deposits from the teacher. • When positioned as passive recipients, students are positioned as adapters to the world as is rather than as shapers of the world to be made. • Learning can emerge from a *problem-posing pedagogy,* designed around the ideas, hopes, doubts, fears, and questions that emerge in a person's relationship with the world—what Freire refers to as "generative themes" (Garcia, 1974). • Additional resources: Frankenstein (1983), Freire (1970/2000), Powell (1995), and Skovsmose (1994)

 Available for download at **resources.corwin/TMSJ-MiddleSchool**

As you look at the interconnectedness and interdependence of each body of work, you can see similarities to that of a tree (Figure 1.3). Each of these bodies of work emphasizes important qualities of an equitable mathematics classroom; they interact with each other as the *roots* of a mathematical experience for students, which allow for the teaching of mathematics for social justice (*trunk*). TMSJ in

Figure 1.3. Standards-Based Instruction, Complex Instruction, Culturally Relevant Pedagogy, and Critical Mathematics Education Are the Roots That Make Effecting TMSJ Possible

turn creates more access to mathematics for students (*branch*), allowing them to see mathematics as a tool that helps them read and write the world (*branch*). TMSJ can also shift the power dynamic between teachers and students (*branch*), which is a reassignment of authority that positions students as doers of mathematics (*branch*). This development of these tools and shift in positionality in the classroom can empower students to improve their lives and their communities (*leaf*). The application of mathematics toward social justice issues that interest students can result in deeper engagement in the mathematics classroom (*branch*) and more engagements with mathematics outside of the classroom (*leaf*). Finally, TMSJ plays a role in building an informed society (*branch*), giving students tools to engage in critical consciousness (*branch*).

Teaching Mathematics for Social Justice

Larnell et al. (2016) observed that descriptions of TMSJ are based on one of two perspectives, either TMSJ as access or TMSJ for critical consciousness. TMSJ for access is about enabling students to advance in the social order as is. This can be seen in the way we develop a space where students can succeed. TMSJ for critical consciousness is about empowering students to alter or improve the present social order (Gholson et al., 2017). This can be accomplished by giving students the tools to challenge injustice through mathematics, like with the lessons in this book. Both of these perspectives are important when considering middle school students, who are at an educational stage where mathematics starts to act as a gatekeeper, and barriers to access can have long-term effects. Simultaneously, as students become more independent and their communities expand and become more complex, their social, emotional, and mathematical growth depends on them developing both tools to process and understand the world around them and agency to respond to injustice. Our interpretation of TMSJ explicitly strives to achieve both, aligning with Freire's (1970/2000) conception of a pedagogy that creates both freedom from oppression and freedom to create culture.

> Generative themes are contexts, topics, problems, and questions students pose. They are considered generative because they contain the possibility of opening again new themes, new problems and questions posed, and new tasks to be fulfilled.

Following on Freire's critical education, Gutstein (2006) argues that TMSJ must be rooted in viewing students as an important part of the solution to injustice: "Students need to be prepared through their mathematics education to investigate and critique injustice, and to challenge, in words and actions, oppressive structures and acts—that is, to 'read and write the world' with mathematics" (p. 4). Units of study in the TMSJ classroom are built on a *problem-posing pedagogy* (rather than a banking pedagogy in which the teacher deposits information into the student's mind), designed around the ideas, hopes, doubts, fears, and questions that emerge in a person's relationship with the world—what Freire refers to as generative themes (Garcia, 1974). The themes—contexts, topics, questions, and so on—are generative because they contain the possibility of opening again new themes, new problems posed, and new tasks to be fulfilled.

Students develop and/or apply mathematics to make sense of the problems that emerge in the generative theme, so as to better understand and act upon the concern or question about the world. In this way, TMSJ involves two types of goals: (1) freedom from oppression through the development of mathematical literacy

Figure 1.4. Instructional Goals When Teaching Mathematics for Social Justice

Mathematics Pedagogical Goals	Social Justice Pedagogical Goals
• Reading the mathematical word • Succeeding academically (in the traditional sense) • Changing one's orientation to mathematics	• Reading the world with mathematics • Writing the world with mathematics • Developing positive cultural and social identities

Source: Gutstein (2006).

and (2) freedom to act upon and impact the world through personal and social transformation. The first is a mathematics goal, and the second, a social justice goal. Gutstein (2006) clarifies elements of these two goals, replicated in Figure 1.4.

These two sets of pedagogical goals are tightly connected. While we aim for students to read the mathematical "word," the traditional school content, we also wish for them to use mathematics to interpret and act upon the world. TMSJ also aims for students to view themselves as doers of mathematics and have a positive mathematical identity that is situated in their cultural and social identity. The connections between the two types of goals are where you can create the possibility for students to find meaning in mathematics, through their exploration, understanding, and response to social injustice.

Through our model of the multiple teaching frameworks that contribute to TMSJ, we aim to emphasize our view that TMSJ is much more than the lessons you might implement in your classrooms. It is about the relationships you build with and among students; the teaching practices that help you do that; and the goals to develop positive social, cultural, and mathematics identities—as authors, actors, and doers.

> TMSJ is much more than the lessons you might implement in your classrooms. It is about the relationships you build with and among students; the teaching practices that help you do that; and the goals to develop positive social, cultural, and mathematics identities—as authors, actors, and doers.

WHY SOCIAL JUSTICE IN MATHEMATICS EDUCATION?

When only focused on teaching mathematics, you know all too well that disconnects can occur between the content and students' passions and lived realities. Contextualizing mathematics instruction in students' experiences of social injustice helps them become more interested in mathematics (Rubel, 2017). Furthermore, TMSJ supports students' use of mathematics to better understand social injustices they recognize in their lives and to be able to act upon those injustices. In doing so, students learn more mathematics. Figure 1.1 provides examples of why the Social Justice Standards are important for students' long-term mathematical growth.

So why bring social injustices into the mathematics classroom? Besides offering critical issues for students to examine and to serve as context for mathematics development and investigation, there are four other answers to this question that resonate with us. Teaching mathematics for social justice can

- build an informed society,
- connect mathematics with students' cultural and community histories,

- empower students to confront and solve real-world challenges they face, and

- help students learn to value mathematics as a tool for social change.

Each of these rationales to TMSJ resonates with the foundations of a public education, effective instructional practices, and a purpose for mathematics education shared by mathematics teachers everywhere.

Build an Informed Society

To create a just society, students must become better informed about not only their own lives but also the lives of others that may be different from their own. It is paramount that you connect students to the injustices expressed by members of their school, community, city, and country—especially to injustices they may be unaware of as experienced by people with different social and cultural experiences than them. Mathematics serves a special role in informing and educating citizens about these issues. By exploring the context of important issues and relating them to mathematics, students become aware of how mathematics may be used to help them better understand the issue, possibly sorting through misconceptions and rhetoric. A student with a meaningful mathematics education is not just academically successful (*academic achievement* in CRP; Ladson-Billings & Tate, 1995), but they are also prepared to make informed decisions in a modern, ever-changing society.

Connect Mathematics With Students' Cultural and Community Histories

Too often, students experience mathematics in schools as something detached from meaningful contexts. Thus, many students perceive mathematics as unfamiliar and unimportant. This leaves many of them with the sense that mathematics is inaccessible and not connected to them or who and what they value. Students bring with them to the mathematics classroom a wealth of informal mathematical knowledge in their everyday cultural and social experiences. Students' informal knowledge and experiences are valuable resources for mathematics teaching and learning.

We know that when classroom experiences and reasoning are meaningfully connected to students' ways of knowing, the learning that occurs—both cognitively and culturally—is powerful and lasting (National Research Council, 2000). By grounding learning in students' own cultural and community histories, you have the opportunity to create both deeper knowledge and greater valuation of students' own culture and to develop their understanding of other cultures—the sort of *cultural competence* called for by Ladson-Billings and Tate (1995) and Ladson-Billings (2021).

Empower Students to Confront and Solve Real-World Challenges They Face

TMSJ helps students to build a *critical consciousness*, identifying issues that are unjust (Ladson-Billings & Tate, 1995), and then to use mathematics as a tool to analyze, critique, and confront those unjust contexts. Your role is to learn about your students to identify generative themes, and thus help them uncover and explore the issues of injustice their families and communities face.

Help Students Learn to Value Mathematics as a Tool for Social Change

The potential of education is to support students to create better lives for themselves and to create a better society for each and every individual. Mathematics, seen as human activity, is a powerful tool to achieve both of these goals (Skovsmose, 1994). When students use mathematics to explore, understand, and respond to social injustices they experience or care about, they learn not only the power of mathematics for social change but also that they are actors on the world with the power to transform inequities and create social change. We want students to recognize that their mathematical power can improve the conditions of not only their own lives but also the lives of others.

CONCLUSION

We believe building instruction from the student's lived experience is paramount for developing the whole, mathematically proficient student. One teacher may demonstrate an algorithm on the board and then require students to repeat the process. Another teacher may implement an inquiry lesson and then have students share their solution pathways. Both of these situations teach mathematics with a specific goal in mind. The first teacher likely has a focus on procedural fluency intended to help students be successful on standardized testing. The second teacher likely has a focus on teaching mathematics for increasing students' problem-solving ability. Both of these classroom lesson analogies exemplify teaching mathematics with good intentions, but the teacher goals are different.

Similarly, you must consider your goal as an educator in a society that promotes justice by and for its citizens. Such a society requires informed and educated officials and citizens who can sift through rhetoric using logic and truth to guide policy, voting, and persuasion. A goal as a mathematics teacher teaching in a society governed by its people should be to engage students in critical inquiry about the world and potential injustices surrounding them, pushing students to imagine and create what a world would look like with justice, fairness, and equality. When you provide students opportunities to construct, revise, add to, and share the story with others, including the next generation, you help empower them to be active agents of change to improve justice in the world.

Similarly, we believe the same is true for mathematics teachers.

> *I never thought about the importance of empowering my students to change their world. Now, having experienced the impact of a social justice teaching demonstration, I am constantly looking for ideas to incorporate social justice topics that affect my students into my mathematics lessons . . . I would suggest that teacher educators require students to not only experience a social justice lesson within their program of study, but also design their own to use in their future classrooms.* —Dacia Irvin, Baker Middle School, Muscogee County, Georgia, Teacher of the Year (2019)

Dacia's comment exhibits how teaching a social justice lesson in the classroom impacted her view of teaching mathematics in a society that promotes justice for all people. She noted how her teaching empowered students to be active agents of change. She identified the need and desire to connect mathematics and social justice concerns. It is evident that her experience teaching mathematics for social justice changed her view of what mathematics may be used for by teachers and students alike. As teachers and students experience new ideas by teaching mathematics for social justice, they begin to develop an orientation that can promote a more just and equitable society.

We are excited to share lessons contributed by educators across the United States that demonstrate their efforts to teach mathematics for social justice and to achieve the four goals we've identified here. As you better understand how children learn mathematics, recognizing the importance of meaningful context and drawing upon students' cultural assets, it is an opportune moment to draw upon the generative themes you learn from your students to engage them in mathematically rich activities and explore, understand, and respond to the social injustices they care about.

It is also an opportune time for you not only to reshape how you teach mathematics but also to use what you teach in mathematics as a tool to empower your students to be agents of change in their society and in their own lives. To put it in Dacia's terms, it is time to empower our "students to change the world."

REFLECTION AND ACTION

1. At the beginning of this chapter, you reflected on your own definition of social injustice and justice. How might you help students develop their own definitions?

2. Visit a colleague and initiate a conversation about why they became a teacher. Does the conversation draw on any of these lenses?

   + Build an informed society

   + Connect mathematics with students' cultural and community histories

   + Empower students to confront and solve real-world challenges they face

   + Help students learn to use mathematics as a tool for social change

3. In the next chapter, you will be asked to imagine your classroom as a beloved community. Teacher positionality plays an important role in this community. What elements of your identity do you share with your students? Which aspects of your identity are unknown to them?

BUILDING AND SUSTAINING A BELOVED COMMUNITY IN THE MIDDLE SCHOOL MATHEMATICS CLASSROOM

Middle school students are developmentally at an age where they are learning a lot about themselves, their peers, their roles in the community, and more. Your students are navigating multiple physical, emotional, and social areas of growth and discovery. As a middle school teacher, you have the privilege to bear witness to and support this navigation.

Young adolescents can learn about themselves and their peers and navigate their own lives situated within their larger contexts.

Depictions of middle school often emphasize students in various cliques following a social hierarchy that establishes and determines their popularity. The middle grades are often seen as a time that one needs to get *through* between elementary school and high school, and middle school students are often perceived to be self-involved, rotten to each other, and exclusive. They are also perceived to be too young or immature to learn about, reflect on, and act on real-world issues. In general, middle school can be a stressful time for students and teachers alike navigating such negative perceptions. However, a different vision of middle school is possible, one that is based on young adolescents learning about themselves and their peers, navigating their own lives situated within their larger contexts. It is the space where students navigate the transition from being children to being adolescents.

> *[Middle school students] deserve an education that will enhance their healthy growth as lifelong learners, ethical and democratic citizens, and increasingly competent, self-sufficient individuals who are optimistic about the future and prepared to succeed in our ever-changing world. (Lounsbury, 2010, p. 53)*

Teaching mathematics for social justice (TMSJ) can be one approach you take to support your students in making sense of their and others' lived experiences and to understand their and others' agency. TMSJ can foster students' *critical literacy* as they uncover, unpack, and articulate assumptions people hold connected to dominant narratives about our world (Vasquez, 2016; Vasquez, Janks, & Comber, 2019). Critical thinking and communication are central skills in this developmental stage.

Young adolescents in middle school are developing their sense of identity, their relationships to one another and the world, and their capacity as change agents. As discussed in Chapter 1, supporting middle school students to examine identity, diversity, justice, and action (see the Social Justice Standards from Learning for

Justice, 2016) is critical in all spaces of learning, including the middle school mathematics classroom.

Middle school students encounter all kinds of (in)justice issues in their daily lives, in literature, and in learning about history. Through engaging in TMSJ, you can create space for students to uncover how mathematics is essential for exploring, understanding, and responding to social, racial, economic, political, and environmental justice issues.

FOUNDATIONS FOR BELONGING IN MIDDLE SCHOOL MATHEMATICS

To change the narrative from middle school being an inevitably miserable passage into adolescence to one that focuses on learning and connection, middle school classrooms must be places where all students and teachers feel that they belong and where they can collaboratively develop. When you promote adaptive academic and interpersonal contexts, students' sense of belonging is enhanced (Anderman, 2003). The Collaborative for Academic, Social, and Emotional Learning (CASEL) defines the characteristics of *belonging* as self-awareness, social awareness, responsible decision-making, self-management, and relationship skills. Each is important in the mathematics classroom, as shown in Figure 2.1.

Figure 2.1. Characteristics of Belonging

Characteristic	Description	How It Can Show Up in the Mathematics Classroom to Foster Belonging
Self-awareness	The ability to accurately recognize one's own emotions, thoughts, and values and how they influence behavior. The ability to accurately assess one's strengths and limitations, with a well-grounded sense of confidence, optimism, and a "growth mindset"	Reactions to and perceptions of numerical data and representations Acknowledgment and reflection of mathematical strengths and growth areas
Social awareness	The ability to take the perspective of and empathize with others, including those from diverse backgrounds and cultures. The ability to understand social and ethical norms for behavior and to recognize family, school, and community resources and support	Openness to learning about people similar to and different from oneself, using mathematics Recognition of family, school, and community assets and how mathematics can support or convey those assets
Responsible decision-making	The ability to make constructive choices about personal behavior and social interactions based on ethical standards, safety concerns, and social norms. The realistic evaluation of consequences of various actions, and a consideration of the well-being of oneself and others	Data, along with non-numerical factors, contributing to decision making that prioritizes the well-being of self and others

(Continued)

Characteristic	Description	How It Can Show Up in the Mathematics Classroom to Foster Belonging
Self-management	The ability to successfully regulate one's emotions, thoughts, and behaviors in different situations—effectively managing stress, controlling impulses, and motivating oneself. The ability to set and work toward personal and academic goals	Awareness of self (contributing and also listening) in mathematical group work
Relationship skills	The ability to establish and maintain healthy and rewarding relationships with diverse individuals and groups. The ability to communicate clearly, listen well, cooperate with others, resist inappropriate social pressure, negotiate conflict constructively, and seek and offer help when needed	Value the humanity of mathematics classmates in working toward common goals

Source: Adapted from CASEL Framework 2022 via CASEL.org

PAUSE AND REFLECT

Take a moment to reflect in the space provided. In which area do your students tend to be the strongest at? What area would you like to support them to grow more?

BUILDING A "BELOVED COMMUNITY" IN THE MIDDLE SCHOOL MATHEMATICS CLASSROOM

Having a sense of community not only makes class more enjoyable but is also essential for social justice mathematics exploration that is rooted in and respects all people's humanity. That is, TMSJ only works if the community is intentionally built around principles of justice and anti-oppression. Stereotyping, bullying, and discrimination pervade dominant cultures, so you must do work to create spaces, including within our mathematics classrooms, that explicitly resist such ways of being.

The *Beloved Community* is one such vision and is explained by The King Center, the memorial institution founded by Coretta Scott King to further the goals of Martin Luther King.

> *Dr. King's Beloved Community is a global vision in which all people can share in the wealth of the earth. In the Beloved Community, poverty, hunger, and homelessness will not be tolerated because international standards of human decency will not allow it. Racism and all forms of discrimination, bigotry and prejudice will be replaced by an all-inclusive spirit of sisterhood, brotherhood, [and siblinghood]. (The King Center, n.d.)*

Your mathematics classroom can strive to be a microcosm of Beloved Community that supports middle school students' social, emotional, and academic development by seeing and recognizing their humanity and interconnectedness *and* what mathematics has to do with that.

Next we explore five facets of building and sustaining a Beloved Community in the middle school mathematics classroom and how they explicitly support social justice mathematics exploration:

1. Reflect on your teacher positionality.

2. Learn about your students as people *and* specifically as young adolescents who do mathematics.

3. Establish mathematics classroom community commitments.

4. Engage students in mathematics community builders and icebreakers.

5. Facilitate temperature checks, in relation to and beyond mathematics.

Reflect on Your Teacher Positionality

To teach mathematics in a way that allows both you and your students a sense of belonging requires you to first (and continuously) deeply reflect on your own identities, experiences, and the positionality you bring to the classroom. Examining your positionality encompasses an analysis of the social positions and power relations you experience as a person in society and as a teacher specifically. Your identities, lived experiences, and positionality shape how you design and implement curriculum, as highlighted by TODOS: Mathematics for ALL (2021) in a discussion of racial identity and racial trauma specifically:

> *Mathematics lessons that focus on understanding social and racial injustices are an important piece of the broader struggle for justice. However, if we as teachers simply take an activity and implement it in our classrooms without first doing the self-reflective work to understand how we all are impacted by racial trauma, then we may not be able to engage with the lesson in ways that are positively impactful for students. It takes time to do the hard self-reflective work of understanding how we are all impacted by racial*

trauma and then take steps to heal from it (Menakem, 2014)—time that our communities need mathematics teachers to take. (p. 11)

Without doing this work *before* diving into social justice mathematics lessons (SJMLs) and continuously throughout your career, you can unfortunately do more harm than good in implementing SJMLs. There are many resources available to you to further your reflection on your identities, lived experiences, and positionality—for example, through workshops by organizations such as the National Equity Project and books such as *Difference Matters: Communicating Social Identity* by Brenda Allen. We especially recommend seeking to learn from texts, webinars, and workshops, designed by people (and compensating the labor of people) who have marginalized identities that you do not share, in order for you to expand your awareness. We recommend you seek resources to learn about intersectionality, conceptualized by Kimberlé Crenshaw, which views race, gender, class, and other social identities not as distinct facets of identity, but rather how they intersect to shape our experiences.

Reflecting on your positionality is an essential process that should inform how you share about who you are with your students. Sharing about yourself with students humanizes you in their eyes and helps support an understanding that who we are as people is relevant to who we are as doers of mathematics, especially mathematics that explores the social and political world. Of course, you should only share about aspects of your identity that you are comfortable sharing.

There are various ways for you to share about who you are with your students. You may consider sharing about your journey through mathematics education; for example, if you were tracked into the advanced mathematics class when you were in seventh grade, how do you make sense of that tracking in relation to your identities? Another way of sharing about yourself is to use a Numbers of You framework, explaining how different numbers tell important parts of your life story (e.g., a significant year in your life, your favorite number, how many years you have been teaching) while also communicating that numbers can never tell the whole story.

You could share about yourself prior to diving into Lesson 7.1 (*Hey Google, Who's a Mathematician?*), in which students examine the disproportionate amount of elder white men considered famous mathematicians and the message this sends to those mathematics teachers and students who are not elder, white, and/or men. Would students find someone who looks like you in that image search? What does that mean to you as a teacher of mathematics?

Sharing about yourself can help in fostering a classroom culture in which you and your students alike do not make assumptions about or stereotype people and all acknowledge that everyone in a classroom community needs to seek to understand one another's stories, which are unique yet also situated in a complex and inequitable social world.

Learn About Your Students as People *and* Specifically as Young Adolescents Who Do Mathematics

Students' mathematical identity includes how students see themselves as doers of, or learners of, mathematics and relationships that students need to develop as successful mathematics learners (Aguirre et al., 2013). For your students to develop an identity in mathematics, they need to understand how they fit into their classroom, their community, and the mathematics community at large. This begins with grounding themselves in their own community and heritage and building on that to develop mathematical skills. Thus, students need to see themselves represented and feel valued in the larger mathematics community. It is essential for you to learn about the funds of knowledge your students bring to the classroom—learning about their personal, familial, and cultural assets and ways of knowing. The TEACH Math project (Turner et al., 2015) offers a community exploration module that we encourage you to incorporate into your practice (https://teachmath.info/modules/community-exploration-module/).

PAUSE AND REFLECT

What are the personal, familial, and cultural assets your students bring to the classroom?

Providing opportunities to examine the mathematics and statistics of the community in which you teach communicates to students that mathematics is all around us and the mathematics of their community matters. Through Lesson 6.4 (*Middle School Math to Explore People Represented in Our World and Community*), you can create space in your classroom to learn about demographics in your local community if it were a "village" of 100 people.

Providing opportunities for your students to individually share with you about themselves also opens up space to build meaningful teacher–student relationships. One way to do this at the beginning of the school year is through a Student Information Sheet or a Student Survey. We encourage you to consider prompts like those in Figure 2.2 that are relevant for your local contexts.

After you have distributed and reviewed the Student Information Sheet or Survey responses, it is essential to communicate to your students that you have read them, you will remember what they shared, and you take what they shared to heart. Individually, you can follow up with particular students regarding information they shared about themselves. As a class, you can share aggregate data in various data representations, such as sharing language data on the class. (See Lesson 6.4, *Middle School Math to Explore People Represented in Our World and Community* in this book for an activity on comparing class data to data about the world.)

Students may create name tents with pictures and descriptions of themselves on the outside of the tent that may be seen or observed by others. On the inside of the name tent, students may write characteristics of themselves that are hidden or not readily seen by others. Students may choose to disclose and discuss whatever parts of their name tents they feel comfortable sharing during class.

Another beginning-of-the-year assignment you can revisit at the end of the year is a "math autobiography" (Gutstein, 2006; Raygoza, 2016), in which students reflect on their experiences doing mathematics over time, their perceived strengths and weaknesses as a mathematics student, why they feel mathematics is important to learn, and what their goals are for mathematics this year (see Figure 2.3).

You can also make space for students to tell the story of their recent mathematics class experiences; mathematics in their lives during the summer; and the mathematics in the world they wonder about right now (numbers, relationships, representations, etc.). Check out the *Woke Math* blog for a phenomenal post on telling stories in mathematics class, in which Ethnic Studies Mathematics Teacher Shraddha Shirude (2019) writes, "Methods are important. However, we do not need to *center* methods in class. We need to center *stories*. By centering stories in mathematics class, we center humanity. When we center humanity, we bring life back to mathematics."

Figure 2.2. Sample Student Information Sheet

Student Information Sheet

Thank you in advance for sharing about yourself that which you feel comfortable! You have the right to pass on any questions.

My name is:

In class, I prefer to be called / nickname:

My birthday is:

My gender pronouns are:

Grade in school:

Home phone number:

Family member's name(s) and their relationship to you:

I am from or local to:

The language(s) I speak with my family is/are:

I am talented at:

I would like to get better at:

My favorite music genre, group, or song is:

My favorite food is:

I am . . . (you decide what to put here!)

In the future, I would like to:

I come to school because:

In the past, I have had good experiences in school when:

I have had bad experiences in school when:

Something I would like my teacher to know about me is:

Some things I balance my time with (outside of school) are:

I would like to change the world by:

The questions I have about this class are:

Is there anything else you would like to share that may help me support you as your teacher?

online resources ⬆ Available for download at **resources.corwin/TMSJ-MiddleSchool**

Figure 2.3. Sample Math Autobiography

MATH AUTOBIOGRAPHY

Use the following guidelines to write a six-paragraph letter to me explaining your journey and experiences with mathematics. This letter is an opportunity for you to explore your identity as a mathematics student (and growing mathematician!). Please neatly write or type your final letter. Stick to the letter format below.

[Date]

Dear [Teacher Name],

Paragraph 1: My name is [name]. I am [age] years old, and I am a [grade]th grader at [our school name]. Add a few more sentences about yourself here!

Paragraph 2: What do you think about math? What kinds of math do you do outside of school, on your own, with your friends, or with your family? Do you like/love/enjoy math? Why or why not? Explain.

Paragraph 3: What has doing math at school been like for you in the past? How did the last school year of math go for you? Do not only write about your grades but your experiences learning.

Paragraph 4: Why do you think math is important for you to learn? Think of all the reasons you can. Do you believe that math can be used to understand and change the world? Why or why not?

Paragraph 5: What are your strengths and weaknesses as a math student?

Paragraph 6: What are your goals for math class this school year? List all of them and explain why you are reaching for those goals. Explain who will help you to reach your goals.

[Include anything else at the end of the letter that you would like to add.]

Sincerely,

[Your name or signature]

 Available for download at **resources.corwin/TMSJ-MiddleSchool**

Establish Mathematics Classroom Community Commitments

Commitments matter because each class is a sacred community of people. To learn and grow together and be vulnerable and courageous, everyone in a classroom community needs to agree to ways of being together to ensure that the classroom is a nurturing, respectful, and academically challenging space. Classroom community commitments go beyond rules or classroom management tools; they help students create community and learn.

It is critical that you take time to establish commitments—ground rules, norms, or ways of being—so that everyone can be present, mindful, safe, and brave in ways that are specific to your mathematics classroom and all the people within it who make up the class community (students, you, co-teachers, assistants, aides, student teachers, family volunteers).

There are different ways to create classroom community commitments. You may present a set of norms to students that they provide feedback on, specify, and add to (see examples of commitments in Figure 2.4), or you can co-create classroom

Figure 2.4. Examples of Middle School Mathematics Class Community Commitments

For Middle School Classrooms Generally	Commitments Specific to Doing Mathematics
We (do our best to) start and end on time.	Mathematical dialogue and debate is wonderful. No one's humanity is up for debate (Doxtdator, 2018).
Be present. Bring your mind and heart to our shared endeavors.	Everyone is a mathematician. Confusion and uncertainty are part of what it means to do math (Horn, 2012).
Be comfortable with being uncomfortable as you grow and learn.	Listen deeply to each other's mathematical ideas.
Make space, take space. Take turns! You have a right to pass.	Encourage, ask a question, or share an idea; don't just share answers.
One speaker at a time. Try not to interrupt, but if you do, apologize.	Ask critical questions, of mathematics and the world.
Use personal pronouns and gender-conscious language.	

norms with students. Student voice and reflection on norms are critical, so that the norms center what matters most to students. You can invite students to create a list of their rights and demands for learning in mathematics class. After establishing norms, you may consider having small groups make artistic representations of each of the norms to decorate the classroom.

Be sure to reflect on norms over time, individually and as a group. Be flexible in adapting them. And make space to process how they are going.

When implementing a SJML, we suggest explicitly revisiting classroom commitments, and even crafting specific norms for the lesson at hand, in relation to the kind of collaborative work and contexts you will be exploring (see *Context Matters* in Chapter 3).

Engage Students in Mathematics Community Builders and Icebreakers

Community-building activities foster trust in the classroom and remind everyone that the classroom community consists of fellow *people*, not just fellow students or learners. It is crucial to establish classroom commitments, as discussed earlier, before diving into community builders, so that the class community can be invited to participate in the community builders with the commitments in mind.

The community builders you incorporate can be related to mathematics or not. Figure 2.5 shows examples of icebreakers that include a mathematical framing.

Figure 2.5. Examples of Middle School Mathematics Class Icebreakers

Community Builder	Description	How It Supports Building Community in the Mathematics Classroom
Show and Share	Show and share about an object that represents your journey as a mathematics student.	Storytelling about past mathematics experiences can create space for new, humanizing stories to be co-created.
My Variables	Share three variables about yourself.	Mathematics can shed light on our diversity.
Concentric Circles	Form two concentric circles, with students facing one another. Discuss a specific get-to-know-you prompt in pairs, and then rotate so that students are with new partners for new prompts.	One-to-one connection can break down walls between students who may not know each other well yet, making it more likely they will feel comfortable diving into mathematics group work together.
My #1 Number	Talk about an important number in your life and why it means a lot to you.	Mathematics builds collective recognition of the significance of numbers in all of our lives.
Community Resource Mapping	Create and share a map of community assets.	Seeing and celebrating the assets in the greater local community can allow students to learn about one another's lives and cultural practices outside the classroom.

While community builders are most often integrated at the beginning of the school year to help students get to know one another, continuing community builders throughout the school year is essential for various reasons:

1. The composition of your class community may change, with new students joining the class or students leaving the class. Many teachers use community-building activities to help create trust in the formation of new groups multiple times in each quarter.

2. Community building is work that is never completed. Even if the class composition does not change, it requires work to maintain community. Engaging in periodic community builders as a whole group and among small groups every time students move or form new groups can allow students to get to know one another as people (who are mathematicians) and make the social justice mathematics learning experience more valuable, vulnerable, and fruitful.

3. Student identities develop exponentially in the middle grades, and navigating peer relationships among this can be challenging, so creating space to break the ice is crucial.

You can also engage in whole-group community building by making room for student announcements and celebrations at the beginning of class and for appreciations (for one person, a small group, or the whole class) at the end of class.

Facilitate Temperature Checks, in Relation to and Beyond Mathematics

To build and sustain a Beloved Community, you can engage students in a "temperature check" at the beginning of class. Temperature checking is a trauma-informed practice that can help determine how students are doing individually and allow teachers to have an overall sense of the social-emotional "temperature" in the room. If you use temperature checks, you can reduce interactions among students that lead to student isolation, frustration, and insensitivity. Temperature checks can also support your students to feel more comfortable, warming up to speak. Some examples of temperature checks are as follows:

- A rose, thorn, and bud (i.e., something sweet/wonderful/positive, something prickly/challenging/negative, and something blossoming/emerging/growing)

- Represent your week in an emoji or hashtag

- A joyful practice that is energizing you (as a student or as a human in the world)

Try to avoid temperature checks that make assumptions about economic status (e.g., assuming students have traveled over a school break), religion (e.g., assuming students celebrate particular holidays), and other facets of students' lived experiences and identities.

The tool of temperature checks is an important one when you consider embarking on SJMLs with middle school students. Exploring issues of social (in)justice will bring up human reactions that you must prepare your classroom community for so that it can be a safe, vulnerable, and brave space. Including temperature checks coming into and as you explore social issues allows you to be more sensitive to students' experiences. In advance of temperature checks, be sure to become knowledgeable about resources at the school or community level for students who disclose trauma. Becoming aware of trauma and resources for healing will also help you understand when to not proceed, or if and how to pause, when engaging in TMSJ.

Going beyond a content or trigger warning for an SJML, a temperature check can allow you to be more sensitive to and understanding of students' experiences, the individual and collective trauma they have experienced, and their mental health—all in relation to the specific topic you may explore in class together. Depending on what you learn within temperature checks, it may be necessary to discuss your lesson ahead of time with mental health practitioners on your campus and/or support individual students to access mental health resources on campus. For further ideas on dealing with emotional or sensitive conversations within SJMLs, see Chapter 4. Continuously educating yourself on trauma-informed and healing-centered pedagogy is critical to your efforts to not exacerbate harm through your SJMLs.

REFLECTION AND ACTION

This chapter explored how you can create the foundations for belonging in a Beloved Community in your middle school mathematics classroom, rooted in a vision of a middle school mathematics classroom that is based on young adolescents learning about themselves and their peers, navigating their own lives situated within their larger contexts. To do this work, you must continuously (1) reflect on your positionality as a person and teacher; (2) learn about your students as people *and* specifically as young adolescents who do mathematics; (3) establish mathematics classroom community commitments; (4) engage students in mathematics community builders and icebreakers; and (5) facilitate temperature checks, in relation to and beyond mathematics.

Take a moment to pause and reflect on your instructional practices that relate to this chapter. In which of these areas have you already done substantial work? Which of these areas are newer to you? What next steps will you take to create foundations for belonging and Beloved Community in your middle school mathematics classroom?

Consider one or more of the following steps that can help you take action in creating a Beloved Community:

1. Pick a community-building activity, temperature check, or identity depiction task to complete with your class.

2. Consider how your task may intersect with a lesson from Chapters 6–10. Incorporate a lesson from a later chapter with careful attention to how this task may shed greater light on your students' lived experiences.

3. Visit your students' homes, community, places of worship, and extracurricular events. Be purposeful in getting to know those attending and working alongside them in their efforts.

The following chapter moves more specifically into preparation to implement middle school SJMLs.

FOSTERING A CLASSROOM TO TEACH MATHEMATICS FOR SOCIAL JUSTICE

CHAPTER
3

As we shift from a discussion about middle school students' essential characteristics in the previous chapter, we acknowledge that teaching mathematics for social justice (TMSJ) involves more than a series of lessons taught at opportune times throughout a school year. TMSJ must permeate the classroom culture, helping students see, recognize, and value how mathematics allows them to understand and critique the world through mathematics. We can help empower students to take more personal ownership of current issues as they advocate for change using mathematics as a tool in their efforts.

As you begin to consider these recommendations for what is essential in implementing middle school social justice mathematics lessons (SJMLs), this chapter is organized into four sections: *Content Matters*, *Context Matters*, *When Matters*, and *How Matters*. Each section guides your consideration as you develop a plan to justify the use of SJMLs in your mathematics classroom and build upon existing structures in middle schools (e.g., interdisciplinary teams) that could help respond to potential backlash or pushback.

- *Content Matters* outlines the importance of making sure lessons focus on essential content, emphasize content standards and mathematical practices, and are part of a coherent learning experience for students.

- *Context Matters* guides your interdisciplinary team and helps you navigate your current school or district setting as you consider SJMLs that might involve topics that may be controversial in some communities, might result in student responses leading to advocacy or other action, or might raise parent or community stakeholders' concerns or opposition.

- *When Matters* offers recommendations for when to teach the lessons and ways to infuse them into a unit of study.

- *How Matters* provides an overview of the SJML template used for the lessons in this book and provides three lesson models (types). Suggestions for how you might facilitate the SJML learning experience while considering students' responses to traumas associated with various social justice topics are also included.

We will walk you through each of these four ideas and encourage you to consider them as you begin to plan to implement a SJML. By way of previewing each of these sections, Figure 3.1 names some questions to ask yourself or your team to start to consider designing and implementing lessons that address social injustices.

Figure 3.1. Guiding Questions for What Matters

Guiding Questions for What Matters

Content Matters

When considering any SJML, ask yourself:

1. How will the lesson contribute to the learning goals for my class?
2. How does the lesson contribute to developing students' deep understanding of the mathematical standards for the course?
3. How does the lesson connect to an issue that is relevant to my students?
4. How does the SJML allow students to use mathematics as a sociopolitical tool of analysis?

Context Matters

When considering any SJML, ask yourself:

1. What is my purpose for including the social justice topic as part of the lesson? Consider the overarching goals for your class, alignment to content standards, connections to students' lived experiences, possible biases and alternative perspectives of yourself and others, and how you will facilitate the learning and conversations to allow students' voices to be shared and heard.
2. What do I know about how this topic intersects with local concerns and interests? How might this topic be received in my local setting?
3. How does the SJML contribute to building students' identity and agency?
4. Who is on the interdisciplinary team that could be allies ready to support you?
5. What interdisciplinary opportunities exist among your grade-level team?
6. How can you support students who might experience trauma as a result of discussing social justice topics?

When Matters

When considering any SJML, ask yourself:

1. How might this SJML contribute to the goals for this particular unit or course?
2. What current events are relevant to this SJML?
3. How does the SJML build on previous lessons or preview future lessons?
4. What might your students already know about the mathematical and social justice topic of this SJML?

How Matters

When considering any SJML, ask yourself:

1. What instructional strategies will I use to engage all students in the lesson, mathematics, and social justice topics?
2. What questions will I use to facilitate the learning process for both the mathematics and the social justice issue?
3. How might my students react to this SJML, and how will I prepare for that?
4. How will I assess my students' ability to apply the mathematics they are learning in the context of the lesson?
5. How will I assess the degree to which they've met my mathematical goals and social justice goals?

online resources Available for download at **resources.corwin/TMSJ-MiddleSchool**

CONTENT MATTERS

We applaud you for your decision to use the SJMLs in your classroom and have intentionally designed the included lessons to address the middle school mathematics content standards in your state or region. We recognize that each lesson's content plays a major role in helping students make sense of both mathematics and their world by using mathematics in an authentic and empowering way. When selecting (or designing) a SJML, make sure that the lesson's content contributes to helping you achieve

1. the overarching goals you have established for the class (i.e., developing students who critically think as doers of mathematics),

2. mathematical goals related to the content standards as well as the mathematical practices, and

3. a deeper understanding of issues relevant to your students' lived experiences—generative themes—and how they might advocate for change.

As you think about how the content of each lesson contributes to the mathematics story for your students, the careful selection of lessons and associated tasks is critical. Tasks that have a high cognitive demand involve making connections, analyzing information, and drawing conclusions (Smith & Stein, 2018). Lessons that involve such tasks help students deepen their understanding of mathematical concepts as they make connections and see the why and how of mathematics. Reflect on whether the mathematics concept might be difficult to introduce while making connections to social justice topics. Sometimes you may need to decide whether it's best to engage in social justice topics at the beginning, middle, or end of, or throughout, a unit.

Planning for a SJML requires you as a teacher to not only be thoughtful about the design and use of mathematics but also think deeply about the context of social injustice. Preferably, the issue or concern of the injustice is a problem posed by students in your class or a concern they have about their world. In this sense, the topic can serve as a generative theme and drive your students' interest. Take care that the social issue does not just serve as a superficial application of mathematics to a different context; rather, that context is being used to help students learn about mathematics as well.

You may be feeling nervous because you don't feel like an expert in certain topics—or may not feel expert enough. As part of your preparation, you should learn all you can, and not just from media and alternative media sources. We strongly encourage you to talk with people closely involved—they will provide the most insight. However, it is certain you will also not know everything there is to know or may stumble on sensitive issues. Position your students as collaborators and ask that you learn together and that they will teach you. In fact, this position doesn't have to be so different from your relationship with your students when it comes to learning mathematics, generally. Having examined both the mathematics content and social justice content, implementing a SJML with a TMSJ stance implies that at the forefront of your planning, you

- design lessons that help students develop self-efficacy in their ability to critically think as doers of mathematics (mathematical identity);

- provide opportunities for students to use and connect the mathematics they know (agency) to further grow their mathematical skills;

- seek out opportunities to infuse topics that are relevant to your students' background, culture, and lived experiences (access); and

- guide students as they identify social injustices or disparities and support their efforts to become agents of change (empowerment and agency).

Taking a TMSJ stance provides a framework to thoughtfully plan lessons that provide students with meaningful access to mathematics, help them better understand their world, and help them develop agency and empowerment. As you prepare to teach a SJML, consider the age appropriateness of both the mathematical and social injustice content. Likely your state standards document can serve as a guide to the appropriateness of the mathematics. Our recommendation is that you speak with a few trusted friends and possibly experts on the age appropriateness of the social content. Some information, whether written, audio, or video, may be challenging for young adolescents to fully understand. Some information or discussions in fact may be traumatizing, due to students' personal experiences with injustice (see Chapter 2). Be prepared to respond to this; an important first step is to validate feelings and experiences. And while it can be beneficial for the student to better comprehend the situation through participation in the SJML, it should not be required. Finally, be sure to follow up with a school counselor or specialist to ensure the student has an opportunity to find support.

CONTEXT MATTERS

As you consider the context and setting of your classroom, it is important to assess the climate and culture of your school and district in regard to the teaching of controversial topics, especially those that people might not see as traditionally fitting into a mathematics classroom. Topics that seek to help students gain an understanding of inequities in their school, community, or society, and teach students to advocate for change, may draw the attention of those who wish to avoid controversy or seek to maintain the status quo. This supports the importance of being mindful in contextualizing mathematics in sociopolitical terms. Topics themselves are not inherently controversial; however, the local sociopolitical context, as well as ill-planned dialogue, can create controversy around a topic. So when we name a topic as controversial, we really mean to indicate that it has the potential to create controversy or disrupt a space of comfort or the status quo. We advise that the overall goal of carefully assessing your context is to help plan for and anticipate possible opposition and backlash that might come during or after the lesson (e.g., parent complaints to administrators, students forming some type of protest). Thus, as you plan for a SJML, consider the recommendations in Figure 3.2 when pursuing the topics of interest to your students, particularly those that may be controversial in your community.

Figure 3.2. Considerations for Pursuing Controversial Topics in Middle Grade Classrooms

Considerations for Pursuing Controversial Topics in Middle Grade Classrooms

1. Check to see if your school or district has policies or procedures for teaching controversial topics (i.e., rights of transgender students, politics, or religion).

2. Be mindful of social justice topics that are recent hot-button issues in your school, district, or local community.

3. Send a monthly newsletter that includes mathematics and social justice topics that you plan to discuss in the classroom.

4. Look for interdisciplinary cross-curricular connections, providing the opportunity to collaborate with a colleague who teaches a related topic as part of their curriculum (i.e., voter suppression may be part of the social studies curriculum, or global warming may be included in the science curriculum). These individuals can also provide support when you are trying to determine how you begin to discuss controversial topics.

5. Identify community stakeholders (i.e., parents/caregivers, businesses, religious groups) who work with related issues and consider inviting them to be a resource to you as you plan for the lesson or during the teaching of the lesson.

6. Determine if the social justice issue is one that you need to seek approval for from your administrator before teaching.

> Sending letters home to families at the start of every unit, informing families not only of the math content but the social justice topic as well, promoted strong family engagement. It encouraged more family input, [provided] new ways for me as the teacher to look at aspects of topics, [and] gave support for any teachers fearful of trying to teach with a social justice lens. Having a supportive administrator is helpful. Analyzing results and data allowed us to show improvement that could not be negated.
>
> —Peggy Nayar, Lesson Author

To further assess your instructional context and setting, also consider the purpose, audience, allies, and timing for the SJML you have selected.

Purpose

What is your reason for introducing the social justice topic in your mathematics class? Try to make sure you have considered any personal biases you may hold in regard to the topic and also any personal agendas. For example, if I am a Christian male, I must approach topics related to the religious identity of students and families thoughtfully and reflectively, realizing that my perspectives have been strongly influenced by the social context of my experiences and beliefs, including the community in which I live and how I am responded to in public. I must recognize that my actions—both intentional and unintentional—impact students' experience of the classroom community as well as shape the perspectives and beliefs of each and every student. You may have to answer questions about why you chose this topic to include in your mathematics classroom. Thus, having a clear purpose that you can articulate will be helpful if the time comes that you need to respond to questions and concerns.

Audience

Think about the students in your mathematics classroom and those students with whom they will interact in other classes, extracurricular activities, and outside of school. Controversial topics/issues have a way of reaching others, just like rumors or bad press. Establish classroom norms or commitments (see Chapter 2) so students are comfortable with expressing their perspectives on issues without the fear of personal attacks from their peers (see Figure 3.3 for possible ground rules for SJMLs).

You need to establish a classroom where acknowledgment and respect are a part of the culture as an art of teaching that promotes equity. It is important that you remind students of the importance of seeking to understand the similarities and differences of other people. Students should be required to acknowledge and respect other students' identity groups, backgrounds, cultures, and lived experiences.

In addition, consider the resources (i.e., articles, videos, worksheets, data sets, websites) that you will provide students for the lesson or they might research on their own. You must be thoughtful not only about who your students are but also how they might respond to the social justice topic.

Figure 3.3. Possible Ground Rules for SJMLs

Possible Ground Rules for SJMLs

1. Actively listen and respect others when they are talking.
2. Use your own experiences to share insight into the topic.
3. We will affirm one another's perspective rather than seek to discredit it by presenting our own perspective of their story.
4. Our goal is to gain a deeper understanding and not necessarily always agree.
5. Our tone and body language in addition to our words can show disrespect. And we can provide support (head nods, smiles, etc.) when someone is struggling to speak.

Allies

Administrators, colleagues, parents/caregivers, community members, and students are all people you should consider when seeking allies in the planning process. As you complete Part 1 of the SJML Planner (Figure 3.6 later in this chapter), consider providing your administrator with a brief overview of the lesson or the actual lesson plan. Invite them to visit your class to see it in action and provide you with feedback; this will allow them to hear the discussion firsthand. For topics that connect with other content areas, invite a colleague to be a part of the lesson, possibly serving as a content-expert guest. For example, middle school teachers can think about grade-level team structures and the opportunities that may be available to work together. This partnership creates another way for you to think about bringing social justice topics into the mathematics classroom. Even better, consider parents/caregivers and community members who may work in a field related to the social justice issue. Your students are often overlooked as the expert. It is likely that many of them know a great deal about the social context you are considering for the lesson. They likely know very much about the school and community context, and they may know individuals who might be allies in your planning. You will be less open to criticism when you have collaborated with others.

Timing

They say "timing is everything," so consider when you will teach the lesson. Choosing to introduce a topic/issue immediately after a related incident in your school or district may or may not be the most opportune time for the lesson. For example, it may not be best to teach a lesson on violence against transgender students shortly after a student has been attacked in your school or community. It may be better to wait and allow students to process what has happened. On the other hand, using a current event (of the school, community, or nation) may very well be a powerful way for your students to process the issues. This decision relies strongly on both consultation with your allies and students as well as your expertise.

WHEN MATTERS

This book provides an initial set of SJMLs for teachers to intersperse in their regular curriculum—as part of an already planned unit of study, to achieve larger goals of the course, or to implement as a special learning opportunity, possibly in response to a local event and a problem posed by students expressing a question or concern of injustice about the event—a generative theme. While each of the lessons addresses specific mathematics content standards, the authors have attempted to provide a structure that will allow for flexible use as well as a model that will allow you to modify or develop your own SJMLs in the future. In addition, opportunities to modify local contexts are included when available to help teachers implement lessons in diverse contexts. So as you consider when you might teach a SJML, revisit the previous sections *Content Matters* and *Context Matters* when identifying which lessons you want to use.

Unit Connection

Using a SJML as part of your unit is the first scenario we will discuss. When planning to use a SJML as part of a unit, you should

- consider the mathematical content to determine how it best supports the content standards for a unit, and

- determine if the SJML best fits at the beginning, middle, or end of the unit (Figure 3.4).

Figure 3.4. Considerations for Integrating Social Justice Topics Into a Unit

	Mathematics Content	**Social Justice Topic**
Beginning	The content builds from prior knowledge and will engage reasoning to support the development of new concepts.	The topic may be one that you want to connect to throughout the unit as an instructional resource or performance task/assessment.
Middle	The content may draw upon conceptual understanding but is now leaning more toward building procedural skills and fluency.	SJMLs can be used to provide meaningful practice, thus reinforcing concepts and skills taught during the current or previous units.
End	The content may require application or problem solving requiring both conceptual understanding and procedural skills.	Using SJMLs at the end of a unit can serve as a formative way to assess students' ability to apply their learning to real-world authentic situations.

Course Connections

When the mathematical content is not explicitly included as part of the unit of study, you may consider using a SJML for

- teaching mathematics content standards;

- teaching mathematics practice standards, including mathematical modeling strategies;

- reteaching opportunities, in order to strengthen students' prior knowledge in advance of a unit;

- continued learning opportunities to provide additional practice with concepts and skills in addition to some application where appropriate; and

- opportunities to preview future course content (see Figure 3.5).

Figure 3.5. Considerations for Integrating Social Justice Topics Throughout a Course

Instructional Aim	Mathematics Content	Social Justice Topic
Teaching mathematics content standards	Supports current grade-level content standards and may be considered a replacement lesson for your curriculum	Select topics that are more directly connected to a current experience (i.e., school, community, country/province) that students have generated concern about or have some knowledge about. This may limit the need to fully explain the "context" of the social justice issue and allow more focused time on mathematics
Teaching mathematical practice standards and mathematical modeling strategies	Provides an opportunity to engage students in the work of mathematicians	
Reteaching prior knowledge and skills	Reintroduces select concepts and skills from a prior grade that may need to be reviewed for an upcoming topic	
Providing continued learning and additional practice	Provides additional time and practice of select concepts and skills that are foundational for the course based on students' identified needs. Great opportunity for distributed practice after a unit has been taught	
Previewing future course content	Introduces concepts and skills that will be a part of a future unit in the course. This is especially helpful when students may have limited or no prior instruction for the concept/skill	

Interdisciplinary Learning Experiences

Interdisciplinary learning experiences provide opportunities to organize and make important connections to students' lives and multiple subject areas (Hardré et al., 2013). You can help students see ways social justice issues are organized by connecting their lived experiences to other disciplines, which allows students to acquire a deeper knowledge through explorations of injustices in various contexts. Interdisciplinary learning has focused on assisting students with making connections between disciplines in order to gain more depth of knowledge about social justice topics. In addition, working alongside others from other disciplines can provide you with an opportunity to learn from your colleagues in a safe environment and put social justice topics into practice in multiple subjects.

Special Opportunity

Another scenario is to use SJMLs for special events or learning opportunities for the purpose of connecting mathematics to other content areas, which will contribute to helping students see how mathematics plays a role in analyzing and interpreting a current event. You may consider having a family night that focuses on mathematics that incorporates a SJML. Connecting to other content areas provides students the opportunity to use their mathematics outside of the classroom in ways where they may not be "prompted" to do so as they are in a mathematics classroom.

SJMLs can also be used for special events or learning opportunities in a less formal manner, since the lessons may emphasize a social justice topic and connect to a current event or issue that is immediately relevant to your students. For example, you may choose to use Lesson 9.3 (*Gender Pay Gap*) during Women's History Month in March to highlight the need for continued attention and actions on related issues. When deciding to use a lesson as part of a special event, you should be mindful of the lesson's mathematical content and students' mathematical background. Students may need additional support in the form of activating prior knowledge, preteaching, or scaffolding. Note that for some of these lessons you may be able to adjust the mathematics content for a SJML to align to a current standard.

HOW MATTERS

The SJMLs provided in this book have an inquiry-based, or exploratory, approach and may be organized in one of several instructional models. We encourage you to thoroughly review each lesson, the provided instructional model, and lesson goals (mathematics and social justice) to determine the appropriateness for use in your classroom, make adjustments as needed in order to localize the lesson so that it is relevant to your students, and plan for the appropriate level of guidance and support.

When engaging in social justice lessons, there are times when students will have to acknowledge the systemic and institutional inequities that pervade our society and actively address privilege and oppression (Cochran-Smith, 2004). Traumatic events can have a negative impact on students' schooling and mathematics learning experiences. There might be backlash from students when teaching social justice topics, such as acting out during these conversations. You will need to have insight and flexibility in instructional practices to handle those situations. Allow all voices to be heard, seek diverse perspectives, keep conversations respectful, and determine norms during these conversations.

Trauma can impede the ways in which young adolescents might engage in social justice topics during mathematics. Trauma can create social and emotional impairments that often result in behavioral problems, difficulty controlling emotions, and problems in interpersonal relationships (Wolpow et al., 2009). Because young adolescents are still developing morally, intellectually, and emotionally, their experiences of trauma might heighten their responses to some social justice topics. As a result, you might need to help students deal with emotions that result from examining injustices.

These four sections attending to what matters—*Content, Context, When,* and *How*—are included in the SJML Planner (Figure 3.6; also Appendix F) we developed to help you organize information in your initial preparation for a lesson. Please note that the SJMLs provided in this book do not provide information on each of these sections, nor are they structured in this form. Some of the decisions you'll need to make when planning are very specific to your local and personal contexts. While this template can help with the planning you will need to do when introducing issues of social injustice into the mathematics classroom, be reminded that additional questions are also provided in Figure 3.1 to help you think through the elements of the SJML Planner. In Chapter 4, we highlight instructional tools for your consideration as you establish goals and plan for discourse, questioning, and assessing student understanding. Further, we step through the design, structures, and resources available in each of the lessons in this book.

RESPONDING TO PUSHBACK OR BACKLASH

Planning for a SJML requires you to think about more than the mathematical goals, such as the social justice goals and thus social injustices that might be addressed in the lesson. This chapter was designed to support you in your planning process, especially when dealing with a social justice issue. Here are some recommendations to consider when responding to and dealing with backlash or pushback.

1. *Seek to understand others' perspectives.* Ask questions to gain an understanding of any raised issues or concerns. Make notes of potential power dynamics, privileges, or biases that might be at play. Encourage that all voices should be heard when discussing social justice topics., In other words, both perspectives should be shared and conversations should remain respectful and situate statements using facts, when possible.

2. *Avoid responding from an emotional stance.* Monitor the emotional levels by setting norms and taking breaks during the conversation to handle emotions. There might be backlash from young adolescents. You should use opportunities to help students develop morally, intellectually, and socially when engaging in social justice topics.

3. *Anticipate ways to respond to bias and stereotypes.* You should be careful to avoid singling students out when having discussions about social justice issues (Steele & Cohn-Vargas, 2013). It is essential for you to reiterate that complex social justice topics do not have simple answers.

4. *Don't send lengthy responses through email.* You have very little control over the perceived tone of an email and how parts of your message will be interpreted or used. A timely response is needed and remember that brevity is your friend. For example, "Thank you for your email. When is a good time for me to call to discuss your concerns?"

Figure 3.6. Social Justice Mathematics Lesson Planner

Social Justice Mathematics Lesson Planner

PART I

CONTENT

Essential Mathematics Concepts	Mathematical Practices

Social Justice Issue or Standards (Middle School Outcomes)

CONTEXT

Purpose

Audience

Allies

Timing

WHEN (in unit)

Circle one: Beginning Middle End Special Lesson	**Number of periods:** _____

Benefits

HOW

Circle one: Mathematics Tasks   Three-Act Tasks   Project-Based Learning   Other

OUTCOME/ACTIONS

PART II

Introduction/Engagement:

What will the teacher do?	What will students do?

Investigation/Exploration:

What will the teacher do?	What will students do?

Share and Discuss:

What will the teacher do?	What will students do?

Taking Action:

5. *Share your rationale for integrating a social justice topic into your mathematics classroom.* Use the guiding questions for what matters (Figure 3.1) and the Social Justice Mathematics Lesson Planner (Figure 3.6) as you prepare for the lesson and anticipate outcomes or reactions. Having a written plan to reference will allow you to share some of your thinking as you planned for the lesson.

6. *Pick your battles.* Recognizing areas of compromise and noncompromise in the conversation is important. Refrain from getting caught up in the small nuisances and details. Remember that your goal is to continue to use SJMLs in your mathematics classroom, so retreat to teach another day when the opportunity presents itself.

CONCLUSION

For many of us, the implementation of SJMLs is something we are fairly new to. As you begin to prepare, first recall that much of what will be required of you is already in your repertoire. There are many aspects of TMSJ that you have already implemented, such as organizing your curriculum around the content standards and teaching in ways aligned to Standards-Based Mathematics Instruction. You utilize many strategies to engage your students in meaningful discourse centered on their mathematical ideas. You draw upon students' social, cultural, and academic resources (funds of knowledge) in your instruction. For many of us, the aim to develop a critical consciousness in our students is something we value and are ready to dive into. The SJMLs in this book can be an excellent starting point.

In this chapter, we ask you to consider four elements in preparation: *Content, Context, When,* and *How.* What mathematical goals are important to achieve? What social justice goals can help further students' critical consciousness? Who can be allies? Does the topic reflect the interest of students? Might it simply be too hot at a particular moment in time? And finally, what is the best way to teach it? For many of us, we feel expert in mathematics but are quite wary of trying to teach about social injustice. Our experiences suggest that the best approach is to learn all you can, be open to learning more from your students, and foster a classroom environment that is hard on ideas and soft on people.

REFLECTION AND ACTION

Consider the following strategy for collaborating on the implementation of one of the lessons in this book or for developing your own. Work with your interdisciplinary team and organize your discussion using the following protocol:

1. Individually, write down a topic of injustice that may be of interest to your students on one note card or slip of paper. Create an additional note card for each topic. Spend several minutes identifying topics you'd be interested to build on across different classes.

2. Place all the cards in a bag, then draw one card. Discuss this topic as an interdisciplinary team.

3. Begin the discussion by responding to the *Content* and *Context* questions from Figure 3.1 ("Guiding Questions for What Matters").

4. Use all components of Part 1 from Figure 3.6, the Social Justice Mathematics Lesson Planner, to begin planning a lesson that might be used for this topic. (Here you are beginning to address the *When* and *How* questions in Figure 3.1.)

5. For your topic, identify allies, within and outside of your school, that you need to develop during the planning stage. What information would each of these people be curious to know? What might you learn from them?

6. Pair up and role play with someone: how would you share the topic/lesson idea with an administrator or colleague?

7. Finally, debrief and reflect individually and then with your interdisciplinary team. Which section (*Content*, *Context*, *When*, or *How*) was most difficult to complete?

INSTRUCTIONAL TOOLS FOR THE SOCIAL JUSTICE MATHEMATICS LESSON

As you think about establishing a classroom community based on TMSJ and integrating social justice issues into your classroom, we believe that your instructional toolkit is already filled with strategies to support mathematics teaching and learning or that you know where to go to learn new strategies. In this chapter, we will take a look at four areas—establishing goals, equitable teaching practices, discourse, and assessment—and provide suggestions for you to consider as you plan for the social justice component of the lesson.

PAUSE AND REFLECT

Before you continue reading, take a few minutes to reflect and make note of your instructional practices, facilitation of discourse, and assessment strategies.

ESTABLISHING GOALS

> The goals that you establish for your students directly impact the types of lessons you will design and implement, how you will assess your students, and ultimately how your students will see themselves as mathematically confident and competent critical thinkers and as doers of mathematics.

The goals that you establish for your students directly impact the types of lessons you will design and implement, how you will assess your students, and ultimately how your students will see themselves as mathematically confident and competent critical thinkers and as doers of mathematics. The SJMLs in this book are unique in the middle school mathematics classroom as they explicitly identify social justice outcomes aligned to the lesson, in addition to the mathematics content and practices. Thus, you must set both mathematics *and* social justice goals.

Mathematics Content

What important mathematical ideas do I want students to understand and be able to use?

The lessons in this book are based on the six mathematical domains of Ratio and Proportional Relationships, the Number System, Expressions and Equations, Geometry, Statistics and Probability, and Functions. Lessons aligned at the domain level provide you with flexibility when selecting lessons to meet your school, school system, or state expectations. This also helps you focus on designing learning experiences that allow your students to demonstrate what they know and can do. See Appendix C for a listing of the Essential Middle Grades Concepts and Appendix E for a listing of lessons aligned by domain.

Mathematical Proficiencies, Practices, and Processes

How do I want students to engage in mathematical activity?

There are particular ways of thinking and working that are characteristic of mathematical thinkers. Levassuer and Cuoco (2003) describe these as mathematical habits of mind that enable us to reason about the world from a quantitative and spatial perspective and to reason with our mathematical knowledge and skills to make sense of and solve problems. The Standards for Mathematical Practice (National Governors Association Center for Best Practices & Council of Chief State School Officers [NGACBP & CCSSO], 2010), Strands of Mathematical Proficiency (NRC, 2001), and NCTM Process Standards (NCTM, 2000) capture the expertise mathematics educators seek to establish in the middle school classroom as they develop their students. See Figure 4.1.

Figure 4.1. Mathematical Practices, Proficiencies, and Processes

Standards for Mathematical Practice (SMP) (NGA Center and CCSSO 2010)	
1. Make sense of problems and persevere in solving them.	
2. Reason abstractly and quantitatively.	
3. Construct viable arguments and critique the reasoning of others.	
4. Model with mathematics.	
5. Use appropriate tools strategically.	
6. Attend to precision.	
7. Look for and make use of structure.	
8. Look for and express regularity in repeated reasoning.	
Process Standards (NCTM 2000)	**Strands of Mathematical Proficiency** (NCR 2001)
Problem Solving	Conceptual Understanding
Reasoning and Proof	Procedural Fluency
Communication	Strategic Competence
Connections	Adaptive Reasoning
Representation	Productive Disposition

Social Justice Standards

How do I want to engage students in anti-bias, multicultural, and social justice issues to develop knowledge and skills to reduce prejudice and advocate for collective action?

The Social Justice Standards by Learning for Justice (2016) describe a social justice and anti-bias focus through four domains—Identity, Diversity, Justice, and Action—and can be found in Appendix D. Each domain is composed of anchor standards with grade-appropriate learning outcomes (see Figure 4.2). See Appendix E for a list of lessons correlated to the Social Justice Standards.

Figure 4.2. Social Justice Standards: Outcomes Appropriate at Grades 6–8 (Learning for Justice, 2016)

Social Justice 6–8 Grade Level Outcomes	
Identity 1	I know and like who I am and can comfortably talk about my family and myself and describe our various group identities.
Identity 2	I know about my family history and culture and how I am connected to the collective history and culture of other people in my identity groups.
Identity 3	I know that overlapping identities combine to make me who I am and that none of my group identities on their own fully defines me or any other person.
Identity 4	I feel good about my many identities and know they don't make me better than people with other identities.
Identity 5	I know there are similarities and differences between my home culture and the other environments and cultures I encounter, and I can be myself in a diversity of settings.
Diversity 6	I interact with people who are similar to and different from me, and I show respect to all people.
Diversity 7	I can accurately and respectfully describe ways that people (including myself) are similar to and different from each other and others in their identity groups.
Diversity 8	I am curious and want to know more about other people's histories and lived experiences, and I ask questions respectfully and listen carefully and non- judgmentally.
Diversity 9	I know I am connected to other people and can relate to them even when we are different or when we disagree.
Diversity 10	I can explain how the way groups of people are treated today, and the way they have been treated in the past, shapes their group identity and culture.
Justice 11	I relate to people as individuals and not representatives of groups, and I can name some common stereotypes I observe people using.
Justice 12	I can recognize and describe unfairness and injustice in many forms including attitudes, speech, behaviors, practices and laws.
Justice 13	I am aware that biased words and behaviors and unjust practices, laws and institutions limit the rights and freedoms of people based on their identity groups.
Justice 14	I know that all people (including myself) have certain advantages and disadvantages in society based on who they are and where they were born.
Justice 15	I know about some of the people, groups and events in social justice history and about the beliefs and ideas that influenced them.

Social Justice 6–8 Grade Level Outcomes	
Action 16	I am concerned about how people (including myself) are treated and feel for people when they are excluded or mistreated because of their identities.
Action 17	I know how to stand up for myself and for others when faced with exclusion, prejudice, and injustice.
Action 18	I can respectfully tell someone when his or her words or actions are biased or hurtful.
Action 19	I will speak up or take action when I see unfairness, even if those around me do not, and I will not let others convince me to go along with injustice.
Action 20	I will work with friends, family, and community members to make our world fairer for everyone, and we will plan and coordinate our actions in order to achieve our goals.

TEACHING EQUITABLY

Students who have the opportunity to critically examine issues of social injustice become informed citizens and active participants in a diverse society. However, this only happens with intentional instruction. When topics and voices are expressed that do not reflect the dominant or majority view, students have the opportunity to feel recognized, opening the potential to develop a sense of empowerment. These voices, however, can easily be shrouded by the dominant view, which is assumed to be normal in the local setting. As a TMSJ teacher, you can draw upon equitable teaching practices to deter this from happening.

We wrote in Chapter 1 that Standards-Based Mathematics Instruction (SBMI) is a foundational instructional quality of TMSJ. SBMI is characterized by an emphasis on learning for understanding and discourse-rich learning environments, marked by conjecture, reasoning, and justification. The NCTM publications *Principles to Actions* (NCTM, 2014) and *The Impact of Identity in K–8 Mathematics: Rethinking Equity-Based Practices* (Aguirre et al., 2013) identify research-based instructional practices that exemplify these two qualities of SBMI. *Principles to Actions* provides a set of eight mathematics teaching practices. *The Impact of Identity* names five equity-based instructional practices that support teachers in the use of these eight equitable mathematics teaching practices. Both sets are provided in Figure 4.3.

Bartell wrote, "When a teacher engages in practices that include strategies such as establishing norms for classroom participation, positioning students as capable, attending explicitly to race and culture, and pressing for academic success (for more, see Bartell et al., 2017), more students have access to the mathematics" (NCTM, 2020, p. 68). We hope the lessons in this book provide an opportunity for you to build important connections between your own teaching and these equity-based mathematics teaching practices as you center and create authentic experiences for students while also generating mathematics- and justice-based questions that probe particular issues and situations relative to students' lives and society.

Mathematics Teaching Practices	Equity-Based Practices
1. Establish mathematical goals to focus learning.	1. Go deep with mathematics.
2. Implement tasks that promote reasoning and problem solving.	2. Leverage multiple mathematical competencies.
3. Use and connect mathematics representations.	3. Affirm mathematics learners' identities.
4. Facilitate meaningful mathematics discourse.	4. Challenge spaces of marginality.
5. Pose purposeful questions.	5. Draw on multiple resources of knowledge (math, culture, language, family, community).
6. Build procedural fluency from conceptual understanding.	
7. Support productive struggle in mathematics.	
8. Elicit and use evidence of student thinking.	

Source: Reprinted with permission from Principles to Actions: Ensuring Mathematical Success for All, copyright 2014, by the National Council of Teachers of Mathematics. All rights reserved.

MANAGING DISCOURSE

Although we've established a separate section on discourse, it is a core element in your instruction. Our current understandings of how people learn help us to value the importance of facilitating meaningful mathematics discourse (Mathematics Teaching Practice 4). In fact, Huinker and Bill (2017) identify Mathematics Teaching Practices 3, 5, 7, and 8 as all being in service of supporting student-to-student discourse.

PAUSE AND REFLECT

Review these four practices and describe which equity-based practices you think promote discourse.

Managing discourse also means that you must be prepared to attend to the impact of status on student access to the whole-class and small-group discourse. You may have noted that all of the equity-based practices, but specifically practices 2–5, target discourse. Chapter 1 introduced Complex Instruction (CI; Cohen & Lotan, 2014). Here, we extend that by drawing upon the wealth of knowledge that Horn's (2012) book *Strength in Numbers* contains, which builds on the principles of CI and specifically supports discourse in secondary mathematics classrooms. We will summarize some key strategies in the following section, but more can be found by reading it yourself.

Discourse to Keep Students Engaged

During my instructional planning, how can I design opportunities and structures for meaningful discourse related to the lesson goals that engages my students?

An important part of facilitating a SJML is creating an environment of safe and open discourse among the students and with yourself, the teacher. As discussed in the previous chapter, this discourse often will extend to other teachers, school leaders, families, and community stakeholders. We know that classroom discourse does not happen by accident, nor is it an automatic component of particular lessons. How you position students to speak and be heard in the mathematics classroom is one of the most important responsibilities of a teacher and helps build a student's "sense of *belongingness*—students' innate need to establish close relationships with others" (Horn, 2017). The quality of discourse is also dependent on a classroom community of mutual respect that has been previously established by you.

As you plan to facilitate the SJML, the role of discourse is critical to keeping students engaged in the lesson. Here are three questions you should consider when planning to integrate a social justice lesson into your classroom. Your responses to these questions should provide you with some indication of how engaged your students might be with lessons that involve a social context, especially one that might be of a sensitive nature.

1. How are you being attentive to student concerns or questions about their school or community (or world)?

2. How are you staying connected with students' communities and realities?

3. Why would students see you as an advocate for them and other socially and historically marginalized people in their community?

You may also want to consider the *5 Practices for Orchestrating Productive Mathematics Discussion* (Smith & Stein, 2018) as a resource when planning for classroom discourse. This will help ensure that your plans anticipate students' responses as you monitor their work and select, sequence, and connect students' responses in meaningful and equitable ways to highlight the mathematical and social justice goals in each lesson. Figure 4.4 lists the five practices; we have added additional ideas for you to consider as you plan to use these strategies when integrating issues of social injustice into the mathematics classroom.

Figure 4.4. Benefits of Using Discussion Practices When TMSJ

Strategies to Support Discourse When Planning the SJML	
5 Practices—Mathematics Discourse	**Social Justice Discourse**
Anticipating likely student responses to challenging mathematical tasks and questions to ask students who produce them	***Anticipating*** likely student points of view and asking questions that help them identify specific points of agreement and disagreement
Monitoring students' actual responses to the tasks (while students work on the tasks in pairs or small groups)	***Monitoring*** students' ideas or positions to keep them focused on and grounded in the topic
Selecting particular students to present their mathematical work during the whole-class discussion	***Selecting*** particular students to share their outcomes or decisions with support so that diverse perspectives and voices are represented
Sequencing student responses that will be displayed in a specific order	***Sequencing*** student responses so that different perspectives and opposing points are shared and valued
Connecting different students' responses and connecting the responses to key mathematical ideas	***Connecting*** different student responses and perspectives to help students identify places where compromise might be possible, or if they should "agree to disagree"

Valuing Voice

How can I ensure that mathematical discourse—students actively speaking and being heard—is a prominent and equitable part of each student's experience in our classroom community?

Valuing each and every student's voice is a hallmark of a classroom grounded in TMSJ and equitable teaching practices. As a teacher, your actions and voice set the tone for establishing a classroom community that nurtures acceptance, belonging, and respect. The manner in which you speak and listen to your students in whole-class, small-group, or student-to-teacher settings establishes the model students will follow. Valuing voice is also demonstrated when you acknowledge students' different ways of thinking about ideas, which fosters a middle school student's sense of *autonomy*—the need to behave according to one's interests and values (Horn, 2017). Figure 4.5 suggests ways that you may promote or discourage discourse during a SJML.

Your actions and voice set the tone for establishing a classroom community that nurtures acceptance, belonging, and respect.

Figure 4.5. Teachers' Actions to Facilitate Discourse During a SJML

	Promotes Discourse	Discourages Discourse
Verbal Patterns	Facilitates and referees debate between students in a way that promotes student identity	Allows students to use language and tone that reduce the voice and experiences of other students
	Uses questioning probes and revoicing strategies that ensure equity and student ownership of thought	Moves from one mathematical or social discussion to the next without ensuring student voice was understood by others
	Uses open questioning to facilitate discourse, which allows students to create and maintain ownership and voice	Uses closed questions, which often include one-word responses that funnel students into a preplanned thought pattern
	Uses good "wait time" to promote discourse among students	Answers student questions quickly in group and classroom discussions without seeking student voice
Nonverbal Patterns	Uses tasks with multiple entry points, pathways to a solution, or differing solutions	Uses tasks with one solution or one potential pathway to a solution often found in directions of an activity or scaffolded exercises
	Uses body language such as head nodding, eye contact, and other strategies to affirm student voice	Uses head turns, lack of eye contact, crossing arms, eye rolling, and other body language that devalues student voice, perspective, or thought
	Chooses and uses student thought and perspective to ensure connection to social justice and mathematics	Does not seek equitable participation by calling on students and/or purposefully disregards student solutions, voice, and/or perspective

Students who feel a sense of respect, acceptance, and value from you are more likely to behave similarly among their peers. Modeling equitable speaking and listening skills with students will help establish a classroom community that makes room for discourse about controversial topics as part of lessons. You can be intentional in highlighting when students demonstrate speaking and listening skills that are productive to fostering respectful discourse.

Learn more about "fostering civil discourse" from Facing History and Ourselves (http://bit.ly/2mlfNhF).

Supporting All Students—Emergent Bilinguals and Students With Special Needs

The SJMLs in this book were prepared with a diverse student population in mind. The lesson design, task structure, mathematical and social justice content, and teaching practices were intentionally selected to allow all students in your classroom to explore mathematical concepts and communicate their ideas, regardless of areas of mathematical and cultural strengths and backgrounds. Emergent Bilingual students—a term we prefer over English Language Learners because it assigns competence rather than deficit—benefit from guided activities that allow for personal discovery as well as opportunities to express mathematical ideas with a language that makes sense to them. It is important for you to engage Emergent Bilingual students in student-student and whole-class discourse so that they can practice shifting from natural to academic language structure as well as develop mathematical concepts. In fact, your value of this discourse benefits all students.

Students with specific learning needs are also served well in a discourse-rich, multidimensional classroom in which all students have some mathematical strengths but none are viewed to be strong in everything. When given rich tasks, students need the mathematical (and social and cultural) assets and strengths of all of their peers in order to be successful with the task. Of course, when a required accommodation has been identified, it is still appropriate to use when TMSJ.

Dealing With Emotional or Sensitive Conversations

What sensitive issues or varying views, feelings, and perspectives should I be prepared to acknowledge, support, and validate?

As you plan for your lesson, we encourage you to carefully consider the classroom community when selecting the social justice issue. Earlier in Chapter 2 we discussed establishing a classroom community in which ideas are discussed, not people, and that all of us learn together. Rather than focusing on how a fellow student is wrong, you should have students seek to understand the reasoning that brought them to their conclusion.

In addition to establishing a classroom community focused on ideas and thinking, there are some topics where you will need to plan for a contentious topic, for emotional discussions, or for students to understand diverse perspectives. Figure 4.6 highlights several resources to support these three aims with a brief description. We recommend that you review each resource to learn more.

Figure 4.6. Resources for Creating Safe and Brave Spaces for Conversations on Sensitive and Difficult Topics

Resource	Description
Learning for Justice. *Let's talk! Discussing race, racism and other difficult topics with students.* https://bit.ly/3dhjsV7	This free resource includes strategies to talk openly about social inequalities and discrimination and other sensitive topics. It provides three strategies with specific examples for each: 1. Reiterate → Contemplate → Respire → Communicate 2. Check in with students. 3. Allow time and space to debrief. A wealth of other resources is also included.
Facing History and Ourselves. *Teaching current events educator guide.* http://bit.ly/2kvsdTK	This free resource provides tools and strategies to build your toolkit as you plan to facilitate discussions involving social justice and other sensitive topics. A highlight of the resource is the *Recommended Teaching Strategies*, which is organized in the following two sections: **When you want students to** discuss a contentious topic, **Try this strategy: Four Corners** where students show their position on a specific statement by standing in a particular corner of the room.

Resource	Description
Boaler, J., & Humphrey, C. (2005). *Connecting mathematical ideas: Middle school video cases to support teaching and learning.* Heinemann.	This resource provides insight into Cathy Humphrey's middle school classroom through rich text and video. Middle school topics related to geometry, proof, algebra, and fractions are included along with connections to • building on student ideas, • encouraging class participation, and • using student errors.
Cabana, C., Shreve, B., Woodbury, E., & Louie, N. (2014). *Mathematics for equity: A framework for successful practice.* Teachers College Press.	The historic Railside High is used in this text to provide a framework for successful equitable practices. Space is provided in describing the Railside approach, student experiences, community, and teacher development. Chapters of particular interest include insights into the following: • Detracked, heterogeneous classrooms • Supporting engagement and success • Equitable pedagogy • Reculturing a department

In many lessons that examine social injustices, there is potential that the names and terms we use for people involved can serve to further marginalize them, both in terms of their own sense of identity and how others perceive them. Terms matter for students (people) deprived of representation and can open a door toward the development of a more positive identity. Further, a definition can point to a community, making a person feel less alone (Learning for Justice, 2016). Identity is important.

We view the mathematics classroom as a place where students learn to exchange ideas, listen respectfully, and provide constructive feedback. Our aim is that all students can experience this discourse-rich classroom without fear or intimidation. Many of us have refined the development of this classroom culture for discussions of mathematics. However, many of us feel a small amount of intimidation to bring issues of social injustice into the same climate. We do not feel well prepared to respond to emotional or sensitive issues. We know that many topics will be very meaningful to many of your students. We believe that engaging your students in difficult conversations provides them opportunities to develop critical thinking skills, empathy and tolerance, and a sense of civic responsibility. We encourage you to rely upon the lessons in this book and the teaching strategies in this chapter to plan for these meaningful, difficult, and important conversations.

ASSESSING PURPOSEFULLY

Assessment is a mainstay of the mathematics classroom and plays an important role when integrating social justice issues into the lesson. It is the aspect of your work in which you seek to know the degree to which your students have met your goals. You may want to develop assessments that capture students' understanding of the mathematics (content and practice), the social justice issue, and the interplay of the two. Assessment also connects to two of the motivational features that Horn (2017) mentions we need to consider when planning for a SJML: competence and accountability.

- **Competence** describes the need to be successful in meeting goals and interacting with the environment (Wigfield & Eccles, 2002).

- **Accountability** refers to the structures and routines that oblige students to report, explain, or justify their activities.

Assessment Strategies

Which strategies can I use to monitor my students' understanding of the mathematics and social justice issue?

The use of formative assessment gives you the opportunity to provide students with actionable feedback and support the discourse related to the mathematics content and the social justice issue. Strategies to support formative assessment and connect formative assessment to discourse are shared in Figure 4.7, which is adapted from *A Fresh Look at Formative Assessment in Teaching Mathematics* (Silver & Mills, 2018). We specifically identify strategies that support your actions that positively position your students, build their identity and sense of agency, and support a motivational instructional design.

Figure 4.7. Assessment Strategies to Support Mathematics and Social Justice Discussions

Strategies for Supporting Formative Assessment	Connections Between Formative Assessment Strategy and Discourse
Providing feedback that moves learners forward	• Be strategic about when to tell (e.g., when to show students what to do rather than letting them struggle through and figure it out). • Have high expectations for how students should work with their groups by not rescuing them when they are stuck. • Explore incorrect answers sometimes. • Monitor the room as students are working. • Use what you learned during monitoring to plan for productive discussions.
Activating students as the owners of their learning	• Invite students to share their ideas. • Position students as having the right to evaluate the reasonableness of one another's mathematical ideas. • Position students as authors of mathematical ideas. • Facilitate a growth mindset through discourse.
Activating students as resources for one another	• Have students talk to one another in mathematics class. • Make strategic use of group work. • Use the think-pair-share strategy to give students an opportunity to think individually and to give all students opportunities to discuss their ideas. • Provide students with accountable talk stems (Michaels et al., 2008).

Developing opportunities for students to exhibit a sense of competence relies heavily on the assessment techniques you choose to use. Three assessment techniques that we have found useful in planning to assess students' understanding of the social justice issue are highlighted in *The Formative 5: Everyday Assessment Techniques for Every Mathematics Classroom* (Fennell et al., 2017)—observations, interviews, and Show Me. In Figure 4.8, we have expanded on the brief description of the technique with recommendations as you plan to use observations and interviews as assessment tools.

Figure 4.8. Techniques to Assess Students' Mathematics and Social Justice Understandings

Assessment Technique	Brief Description	Assessing Mathematics Goals	Assessing Social Justice Goals
Observations	Make informal and targeted observations of students engaged in mathematics learning. Constantly gather evidence of student progress as they engage in the mathematics and social justice task.	Take time to record a few "look-fors" for the mathematics part and social justice issue. For closed mathematics tasks, develop an answer key with anticipated answers. For open mathematics tasks, create several solution pathways or key features you expect to see in a solution.	Anticipate multiple perspectives and various points of view. Monitor students' emotions during small- and whole-group discussion.
Interviews	Have brief informal conversations with a student or small group of students that provide a "deep dive" into student thinking and understanding. Continuously monitor student progress and look for opportunities to support students.	Write questions that push and probe students' thinking. Include questions that address anticipated misconceptions and suggestions to build from students' strengths.	Write specific questions that address the lesson's social justice goal or standards. Develop question prompts to connect the mathematics and social justice issue.
Show Me	This is a performance response that requires a student or group to demonstrate their thinking and orally explain their response.	Use this on-the-spot extension of an observation or interview that extends or deepens your understanding of what was observed or heard in an "interview." Help identify a level of progress in understanding of a standard.	Solicit multiple perspectives and various points of view with supporting information. Encourage students to consider "role playing" or storytelling as a means to "demonstrate" their response.

"Misconceptions"

How might I consider what students do not yet understand in ways that view them as sensemakers rather than deficient?

In your role as a mathematics teacher, you are required to structure your curriculum in a way that focuses on what you want students to know (e.g., the standards students need to learn this year). Almost by default, you are positioning students using a deficit perspective. Further, your use of formative, summative, and standardized testing leans toward a focus on students' errors and mistakes. For example, many districts analyze their test data for what young adolescents score poorly on and ask teachers to respond with the assumption that students do not know or did not learn the concept.

While knowing what your students do not yet understand is useful in some areas of your work, there are several rationales to also focus strongly on student conceptions, or funds of knowledge. First and possibly foremost, attending to what students know and can do positions them in a positive manner; rather than lacking or broken in some way, students are viewed as having reason and experience worthy of an adult's attention. Second, by examining social, cultural, and academic (mathematic) funds of knowledge, we can design instruction that is connected to and builds on these assets. And finally, such an orientation means you are learning from the students as the students are learning from you. Both have expertise.

> As our profession is recognizing more and more that mathematics is a social activity, mathematics is not the universal truth we may have once thought it to be. We have learned to value many valid pathways to a solution, and even possibly many valid solutions. Even mathematics is not so cut and dried.

Recognizing the social, cultural, and academic expertise that students bring helps address a challenge many will experience as you implement SJMLs. We have felt a tension in our work teaching mathematics for social justice—how to know what is the "right" answer to the social dilemma or injustice. And further, what is our role in communicating to students the "correctness" of their personal conclusion? In some ways, we're drawn to wish things were more cut and dried, like mathematics. However, as our profession is recognizing more and more that mathematics is a social activity, mathematics is not the universal truth we may have once thought it to be. We have learned to value many valid pathways to a solution, and even possibly many valid solutions. Even mathematics is not so cut and dried.

Yet not every answer to a mathematical problem is correct, nor will every pathway be productive or more broadly applicable. Horn (2017) shares that normalizing mistakes as opportunities to grow and learn addresses a motivational feature by increasing a student's sense of competence. We, the authors, have learned that we can rely on our experiences teaching mathematics in which we draw out student ideas to devise parallel strategies for the discussion of student ideas about a social context or injustice. You should ask students to support their conclusions with evidence. Sometimes rather than building from definitions, ask students to state the moral or ethical grounding (or assumption) that leads to the conclusion they make.

Finally, as we rely on the middle grades outcomes for the Social Justice Standards to identify learning goals for the lesson, rather than aiming for students to have the *right* answer to the social injustice, we can ensure the classroom experiences and conversations serve the social justice goal we set for the lesson. Just as with mathematics, you shouldn't use a task for the purpose of students obtaining correct answers. Rather, the task you use creates a rich opportunity for robust and personally meaningful investigation, reflection, and understanding of important mathematical ideas and social experiences (see Chapter 5), those you've named as the goals for the lesson.

When you focus too strongly on student mathematical errors and misconceptions, students become wary to share ideas in the classroom. The same principle holds true if you silence ideas related to social justice issues; this would be counter to the aims. You must develop a classroom community that values students sharing ideas, mathematical or critical, even when they are in "rough draft" form (Jansen et al., 2017). And then ask your students to shape, debate, and refine them as a part of the lesson. When your mathematics hat is on, you might not let an error in mathematical understanding go—sometimes corrected in the moment, or noted as problematic in the class, likely to be revisited in the future. We suggest a very similar approach to concerns about unjust conclusions about a social context. It is your role as teacher to have some say in identifying where student ideas conflict with experts in the area, or even when there is disagreement among people who are deeply involved in the topic. But rather than viewing your students as incorrect or ignorant, we suggest working to examine students' current conceptions or funds of knowledge, to understand what may lead them to the conclusions they reach. Consider Sara Rezvi and Tyrone Martinez-Black's reflection on Lesson 10.3 (*Water Is Life—Our Collective Past, Present, and Future*) as a way to think about the intersection of the words we use, the context in which we live, and our progress:

> *When we hear the phrase "water is life" from Indigenous activists in Standing Rock and beyond, what questions are brought up for those of us who are settlers in the United States? In American society, water is assumed to be drinkable, potable, hygienic, and perpetually flowing. Yet, as in the cases of Flint, Michigan and Standing Rock, access to water for Black, Indigenous, and other people of Color continues to be severely limited due to human greed, capitalism, legacies of colonization, and white supremacy. The purpose of this lesson series is to take a deep look from a social justice and mathematical perspective that allows students to interrogate their relationship to water. We explicitly use First Nations' perspectives on the relationship of people to water in past, present, and future forms. We refer to this way of knowing as an alternative to Western conceptions of water as property, expendable, and commodifiable. Instead, we direct students to interrogate their relationship to future generations and the Blackfoot nation epistemology of cultural perpetuity.*

As we rely on the Social Justice Standards middle grades outcomes to identify learning goals for the lesson, rather than aiming for students to have the *right* answer to the social injustice, we can ensure the classroom experiences and conversations serve the social justice goal we set for the lesson.

CONCLUSION

If this is your first foray into implementing SJMLs, you may have some trepidation about how things will turn out in your classroom. We report on the experiences of several teachers and lesson authors in Chapter 11. One common theme is the value of planning in advance. While you cannot predict every detail of how a lesson will unfold in your classroom, we've all experienced the power of pausing to consider how students might respond to a prompt or task, how to better structure an activity to ensure equitable discussions, how to encourage responses to ideas rather than people, or how to justify a response through previous knowledge and facts.

Our emphasis is that you already possess many of the skills to implement a SJML effectively. You might learn some additional strategies to support student-to-student discussion about sensitive or emotional topics. Or you might refine your formative assessment strategies to focus on your social justice goals in addition to your mathematical goals. Ultimately, TMSJ presses you to be a more effective teacher of young adolescents. Mathematics doesn't have to be the only thing foregrounded. Students' interests and concerns about their lives, their community, and their world can provide you the rich context to develop informed and active citizens, able to participate in and shape a diverse society.

REFLECTION AND ACTION

Take a moment to pause and reflect on your instructional practices. How do your instructional practices build a sense of belonging and community? What assessment techniques might you need to adjust or learn? How ready are you to facilitate discussions—not just about the mathematics but about topics that might lead to emotional or divisive responses?

One of these steps can help you take a first step toward implementing one of the full SJMLs in this text.

1. Identify one Social Justice Standard from Learning for Justice (see Appendix D) that resonates with you. Select one discourse strategy that can help achieve that standard to integrate into your class.

2. Identify one Social Justice Standard from Learning for Justice (see Appendix D) that resonates with you. Select one formative assessment strategy for that standard to implement in your class.

3. Connect with a colleague who teaches Social Studies or English and discuss some of the techniques they might use to engage students in emotional or sensitive issues. Better yet, visit their classroom to see them in action.

TEACHING THE SOCIAL JUSTICE MATHEMATICS LESSON

CHAPTER

5

Tasks that honor the uniqueness of middle school students as individuals while also allowing them to experience mathematics in personal and socially meaningful ways are tasks that empower students to maximize the utility of mathematics in a variety of ways to understand and critique the world. (NCTM, 2020, p. 14)

Now that you have some background on the purpose, strategies, and pedagogical tools for teaching mathematics for social justice, you are ready to begin planning to teach a social justice mathematics lesson (SJML), a lesson that can begin to help you to achieve the aims set in this opening quote from NCTM's (2020) *Catalyzing Change in Middle School Mathematics* as the purpose of middle school mathematics. You will find that the lessons in this book begin with a problem. Most lessons begin with an opportunity for students to learn more about the problem and its social contexts. Students are encouraged to wonder and ask questions, providing an opportunity for them to create a meaningful and personal connection to the issue. This begins an authentic, sustained inquiry into the task. Although most of our lessons in this book are designed for 2 to 3 days, some could be extended even further. Each lesson closes with students taking action, a form of a public product.

In this chapter, we will provide you with details about the structure of the SJMLs in this book. It is our hope that understanding elements of the structure will not only help you think through the lessons as you plan to use them, but also support you to develop your own SJML. To do so, we introduce the design principles of the SJMLs in this book as well as the organizational components of the lessons. We introduce seven design elements, describe the lesson components, and provide guidance as you plan to implement them.

SOCIAL JUSTICE MATHEMATICS FRAMEWORK

The SJMLs we have included in this book align with seven design principles (see Figure 5.1) we established to maximize the learning opportunities afforded by the intersection of mathematics content and social justice issues. The seven elements of our SJML framework are as follows:

1. Building and Sustaining Beloved Community

2. Equitable Mathematics Teaching Practices

3. Authentic, Challenging Question or Concern

4. Social and Mathematical Understanding

5. Social and Mathematical Investigation

6. Social and Mathematical Reflection

7. Action and Public Product

These seven design elements have similarities to lesson design structures of Project-Based Learning (PBL) in which we place the focus of the PBL on examining a social injustice with mathematics, so that we can take action. Additional details about the seven elements of the framework are provided in Chapter 11, where we offer guidance as you begin to design SJMLs of your own.

Figure 5.1. Framework for the Social Justice Mathematics Lessons in This Book

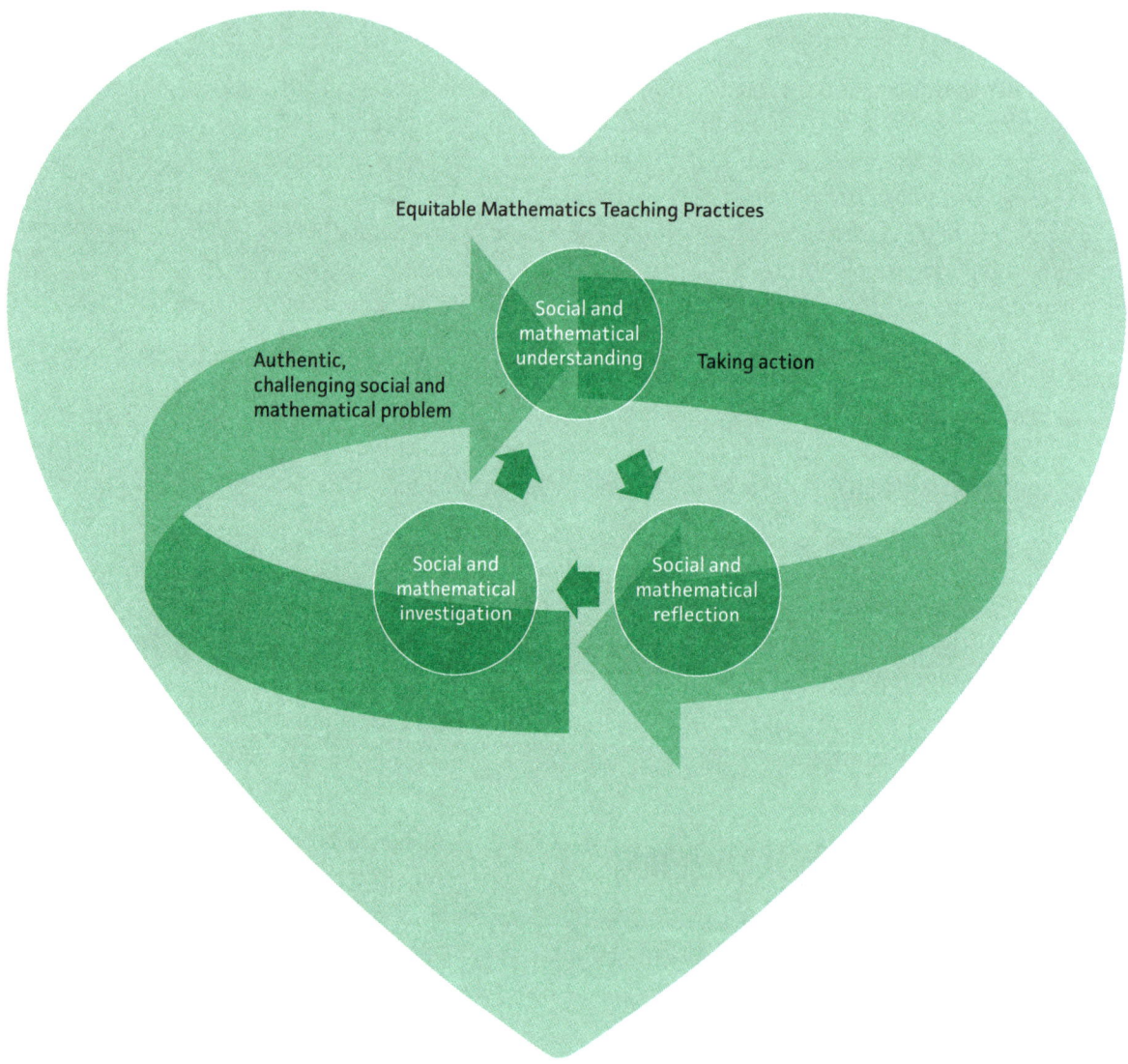

Source: Adapted from Berry et al. (2020).

Element 1: Building and Sustaining Beloved Community

In order for a SJML lesson to be completed effectively and in its entirety, you must create and sustain an environment of belonging, a Beloved Community (see Chapter 2). Your mathematics classroom should strive to be a microcosm of Beloved Community that supports middle school students' social, emotional, and academic development; helps them see and recognize our humanity and interconnectedness; and helps them understand what mathematics has to do with that.

Earlier in this book, we spent time describing this community and provided some tasks, tools, and practices you may use to build this community. The five facets of building and sustaining a Beloved Community in middle school mathematics classrooms that we introduced in Chapter 2 are as follows:

1. Reflect on your teacher positionality.

2. Learn about your students as people *and* specifically as young adolescents who do mathematics.

3. Establish mathematics classroom community commitments.

4. Engage students in mathematics community builders and icebreakers.

5. Facilitate temperature checks, in relation to and beyond mathematics.

Element 2: Equitable Mathematics Teaching Practices

A SJML must be structured in such a way that allows you to enact the equitable mathematics teaching practices introduced in Chapter 4. These practices are foundational to TMSJ and help you establish a classroom culture and environment that is ready to take on the challenge of integrating social justice issues with mathematics. We encourage you to revisit Figure 4.1 listing eight mathematics teaching practices (NCTM, 2014) and five equity-based practices (Aguirre et al., 2013).

The SJMLs selected for this book provide opportunities to build on the social, cultural, family, and community knowledge that your students bring to your class and to challenge spaces of marginality, specifically by centering student experiences and knowledge as legitimate intellectual spaces for investigation of mathematical ideas. Ensuring that the SJMLs provide opportunity for these equitable mathematics teaching practices is a fundamental step toward achieving Gutstein's (2006) mathematics and social justice goals: to read and write the word of mathematics and to develop positive social, cultural, and mathematical identities (see Chapter 1).

> The SJMLs in this book provide opportunities to challenge spaces of marginality.

Element 3: Authentic, Challenging Question or Concern

The SJMLs in this book are grounded in a question or concern that likely could arise from students, allowing for authentic and challenging learning. Queshonda, one of the authors of Lesson 9.2 (*The Mathematics of Toxic Air Emission* by Queshonda and Oluwaseun Kudaisi), shared that "one of the things motivating me to write this lesson is that it is actually the community that I grew up in. As a child, people in the community would discuss the issue of us having bad air.

However, we as children didn't realize that this was a documented reality." The social contexts of the SJMLs in this book include topics of injustice such as ability, econimic inequality, envioronmental isues, health inequality, human rights, representation, and world diversity. These contexts of these lessons can promote students to observe patterns, critique information, learn to ask questions, and reflect.

As noted, the contexts of the lessons here may be less authentic to your students. When possible, you might adjust the lessons to include a local context, data, or resource. Another strategy is to ask students, "Is there a similar concern here?" And encourage them to lead the modification to the SJML. The local, authentic context can serve as a powerful way to increase student engagement and motivation to learn mathematics, understand the social justice issue, and plan and carry out collective action.

Element 4: Social and Mathematical Understanding

When we consider "understanding" in the mathematics classroom, we often think about students involved with reasoning, justification, and proof. Implementing a SJML, you must now clearly identify what students need to know and understand mathematically and socially when engaged in the SJML. These lessons identify and provide opportunity to assess goals of three distinct types:

- *Mathematics Content*—what we want students to know and be able to do.

- *Mathematics Practices*—how we want students to show what they know and can do (NGACBP & CCSSO, 2010).

- *Social Justice Standards*—how we want students to demonstrate their understanding of and response to an issue (Learning for Justice, 2016).

While many of us have expertise in the middle school mathematics content and practices, the context of our social justice lesson goal(s) might be a new landscape. We offer a recommendation: allow yourself to be a learner as you implement these SJMLs. Focus first on the assessment of the social justice goal; don't feel like you have to know how to respond to support students who are not there yet or those who are excelling. Second, as you listen to how your students respond to the social justice learning goal, you likely will better understand the goal yourself. We encourage you to return to the Social Justice Standards, or specifically the middle grades outcomes, developed by Learning for Justice (2016) to consider how you might refine the learning goals you have for your students.

Element 5: Social and Mathematical Investigation

Because the tasks emerge from student questions or concerns, it is also important that the lesson be grounded in a mathematically driven investigation of the social context. By pursuing questions or concerns of your students (Element 3) and implementing tasks that build upon their own reasoning and problem solving, the lessons have strong potential to further positively impact your students' social and cultural identities. We wanted to ensure the lessons also provide an opportunity for positive impact on students' mathematical identities.

We also intentionally designed or selected tasks that support a discourse-rich classroom. As you implement the lessons in this book, plan for how you will *facilitate meaningful mathematical discourse*, Mathematics Teaching Practice 4 (Figure 4.3). The strategies you choose to use are critical to how you will position students as competent learners and doers of mathematics. Discourse instructional strategies are provided in Chapter 4, such as those in Figure 4.4.

Element 6: Social and Mathematical Reflection

Many characteristics of each lesson's implementation promote student reflection. These characteristics include the level of cognitive demand of the task, the questions posed by the students and yourself, and discussion about solution methods or ideas with peers. The SJMLs in this book are designed with characteristics that promote reflection about the mathematics, about the social issue, and about how the two inform one another.

Revisit your responses to the *Guiding Questions for What Matters* (Figure 3.1), along with Figure 3.2, to use a guidepost, especially when dealing with topics that may be considered emotional, sensitive, or controversial. As you consider opportunities for your students to reflect during a SJML, we recommend that you provide your students with individual, pair, small-group, and whole-class time to reflect.

Element 7: Action and Public Product

The final element required in the SJMLs in this book was to include an opportunity for students to take action or develop a public product. It was our determination that unless some form of action was included, a lesson could not fully achieve the potential we hoped to see. We asked that each lesson include a taking action or public product element that aligned with the Social Justice Standards, in which students express empathy, recognize their own responsibility, speak up with courage and respect, make principled decisions about when to take a stand against bias, or plan and carry out collective action.

Our hope is to not minimize the role of mathematics in the lesson but allow students to share how the mathematics has helped shape their understanding of the social justice issue. Thus, lessons in this book focus greater attention on the development of social justice goals because we know you will have several ideas to handle the mathematics.

An end result of investigating social justice issues is often a deeper understanding and awareness that somehow connects to

- *Identity*—how we view ourselves;

- *Diversity*—how we view others and their perspectives; and

- *Justice*—how we view fairness and unfairness, unequal power relations, and the impact of bias.

Each SJML has a "Taking Action" section that provides possible actions for you to consider introducing to your students. The suggestions are just that—a list of possible actions to get you, and the class, thinking. It will be more relevant and meaningful if you and your students develop your own actions as you seek to reduce injustice. It will also help determine the level of involvement with other students, teachers, administration, community, and other stakeholders.

REFLECTIONS ON PLANNING FOR "TAKING ACTION"

by Odesma Dalrymple, Marissa Forbes, Celina Gonzalez, Kristin Komatsubara, Perla Lahana Myers, and Joi Spencer, authors of Lesson 9.1 (*Playing With Data*)

Students become empowered when the world opens up to them and they make sense of phenomena that they may have never previously questioned. For example, when students recognize that their neighbors have greater or lesser access to parks or green spaces, it may awaken deeper curiosities and perhaps actions. Likewise, it may help them gain a better sense of the challenges that they have faced unknowingly. (What is it like for a young person to recognize that their community does not have the same resources as another?)

It has become imperative that young people understand how to decode the complex messages, images, and data that inundate their world daily. This lesson addresses data, including how data are represented; the messages and stories that data tell; and how the data align with who we are—our identities, emotions, interests, and values. The lesson allows students to explore the ways in which the same representation can tell two or even more stories. It is the reader, the learner, the scholar who must do the thinking and make the decision about what the data actually mean.

The students presented their learning of and their own thoughts about social justice when they presented their final representation to attendees of a STEAM Youth and Community Conference (https://bit.ly/2ZRBQkm) and on the University of San Diego S(TEAM)2 Academy website (https://bit.ly/3rs7Dnj).

PLANNING TO IMPLEMENT A SJML

Through planning, teachers can anticipate likely student contributions, prepare responses that they might make to them, and make decisions about how to structure students' presentations to further their mathematical agenda [and social justice agenda] for the lesson. (Smith & Stein, 2018, p. 5)

Lessons that promote mathematical and social discourse provide opportunities for students to engage in different mathematical strategies to solve or understand a social concern. Lessons may also potentially include different mathematical solutions to questions based on a student's perspective of a social concern. For example, in Lesson 10.1 (*Map Projections*), students determine that no two-dimensional projection preserves the integrity of all true spatial properties but that some may have more impact on social perspectives than others by elevating or diminishing specific properties. Thus, student perspectives may fluctuate but are grounded in different mathematical solutions.

How you implement a lesson is not a linear, predictable, or systematic process, as the structure of the SJMLs presented in this book may suggest. Although you may implement lessons as is from the book, lessons may also be adjusted to fit social justice concerns from the local community or questions from your students. Many excellent social justice lessons arise in classrooms in which teachers listen to and are involved with students and have established a culture of facilitating social justice topics focused on mathematical standards. Consider Lesson 7.1 (*Hey Google, Who's a Mathematician?*). This lesson directly encourages the examination of a search engine's images to understand potential bias toward a certain perceived gender and race. This same lesson may be modified to look at potential biases of different topics like what a family looks like or compare different search engine results transitioning the lesson from a traditional sixth-grade lesson on univariate distributions to a seventh-grade lesson comparing two distributions. More reflections from teaching these lessons and recommendations from the lesson authors of this book are provided in Chapters 11 and 12.

Common Structures for All SJMLs

The learning experiences students value and remember are often a result of their teacher's thorough planning and preparation. As you prepare to implement one of the SJMLs provided in this book, we recommend that you start by carefully reading through the lesson materials provided by the lesson authors. Each SJML provides the following information: Lesson Overview, Teacher Notes, and Student Resources.

Lesson Overview

The Lesson Overview provides the basic information about the social justice topic and mathematics content as well as a brief summary of the lesson. This snapshot will help you determine if you should study the lesson in more detail now or flag it to revisit at a later date. The Lesson Overview contains the following information:

- Short introduction to the context of the lesson
- Social Justice Standards middle grades outcomes addressed
- Short introduction to the Deep and Rich Mathematics of the lesson
- Mathematical Content Domain
- Mathematical Practices
- Structure and time of the lesson
- Resources and materials needed

Teacher Notes

Teacher Notes include detailed procedures that step you through how the author(s) carried out the lesson's implementation. Many of the lessons were written in one of the following three instructional models: Teaching With High-Cognitive-Demand Tasks, Three-Act Tasks, or Project-Based Learning (PBL). We recommend that after you review the Teacher Notes once, identify the core structures and then map them to the context of your students and classroom. Identify your mathematical and social justice goals for the lesson. Next, determine how you will assess student attainment of these goals.

As a next step, make any needed modifications to allow you to achieve and assess these student learning goals, which includes adjustments to best match the interests of your students, meet their learning needs, and match your instructional goals. In addition, you may want to revise the lesson's context to localize it to your school setting where appropriate. For example, Lesson 6.1 (*Food Apartheid: Graphing and Understanding Access to Healthy Food*) is set in Washington, DC, and should be adapted to your locality by using an internet mapping website to allow students to explore food access in their community.

After briefly introducing the Student Resources section, we next discuss several common elements of the Teacher Notes in order to highlight some of the resources available to guide you through the lesson planning and implementation:

- Introducing the lesson
- Facilitating the exploration
- Closing the lesson

Although each of these is familiar to us as mathematics teachers, we focus our discussion on the development of the social justice goals.

Student Resources

Many of the lessons provide suggested resources and worksheets. Previews of some of the resources and worksheets are provided in the book to help give context as you read the Teacher Notes. In order to allow you to modify or contextualize the lesson to your local setting, the lesson worksheets are provided in Word format on the book's companion website (resources.corwin.com/TMSJ-MiddleSchool).

Using the Lesson Overview to Plan

Each SJML is conceptualized as a sequence of 2 or 3 days (typically). However, the lesson authors taught in schools with classes of different time lengths. This will be one adaptation for many of you. Each day of the lesson is structured most typically with three phases: an opening, student work time, followed by discussion and conclusion. Most of these are tightly patterned after the structure to launch, explore, and then summarize (Smith & Stein, 2018). Next, we elaborate on each of these, specifically with the development of your social justice learning goals in mind. We discuss strategies to introduce the lesson, make recommendations about facilitating the exploration, emphasize important elements of closing the lesson, and reflect on the opportunities to assess student attainment of the learning goals.

Introducing the Lesson

TMSJ requires you to be attentive to developing your students' social, cultural, and mathematical identities. Many students' attitudes and perspectives are largely shaped by their past experiences in mathematics classes and their own lens of viewing the world. For this reason, many of the lessons introduce students to the social context at the beginning of the lesson. While there are numerous ways to introduce the social context, it is important that this introduction creates an interest in students to further explore. In this way, you have created an intellectual and possibly social or emotional need that will drive student learning. We highlight three options that can be effective to begin the SJML: storytelling, using articles or videos, and mathematics.

One effective strategy for introducing the SJML is the use of storytelling. Delgado (1990) identifies this pedagogical practice as a powerful opportunity for marginalized groups to draw attention to their own experiences and provide insights to others through their own lens. An example of this can be found in Lesson 8.3 (*The Black Vote in America: The Impact of the 1965 Voting Rights Act*). Chuck Munter and Cara Haines use storytelling to help connect students to the Black struggle struggle for representation in governance.

Articles or videos are another method that provides students the opportunity to connect to the social context of the lesson by drawing upon students' social, cultural, and academic funds of knowledge (González et al., 2001). For example, they may recall and build upon a conversation with family about a current topic or may draw upon skills learned in an English Language Arts class to read, analyze, and summarize print or video. Students may be introduced to an injustice through a video such as in Lesson 8.2 (*National Team Pay Investigation*). Lauren James, a teacher who field-tested Lesson 9.1 (*Playing With Data*), described the strength of using an image to connect to a city in Africa and wealth distribution:

> *I thought the image of Dar es Salaam was a good entry into a conversation about wealth distribution and access to resources. All students were looking at the same image and all students (in my classes) reached the conclusion that it wasn't fair that some people who may be wealthier had more green space, bigger homes, and even swimming pools, while some people (who may be less wealthy) live in communities that are more densely packed.*

Another route that some authors have taken to the opening of the lessons is to focus on the mathematics of the lesson first, allowing the mathematics to unearth or highlight the social concern. Gloria Ladson-Billings (2019) shared a story of a lesson taught by Bill Tate, currently President of Louisiana State University in

For additional strategies to develop your students' literacy skills, consider *Developing Literate Mathematicians: A Guide for Integrating Language and Literacy Instruction into Secondary Mathematics* (Hoffer, 2016).

Baton Rouge. He offered students $\frac{1}{14}$ and $\frac{1}{4310}$ for comparison mathematically with no context to students. As students began to see how different the two fractions were, they were introduced to the context of these fractions: $\frac{1}{14}$ of slaves in Massachusetts were declared insane, while $\frac{1}{4310}$ of slaves in Georgia were declared insane (Deutsch, 1944). Students were then asked to speculate why they believed the fractions were so different. Before the Civil War, several Southern states had laws to restrict or deny mental health care for Black people, such as a statute in Virginia that "decreed that 'no insane slave should be received or retained in either (the Eastern or Western Lunatic) Asylum so as to exclude any white person residing in the state'" (p. 480). In Dr. Tate's lesson, he drew in the learner's interest through the mathematics, opening the opportunity to consider the context of a social injustice and possibly further pursue the mathematics.

Facilitating the Exploration

Your primary role while students work is to facilitate student discourse, question to probe and push students' thinking, and use formative assessment strategies to learn students' level of understanding of both the mathematics and social justice goals. Chapter 4 offers several strategies to facilitate student-to-student discourse. The lessons in this book include both deep mathematical and social justice-related questions to support you in this work. The questions appear throughout each part of the lesson materials or are listed at the end. Although some questions may be used at the beginning of a lesson, the exploration phase of a lesson is the best opportunity for teachers to pose many of these questions.

During the opening of a lesson, your students may be reluctant to openly share their experiences around sexism, drugs, racism, experiences of bullying, religious intolerance, and more out of fear of how they may be seen by their peers. Similarly, students are less apt to share their mathematical thoughts openly without opportunities for self-reflection and peer checking. Using the deep mathematical and social justice questions in small groups allows students to grapple with their thoughts and how their responses might be perceived by their peers at a smaller level. In addition, you may purposefully group students in ways that provide support for student sharing of ideas.

During the exploration portion of a SJML, you should monitor student progress (see Figure 4.4) both mathematically and socially. Consider

- purposefully monitoring student explorations and facilitating these pathways toward meaningful mathematical and social discoveries,

- completing brief notes on students' mathematical pathways to answering the social justice issue or concern, and

- looking for students who have stories to tell that relate to the SJML and providing them with the opportunity to share with others outside their small group.

Closing the Lesson

Student voice and choice in the opening and middle of the lesson are important components of ensuring equitable participation and learning. However, the closure or summary of the lesson is arguably the most important time in which you select, sequence, and connect ideas from students. During this time, students begin to reflect on their personal or small-group ideas about the context and the mathematics, connect to others' ideas, and solidify their understandings. As you help students to identify connections to both the mathematical and social justice goals of the lesson, you can also be thoughtful to positively impact student identity. When possible, you should

- use the stories from students to highlight the mathematical and social justice question and exploration,

- use the notes you keep while monitoring to purposefully select students to share their stories as they relate to the SJML and are meaningful to highlighting the mathematical discoveries, or

- identify student ideas to be shared with the whole group in ways that promote the understanding of the main goals of the lesson.

Final Thoughts on Planning to Implement

It is our belief that a well-planned lesson is an important quality of effective, student-centered instruction. A carefully thought-through plan provides the greatest opportunity for powerful improvisation. When you have a strong sense of the learning goals, have considered at least one pathway to those goals, and have predicted how students may engage in that pathway, you can make sound decisions based on unpredicted student interests, experiences, or questions. Establishing goals and strategies to assess ensures you maintain a focus as the pathway is changed.

Each lesson in this book offers an overview, implementation guide, and student resources. We hope that these planning resources provided by the SJMLs in this book offer sufficient structure for you to design a well-planned lesson. These SJMLs are intended to be contextually relevant to your students and allow you to attain the mathematical goals for your middle school mathematics course.

LAST WORDS BEFORE YOU GO TEACH

We hope that the previous chapters have laid the groundwork for you to further advance your efforts to TMSJ. We have attempted to provide resources and connect to strategies you already know well in order to create some comfort—both competence and confidence—to implement some of the SJMLs in this book, modified to fit your student and school context. We are hopeful that shortly you will be ready to create your own SJML; we offer some recommendations in Chapter 12.

As you begin to implement the lessons in this book in your own classroom, consider the following chart of instructional strategies for teaching mathematics for social justice (Figure 5.2) to guide your preparation and implementation of the SJMLs in this book.

Figure 5.2. Social Justice Teaching Strategies

Teaching Mathematics for Social Justice—Instructional Strategies		
Preconditions for Teaching for Social Justice	**Characteristics of Teachers Who Teach for Social Justice**	**Social Justice Teaching Strategies**
Recognize and validate students' perspectives.	Incorporate student mathematical strengths and varied perspectives.	Engage students in the varied perspectives of other students.
Appreciate varied perspectives in school.	Demonstrate high expectations of each and every student.	Engage students in actionable social change efforts.
Value teacher–student relationships.	Facilitate discussion between students that ensures opinions are valued.	Legitimize students' real-life experience.
Value the stories and lived experiences of others.	Form emotional affiliation with each and every student.	Provide storytelling of others to shape and describe varying perspectives.
Provide space for authentic student voice.	Exhibit a genuine caring attitude toward each and every student.	Use investigative learning processes.
Ensure security for marginalized adolescents.	Engage with the community.	Provide real and meaningful opportunities to engage with data and contextual situations.
Give opportunity for students to authentically speak.	Listen actively and synthesize student voice.	Include content relative to students' lives—social and cultural experiences.

This chart can be also used as an opportunity to reflect and self-assess aspects of your instruction and relationship with students.

- Which of these do you think will come most easily for you?

- Which strategy do you need to learn more about?

- How might you build on your existing strengths as a mathematics teacher concerned with social injustices to take on more of these strategies?

- Which of these are you going to keep an eye out for as you look through the SJMLs in this book?

It is our hope that in looking at Figure 5.2 more than one time, you will also add instructional strategies and teaching qualities in each of the categories. At this stage in our careers, we do not think we have learned all there is to know about teaching mathematics well. Each of these instructional strategies or qualities provides a focus area where each of us as teachers can make a commitment to learning that becomes a career-long aim.

CONCLUSION

With this chapter, we close our introduction to TMSJ and send you forward to examine the SJMLs that come next. In this chapter, we discussed the seven design principles present in each of these SJMLs and identified how each of the lessons in this book is organized. It is our sense that understanding both of these aspects of our book will be useful for your implementation. However, more importantly to us, we hope that these design principles and lesson structures will propel you into the development of high-quality mathematical investigations into issues of social injustice that your students are passionate about.

REFLECTION AND ACTION

It's time to dive in and teach a SJML. For some of us, it is a significant step—a sort of leap of faith that our classroom won't go crazy, or that we won't get a parent or principal complaint. Consider the following as steps to help you get started to further enhance the work you are already doing.

1. Return to Figure 1.2, which identifies four frameworks of equity-driven mathematics teaching practices. Commit to enhancing your instruction in alignment with the ideas of one or more of these frameworks. Teaching is, for us, the most important aspect of TMSJ.

2. Review the SJML lessons in Chapters 6–10 and select one to teach in your class. Use the *Guiding Questions for What Matters* (Figure 3.1) as you consider the lessons and do your planning.

3. Invite a colleague/critical friend to watch you teach the lesson and help you reflect on the lesson.

As you begin or continue your journey to integrate social justice issues into your mathematics classroom, we invite you to share your story. Together we can all make a difference in the lives of our children. A facebook group—Mathematics Lessons to Explore Injustice (https://www.facebook.com/groups/178098736933840)—offers a space to share your experiences implementing one of the lessons in this book. We invite you to share your stories with us through Facebook and on Twitter @SJMathematics. Please consider responding to any of the following questions or contribute anything else you would like to share. The lesson authors have shared responses to these questions as well, which we provide in Chapter 11.

1. How did you modify the lesson for your context?

2. What was most valuable about your (and your students') experience when implementing a TMSJ lesson(s)?

3. What challenges have you been confronted with in the classroom, school, or community when TMSJ, and how did you overcome these challenges?

4. What bits of advice would you offer anyone implementing this or other TMSJ lessons?

PART II

SOCIAL JUSTICE MATHEMATICS LESSONS

CHAPTER 6

THE NUMBER SYSTEM

Middle school students should be able to apply and extend previous understandings of multiplication and division of whole numbers to fractions. Students computing fluently with multidigit numbers, using common factors and multiples, offer great opportunities to extend number standards to social injustices and real-life problems. Using contextual situations that relate to social experiences provides wonderful opportunities for students to use reasoning and sensemaking to build procedural fluency from conceptual understanding. Additionally, these situations can also help make mathematics more accessible by creating the need for students to label quantities and explain what a quantity means, and they provide teachers with an excellent opportunity to include lessons that attend to social injustices. The following lessons are some examples of teachers who have done just that.

LESSONS

Lesson No.	Lesson Title	Mathematics Focus Areas	Grade	Social Justice Topic	Authors
6.1	*Food Apartheid: Graphing and Understanding Access to Healthy Food*	• Number	6	Health Inequality	Becky Evans, Emmalee Bielenberg, Julia Novosad, and Cassie Ruettiger
6.2	*Cor(o)ner Stores and Food Apartheid*	• Number • Ratio and Proportion	6	Health Inequality	Michelle Cody and Kari Kokka
6.3	*Billionaire Power*	• Number	7	Economic Inequality	Dr. Natalie Odom Pough and Y. Rhoda Latimer
6.4	*Middle School Mathematics to Explore People Represented in Our World and Community*	• Number • Ratio and Proportion • Statistics and Probability	7	World Diversity	Mary Candace Raygoza, Eva Thanheiser, Courtney Koestler, Jeff Craig, and Lynette Guzmán

LESSON 6.1 FOOD APARTHEID: GRAPHING AND UNDERSTANDING ACCESS TO HEALTHY FOOD

Becky Evans, Emmalee Bielenberg, Julia Novosad, and Cassie Ruettiger

HEALTH INEQUALITY

Many times, fast-food restaurants and convenience stores serve as locations for meals instead of full-service grocery stores. The city of Lincoln, Nebraska, has a wealth of resources for families to gain access to food such as the local food bank, weekly food backpack distribution, and other programs. However, none of these supplemental programs address why almost half of students' families may struggle to gain access to fresh food. *Food apartheid* refers to how communities have differential access (whether because of distance and/or insufficient public transportation) to healthy foods. The purpose of this lesson is for students to understand that all people do not have equal access to food within their community. The location of where one lives may impact their access to fresh and healthy food, which may have other lasting effects. The lesson will enable students to explore where in their city grocery stores and fast-food restaurants are located and consider how a person's relative location to these resources may impact their actions.

DEEP AND RICH MATHEMATICS

In this lesson, students will use coordinate graphing in all four quadrants to make sense of their local community. Each student will be responsible for graphing several points throughout the plane. Then, students will collectively look at all their plotted points to identify patterns and make observations regarding the data. They will use their observations to construct their own arguments as to if or how food apartheid impacts their community. Determining the distance between two points using students' own language and strategies will lead into a discussion of magnitude and absolute value. This will enable them to explore the potential consequences of lack of access to a full-service grocery store.

ABOUT THE LESSON

This lesson uses a launch–explore–summarize instructional model and is intended to take approximately 90 minutes to complete across two class periods.

Lesson 1: Students will use coordinate graphing in all four quadrants to make sense of their community. Describe the distribution of grocery stores in the city. Think about which regions look like they could be negatively impacted by food apartheid and begin to consider solutions to this problem.

Lesson 2: Compare distances from your "home" point to different grocery stores and fast-food restaurants. Discuss student-generated methods for finding the distance between two points and relate those methods to absolute value.

Resources and Materials
- Teacher Resource 1: *Image of a Complete Grocery Store Map* (Day 1)

- Teacher Resource 2: *Grocery Store Coordinates*

- Desmos activity: *Map of Lincoln*, to be shared with students (https://bit.ly/3luWAWD)

- Video: "Food Deserts in DC: Let's Talk," from NPR (https://bit.ly/31l2gM1)

- Blog post: "Food Apartheid: Racialized Access to Healthy Affordable Food" by Nina Sevilla, *NRDC Expert Blog*, April 2, 2021 (https://on.nrdc.org/3rteFYP)

- Website: U.S. Department of Agriculture (USDA) Food Access Map (https://bit.ly/32WfX4z)

LESSON 1 FACILITATION

Where Are the Grocery Stores?

Launch (15 minutes)
- Begin the lesson by drawing on students' funds of knowledge. Ask: *What is a food apartheid?* Engage students in a think–write–pair–share to allow for individual processing time prior to opening up for sharing ideas to the whole group. Record students' ideas in a visible location (but do not provide feedback on the accuracy of their response).

- Tell students, *Today we will use what you know about graphing within the coordinate plane to explore our community and determine if a food apartheid exists where we live.*

- Show the video "Food Deserts in DC: Let's Talk" (https://bit.ly/31l2gM1). Facilitate a short discussion about the video. Potential discussion questions include these:

  + *How does food apartheid impact those who do not have access to transportation?*

  + *What might the impact be on these individuals and their families?*

- Include conversation on problematizing the term *food desert*, such as described in the *NRDC Expert Blog* post, "Food apartheid: Racialized access to healthy affordable food," by Nina Sevilla, April 2, 2021 (https://on.nrdc.org/3rteFYP).

- Engage students in a whole-class conversation about graphing points in the first quadrant (fifth-grade standard). Select one point from Quadrant I to graph as a class (each point represents a grocery store). Emphasize that mathematicians have agreed on the convention that the first coordinate is related to the horizontal axis and the second coordinate is related to the vertical axis. Ask: *How does the location of a point change if we move horizontally then vertically? Vertically then horizontally?*

Explore (20 minutes)

- Tell students that their job is to answer this question: *How do we experience food apartheid in our city?*

- Assign students into small groups of three to four.

- Distribute physical or digital versions of the city map and a list of several coordinates to each student.

- With each group, assign each student several coordinates/grocery stores to plot on the map of the city (make sure that the coordinates are spread throughout multiple quadrants). Students can ask those within their small group to verify the location of their points.

Summarize (10 minutes)

- Engage students in a whole-class discussion about the distribution of grocery stores across the city. (If you are using Desmos to facilitate this lesson, overlay each student's points onto a single map.) Use the following questions within the discussion:

  + *What do we notice is the same about all the coordinates in Quadrant I?*

  + *What about Quadrant II? Quadrant III? Quadrant IV?*

In this discussion, make sure the following mathematical concept is clear: In the first quadrant, x- and y-values are positive. In the second, x is negative and y is positive. In the third, both x- and y-values are negative. In the fourth, x is positive and y is negative.

+ *What do you notice about the distribution of the grocery stores in the city?*

Prompt students to use precision in their descriptions of the map. Use mathematical language to describe the relative distance between points and in relation to the origin. Potential methods include describing the space between coordinates, estimation, or precise measurement. Record students' observations onto the whole-class map.

+ *Based on our graph, how do you think people experience food apartheid in our city?*

+ *Where might you recommend building a grocery store? Why?*

LESSON 2 FACILITATION

Home to the Grocery Store

Launch (5 minutes)

- Begin class by connecting to prior learning. Say:

  Yesterday we looked at a map of our city and you plotted points that represent grocery stores to decide how our city experiences food apartheid. Today we will look at some additional points on our map to consider the potential impact on those living in areas with restricted access to full-service groceries.

- Arrange students into the same small groups. Display/distribute the Quadrant I fast-food graph. See the Desmos activity, *Map of Lincoln*.

- Ask:

  + *What do you think these red points represent?*

  + *What relationships do you notice between these points and the grocery stores you located yesterday?*

  Allow students time to discuss and then share out ideas with the large group. Record observations on the board.

Explore (20 minutes)

- Say: *Today we are going to put ourselves into someone else's shoes. Imagine that you live at Home ___ and that you do not have reliable access to a car* (each group will be provided a different home point within Quadrant I or Quadrant IV). *You rely on your bike or walking to get from place to place.*

How far would you need to walk to get to a fast-food restaurant? How far would you need to walk to get to a full-service grocery store?

+ Tell students their job is to determine the distance between both their assigned home and the nearest fast-food restaurant and their home and the nearest full-service grocery store. Consider providing students with the list of coordinates for grocery stores shown in the Desmos activity, *Map of Lincoln*. This list will provide students with the names of the fast-food restaurants in their quadrant as well as the coordinates.

+ Provide students with the following information: One unit on the coordinate plane is equivalent to .25 miles. Students can provide their answers in either whole units or miles. They must travel in straight lines (stay on the sidewalk). Students can only move in the four cardinal directions when creating their routes and finding the distance.

• Monitor students' conversations. Record any important topics you overhear a team say that may be beneficial for the whole class to hear. When working within their small teams, students will need to determine the following:

  + The closest fast-food restaurant given a straight path

  + The closest full-service grocery store given a straight path

Summarize (20 minutes)

• Engage in whole-class discussion about the distribution of fast-food restaurants.

  + *Where are they located? How close are they to each other?*

  + *Are there grocery stores near them?*

  + *Do most people live closer to fast-food restaurants or grocery stores?*

• Engage students in a whole-class discussion related to their discoveries. During the conversation, be sure to ask the following questions:

  + *How did you determine the distance between the two points/locations?* (Allow students to share different approaches; that is, students may count by single units, add together distances to find a total, etc.)

• Make sure to include a discussion of how moving "left" or "down" within the coordinate plane may indicate a negative number. However, this distance represents the absolute value, or the magnitude, of that distance.

  + *What was the closest fast-food restaurant to your assigned home? Closest full-service grocery store? How did you know?*

  + *Why might someone choose to go to a fast-food restaurant instead of a grocery store?*

+ *What could be some consequences of living closer to fast-food options than a full-service grocery store?*

+ *What might be some ways to eliminate food apartheid in our community or others? What might be some challenges to making these changes?*

TAKING ACTION

Option 1: Each class will likely develop their own plan for action, specific to the culture and needs of their community. Some possibilities include the following:

- Write a letter to your local government officials, advocating for the government to incentivize grocery stores to open new locations in communities without much access to grocery stores.

- Contact local grocery stores and suggest that they provide free or low-cost delivery services of fresh groceries to people living beyond walking distance of the store.

- Encourage students to do additional research related to the following:

    + The impacts of limited access to fresh foods

    + How communities are or are not addressing food apartheid

Option 2: Engage students in using the USDA Food Access Map (https://bit.ly/32WfX4z) to determine if there are areas within your local community that are actually designated as "food deserts." Students can compare this with the conclusions they arrived at during the lesson.

ONLINE RESOURCES

 Available for download at **resources.corwin/TMSJ-MiddleSchool**

▼ *Teacher Resource 1: Image of a Complete Grocery Store Map (Day 1)*

▼ *Teacher Resource 2: Grocery Store Coordinates*

LESSON 6.2 COR(O)NER STORES AND FOOD APARTHEID

Michelle Cody and Kari Kokka

HEALTH INEQUALITY

The two activities in this lesson explore health inequities related to *food apartheid* in under-resourced communities. Students are invited to analyze the availability of healthy foods and the distance needed to travel to purchase groceries. Both lessons invite students to share their feelings about living in a community without many healthy food options, using healing-informed social justice mathematics practices (Kokka, 2019) that incorporate radical healing (Ginwright, 2016) and trauma-informed care practices (Harvey, 1996).

DEEP AND RICH MATHEMATICS

In Lesson 1, students will engage with the number line to learn addition and subtraction of signed numbers and use the magnitude of distance from home to learn absolute value. In Lesson 2, students will use rate, ratio, and proportion to find one's travel rate to analyze the time to travel to and from grocery stores. The activity considers individuals with different travel times of walking, running, biking, skateboarding, and using a wheelchair or walker. Conceptual understanding is encouraged through the use of open-ended questions that invite multiple solution pathways with multiple representations of the mathematics. Extensions are also offered to adapt the activity for high school mathematics, such as graphing distance versus time and velocity versus time graphs for various travel scenarios, including taking the bus.

ABOUT THE LESSON

This lesson uses a launch–explore–summarize instructional model and is intended to take approximately 180 minutes to complete across two class periods.

Lesson 1: *Cor(o)ner Store: Where Healthy Food Comes to Die.* Students use the number line to add and subtract signed numbers and understand absolute value while exploring traveling to multiple stores to purchase healthy foods in a community without nearby access to grocery stores.

SOCIAL JUSTICE OUTCOMES

- I know that all people (including myself) have certain advantages and disadvantages in society based on who they are and where they were born. (Justice 14)

- I will work with friends, family and community members to make our world fairer for everyone, and we will plan and coordinate our actions in order to achieve our goals. (Action 20)

ESSENTIAL MIDDLE GRADES CONCEPTS

- **Number**—Grade 6: Apply and extend previous understandings of numbers to the system of rational numbers.

- **Ratio and Proportion**—Grade 6: Understand ratio concepts and use ratio reasoning to solve problems.

MATHEMATICS PRACTICES

- Construct viable arguments and critique the reasoning of others.

- Model with mathematics.

Lesson 2: *Grocery Trek.* Students use rate, ratio, and proportion to analyze how long a trip would take to travel to multiple stores to purchase healthy foods in a community without nearby access to grocery stores. Students measure their own travel rate and consider multiple means of movement, such as walking or using a wheelchair, walker, bike, or skateboard.

This lesson is inspired by the Chicago Grassroots Curriculum Taskforce Community Tour (Pulido et al., 2013).

Resources and Materials

- Worksheet 1: *Cor(o)ner Store: Where Healthy Food Comes to Die* (1 per student)

- Worksheet 2: *Grocery Trek* (1 per student)

- Worksheet 3: *A Call to Action* (optional, 1 per student)

- Student Resource 1: *Task Card*

- Teacher Resource 1: *Grocery Trek Optional Scaffolds and Extensions*

- Teacher Resource 2: *Additional Lesson Resources*

- Poster paper (bigger than 11.5 × 17 inches)

- Markers

- Rulers

- Graph paper

- Timer (1 per group)

- Measuring tape (1 per group)

- Video: "Verify: Living in a Food Desert" (https://bit .ly/3dcIRzh)

- Video: "Food Deserts in DC: Let's Talk," from NPR (https:// bit.ly/31l2gM1)

- Article: "'Food Apartheid,' Not 'Desert,'" University of Texas at Austin Campus Environmental Center (https://bit .ly/3G947SL)

- Website: Worldometer, "United States Population" (https:// bit.ly/3xNEvIc)

- Website: U.S. Department of Agriculture Economic Research Service, "Food Access Research Atlas" (https://bit .ly/3lr4wbm)

- Video: "Why Is the 1% So White?" from Francesca Ramsey of MTV Decoded (https://bit.ly/3Id0ixP)

- Documentary series: "Race: The Power of an Illusion" (https://www.racepowerofanillusion.org)

- Article: "The Future of Healing: Shifting From Trauma Informed Care to Healing Centered Engagement" by Shawn Ginwright, May 31, 2018 (https://bit.ly/3Dh3Vze)

- Article: "Why Do Corner Stores Struggle to Sell Fresh Produce?" by Sam Bloch, *The Counter*, February 21, 2019 (https://bit.ly/3xOJNTJ)

- Article and video: "West Oakland Food Desert Blooms With a Single Produce Market," from KTVU (https://bit.ly/3dcOldl)

- Video: "Grown in Oakland," from *The Atlantic* (https://bit.ly/3xL5nIY)

This lesson was written using the following resources: Ginwright (2016, 2018), Harvey (1996), Kokka (2019, 2020), and Pulido et al. (2013).

Note: You can adjust the map and locations to reflect authentic locations in your local community.

LESSON 1 FACILITATION

Cor(o)ner Store: Where Healthy Food Comes to Die

Launch (10 minutes)

- Place the students into small groups of three or four.

- Pass out Worksheet 1 (*Cor(o)ner Store: Where Healthy Food Comes to Die*).

- Use one of the following options to begin the lesson:

  + Have the students start with the *Do Now*: "Which One Doesn't Belong" (5 minutes). Invite students to identify which one of the images does not belong. Make sure that they write their response and explanation.

  + Use the following prompt to activate prior knowledge: *In your own words, define absolute value and create a visual representation.* Have students share their written example of visual representation with their table partners.

+ Facilitate a *Brain Trust* by saying, *List as many places as you can where you believe you can find healthy/fresh food.* (2 minutes)

- At their tables, tell each student to silently read one of the vocabulary words on the paper. They will be responsible for being the expert on that new vocabulary word to discuss the definitions with their classmates.

- Prior to showing the video, allow students to ask questions about the vocabulary words to collectively develop the meaning of the terms. Read the title of the video and invite students to craft questions prior to watching the video.

- Have the students watch the video "Verify: Living in a Food Desert" (watch 0:00–3:33).

- Pose the following questions to begin a brief discussion about the video:

  + *What are some of your first thoughts about what the video refers to as "food deserts"?*

  + *Why would they exist?*

  + *What are some potential challenges people of these communities might face?*

  + *How do you feel about this? Record your thoughts and feelings.*

Together, review the article "'Food apartheid,' not 'desert'," which problematizes the deficit labeling of communities with poor access to healthy food as "deserts" (https://bit.ly/3G947SL).

Explore (60 minutes)

- Tell students to read "Akeelah's Search for Fresh Seafood" (Worksheet 1) and then write their responses to the questions about Akeelah's access to healthy food. (3 minutes)

- Have students share their thoughts with their table partners, and then have a whole-class discussion to help students share their thoughts.

- Prompt students: *Look at the recipe card and think about which of the ingredients Akeelah might already have in her home, and which ones she might need to go to the store to get.*

- Hand out the Task Card and explain to the students that they are going to create a visual representation of Akeelah's journey to make gumbo for her Nana.

- Remind students that as they complete the mathematics task, they should also discuss what Akeelah might be feeling or experiencing through her travels.

- As students are working, walk around and monitor group progress, ask questions to support thinking, and probe for conceptual understanding.

- During the last 10 minutes, allow students to engage with other groups by doing a gallery review.

- Students should focus on the mathematical reasoning *and* the thoughts that other groups have around Akeelah's trip and how it might affect her community and others' communities.

We do not use "gallery walk," since not everybody walks. We choose *gallery review* to use inclusive language.

Summarize (20 minutes)

- Give students time to reflect on the process and ask how Akeelah's trip might affect her health, nutrition, quality of life, and feeling about her neighborhood.

- Have the students work with their groups and reflect on the last question: *Why do you think the authors called the activity "Cor(o)ner Store"? Use the definitions from page 1 to help support your connections.*

- Give students a chance to share their ideas with the whole group.

- At the end of class, explain to the students that they are going to create a social media campaign around what they have learned about food apartheid.

- **Note:** You may need to talk with students' caregivers in order for students to use social media. Posters can be used as well. This is explained in the Taking Action section.

LESSON 2 FACILITATION

Grocery Trek

Lesson 2 was designed to be inclusive for people who may move and travel in various ways. Please be mindful if students and/or their family members use additional travel means, such as a wheelchair.

Launch (15 minutes)

- Begin by reminding students that the video "Verify: Living in a Food Desert" mentioned that 1 in 10 Americans live in what is referred to as a *food desert*. The video estimates this to be 30 million Americans. Check the Worldometer website (https://bit.ly/3xNEvIc) to see how many people are currently living in the United States to answer question 2 of the *Do Now/Warm-Up.*

- Start the *Do Now/Warm-Up.* (You can write the questions on the board, print them out, or project them.)

According to the Worldometer website, how many people are currently living in the United States? (Display the Worldometer website: https://bit.ly/3xNEvIc.)

You might recall that the video from last class, "Verify: Living in a Food Desert," mentioned that 1 in 10 people in the United States live in a food desert. The video was made in 2017. If we assume in 2022 (or whatever year you are in) that 1 in 10 still live in a food desert, how many people are currently living in a food desert?

- Have students watch the video "Food Deserts in DC" by NPR (https://bit .ly/31l2gM1). (3–4 minutes)

  Note: Before you show "Food Deserts in DC," ask students to think about this question as they watch:

  + *What do you notice is similar about Toni's situation in this NPR video and Jeretha's situation in the Verify video from yesterday?*

- After watching the video and discussing Toni and Jeretha's situations, have students explore the U.S. Department of Agriculture Economic Research Service website (https://bit.ly/3lr4wbm), which includes the Food Access Research Atlas. Click on "Enter the Atlas." If your school does not have devices, you can print out a section relevant to your teaching context for students to view in partners or small groups. If your printing access is limited to black and white, you may need to project the image because the colors have different meanings.

- As students explore areas of the map (you can use your own local context), ask them these questions:

  + *What do you notice about the areas of the map that indicate low access to grocery stores?*

  + *How does this make you feel?*

  + *What actions can you take so that everyone has access to healthy foods?*

- Consider what would be most appropriate to share with students given the social context in which they live. For instance, for students of relatively privileged backgrounds, you may want to have them analyze the availability of healthy foods in their own community in comparison to a less-resourced community. Consider how you can do this in a way that neither encourages students to take on a "savior" mentality of wanting to help neighbors living in a less-privileged area nor reinforces negative implicit biases they may hold about residents of a nearby less-resourced community (Kokka, 2020). For instance, students may incorrectly believe that residents of the less-resourced community somehow brought their circumstances on themselves or have not worked hard enough, rather than understanding systemic oppression or the effects of federal policies such as redlining, racial profiling, inequitable school funding structures, and others. Students in an under-resourced school may already face challenges of living in areas without access to healthy foods. You may need to engage students in discussions of such systemic inequities so that students of historically marginalized backgrounds do not experience internalized oppression. Here are some suggested resources for students of all backgrounds:

+ Video: "Why Is the 1% So White?" from Francesca Ramsey of MTV Decoded (https://bit.ly/3Id0ixP)

+ Documentary series: "Race: The Power of an Illusion" (https://www .racepowerofanillusion.org)

+ Article: "The Future of Healing: Shifting From Trauma Informed Care to Healing Centered Engagement" by Shawn Ginwright, May 31, 2018 (https://bit.ly/3Dh3Vze)

- Remember to highlight students' strengths and deep knowledge of their own communities. Moreover, it's essential to discuss the strengths, joy, and contributions of all students' communities. Consider your own unique school and community context and think about your students' and your own intersectional identities when engaging in this lesson.

Explore (55 minutes)

- Pass out Worksheet 2 (*Grocery Trek*) and invite students to read the scenario. You may want to recreate this using a map of locations specific to your own context.

- Invite students to share their thinking with a partner or teammate. Engage in a whole-class discussion for students to share their thoughts. Encourage multiple solution strategies and ask students to come up with as many different strategies as they can.

- For Part 2, allow students to spread out in their teams to find each other's travel rates. Since not everyone may walk, invite students to also consider other means of transportation (e.g., using a wheelchair) as indicated in question 7. Question 3 in Part 2 may take some time. Rotate to the groups to check in, and you may use the Teacher Resource, *Grocery Trek Optional Scaffolds and Extensions*. Scaffolds can at times decrease the cognitive demand of the task, and they are therefore provided separately for you to choose whether to use them.

- When groups have finished question 3 in Part 2, invite them to continue working in their groups on questions 4 and 5. Make sure you ask groups to justify their thinking as you circulate. There are many "correct" answers as well as solution pathways to question 5.

- If time permits, invite students to do questions 6–10. Otherwise, invite students to move to Parts 3 and 4.

Summarize (20 minutes)

- When all groups are ready, bring the class together to discuss their strategies and justifications. Question 5 is the main question to focus on, so begin with a whole-class discussion to answer one of the following central questions:

  + *Taking into consideration the time you will spend at each store, will you have enough time to go with Akeelah to help her carry the groceries home and still make it to Leola M. Havard Early Education School by 12:30 p.m.?*

  + *How long can you spend at each store and how fast would you need to travel to get the groceries and make it to the school on time?*

  Note: This task is designed to be inclusive of those who use wheelchairs or other means of transportation; therefore, the term "travel rate" is used instead of "walking rate." Question 10 (optional) invites students to try it out, if feasible for your class. This may require caregivers' consent.

- Use the following questions to facilitate a whole-class discussion:

  + *What other factors do you need to consider in addition to the travel rates and total travel time?*

  + *What accessibility considerations should you make?*

  + *In this scenario, you and Akeelah needed to travel to multiple stores because there was not an affordable grocery store nearby with ample food to sell. Question 12 asked about your feelings about needing to travel and visit multiple stores. How may it affect you, Akeelah, and others in the community?*

  Note: This may be a time to discuss with students the differences between needing to travel this distance, and without a car or public transportation, versus wanting to purchase food from multiple stores.

- Invite groups to share at the board or with the document camera how they approached the problem and their justifications. Invite classmates to ask questions and provide justifications. Encourage multiple solutions and solution strategies. Use the turn-and-talk routine throughout to increase opportunities for participation and engagement.

- If you have extra time or multiple days for this, you can even have each group create a poster for a gallery review discussion.

- Make sure to leave time to discuss Part 4, the reflection and call to action. Ask students for their reactions to the article they read the previous night. Invite groups to share their plans for their social media campaign as described next in the Taking Action section.

TAKING ACTION

Consider having students take action as follows:

1. Read an article and watch a video about a community working to turn corner stores into markets. Here are some videos and articles to choose from:

   + Article: "Why Do Corner Stores Struggle to Sell Fresh Produce?" by Sam Bloch, *The Counter*, February 21, 2019 (https://bit.ly/3xOJNTJ)

   + Article and video: "West Oakland Food Desert Blooms With a Single Produce Market," from KTVU (https://bit.ly/3dcOldl)

   + Video: "Grown in Oakland," from *The Atlantic* (https://bit.ly/3xL5nIY)

2. Create a social media (or poster) campaign explaining what they know about food apartheids and their effects on communities.

3. Write a reflection around how Akeelah might feel about her access to healthy foods and her journey, and how that might affect her view of the community. Consider what actions can be taken to support Akeelah's community (this is at the end of Lesson 1 as a scaffold).

We have decided to add the social media campaign piece to allow students to create advocacy on their own. Students will present their work in a gallery review so they can honor each other's creativity around the mathematics learning.

ONLINE RESOURCES

 Available for download at **resources.corwin/TMSJ-MiddleSchool**

▼ *Worksheet 1: Cor(o)ner Store: Where Healthy Food Comes to Die*

▼ *Worksheet 2: Grocery Trek*

▼ Worksheet 3: A Call to Action

Taking Action

1. Read the article and watch the 3-minute video: KTVU "West Oakland Food Desert Blooms With a Single Produce Market" by Bob Roth (https://bit.ly/3dcOluH).

2. Identify what in their plan might work for communities impacted by food apartheid. What would you change and what else could you do? What more could the city do? What could you do in your own community (with classmates, teachers, family, and community members)?

3. Use what you learned in class and with the videos and this article to create a social media campaign to help areas impacted by food apartheid gain affordable access to fresh foods (Twitter, Instagram, TikTok, a poster, etc.). Make sure your teacher and/or caregiver support your use of such tools.

▼ Student Resource 1: Task Card

The Task Card

Objective: To create a visual representation of YOUR group's THINKING around the math.

Akeelah gets on the T-train by her house at **La Salle and 3rd Street**. She takes the train **SOUTH 6 blocks** to get seafood at Let's Eat BBQ and More at Revere Street and 3rd Street. She buys shrimp there, but they have no crab. She gets back on the train **NORTH 38 blocks** to go to Safeway at 4th Street and King Street to buy fresh crab meat on sale. Safeway is out of okra, so she gets back on the train and takes it **48 blocks SOUTH** to Duc Loi Supermarket near Carroll and 3rd Street to buy the okra she needs. Finally, she takes the train **NORTH** back to her house, but the train breaks down after **13 blocks** at the Oakdale Street stop.

Don't forget: You are using a vertical NUMBER LINE. North of zero are positive numbers and south of zero are negative numbers.

Math Goal Checklist (What you are asked to produce by the end of the task)

1. Summarize.
2. Draw a diagram showing her TRIP throughout the day.
3. Label the stop by her house as ZERO, because that is where she started.
4. Label all of the distances on the trip in numbers and what she purchased: crab, okra, and shrimp.
5. Use a different color for each part of her trip.
6. Identify where she ended her trip in relation to her house, which is ZERO.
7. Make sure you label the positive and negative numbers on the number line.
8. Write a number sentence (expression and/or equation) for each trip segment.

Mathematical Thinking (What mathematical reasoning will you uncover?)

1. How many total blocks would she need to travel?
2. Akeelah traveled the longest distance between which two stores?
3. What other mathematics did you notice when mapping out Akeelah's journey?
4. Summarize Akeelah's journey for the poster. Make sure people can understand what is being asked.

Social Justice Thinking

1. How might Akeelah's journey affect her?
2. Akeelah is young. Who in her community might be more affected by food apartheid? Use your knowledge from your own community and what you learned in the video to help support your thinking.
3. Using the videos and definitions, write three facts about food apartheid.

▼ Teacher Resource 1: Grocery Trek Optional Scaffolds and Extensions

Grocery Trek Optional Scaffolds and Extensions

OPTIONAL SCAFFOLDS FOR PART 2

Method A: Travel a fixed distance and time it.

Time how long it takes you to walk or travel via a wheelchair a fixed distance to find your walking rate. Measure an agreed-upon distance using a measuring tape and time how long it takes you to walk that distance. Your walking or travel rate should be expressed in feet per second (*or how many feet you travel in one second*). As your team is working, take note of how you round any of your estimates. How might different rounding choices affect your rate?

a) What is your walking or travel rate (in feet/second) (or how many feet you travel in one second)? (Don't forget to take the mean, or average, after doing this 2 to 3 times.)

b) What are your teammates' mean travel rates (in feet/second) (or how many feet do your teammates travel in one second)? What might account for differences in rates? How might others travel (e.g., with a walker, crutches, wheelchair)?

What is the distance being traveled? _____ (make sure to include units)

Time (seconds) to travel the distance	Name	Name	Name	Name
Attempt 1				
Attempt 2				
Attempt 3				
Average or mean time (seconds)				
Average or mean travel rate (include units, e.g., feet/second)				

▼ Teacher Resource 2: Additional Lesson Resources

Additional Lesson Resources

STEPS TRACKER EXTENSION ACTIVITY

The "steps tracker" extension may be engaging, yet there are concerns with this portion since not everyone travels via walking. If you choose to do this activity, here are some helpful resources:

- Article: News-Medical, "How Do Wearable Fitness Trackers Measure Steps?" by Cashmere Lashkari, B.Sc. https://www.news-medical.net/health/How-do-wearable-fitness-trackers-measure-steps.aspx
- Article: Lifewire, "How Does Fitbit Track Steps?" by Kat Aoki, December 2, 2020. https://www.lifewire.com/how-does-fitbit-track-steps-4774295
- Article: Healthline, "How to Calculate Stride Length and Step Length," https://www.healthline.com/health/stride-length
- Article: Wikihow, "How to Measure Stride Length," https://www.wikihow.com/Measure-Stride-Length#~:text=When%20you%20know%20your%20total,0.89%20metres%20(2012.9%20ft)

Videos and Articles

- YouTube video: The Obama White House, "Eliminating Food Deserts in America," https://www.youtube.com/watch?v=SMy-iWjTBQS&t=4s
- Article: NPR, "How to Find a Food Desert Near You" by Nancy Shute, March 13, 2013. https://www.npr.org/sections/thesalt/2013/03/13/174112591/how-to-find-a-food-desert-near-you
- ArcGIS story map: "Food Deserts in America," https://www.arcgis.com/apps/MapSeries/index.html?appid=226602 4e097e4a0d86a6ac028a183dff
- Website: IMPAQ, "Barriers to Food Access Locator," https://impaqint.com/barriers-food-access-locator

LESSON 6.3 BILLIONAIRE POWER

Natalie Odom Pough and Y. Rhoda Latimer

ECONOMIC INEQUALITY

This lesson focuses on economic inequality, particularly a living wage and consumerism. Students are seeking ways to be successful throughout school and life, and within their community. As they obtain success, students must understand the adage, "To whom much is given, much is required." It is posited that many of the world's issues could be eradicated by some of our most wealthy individuals. It is important for younger generations to explore and troubleshoot this option as well as use mathematics to support their justifications for moving in that direction. Students need to think critically about the system that has been created and why/how it is maintained through capitalistic beliefs and practices.

DEEP AND RICH MATHEMATICS

Students will develop a deep understanding, model multiple representations, and communicate relationships of large numbers and number systems. Students will also access deep and rich meanings by selecting appropriate methods and tools for reasonable estimation, proportional reasoning, and computational fluency while solving real-world problems.

ABOUT THE LESSON

This lesson uses a launch–explore–summarize instructional model and is intended to take approximately 225 minutes to complete across three class periods.

Lesson 1: Students investigate the difference between 1 million and 1 billion as they deepen their understanding of the value of economic wealth statuses.

Lessons 2 and 3: Students are introduced to the income inequalities that have been heightened due to the COVID-19 pandemic and billionaires' economic responsibilities.

SOCIAL JUSTICE OUTCOMES

- I know that all people (including myself) have certain advantages and disadvantages in society based on who they are and where they were born. (Justice 14)

- I will work with friends, family and community members to make our world fairer for everyone, and we will plan and coordinate our actions in order to achieve our goals. (Action 20)

ESSENTIAL MIDDLE GRADES CONCEPTS

- **Number**– Grade 7: Solve real-world and mathematical problems involving the four operations with rational numbers.

MATHEMATICS PRACTICES

- Make sense of problems and persevere in solving them.

- Reason abstractly and quantitatively.

- Construct viable arguments and critique the reasoning of others.

Resources and Materials

- Book: *How Much Is a Million?* by David M. Schwartz (1985) (1 copy)

- Play money: $100 bill (1 per student)

- Worksheet 1: *Proportional Reasoning* (1 per student)

- Worksheet 2: *Top 10 Billionaires* (1 per student)

- Worksheet 3: *The Billionaire Complex* (1 per student)

- Article: "Uber, Other Gig Companies Spend Nearly $200 Million to Knock Down an Employment Law They Don't Like – and It Might Work," by Faiz Siddiqui, *Washington Post*, October 26, 2020 (https://wapo.st/3EvyhiA)

- Article: "Uber and Lyft Used Sneaky Tactics to Avoid Making Drivers Employees in California, Voters Say. Now, They're Going National," by Faiz Siddiqui and Nitasha Tiku, *Washington Post*, November 17, 2020 (https://wapo.st/3plyEGm)

- Computer with internet access

LESSON 1 FACILITATION

Proportional Reasoning

Launch (20 minutes)

- As a class, read *How Much Is a Million?* by David M. Schwartz.

- After reading the book, provide each student a $100 bill of play money. Have the students calculate how much money is in the room.

- Have 12 students stand up and discuss how much money is represented. Use the following questions to extend the discussion into monthly bills and payments.

  + *Do we have enough money to cover $800 rent and a $300 monthly car payment? If so, do we have any money left over?*

  + *Is there enough money represented to take the entire class to a matinee movie ($7 per person)?*

  + *If each person had a $100 bill, how many people would need to be in our classroom so that we have $1 million represented?*

Explore (30 minutes)

- Distribute Worksheet 1 (*Proportional Reasoning*).

- Tell students that as they complete each part of the worksheet, they should work with a partner to check their work.

- Encourage students to create visual representations of the conversions they created in the launch.

Summarize (15 minutes)

- At the conclusion of the assignment, ask: *Which part of the task helped you gain a better understanding of how large one billion is?*

- Allow students to discuss in their groups and then share out with the whole class.

- Tell students that tomorrow they will explore the top billionaires in the world.

LESSON 2 FACILITATION

Top 10 Billionaires

Launch (15 minutes)

- Organize students into groups and begin class with the think–pair–share instructional strategy by having students connect to the prior day's lesson and activity. Using the following prompt, have students calculate and discuss their findings:

  + If the cost of a school lunch is $ _____, how many children can a billionaire feed for the entire school year (180 days)?

- Select a group to explain their reasoning and allow for whole-class discussion about the process and answer.

- Transition into today's exploration. Say:

  + *Today we will explore the Forbes US Billionaires list to gain a better understanding of the economic inequalities, economic power, and economic responsibilities of some of the wealthiest people in America.*

Explore (40 minutes)

- Distribute Worksheet 2 (*Top 10 Billionaires*) and tell students that as they complete the worksheet, they will analyze who the individuals are, the source of their money, how much they make each day, the minimum wage earned by their entry-level employee, how the billionaires' daily salary could positively impact a smaller community-based nonprofit organization, and the philanthropic efforts of their family and/or organization.

- As students work in groups, monitor their progress by asking questions about their responses for the various parts.

Summary (15 minutes)

- Allow students to discuss the graph in Part 3 (*Income Inequality Soars in America*) of Worksheet 2. Students may provide written or verbal responses to question 5, the summary question on Worksheet 2. This activity allows students to see the longevity of income inequality. Although COVID-19 has exacerbated the issue, it has been a problem in America for a long time, as displayed in the graph.

- If time permits, use the questions provided in Part 4 to have a small-group or class discussion about the political cartoon.

  Optional: Consider assigning Part 4 for homework.

LESSON 3 FACILITATION

California's Prop 22 (2020)

Launch (25 minutes)

- Read the details around Proposition 22 (Prop 22) and the issues surrounding it.

  + Provide a summary or have students read the *Washington Post* article "Uber, Other Gig Companies Spend Nearly $200 Million to Knock Down an Employment Law They Don't Like – and It Might Work," by Faiz Siddiqui (https://wapo.st/3EvyhiA).

  + If time permits, read the *Washington Post* article "Uber and Lyft Used Sneaky Tactics to Avoid Making Drivers Employees in California, Voters Say. Now, They're Going National," by Faiz Siddiqui and Nitasha Tiku (https://wapo.st/3plyEGm).

  + It may benefit you to develop a summary of both articles or to provide the readings for homework.

- Distribute Worksheet 3 (*The Billionaire Complex*) and tell students that today they will gain an understanding of how California's Prop 22 impacts app-based drivers.

- Break students into two groups and assign each group the article corresponding to their group number. Tell them that as they read, they should focus on understanding one side of Prop 22.

  + Group 1: No—the vote that benefits drivers

  + Group 2: Yes—the vote that benefits the companies

Explore (40 minutes)

- In small groups, have students randomly select a number between 1 and 10. The selected number represents the placement of the billionaire on the *Top 10 US Billionaires* list that the group will research.

- Tell students that each group is responsible for using the information provided in Part 2 to research the billionaire and the source of their wealth.

- As you monitor students' progress, remind them that the main focus is to gain an understanding of the individual and how they were able to achieve billionaire status.

Summary (15 minutes)

- Say to students:

  The COVID-19 pandemic has exposed a number of inequalities throughout the United States of America. The chart presented in Part 3 provides the data that demonstrates how much the wealth inequalities have grown during the pandemic. You may be aware of the number of Americans who have suffered during the COVID-19 pandemic, and additional awareness should be placed on the number of people who have profited off the pandemic in various ways.

- Instruct students to review the data provided and discuss questions 1–3. Tell them to write a summary of their discussion and be prepared to share out.

- Use purposeful selection and sequencing to call on students to share their summary.

- Use question 4 to facilitate a closing discussion to help students make connections between local businesses that may have closed and/or the expansions or growing presence of these companies due to this growth.

TAKING ACTION

Encourage students to follow through and do some of the actions they suggested in question 4 of Lesson 3. Actions might include the following:

- Have students create public service announcements (e.g., social media blasts) to share key information about the benefits of shopping locally.

- Encourage students to share their presentations with local agencies to advocate for resources for local businesses.

- Invite a guest speaker (local commerce or business bureau, small business owner, etc.) to share with students ways they can help grow the local economy.

ONLINE RESOURCES

 Available for download at **resources.corwin/TMSJ-MiddleSchool**

▼ Worksheet 1: Proportional Reasoning

Proportional Reasoning

Part 1

1. Write out one million in standard form.

2. Write out one billion in standard form.

3. Find the difference between one million and one billion.

Part 2

4. If I were to pay you $1/day, how many days would you have to work to earn $100,000?
   a. What can you do with the value to determine how long you would have to work to get to one million dollars?
   b. What can you do with the value to determine how long you would have to work to get to one billion dollars?

5. If I were to pay you $5/day, how many days would you have to work to earn $100,000?
   a. What can you do with the value to determine how long you would have to work to get to one million dollars?
   b. What can you do with the value to determine how long you would have to work to get to one billion dollars?

6. As of 2020, the average minimum wage in America is $7.25 per hour. If I were to pay you $7.25/day, how many days would you have to work to earn $100,000?
   a. What can you do with the value to determine how long you would have to work to get to one million dollars?
   b. What can you do with the value to determine how long you would have to work to get to one billion dollars?

Part 3

7. The average person works a 40-hour workweek (8 hours/day for 5 days/week). If you were making minimum wage ($7.25 per hour) and assuming you had no bills to pay, how many hours would it take for you to become a billionaire? Feel free to convert your answer into years to best explain how to achieve this goal.

▼ Worksheet 2: Top 10 Billionaires

Top 10 Billionaires

Part 1

1. Look at the Top 10 billionaires in the world. What percent are from the United States? This information can be found at the *Forbes* website "Real Time Billionaires" (**https://bit.ly/3lw2Aym**).

Part 2

2. Who are the top 10 US billionaires presented on the current *Forbes* list? Complete the table below with the following information. Discuss the similarities and differences between these individuals.

TOP 10 US BILLIONAIRES

	Name	Source	Net Worth & Daily Rate	Age	Race & Gender
1			Net Worth:		
			Daily Rate:		
2			Net Worth:		
			Daily Rate:		
3			Net Worth:		
			Daily Rate:		
4			Net Worth:		
			Daily Rate:		
5			Net Worth:		
			Daily Rate:		
6			Net Worth:		
			Daily Rate:		
7			Net Worth:		
			Daily Rate:		
8			Net Worth:		
			Daily Rate:		
9			Net Worth:		
			Daily Rate:		
10			Net Worth:		
			Daily Rate:		

▼ Worksheet 3: The Billionaire Complex

The Billionaire Complex

Part 1

Article 1: "Uber, Other Gig Companies Spend Nearly $200 Million to Knock Down an Employment Law They Don't Like—and it Might Work" (**https://wapo.st/3FvyhiA**)

Article 2: "Uber and Lyft Used Sneaky Tactics to Avoid Making Drivers Employees in California, Voters Say. Now, They're Going National" by Faiz Siddiqui and Natasha Tiku (**https://wapo.st/3plyEGm**)

In small groups, discuss the following questions:

- The campaign for No on Prop 22 was outspent by Yes on Prop 22 $200,000,000 to $20,000,000. Why would companies like Uber, Lyft, etc. spend so much money to campaign for limiting the benefits and compensation of drivers?

- How does maintaining drivers as contractors rather than employees benefit these companies?

- How could the $200,000,000 spent on the campaign be proportioned to benefit the drivers? (Show your answer using proportional reasoning and unit conversions.)

- Although the passing of Prop 22 in November 2020 impacted California drivers, Uber made a national announcement in April 2021 that stated there would be a $250 million incentive program to boost driver earnings and encourage drivers to return to the company. How do these two actions conflict with one another?

Part 2

Your teacher will place your class into 10 small groups. Each group will randomly select a number between 1 and 10. That number will represent the placement of the billionaire on the Top 10 US Billionaires list we previously created. Your group will then research the billionaire and the source of their wealth.

In your small groups, research the following:

About the Billionaire	About the Billionaire's Organization/Corporation
1. Full name	1. Full name
2. Hometown	2. Location of headquarters
3. Education	3. Number of locations in the US and worldwide
4. Average salary	4. Size of organization/corporation
5. Net worth	5. How is revenue generated?
6. Philanthropic efforts	6. Net worth
7. Headshot	7. Philanthropic efforts
8. Any childhood information (parents/caretakers, siblings, etc.)	8. Company's logo and slogan

LESSON 6.4 MIDDLE SCHOOL MATHEMATICS TO EXPLORE PEOPLE REPRESENTED IN OUR WORLD AND COMMUNITY

Mary Candace Raygoza, Eva Thanheiser, Courtney Koestler, Jeff Craig, and Lynette Guzmán

WORLD DIVERSITY

By shrinking our world into a "village" of 100 people and learning about different aspects of their lives (e.g., "nationalities," languages spoken, religions practiced [or not!], access to drinking water, living in poverty), "we can find out more about our neighbors in the real world and the problems our planet may face in the future" (Smith, 2011, p. 7). Exploring a global context through this lens can encourage students' critical inquiry about all of humanity's access to basic human needs and distribution of resources, recirculate stories about ourselves and our neighbors, and open up conversations to learn from and with each other. Furthermore, engaging with these data can affirm and/or challenge assumptions about who lives in our world and facets of their lives, empowering everyone to be more aware of and connected to the global community. Together, we resist American individualism and United States–centric curriculum and stories about humanity.

Critical literacy is a way of seeing ourselves and our world; it is an approach we can take to ask questions to uncover and unpack assumptions we have that are connected to dominant narratives about our world. In this lesson, we can use this approach to

- unpack our own assumptions,

- unpack authors' assumptions in the way that they present data, and

- move beyond a "single story" to get at important counternarratives.

We might ask: What kinds of **critical questions** can we ask about representation in our lives in and beyond this lesson?

SOCIAL JUSTICE OUTCOMES

- I can accurately and respectfully describe ways that people (including myself) are similar to and different from each other and others in their identity groups. (Diversity 7)

- I am curious and want to know more about other people's histories and lived experiences, and I ask questions respectfully and listen carefully and non-judgmentally. (Diversity 8)

- I know that all people (including myself) have certain advantages and disadvantages in society based on who they are and where they were born. (Justice 14)

ESSENTIAL MIDDLE GRADES CONCEPTS

- **Number**—Grade 7: Apply and extend previous understandings of operations with fractions.

- **Ratio and Proportions**—Grade 7: Analyze proportional relationships and use them to solve real-world and mathematical problems.

- **Statistics and Probability**—Grade 7: Draw informal comparative inferences about two populations.

MATHEMATICS PRACTICES

- Make sense of problems and persevere in solving them.

- Construct viable arguments and critique the reasoning of others.
- Model with mathematics.
- Attend to precision.

DEEP AND RICH MATHEMATICS

This lesson prioritizes mathematical ways of comprehending problems. This includes starting problems using mathematics and mathematical language, making sense of mathematical relationships through decontextualizing and contextualizing problems, and utilizing mathematics to reconsider perspective. To accomplish these goals, students should be constantly reminding themselves that imagining the world as 100 people is a lens for framing any problems they identify and analyze. Students use proportional thinking to reckon with large numbers and practice their fluency in going between relative and absolute amounts in order to maintain the complexity and significance of the problems they are identifying and discussing.

ABOUT THE LESSON

The lesson uses a launch–explore–summarize instructional format and is intended to take 180 minutes to complete across two class periods.

Lesson 1: Students build community, learning about similar and diverse characteristics of the class. Together, students explore statistics about the world.

Lesson 2: Students compare data about the world to data about their own community.

Note: Taking Action is included in Lesson 2.

Resources and Materials

- Teacher Resource 1: *Additional Resources*

- One or more of the following:

  + Book: *If the World Were a Village: A Book About the World's People, Second Edition* by David J. Smith (2011), published by Kids Can Press Ltd.

  + Website: "100 People: A World Portrait" statistics (https://bit.ly/3G48v5w)

  + Video: "100 People: A World Portrait," from 100 People (https://www.100people.org/the-100-people-project-an-introduction/)

  + Website: Ed-Data (https://www.ed-data.org/)

  + Website: CIA "The World Factbook" (https://www.cia.gov/the-world-factbook/)

For additional perspectives on teaching this lesson, we invite you to explore the following articles written about a variation of this lesson, by each of the lesson authors in Guzmán and Craig (2019), Raygoza (2016), and Thanheiser and Koestler (2021).

LESSON 1 FACILITATION

Exploring Data About the World

Opening Circle (15 minutes)

Community-Building Activity: "I Love My Neighbor Who . . ."
This lesson starts with a community-building activity designed to (1) support students to get to know one another better to build trust and community and (2) challenge assumptions about one another and identify similarities and differences with one another.

- Have students circle up, seated in chairs. One person starts in the middle (you could model the first prompt and participate, too). The person in the middle says *I love my neighbor who . . .* and fills in the sentence with a trait or experience that is true for them (e.g., *I love my neighbor who speaks Spanish*). For whoever else it also applies to, they must move to a new chair (think musical chairs). The person who does not find a chair is the new person in the middle, and the process is repeated.

- Prior to the activity, as you introduce the instructions, have students (1) revisit classroom norms together to ensure everyone is respectful and also knows they can choose not to participate or can pass on particular prompts, and (2) brainstorm possible topics to include so that they are not caught off guard thinking on the spot and so that a range of topics comes up (e.g., birthday month, hair color or texture, favorite music genre, religion, languages). After students suggest topics, offer some suggestions as well.

  Note: The activity will need to be designed around and accommodate for everyone's physical abilities.

- Close with a reflection by inviting students to share what they appreciated learning about one another. Also consider inviting students to speak to the sentence starter: *We should not make assumptions about people because . . .*

Launch (30 minutes)
- Transition by introducing the idea of shrinking the world's population down to a village of 100 people, which represents who lives within it.

  + One option is to do this by reading page 7 of the book *If the World Were a Village* by David J. Smith. Having the book is not required for this lesson.

- + You may wish to use a globe or map to provide additional context if necessary.

- Before revealing data about the world village, use the following questions to have students make predictions:

  - + *Of the 100 people representing the world, how many of those people do you think would live on each of the continents?*

  - + *Of the 100 people representing the world, how many people do you think would speak* [choose a language]?

  Note: You may also engage students in wondering about other characteristics of the world's village. These predictions are meant to serve as prompts for critical literacy and critical quantitative literacy. They cause students to grapple with what they think about the world that might not be reflected in reality or in data. This process is about revealing assumptions that we develop without a clear source other than our personal interactions and relationships and how those assumptions should be challenged in the face of evidence.

- After students make predictions, explore the characteristics of the "village" by continuing to read the book. There are also many videos available for free online that reveal the world data in a similar fashion, such as the video on the "100 People: A World Portrait" website.

- **Notes:**

  - + Supplementary websites noted in the Teacher Resource are available online at **resources.corwin.com/TMSJ-MiddleSchool**.

  - + Consider exploring statistics about nationalities, languages, and religion; you can also explore other statistics based on student interest.

  - + In many resources, gender is reported as a binary using sex markers of male and female. This can be problematic because it does not unpack the diversity of gender (or sex). This is a great opportunity for critical literacy, pointing out that data sets often assume this binary and problematizing why.

- Zooming in on languages: Use the following questions to explore the language data with students.

  - + *What do you notice and wonder about the data?*

    - Possible Student Noticings or Wonderings

      - The Languages data are presented in a way that assumes people only speak one language.

      - How many people are multilingual? Why are they not represented?

      - There are "almost 6000 languages" (Smith, 2011, p. 10) in the village. What are all of the languages? Is sign language included?

- Chinese, Hindi, Spanish, Arabic, Bengali, Portuguese, and Russian are the most common languages spoken in the "village," in addition to English. Where are these languages typically spoken?

+ *How could we represent the data?*

Explore (35 minutes)

- Tell students that they will explore data from other pages of the book focusing on other characteristics of the world, such as health, education, or wealth, and create their own representations of the data on posters to share with their classmates. Ideally, each group picks a different characteristic so they can all teach each other something about the world. Share that posters should include the following:

    + Titles, labels, and keys

    + A table that includes decimals, fractions, and percentages of the population with the examined characteristics

    + At least two different kinds of representations (hundreds chart, bar graphs, pie charts, etc.) of the same data set

        - Make connections between and among the different representations.

        - Use the same color across representations to represent the same parts of the whole.

You can display a sample arrangement of a poster such as this one:

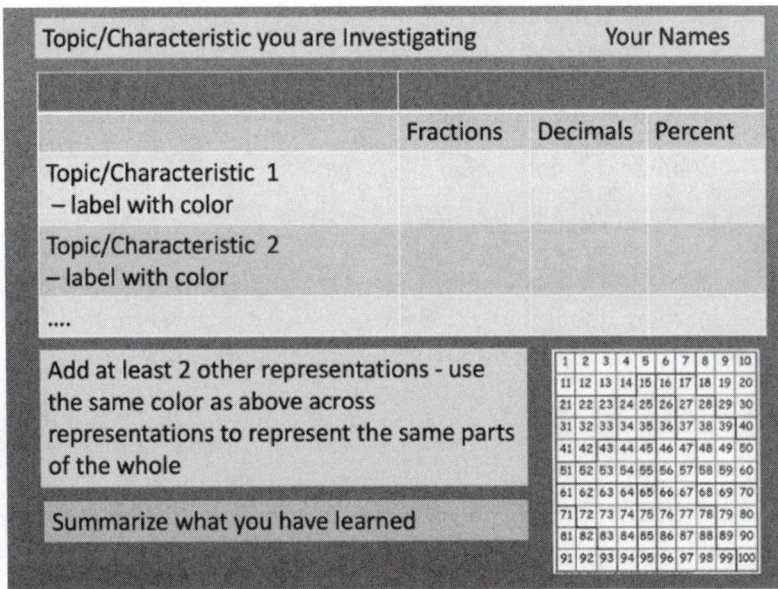

- Have students brainstorm multiple ways of representing data (e.g., conventional ways such as bar charts, pie charts, and pictograms), and more unconventional ways such as hundreds charts [10 × 10 grids].

- Organize students into groups of four.

- Provide independent think time to explore the topic and sketch various representations for the topic. Then allow small groups to share their ideas and to decide on several representations to include on a joint poster.

- As you monitor groups working, prompt students to think about whether another person would understand what they are communicating, without them adding any clarification beyond what is displayed on their poster. Have students ask themselves the following questions:

  + *What is the purpose of the titles, labels, and keys? How do they help someone understand the represented data? If your graph does not have a title, what would someone not know about your data that they should?*

  + *How do different types of data representations help tell the story of who would live in the world village?*

In an effort to use more inclusive language, we use the term *gallery review* for what is commonly referred to as a "gallery walk."

- After all posters are created, have students conduct a gallery review and reflect on how various groups' representations communicate the data. Ask: *What do different representations communicate?*

- During the gallery review, support students in engaging in discourse about how different representations communicate the same data in different ways. For example, a pie chart can show the data as a whole, while a bar graph can highlight the discrete pieces of data.

Summarize (10 minutes)

- Engage students in a reflection by using prompts such as these:

  + *What do you notice about the representations? What do you wonder?*

  + *What is still confusing?*

Continue to create space for students to share (1) world data they appreciated learning about and (2) ways in which their assumptions about the world were challenged through data exploration.

  + *What was surprising to you about the data? Why do you think it surprised you?*

  + *What questions do you have about the data or the issues related to the data?*

LESSON 2 FACILITATION

Comparing the World to Your Local Community

Launch (5 minutes)

- Tell students that today they will work with their group to explore more localized data about their country, city, town, school district, or school. Instead of imagining the entire world represented as 100 people, today they will explore their local region (country, city, town, school district, or school) as 100 people. For example, today students may investigate rates of access to safe drinking water in their hometown or languages spoken in their hometown. They ultimately will compare these data to the world data. The idea today is to explore a local context and understand how the village of 100 people would differ between the local context and the world.

- Provide data about a local context that you and your students are interested in.

- Here are some examples: school district data, language data from school sites, state online databases (e.g., California state data from Ed-Data at https://www.ed-data.org/), country data (e.g., from The World Factbook at https://www.cia.gov/the-world-factbook/), or city data from the U.S. Census or your city's website.

Explore (35 minutes)

- Have students explore several of the characteristics they explored in Lesson 1 if the data are available. If not, brainstorm people or organizations they could contact about obtaining the information or discuss whether the information is even gathered.

- Have students create posters in groups about the country, city, town, school district, or school in the same format as the poster they created about the world the day before. Then, students use their representations of the local data to compare with the world data, if possible. Ask students what they notice and wonder about the data.

- Use the following prompts to facilitate a discussion about the data:

  + *Name any ways in which the world and the local data are similar.*

  + *Name any ways in which the world and the local data are different.*

  + *Are there any categories of data that you think are not accurate? How do you know?*

  + *How would you rate your level of confidence in the data for each of the categories? How do you know if you are correct?* **Note:** Students assess the validity of data and become self-aware of the relative strengths and weaknesses of their intuition and expertly generated data.

Summarize (20 minutes)

- Begin the summary by having students reflect on the following prompts:

    + *What do you notice about the representations?*

    + *What do you wonder?*

    + *What is still confusing?*

- Use the following questions to engage students in a discussion about the data explored during the lesson:

    + *What is surprising about the data?*

    + *What did not surprise you?*

    + *What questions do you have about the data or the issues related to the data?*

Note: The two following questions are especially important to get at the critical literacy part of the lesson:

    + *How do your different representations communicate the data differently?*

    + *What assumptions went into the presentations of the data we explored?*

TAKING ACTION (20 MINUTES)

Depending on what is salient to your students in exploration of local context data, you may encourage your students to take action by

- researching the languages spoken in their school by students and families and then write letters to the principal to advocate for support; or

- identifying people (e.g., family members, politicians) or organizations (e.g., their own school) that may hold assumptions about who is represented in different spaces and to think about bringing attention to those assumptions with data.

CLOSING CIRCLE REFLECTION (10 MINUTES)

Have students discuss the following question in their groups and then with the whole group:

- As we engaged in this lesson learning about the global and local community, what do you appreciate about working with one another here in our classroom community?

ONLINE RESOURCE

 Available for download at **resources.corwin/TMSJ-MiddleSchool**

▼ *Teacher Resource 1: Additional Resources*

Additional Resources

The table below indicates which characteristics of the village are mentioned in each video option.

Link	Where From	Lan- guages	Gen- der	Age	Reli- gion	Shel- ter	Food	Wa- ter	Toi- let	Liter- acy	Educa- tion	Income/ Wealth	Elec- tricity	Cell Phone	In- ter- net	Health
"If the World Were a Village of 100 People: A Story About the World's People" — YouTube https://www youtube.com/ watch?v=FtYjUv2x65g	x	x														
"The 100 People Proj- ect: An Introduction" — 100 People: A World Portrait https://www.100people .org/wp/the-100-people- project-an-introduction/	x		x		x			x		x	x		x			
"If the World Were a Vil- lage of 100 People (2019 Edition)" — YouTube https://www youtube.com/ watch?v=aLjlFoOTJIY	x			x	x	x	x	x			x	x	x			
"If The World Was 100 People" by Jay Shetty — YouTube https://www youtube.com/ watch?v=LXqOdSnoN8g	x	x	x		x	x		x		x	x	x	x	x	x	

CHAPTER 7

RATIOS AND PROPORTIONAL RELATIONSHIPS

Ratios and proportional relationships extend students' work in measurement and multiplication and division from elementary grades using comparisons. Ratios and proportional relationships are fundamental to understanding high school mathematics and science and offer many opportunities for students to connect mathematics with everyday life. Students make connections between part to part and part to whole relationships, developing conceptual understanding of percentages and unit rates and applying this understanding to daily life by comparing the demographics of students in honors classes in tracked schools or other contexts that fit their lived experience. Teachers may use publicly accessible data sets and websites that provide descriptive statistics to study demographics, economy, access, opportunity, family finances, and other topics. Teachers should support students in the middle grades by having students include and make sense of context when creating, comparing, and explaining ratios and proportional relationships and be required to include units of analysis with precision. Ratios and proportional relationships provide an excellent opportunity for teachers to teach mathematical and social justice goals, as you will see in these lessons.

LESSONS

Lesson No.	Lesson Title	Mathematics Focus Areas	Grade	Social Justice Topic	Authors
7.1	*Hey Google, Who's a Mathematician?*	• Ratio and Proportion	6, 7	Representation	Allyson Lam
7.2	*The True Cost of That $29 T-Shirt in the Store Window*	• Ratio and Proportion	6, 7	Human Rights	Bethany Chan, Debasmita Basu, Rebecca Ellis, Frances K. Harper, and Jennifer Ruef
7.3	*Majority and Power*	• Ratio and Proportion	6, 7	Civil Rights/Government Laws	Jennifer A. Wolfe and Farshid Safi
7.4	*Smoking and Vaping: Targeting of Marginalized Communities by the Tobacco Industry*	• Ratio and Proportion	7	Health Inequality	Nichole Campbell and Peggy Nayar

Lesson No.	Lesson Title	Mathematics Focus Areas	Grade	Social Justice Topic	Authors
7.5	*Health, Race, and Ratios*	• Ratio and Proportion • Statistics and Probability	7	Health Inequality	Travis Weiland and Melissa A. Gallagher
7.6	*Health Inequalities: COVID-19 and Other Health Conditions*	• Ratio and Proportion	6, 7	Health Inequality	Tashana Howse and Kendrick Savage

SOCIAL JUSTICE OUTCOME

- I can explain how the way groups of people are treated today, and the way they have been treated in the past, shapes their group identity and culture. (Diversity 10)

ESSENTIAL MIDDLE GRADES CONCEPTS

- **Ratio and Proportions—** Grade 6: Understand ratio concepts and use ratio reasoning to solve problems.

- **Ratio and Proportions—** Grade 7: Analyze proportional relationships and use them to solve real-world and mathematical problems.

MATHEMATICS PRACTICES

- Reason abstractly and quantitatively.

- Construct viable arguments and critique the reasoning of others.

- Attend to precision.

All students can be good at mathematics. Institutions that fail to represent all of our students must not only be brought to light, but pushed to reconsider the ways in which their systems exclude students from mathematics spaces. Students may show what they are learning by doing a similar Google Image search with their families and sharing their presentations with them.

LESSON 7.1 HEY GOOGLE, WHO'S A MATHEMATICIAN?

Allyson Lam

RACE, ETHNICITY, AND REPRESENTATION

Mathematics belongs to everybody. So why is it that a quick Google Image search of "mathematicians" brings up a disproportionate amount of old white men? What message does this send to those of us and our students who are not old, white, or men? And how can we use mathematics to deconstruct and speak out against an algorithm that erases the brilliant mathematicians who represent us?

DEEP AND RICH MATHEMATICS

Students will use data from a Google Image search to create ratios, percentages, and graphs describing representation of mathematicians by racial groups, comparing these data to the racial makeup of their city or school. While students have different sample sizes, comparison of their data will show how proportionality is roughly reflected in their representations.

ABOUT THE LESSON

This lesson uses a launch–explore–summarize instructional model and is intended to take approximately 180 minutes to complete across three class periods.

Lesson 1: Students explore the way mathematicians are represented in a Google Image search and use ratios and percentages to describe their findings around the racial makeup of the search results.

Lesson 2: Students compare their findings to the racial makeup of their school community or city and draft visual presentations to send to Google with their reflections.

Lesson 3: Students learn about mathematicians who represent an inclusive community and share about them with their peers and Google representatives through their finalized presentations.

Resources and Materials

- Tablets or laptops with internet access (1 device for every 2 students)

- Worksheet 1: *Who's a Mathematician?*

- Data on the racial makeup of your school or city

  + If your school or city is predominantly white, you may choose to use demographic data for the United States or the world, as white people are overwhelmingly represented in this Google Image search, and the purpose of this lesson is to advocate for the inclusion of people of Color in mathematics spaces.

Websites

- Google Image search (https://images.google.com)

- Not Just White Dude Mathematicians, by Dr. Annie Perkins (https://bit.ly/3lxaDuN)

- Mathematically Gifted and Black, by Dr. Erica Graham, Dr. Raegan Higgins, Dr. Candice Price, and Dr. Shelby Wilson (https://mathematicallygiftedandblack.com)

- Lathisms, by Dr. Pamela E. Harris, Dr. Alicia Prieto-Langarica, and Dr. Luis Sordo Vieira (https://www.lathisms.org)

- Mathematicians of the African Diaspora, originated by Scott Williams and supported by the National Association of Mathematicians and the Educational Advancement Foundation (https://www.mathad.com/home)

- Indigenous Mathematicians, founded by Kyle Dahlin, Rebecca Garcia, Ashlee Kalauli, Marissa Loving, and Kamuela Yong (https://www.indigenousmathematicians.org/)

LESSON 1 FACILITATION

Hey Google, Who Gets Represented?

This lesson primarily focuses on the overrepresentation of white people in mathematics compared to other racial groups; however, this does not discount the significant exclusion of other underrepresented groups such as groups marginalized by gender identity, sexuality, ability, or age. You should be prepared to acknowledge and validate student observations around these issues.

Launch (15 minutes)

- Begin by instructing pairs of students to go to images.google.com and search for "mathematicians." Ask, *What do you notice? What do you wonder?*

- Allow students to share their responses. Use the following questions to facilitate discussion among the students:

    + *Are these an accurate representation of mathematicians in our community or in the world? Why or why not?*

    + *Evaluate how well these results reflect the diversity of mathematicians in the world today. How do you think these representations came to be Google Image search's top results?*

- Remind your class that mathematics belongs to everyone, which is why you believe that all of them can be good at mathematics!

If the idea of everybody being capable of mathematics is new for your students, consider taking time before implementing this lesson to do the following:

- *Discuss* your class's beliefs about what makes somebody good at mathematics and whether these are innate or learned abilities.

- *Identify* and address barriers to success in math, brainstorming ways your class is working to overcome these barriers.

- *Reflect* on your teaching practices and the ways you can implicitly and explicitly teach students that mathematics belongs to every single one of them.

Explore (25 minutes)

Teacher Preparation: You may choose to modify Worksheet 1 (*Who's a Mathematician?*) and use different racial group categorizations based on identifying nomenclature considered most dignifying in your geographic region and to be as inclusive as possible.

- Begin by instructing students to use Worksheet 1 (*Who's a Mathematician?*) to record their tally count of how many mathematicians appear on a Google Image search by racial group. Say,

    + *You will have 3 minutes to count as many mathematicians from the Google Image search as you can scroll through and record, with as much accuracy as can be expected given the time constraint.*

- Acknowledge the limited nature of this activity:

    + *While we are trying our best to accurately count each person represented, our assumptions may be incorrect, particularly because of the limited information we have on each person and time constraints for this activity.*

- At the end of 3 minutes, each pair of students will have a unique amount of data, based on the number of mathematicians they were able to record. This will allow for students to have different data based on sample sizes, which may tell similar, though not identical, stories.

- Instruct students to use the next 20 minutes to complete the worksheet tasks by responding to these prompts:

  + *For each racial group, describe its representation based on your data as a ratio and a percentage.*

    - Students may choose to use a ratio that describes that racial group compared to all racial groups, or that racial group compared to another racial group.

  + *Create a visual that illustrates your findings.*

    - Students may choose to make one visual that combines all of their data, or a separate visualization for each racial group (e.g., tape diagram, bar graph, percent bar, fraction bar).

- Use the following prompts as you monitor student pairs:

  + *Identify the racial group that you belong to.*

  + *Is your racial group represented fairly in the Google Image search?*

  + *Why do you think this is the case?*

  + *What impact might this have on your community?*

Summarize (20 minutes)

Begin by allowing students to do a gallery review to see the ways other pairs of students represented and visualized the data.

- Facilitate a discussion, asking students to share commonalities between groups as well as unique representations they noticed from other groups.

- Notice how pairs had different amounts of data, since they each recorded as much as they could in 3 minutes, yet their data tell similar stories. Make connections to proportionality.

- Invite students to share how they feel about the way mathematicians are represented, and what impact this can have or has already had on communities of Color.

- Encourage students to trust their brilliance and mathematical prowess over the representation seen during this class, and to be prepared to take action during the next lesson!

In an effort to use more inclusive language, we use the term *gallery review* for what is commonly referred to as a "gallery walk."

LESSON 2 FACILITATION

Hey Google, Do You See Us?

Launch (10 minutes)

- Begin by reviewing your class findings from Lesson 1 (*Hey Google, Who Gets Represented?*).

  + Highlight several of the ratios, percentages, and visual representations students created.

  + Discuss with students why mathematicians were represented this way on Google.

- Provide a brief explanation of search algorithms, as necessary, including how Google assigns relevancy and prioritization to search results:

  + *Google's search engine involves a series of algorithms that use your search history, location, and search settings to show you information that would be most relevant to you.*

  + *It is possible that two different users would see slightly different images given the same image search!*

  + *Many data scientists have pointed out the dangers of algorithms that reflect information that reinforces what the user has already seen.*

  + *Consider referencing Data Feminism by Catherine D'Ignazio and Lauren F. Klein or Weapons of Math Destruction by Cathy O'Neil to inform your facilitation of this discussion.*

- Use the following questions to facilitate a brief discussion:

  + *How do search algorithms empower or disempower specific racial groups?*

  + *Who is paying the greatest cost and who is receiving the greatest benefit from Google Image search algorithms?*

  + *What would it look like for Google to represent mathematicians who reflect a racially inclusive community?*

Explore (30 minutes)

- Begin by sharing the racial demographics of your school community. Use data specific to the school, neighborhood, or city.

  Note: *If your school or city is predominantly white, consider using the racial makeup for the United States or the world, as white people are overwhelmingly represented in this Google Image search and the purpose of this lesson is to advocate for the inclusion of people of Color in mathematics spaces.*

- Use the following questions to facilitate a discussion:

  + *How do the percentages of each racial group in our community compare to that of the Google Image search?*

  + *Which contrasts in percentages are most striking to you and why?*

- Task students with making a visual presentation (on Google Slides or a similar medium) comparing the racial makeup data of the prior day's Google Image search and their community data. Tell them the following:

  + *Presentations should also include graphs or images from the previous day, comparisons with today's racial data, a reflection based on your class discussions, and a demand for change based on the right of all people to belong to mathematics spaces.*

  + They will have time during the next class period to finish their presentation.

- Use the following questions as you monitor student work, assess student understanding, and review their presentation:

  + *For each racial group, describe its representation based on your data as a ratio and a percentage. Create a visual that illustrates your findings (e.g., tape diagram, bar graph, fraction bar, etc.).*

  + *Identify the racial group that you belong to. Is your racial group represented in the Google Image search? Why do you think this is the case? How might this affect your community?*

  + *How do search algorithms empower or disempower specific racial groups?*

  + *How do the percentages of each racial group in our school compare to that of the Google Image search?*

  + *What would it look like for Google to represent mathematicians who reflect a racially inclusive community?*

Summarize (20 minutes)

- Combine pairs into groups of four and allow each pair to share their presentation with their group.

- Bring the entire class together to reflect on class discussions from both lessons. Encourage students to share what they learned, what they hope Google takes away from their presentations, and how this process felt for them.

LESSON 3 FACILITATION

Hey Google, We Can All Be Mathematicians!

Launch (5 minutes)
- Share with students about a mathematician who is not represented in Google's Image search. (See a list of websites in the *Resources and Materials*.)

    + Consider sharing about somebody who is Black, Indigenous, a person of Color, a woman, gender-nonconforming, part of the LGBTQIA+ community, neurodiverse, differently abled, young, and/or a person belonging to another underrepresented community.

    + Feel free to share about a mathematician from your own life.

Explore (35 minutes)
- Tell students to choose a mathematician, one they know of or from the listed website, to research and answer the following questions:

    + *Who were/are they?*

    + *What did they contribute to mathematics?*

- Instruct students to add one additional slide to their presentations: a suggestion that Google include the mathematician they researched in their Google Image search results. The slide should include a picture of the mathematician (if available) as well as important information about who they are and what they contributed.

Summarize (20 minutes)
- Select a few students to share the mathematician they researched and then tell students that throughout the year others will have a chance to present.

    Extension Option: Continue the conversation throughout the year by taking a few minutes each week for one of the following activities:

    + Have a student share about the mathematician they researched.

    + Discuss current events involving the work and leadership of underrepresented mathematicians.

- Facilitate a discussion on possible actions students can take to share their findings about mathematicians who more closely reflect an inclusive community and how algorithms perpetuate underrepresentation of marginalized groups.

TAKING ACTION

- Encourage students to think through how they can share their presentations with others in the school, community, and district.

- Send a letter to Google.

  + Have students compose a letter to Google, summarizing their findings and reflections on representation, belonging in mathematics communities, and how algorithms perpetuate underrepresentation of marginalized groups to each other and Google representatives.

  + Collect all presentations. Then curate the presentations into one slide deck to send to Google. Ensure that privacy settings allow for somebody outside of your school district to view your file.

    - As there is no public-facing Google contact for these concerns, teachers will need a LinkedIn account to contact an appropriate Google representative. To create a free LinkedIn account, first find Google's page on http://LinkedIn (https://www.linkedin.com). Click "People" to search for employees who work in "K–12 Google for Education" or "Google Search and Assistant Product Inclusion." Identify an employee to connect with, whom you can then message. Connecting must be confirmed by the other party, so it is best to prepare this contact before implementing the lesson.

Institutions that fail to represent all of our students must not only be brought to light, but pushed to reconsider the ways in which their systems exclude students from mathematics spaces.

ONLINE RESOURCE

 Available for download at **resources.corwin/TMSJ-MiddleSchool**

▼ *Worksheet 1: Who's a Mathematician?*

Who's a Mathematician?

Instructions: Use this chart to record your tally count of how many mathematicians appear on a Google Image search by racial group. For each racial group, describe its representation based on your data as a ratio and a percentage. Then, create a visual that illustrates your findings.

Assumed Racial Group	Tally Count	Ratio	Percent	Graph(s) / Diagram(s)
White / European American				
Black / African American				
Asian / Asian American / Pacific Islander				
Hispanic / Latina/o				
Indigenous / Native American				
Middle Eastern / North African				
Mixed / Two or more identities				
Other / Unsure				

SOCIAL JUSTICE OUTCOMES

- I can recognize and describe unfairness and injustice in many forms including attitudes, speech, behaviors, practices, and laws. (Justice 12)

- I will work with friends, family and community members to make our world fairer for everyone, and we will plan and coordinate our actions in order to achieve our goals. (Action 20)

ESSENTIAL MIDDLE GRADES CONCEPTS

- **Ratio and Proportions—** Grade 6: Understand ratio concepts and use ratio reasoning to solve problems.

- **Ratio and Proportions—** Grade 7: Analyze proportional relationships and use them to solve real-world and mathematical problems.

This lesson gives students a chance to critique the world and explore their own identities, without teachers or other adults imposing their own beliefs.

LESSON 7.2 THE TRUE COST OF THAT $29 T-SHIRT IN THE STORE WINDOW

Bethany Chan, Debasmita Basu, Rebecca Ellis, Frances K. Harper, and Jennifer Ruef

HUMAN RIGHTS

To lower their production cost and maximize their profit, companies often establish their manufacturing divisions in cheaper and less regulated locations such as Bangladesh, China, Mexico, Cambodia, and other developing countries. The working conditions of the factories, commonly known as sweatshops, are often hazardous and abusive with long working hours and inadequate pay. For example, despite fashion being a $29 billion USD industry, people working in garment factories in Bangladesh are only paid $0.35 USD an hour, which forces them to work for 14–16 hours a day to pay for their daily necessities. In 2013, one such factory on the outskirts of Dhaka, Bangladesh, collapsed, trapping and killing more than a thousand of its employees. The investigation suggested that the factory was under scrutiny because of evidence of unsafe conditions, but no steps were taken to improve them. The goal for this lesson is for students to use mathematics in this social justice context to raise their consciousness about the exploitation being practiced within the four walls of the sweatshops. This lesson gives students a chance to critique the world and explore their own identities, without teachers or other adults imposing their own beliefs.

DEEP AND RICH MATHEMATICS

This lesson is designed for students to apply prior knowledge of percentages and proportions to construct and operate on mathematical models based on real-life scenarios from the clothing industry. Students will use critical thinking skills and problem-solving strategies as they apply and attain mathematical content knowledge. For this reason, teachers should be prepared to share just-in-time information about problem-solving methods others may have used. This preparation will support teachers in recognizing solution strategies and potential challenges in mathematical understandings and preparing to ask focusing questions to guide students' work. Teachers are invited to scaffold the lesson based on anticipation of their students' prior knowledge and experience.

ABOUT THE LESSON

This lesson uses a launch–explore–summarize instructional model and is intended to take approximately 120 minutes to complete across two class periods.

Lesson 1: Students use ratios, proportions, and percentages to estimate the allocation of money to make a t-shirt.

Lesson 2: Students use ratios, proportions, and percentages to analyze the actual profit allocation of a t-shirt and consider what a fair allocation would be.

Resources and Materials

- Worksheet 1: *How Should Money Be Allocated?* (1 per student)

- Worksheet 2: *How Is Money Allocated?* (1 per student)

- Worksheet 3: *What Is Fair?* (1 per student)

- Highlighters, colored pencils, or crayons

- Video: "True Cost Clothing Industry" (https://francesharper.com/clothing-industry-video/)

- Website: Educational Video Center (https://evc.org/)

- Website: Two Dollar Challenge (http://twodollarchallenge.org/our-story/)

MATHEMATICS PRACTICES

- Make sense of problems and persevere in solving them.

- Reason abstractly and quantitatively.

- Construct viable arguments and critique the reasoning of others.

- Model with mathematics.

- Look for and make use of structure.

- Look for and express regularity in repeated reasoning.

LESSON 1 FACILITATION

How Do You Think Money Is Allocated?

Launch (25 minutes)

- Assign stores to students to research the cost of clothing items. Students can bring in advertisements from clothing retailers or look online. Have students share what they found.

  Note: *If students find an item that costs around $29, you might highlight that item for the lesson later on.*

- Have students brainstorm ideas about *what the money pays for* when they purchase an item of clothing. Students might find it helpful to pick one of the items of clothing that they found in advertisements or online. The point of this activity is to get students thinking about the production process, costs, and so on.

- Elicit some ideas from students and then show the video, "True Cost Clothing Industry" (https://francesharper.com/clothing-industry-video/).

- Ask students to reflect on what they noticed or wondered as they watched the video. Then have them share out with the class and make sure students understand the eight different categories involved in the production process. Consider asking the students the categories they remember and writing the list publicly on the board. Thus, students can refer to the list when answering question 1 on Worksheet 1 (*How Should Money Be Allocated?*).

Explore (35 minutes)

- Provide students with Worksheet 1. Ask:

  + *Who and what is involved in the t-shirt-making process?*

  + *How do you think the money is allocated among the eight categories?*

- Tell students to individually complete question 2, which requires them to color/shade the part of the dollar bill that correlates to each category.

- After they complete question 2 and distribute the $100 bill among eight categories, put students in groups of four.

- Ask the students to share their strategies with their group members and discuss the rationale behind the money allocations they choose.

- From each group, ask one student to volunteer for the role of group leader to share what they discussed in their group during the whole-class discussion.

Summarize (10 minutes)

- Facilitate a whole-class discussion. Ask students:

  + *If the $100 bill represents the retail price of a product, how do you think that the $100 bill is distributed among the eight categories? Why?*

For students who divide the bill nonuniformly, ask:

  + *Who should get the maximum portion of the money? Why?*

  + *Who should get the minimum portion of the money? Why?*

LESSON 2 FACILITATION

How Is Money Allocated? What Is Fair?

Launch (10 minutes)

- Distribute Worksheet 2 (*How Is Money Allocated?*). Lead a brief discussion about how money for an item of apparel is typically distributed.

  Note: *Values are included on the worksheet.*

  - ☐ *Materials: 12%*   ☐ *Retailer: 58%*
  - ☐ *Factory profit: 4%*   ☐ *Intermediary costs: 4%*
  - ☐ *Transport: 8%*   ☐ *Brand profit: 12.4%*
  - ☐ *Overhead cost: 1%*   ☐ *Workers: 0.6%*

- Individually, or in pairs, have students color in the $100 bill using the new breakdown.

Explore (35 minutes)

- Divide students into groups of four and assign each group to one of the two group types (see the following table). Students in each group type will focus on some of the percentages given in the Lesson 2 Launch and will calculate how the cost of a $29 t-shirt is distributed across the different categories. Students should each take on one conversion individually and then share their strategies with the other group members.

Group Type 1	Group Type 2
Retail	Intermediary
Profit to brand	Factory profit
Material cost	Overhead cost
Transportation	Payment to the workers

- After students share their strategies within their group, pair up group types and encourage them to compare the amount of money distributed among the eight categories.

- Begin a whole-class discussion by asking, *How does this breakdown differ from what was imagined on Worksheet 1?*

- Use the following questions to further assess students' understanding:

  + *Which among the eight categories received the highest percentage of the retail price of the t-shirt? How much is allotted to that category?*

  + *Which among the eight categories received the lowest percentage of the retail price of the t-shirt? How much is allotted to that category?*

> + *How many times more did the retailer earn compared to the workers who are employed to produce the t-shirt?*
>
> + *How do the numbers look when we are talking about portions of $29 instead of percentages?*

- Tell students to return to the groups they were in for Worksheet 1 and distribute Worksheet 3 (*What Is Fair?*).

- Have students work in groups as they complete the worksheet and record their reasoning for each question.

- As students work, encourage them to think about both the context and the mathematics as they create their responses.

- Use purposeful selection and sequencing to have students share the key ideas from the questions on Worksheet 3 during a whole-class discussion. Pay particular attention to the students' answers to question 4, as this will guide the transition to the summary portion of the lesson.

Summarize (15 minutes)

- Use the following questions to facilitate a closing discussion:

  + *Other than workers' wages, what other factors impacting the workers should companies consider? What benefits can companies get from being ethical?*

    - Possible Answers: A *positive company image attracts more customers/customer loyalty. It creates a company culture that values being ethical and socially responsible. These companies provide jobs to the local community by not outsourcing.*

- Continue the discussion by saying, *Some students have suggested buying only from ethical companies, but that can get expensive. For families that cannot afford to pay more money for ethically made clothing, what are other actions they can take to fight against unethical companies?*

- Have students brainstorm ways they can fight (not support) sweatshops and the unfair treatment (monetarily, physically, and mentally) of workers in the clothing industry.

TAKING ACTION

- Have students research current companies that sell sweatshop-free clothing (e.g., Patagonia) and create a brief presentation to share with your community (e.g., other students at the school, community members, family).

  + Consider the following questions: *Beyond working conditions, what other questions do you have about these companies? Do sweatshop-free clothing companies distribute money differently and/or fairly?*

- Brainstorm individual and family actions to take in relation to ecojustice (e.g., buy used clothing, buy local, wear clothing until it wears out, repair clothing when possible, make your own clothes).

- Create a video or podcast in which you interview sweatshop workers, organizations, or social justice members who have worked toward gaining more rights for workers. Share your video or podcast with the community. See the Educational Video Center (https://evc.org/).

  + Another possible topic: history of labor laws

- Create a list of questions to ask and find out if local businesses or community members endorse sweatshop-free products. If they don't, brainstorm ways to appeal to the community on the importance of supporting sweatshop-free products.

- Challenge families, community members, other teachers, and others to the Two Dollar Challenge (live on $2 for 5 days; http://twodollarchallenge.org/our-story/) as a way to raise awareness of global poverty and raise money for an organization working for social justice.

ONLINE RESOURCES

 Available for download at **resources.corwin/TMSJ-MiddleSchool**

▼ Worksheet 1: How Should Money Be Allocated?

Name: _____

How Should Money Be Allocated?

Imagine that you buy an item of clothing from the store for $100. How should the money be allocated?

1. Who and what is involved in the t-shirt-making process? Create a key matching a color to each category involved in the production process that we discussed as a class.

   Key:

2. The $100 bill below has been subdivided into 100 equal sections, where each section is worth $1. How do you think the money is divided among the eight categories?

   a. Divide the bill among the categories by coloring in the sections on the $100 bill.

   $100

3. Discuss how you divided your bill with your group.

   a. Are you in agreement with what others said?

   b. What differed?

   c. If you could change anything in your drawing, what would you change?

   Summarize your discussion here.

▼ Worksheet 2: How Is Money Allocated?

Name: _____

How Is Money Allocated?

1. Your teacher has now shown you how money from an item is typically divided. On the $100 bill below, show how the money is actually allocated.

   ☐ Materials: 12%   ☐ Retailer: 58%

   ☐ Factory profit: 4%   ☐ Intermediary costs: 4%

   ☐ Transport: 8%   ☐ Brand profit: 12.4%

   ☐ Overhead cost: 1%   ☐ Workers: 0.6%

   $100

2. If a shirt costs $29, and not $100, how much money is allocated to each category involved in the production process based on the percentages in Question 1?

3. Compare and contrast how the money is allocated in your models (Worksheet 1) versus in the model above (Question 1).

▼ Worksheet 3: What Is Fair?

Name: _____

What Is Fair?

In Worksheet 1, you modeled how you think money is divided.

In Worksheet 2, you modeled how money is typically divided.

Now, think about what is *fair*. How should money be allocated?

1. Divide $100 among the eight categories in a way that you think is fair. Color in the bill and label the key.

   $100

   ☐ Materials: _____%   ☐ Retailer: _____%

   ☐ Factory profit: _____%   ☐ Intermediary costs: _____%

   ☐ Transport: _____%   ☐ Brand profit: _____%

   ☐ Overhead cost: _____%   ☐ Workers: _____%

2. Why do you allocate the money in this way?

3. How does this new allocation differ from the previous two versions?

4. What questions do you have about allocating money among the eight categories involved in the production process?

LESSON 7.3 MAJORITY AND POWER

Jennifer A. Wolfe and Farshid Safi

CIVIL RIGHTS/GOVERNMENT LAWS

The United States government consists of three branches—legislative, executive, and judicial—whose powers are outlined in the U.S. Constitution. In this lesson, students begin to develop a shared understanding of the impact on the balance of power and the potential effect on their communities by analyzing changes in ratios and proportional relationships in various government agencies. Using mathematics, students are able to see how disproportionate representation can have major effects in communities.

DEEP AND RICH MATHEMATICS

This lesson focuses on engaging and empowering students to explore and extend the connections and progression of mathematical ideas related to ratio and proportional thinking while leveraging sensemaking and reasoning contextually and mathematically. The series of tasks intentionally creates coherent links between students' mathematical reasoning through thinking about a progression connecting fractions, ratios, and proportions. Furthermore, the notion related to how we explore proportional thinking as change occurs is also central to this lesson. Students will engage in tasks that require them to analyze the effect on ratios and proportions by changing values by small, fixed values while considering large-scale changes that have intended and/or unintended consequences mathematically and contextually.

ABOUT THE LESSON

This lesson uses a launch–explore–summarize instructional model and is intended to take approximately 240 minutes to complete across four class periods.

> **Lesson 1:** Students explore the changes in ratios and proportional relationships regarding possible expansions in the number of Supreme Court justices, including the impact on the balance of power and the potential effect on their communities.

Lesson 2: Students explore the changes in ratios and proportional relationships regarding possible expansions for the number of representatives in the Senate, including the impact on the balance of power and the potential effect on their communities.

Lesson 3: Students explore the changes in ratios and proportional relationships regarding possible expansions for the number of representatives in the House of Representatives, including the impact on the balance of power and its potential effect on their communities.

Lesson 4: Students continue to explore concepts of ratios and proportions and extend and apply their knowledge within a new context, specifically to grant the District of Columbia and Puerto Rico statehood and subsequent inclusion in both houses of Congress.

Resources and Materials

- PowerPoint resource to support lesson facilitation

- Video: "All About the Supreme Court," from *History Channel* (https://bit.ly/32L70La)

- Video: "What Is the Legislative Branch of the U.S. Government?" from *History Channel* (https://bit.ly/3rx103a)

- Laptops/mobile devices and online resources for group research, collaboration, and reflection

- Website: Supreme Court, "About the Court" (https://bit .ly/3rBYSqZ)

- Website: U.S Senate, "Party Division" (https://bit .ly/3ElHeLo)

- Online collaboration resource: Mentimeter (https://www .mentimeter.com/) or Google Slides

LESSON 1 FACILITATION

Majority in the Supreme Court

Launch (20 minutes)

- Organize students into groups of four. Take a moment to revisit your co-constructed classroom norms for interaction and agreed-upon ways of listening and responding to one another (e.g., sentence stems for rough-draft math talk; Jansen, 2020).

- Share the following prompt and allow 3 minutes for individual work:

  + *In your own words, provide a definition for the concept of "majority." Feel free to use example(s) to describe your definition.*

- Have students discuss within their group with a shoulder partner for 5 minutes. Use a Learning to Listen, Listening to Learn Protocol and rough-draft math talk sentence stems (Jansen, 2020) to discuss definitions.

LEARNING TO LISTEN, LISTENING TO LEARN PROTOCOL

- Each person is given equal time to talk, even if they don't use it all.

- Each person has 1 minute to share their definition and ideas. The listener does not interpret, paraphrase, analyze, give advice, or break in with their ideas. The listener must remain quiet, attentively listening to their partner, until the timer goes off.

- After the talker has shared, the listener will

  + thank their peer for sharing their rough-draft definition of *majority.*

  + use the following sentence stems:

    - *What I heard you share was . . .*

    - *When you shared . . . it made me think about . . .*

    - *I used to think . . . but now I'm wondering . . .*

- Some anticipated responses include these: *more than $\frac{1}{2}$, one more than half, depends on the number we are talking about, easy to notice but tough to describe,* and so on.

- Facilitate a class discussion about the contexts, nature, and mathematical aspects of definitions brought up. Look for any connections to students, schools, and communities. Record these definitions, ideas, and connections on a shared Google slide (online) or whiteboard/poster paper (face-to-face in class) so that the learning community can revisit them throughout the lesson.

Explore (30 minutes)

- Display a picture of the Supreme Court from the PowerPoint resource.

- Say: *Consider the Supreme Court of the United States.*

  + *What do you notice?*

  + *What do you wonder?*

- Tell students they have 2 minutes to jot down two things they notice and two things they wonder.

- Have students use the Turn-Taking Protocol (Hackett et al., 2020) to share the two things they noticed and the two things they wondered in their groups. (approximately 8 minutes)

TURN-TAKING PROTOCOL: ONE AT A TIME

- Person 1 shares their ideas.

- Other group members listen, but do not question or comment on what is being said.

- When Person 1 finishes, Person 2 reports their ideas as groupmates listen.

- Repeat until all group members have reported.

- Have students add what they noticed and what they wondered into an online resource (i.e., https://www.mentimeter.com/ or Google Slides) to generate a word cloud, then briefly discuss. (approximately 10 minutes)

  + Show the video, "All About the Supreme Court" (https://bit .ly/32L70La), regarding the history and composition of the Supreme Court.

  + Using their laptops or mobile devices, have students work in groups to research and gather data regarding the current and historical make-up of the Supreme Court. We recommend starting with an exploration of the Supreme Court website, "About the Court" (https://bit.ly/3rBYSqZ).

- Ask: *What are some takeaways that you have in considering the current make-up of the Supreme Court?*

  Anticipated responses include the following: *odd number, even number, mostly older, mostly white, mostly men, seated/standing, glasses, time serving on the Supreme Court,* and so on.

- Use the following questions to have groups focus on how an understanding of ratios and proportions can empower them to think through both the mathematical consequences *and* the contextual/life implications. (approximately 15 minutes)

1. *How much power does any one individual have to impact decisions? Explain your reasoning.*

2. *How easily can major decisions be influenced by any one individual? What are the challenges and affordances of this dynamic? What happens when the ratio shifts from 4 to 5, to 4 to 6, or to 4 to y?*

3. *How does adding two to four seats to the Supreme Court shift power? Mathematically? Contextually? Conversely, what if the two seats added are split across the two groups so the power shifts from 4 to 5, to 5 to 6?*

4. *How would the power of any one individual change if the number of Supreme Court judges were different from the current number? Fewer? More? Explain your reasoning.*

Summarize (10 minutes)

- Use the following prompts for a class discussion to extend students' current thinking and understanding and engage in student-led summarized learnings.

  1. Discuss the mathematical as well as sociopolitical implications related to the notion of majority in the Supreme Court. Look for any connections to students, schools, and communities. Connect the ratio and proportion of the U.S. population vs. the current composition of the Supreme Court. Ask: *Who is not represented in this picture? Do you see yourself in this picture?*

  2. Ask: *Is it representative? Is it designed to be a representative body? What are the historical/institutional/political perspectives?* For example, students may be interested to learn more about the history of the Supreme Court and how it started with six justices.

  3. Ask: *How does an understanding of mathematics related to ratios and proportional thinking empower you to make sense of this context?*

- Open up the discussion with student-generated questions, challenges, and curiosities for future exploration.

Sociopolitical refers to the combination or interaction of social and political factors.

LESSON 2 FACILITATION

Majority and Power in the Senate

Launch (5 minutes)

- Organize students into groups of four with a shoulder partner. Take a moment to revisit your co-constructed classroom norms for interaction and agreed-upon ways of listening and responding to one another (e.g., sentence stems for rough-draft math talk; Jansen, 2020).

- Say:

In our last class, we talked about the concept of majority and explored it in the context of one of the three major separate branches of government: the judicial branch, specifically the Supreme Court. We explored the significance of connecting the mathematics and the context to gain additional perspective in exploring the situation.

Now we are going to turn our attention to another major branch of the government: the legislative branch, or Congress. Congress consists of two governing bodies: the House of Representatives and the Senate. This branch of government makes laws and approves presidential nominees. There are two senators from each state and the number of representatives in Congress is based on the population.

Explore (40 minutes)

- Show the visual representation of the Senate (see the PowerPoint resource).

- Say: *Consider the Senate of the United States.*

  + *What do you notice?*

  + *What do you wonder?*

- Tell students they have 2 minutes to jot down two things they notice and two things they wonder.

- Have students use the Turn-Taking Protocol (Hackett et al., 2020) to share the two things they noticed and the two things they wondered in their groups. (approximately 8 minutes)

  + Show the video, "What Is the Legislative Branch of the U.S. Government?" (https://bit.ly/3rx103a), to explore the history and composition of Congress.

  + Using their laptops or mobile devices, have students work in groups to research and gather data regarding the current and historical make-up of the Senate. We recommend starting with an exploration of the U.S. Senate website, "Party Division" (https://bit.ly/3ElHeLo).

- Ask: *What are some takeaways that you have in considering the current make-up of the Senate?*

- Use the following questions to have groups focus on how an understanding of ratios and proportions can empower them to think through both the mathematical consequences *and* the contextual/life implications. (approximately 20 minutes)

1. *How much power does any one individual have to impact decisions? Explain your reasoning.*

2. *How easily can major decisions be influenced by any one individual? What are the challenges and affordances of this dynamic?*

3. *How does adding two to four seats to the Senate shift power? Mathematically? Contextually?*

4. *How would the power of any one individual change if the number of senators were different from the current number? Fewer? More? Explain your reasoning.*

Summarize (15 minutes)

- Use the following prompts for a class discussion:

  + *Discuss the mathematical as well as sociopolitical implications related to the notion of majority in the Senate. Look for any connections to students, schools, and communities.*

  + *How does an understanding of mathematics related to ratios and proportional thinking empower you to make sense of this context?*

- Open up the discussion with student-generated questions, challenges, and curiosities for future exploration.

LESSON 3 FACILITATION

Majority and Power in the House of Representatives

Launch (10 minutes)

- Organize students into groups of four with a shoulder partner. Take a moment to revisit your co-constructed classroom norms for interaction and agreed-upon ways of listening and responding to one another (e.g., sentence stems for rough-draft math talk; Jansen, 2020).

- Say:

In our last class, we talked about ratio and proportion and the sociopolitical implications related to notions of majority in one of the three major separate branches of government: the legislative branch, with particular focus on the Senate. We explored the significance of connecting the mathematics and the context to gain additional perspective in exploring the situation.

Now we are going to turn our attention to the other governing body within the legislative branch of government, the House of Representatives. There are a total of 435 seats in the House of Representatives. The number of representatives for each district in each state is based on population as measured by the

U.S. Census, with each district entitled to one representative. The House of Representatives has the power to impeach federal officials and elects a president if no candidate receives a majority of electoral votes.

Explore (35 minutes)

- Show a visual representation of the House of Representatives (see the PowerPoint resource).

- Say: *Consider the House of Representatives of the United States.*

  + *What do you notice?*

  + *What do you wonder?*

- Tell students they have 2 minutes to jot down two things they notice and two things they wonder.

- Have students use the Turn-Taking Protocol (Hackett et al., 2020) to share the two things they noticed and the two things they wondered in their groups.

- Tell students that after sharing they will work in their groups to jot down rough-draft ideas to the following questions: (approximately 30 minutes)

  1. *How much power does any one individual have to impact decisions? Explain your reasoning.*

  2. *How easily can major decisions be influenced by any one individual? What are the challenges and affordances of this dynamic?*

  3. *How does adding two to four seats to the House of Representatives shift power? Mathematically? Contextually?*

  4. *How would the power of any one individual change if the number of representatives were different from the current number? Fewer? More? Explain your reasoning.*

Summarize (15 minutes)

- Use the following prompts to lead a class discussion:

  1. *Discuss the mathematical as well as sociopolitical implications related to the notion of majority in the House of Representatives. Look for any connections to students, schools, and communities.*

  2. *How does an understanding of mathematics related to ratios and proportional thinking empower you to make sense of this context?*

- Conclude the discussion with student-generated questions, challenges, and curiosities for future exploration.

LESSON 4 FACILITATION

How Does Power Shift If We Add States?

Launch (5 minutes)

- Organize students into groups of four with a shoulder partner. Take a moment to revisit your co-constructed classroom norms for interaction and agreed-upon ways of listening and responding to one another (e.g., sentence stems for rough-draft math talk; Jansen, 2020).

- Say:

This week, we've talked about ratio and proportion and the sociopolitical implications related to notions of majority in two major branches of government: the Supreme Court and Congress. We explored the significance of connecting the mathematics and the context to gain additional perspective in exploring the situation.

Proposals have been made to grant the District of Columbia and Puerto Rico statehood and subsequently have them included in both houses of Congress. In your small groups, you will collaborate and explore your rough-draft ideas on the effects of this proposal.

Explore (25 minutes)

- Encourage students to work inside a shared *document* to capture their collective ideas in addressing the following questions: (approximately 25 minutes)

    1. *How does adding two new states (DC and Puerto Rico) impact the balance of power in the House of Representatives? Justify your mathematical reasoning (through understanding of ratio/proportions).*

    2. *How does adding two new states (DC and Puerto Rico) impact the balance of power in the Senate? Justify your mathematical reasoning (through understanding of ratio/proportions).*

    3. *Compare and contrast the significance of mathematical reasoning with ratio/proportions and the implications socially, politically, economically, environmentally, educationally, and so on.*

    4. *In this scenario, how is power shifted or redistributed? Is it fair? Who or what is positioned to have power now? Has this changed? How and why? What impact does this shifting and change have on a particular policy or law affecting historically marginalized people?*

Summarize (30 minutes)

- Tell students to collaborate on a shared document (i.e., Google slide/Jamboard) or poster to record their collective responses to the following reflective summary questions:

  1. *How did our explorations through ratios and proportional relationships help you think about how people and communities are positioned and who has power?*

  2. *How have our explorations through ratios and proportional relationships helped you better understand notions of power, positioning, and privilege in your communities?*

  3. *What role does any one individual play in determining the majority across the different contexts we explored across the four lessons (Supreme Court, Senate, House of Representatives, and the possibility of adding two new states—District of Columbia and Puerto Rico—to the Union)?*

  4. *How easily can voices be marginalized in small sample-sized groups?*

  5. *What happens when the power rests in any one individual?*

  6. *What are the connections to fairness and democratic ideals?*

  7. *What are the contributing factors to checks/balances when one group elects/selects significant members of the other group that are designed to serve in a checks-and-balances manner?*

  8. *How can you apply your understanding of ratios and proportions in the contexts we just explored this week to other contexts? What actions will you take?*

- Tell students to create a video reflection (i.e., via Flipgrid) to the following prompts:

  + *What aspects of this task sequence (our collective collaboration over the last four lessons in learning about applications of ratios and proportions) have resonated with you? Personally? Mathematically?*

  + *How have these experiences caused you to reflect on the role of mathematical sensemaking in raising awareness of issues related to power and privilege in society?*

TAKING ACTION

Have students consider the following prompts and the actions they can take to bring awareness to the issue:

- In your community, where do you see instances of such disproportionate representation? (For example, describe the composition of your local school board. How many members are there, and is the board representative of your community?)

- Describe the composition of the teachers in your school district. What are the implications and connections?

- How can you use mathematical reasoning skills to advocate for greater, more proportionate representation in the teaching force and/or school board?

ONLINE RESOURCE

 Available for download at **resources.corwin/TMSJ-MiddleSchool**

▼ *PowerPoint resource to support lesson facilitation*

Majority and Power

- I can recognize and describe unfairness and injustice in many forms including attitudes, speech, behaviors, practices, and laws. (Justice 12)

- I am concerned about how people (including myself) are treated and feel for people when they are excluded or mistreated because of their identities. (Action 16)

- I will speak up or take action when I see unfairness, even if those around me do not, and I will not let others convince me to go along with injustice. (Action 19)

ESSENTIAL MIDDLE GRADES CONCEPTS

- **Ratio and Proportions—** Grade 7: Analyze proportional relationships and use them to solve real-world and mathematical problems.

MATHEMATICS PRACTICES

- Make sense of problems and persevere in solving them.

- Use appropriate tools strategically.

- Look for and express regularity in repeated reasoning.

LESSON 7.4 SMOKING AND VAPING: TARGETING OF MARGINALIZED COMMUNITIES BY THE TOBACCO INDUSTRY

Nichole Campbell and Peggy Nayar

HEALTH INEQUALITY

In this lesson, students investigate data about strategies that tobacco companies use to target young Black people and other marginalized groups. They will make public service announcements in the form of movie trailers or commercials that share facts through tables, graphs, and equations of proportional relationships created by students. Truth.com is the major source of data on smoking and vaping for this lesson (https://www.TheTruth.com) because of the ways the website informs young adolescents.

DEEP AND RICH MATHEMATICS

This lesson uses proportional reasoning to determine whether a fact represents a proportional relationship. Students research a "Truth fact" (from TheTruth.com), create an equation based on $y = kx$, and complete an xy table and graph. Students determine whether their Truth fact has a proportional relationship based on the patterns and relationships of the equation, table, and graph. They are empowered to mathematically analyze and discuss social problems associated with the ways in which companies target marginalized people to buy and use smoking products. Students also analyze other risks associated with smoking and vaping, including impacts on short- and long-term health, finances, and the environment.

ABOUT THE LESSON

This lesson uses a task–question–evidence instructional model. It is intended to take 100 minutes to complete across two class periods.

Lesson 1: Students will model Truth facts with linear equations. They will graph the given equation and plot data points on the same *x-y* axis.

Lesson 2: Students will discuss similarities and differences between graphs created in the previous lesson and use them to discuss related vocabulary. Students will find the constant of proportionality for data the class has worked with.

Resources and Materials

- PowerPoint resource to support lesson facilitation

- Teacher Resource 1: *Vaping Facts With Numbers From TheTruth.com*

- Teacher Resource 2: *Proportional Relationships Project*

- Student PowerPoint: *Truth Fact Activity*

- Worksheet 1: *Constant of Proportionality Guided Notes* (1 per student)

- Worksheet 2: *Constant of Proportionality* (1 per student)

- Video: "Read Between the Lies: Black Lives/Black Lungs," from Lincoln Mondy (https://bit.ly/3DjfAxr) and others from Truth.com (https://www.TheTruth.com/)

LESSON 1 FACILITATION

Smoking and Vaping and Linear Equations

This lesson is intended to make students aware of how tobacco and vaping companies specifically target marginalized people. Students will learn how mathematics can be used to better understand the world they live in and to promote a sense of social justice.

Launch (5–20 minutes)

- Consider showing the video, "Read Between the Lies: Black Lives/Black Lungs," from Lincoln Mondy (https://bit.ly/3DjfAxr) to introduce the lesson. (15 minutes)

  Note: *The video contains one word of profanity.*

- Show PowerPoint slides 3 to 4 and begin a brief discussion.

  + Slide 3 (advertisement images): *What do you notice? What do you wonder?*

  + Slide 4 (vaping devices): *What do you know about vaping?*

- Use the following questions to extend the classroom discussion:

  + *How do you feel about the ads that we looked at?*

  + *Why did they use those images?*

  + *How does this impact people of Color?*

+ *Who is the target audience for these images? Why?*

+ *What is the ad not telling us?*

Explore (35 minutes)

Preparation: Adjust the number of students allowed to select each fact based on the total class size. There are six facts in this lesson. The Lesson Facilitation PowerPoint slide deck was set up for a class size of 24 students.

- Arrange students into groups.

- Consider having students count off from 1 to 6 or use a random selection process to have students select one Truth fact about smoking or vaping.

- Provide students with the Student PowerPoint (*Truth Fact Activity*). Use Lesson slide 5 to walk through the activity and describe how students should submit their completed work.

- Monitor student progress throughout the activity.

Summarize (8 minutes)

- Have students share their completed slide deck with their small group and then the whole class.

- Use the following questions to promote class discussion:

    + *What do you know about Juuls?*

    + *How do you feel about the facts that we looked at?*

    + *Which fact do you think impacts you most directly? What about your family? What about your community?*

    + *Who makes money off this?*

    + *What else could be done with that money?*

    + *How does this impact people of Color more than others?*

    + *How does this impact underrepresented groups more than others?*

    + *Why would tobacco and vaping companies target underrepresented groups and communities of Color? Why would they not target white communities?*

    + *Why is it important to discuss what we are learning with others?*

    + *What can we do about the damage that is being done to people and our environment?*

LESSON 2 FACILITATION

Smoking and Vaping and Proportional Relationships

Launch (5–7 minutes)

- Show slide 13 and use the following questions to activate prior knowledge:

  + *What is the same? What is different?*

  + *What are the requirements for proportionality?*

  + *Which one has a straight line that goes through the origin?*

  + *How do we find the rate of change (or slope) of these lines?*

- Here are some possible responses:

  + As students share that *they are both straight lines*, connect that with one of the requirements for being proportional.

  + When a student notes that *the blue line goes through the origin and the red one doesn't*, link this to the other requirement of being proportional.

  + The discussion should be led toward the left image being proportional because it has a straight line that goes through the origin (0,0).

  + The right one is nonproportional because the line will not go through the origin.

  + We use rate of change rather than slope in this lesson. You may need to define for or remind students that for lines, the rate of change is the same as the slope of the line.

Explore (70 minutes, or possibly 2 days)

- Provide students with a copy of Worksheet 1 (*Constant of Proportionality Guided Notes*).

- Use slides 14–18 to emphasize the rate of change as an important characteristic of proportional relationships, and connect the rate of change in these cases to the constant of proportionality.

  Slide 15:

  + Link change in y to change in x as the rate of change of 2200.

  + Point out to students that, in this case, **the rate of change is a ratio of $\frac{y}{x}$.**

Slides 16 and 17

+ Link change in *y* to change in *x* as the rate of change of 15.

+ Notice that in the chart, the change in *x* is not always 1.

+ Remind students that the rate of change is a ratio of $\frac{y}{x}$.

Slide 18:

+ Use the phrase, *When the requirements for proportionality are satisfied, the rate of change is the same as the constant of proportionality (k).*

+ It is the same ratio of $\frac{y}{x}$. Therefore, $k = \frac{y}{x}$.

- Consider asking the following questions to facilitate the discussion and develop student understanding:

    + *How do the x and y values change?*

    + *What is the rate of change?*

    + *What is the ratio?*

    + *What is the change in y?*

    + *What is the change in x?*

    + *The rate of change is a ratio of what?*

    + *What does constant mean?*

    + *What stayed constant?*

    + *What does k equal?*

    + *What is another term for "rate of change"?*

    + *What does proportional mean?*

    + *What is the equation for k?*

- Distribute Worksheet 2 (*Constant of Proportionality*), and tell students to complete each part individually before checking in with their group.

- Monitor students' progress throughout the activity and use the previous questions to further probe students' thinking.

- Select groups to share their solutions.

Summarize (8 minutes)

- Use slide 19 and Worksheet 1 (*Constant of Proportionality Guided Notes*) to have students write out the summary statement, then turn and talk within their group. Say:

    + *For proportional relationships, the rate of change is the ratio of y to x. Constant means the same. When the relationship is proportional, the term "constant of proportionality" is the same as "rate of change." Constant* *of proportionality is represented by the letter k. $k = \frac{y}{x}$ is an equation that stands for how a proportional graph changes. In the example of 1 tree killed for 15 packs of cigarettes, we could write the equation as $y = 15x$.*

- Use the following questions to promote discussion and connect the mathematics and the context of smoking and vaping:

  + *What facts are you learning about smoking and vaping that you want to represent in tables or coordinate planes?*

  + *How do you feel about what you are learning?*

  + *What will happen if these trends continue?*

TAKING ACTION

Consider the following actions:

- Students can create movie trailers and commercials as forms of public service announcements. See Teacher Resource 2 (*Proportional Relationships Project*) for an example of guidelines for public service announcement projects.

- Students can write letters to tobacco and vaping companies as well as to business owners and politicians.

- Encourage students to share with loved ones and their community what they learned during the lesson.

ONLINE RESOURCES

PowerPoint resource to support lesson facilitation

 Available for download at **resources.corwin/TMSJ-MiddleSchool**

▼ *Teacher Resource 1: Vaping Facts With Numbers From TheTruth.com*

▼ *Teacher Resource 2: Proportional Relationships Project*

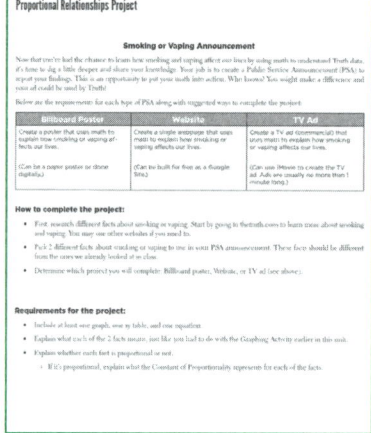

▼ *Student PowerPoint: Truth Fact Activity*

Graphing Activity

Each person will select one Truth fact about smoking or vaping with an equation that represents the fact. There is a maximum of 4 students per fact.

You will need to complete the following on your slide:

* Use the equation to complete the xy table.
* Plot the ordered pairs on your graph paper.
* Complete the summary statement by telling us what your Truth fact is about.

Be ready to share your work with the class.

▼ *Worksheet 1: Constant of Proportionality Guided Notes (1 per student)*

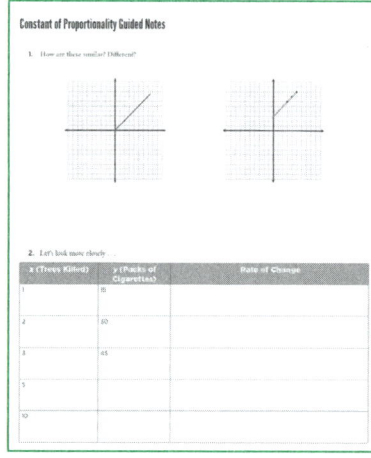

▼ *Worksheet 2: Constant of Proportionality (1 per student)*

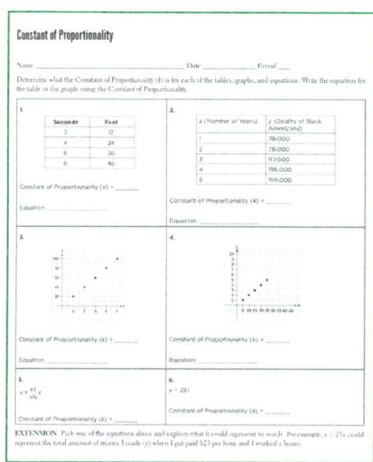

▼ *Lesson slides*

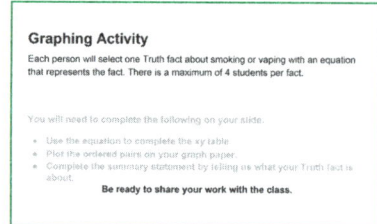

Graphing Activity

Each person will select one Truth fact about smoking or vaping with an equation that represents the fact. There is a maximum of 4 students per fact.

You will need to complete the following on your slide:

* Use the equation to complete the xy table.
* Plot the ordered pairs on your graph paper.
* Complete the summary statement by telling us what your Truth fact is about.

Be ready to share your work with the class.

LESSON 7.5 HEALTH, RACE, AND RATIOS

Travis Weiland and Melissa A. Gallagher

HEALTH INEQUALITY

This lesson focuses on health inequality. We focus on having students interrogate data related to the COVID-19 pandemic; however, this situation is merely highlighting long-standing inequities in health and health care. Health is deeply intersectional with other factors such as socioeconomic status, location, education, race and ethnicity, and gender and is impacted by systemic racism. The goal of this lesson is to present students with real data from the pandemic and support them in reasoning proportionally to read the world with mathematics and create opportunities for students to take action.

CONTENT NOTE

Because of the broad and unequal impacts of COVID-19, it is likely that your students have all had different lived experiences related to the pandemic, some of which might be very difficult such as the death of a close loved one. This does not mean the topic should be avoided; however, it is important for teachers to take care in starting such a lesson where there may be strong emotions for the students. One possible starting point could be to create time and space for students to share their experiences with the pandemic in small groups or in writing. It is also important for the teacher to publicly acknowledge students' emotions related to the issue, as this is not a neutral, abstract topic to discuss.

DEEP AND RICH MATHEMATICS

A deep understanding of ratios, percentages, and proportional reasoning is foundational for later mathematics and mathematical literacy. For example, counts are often presented to make the case that there are no inequities (e.g., more white people have died from COVID-19 than Black people); however, these counts are not an accurate way to look at the impact on different communities because they do not take into account that each group makes up a different percentage of the total population. In order to examine issues of inequity, the data need to be converted into ratios before they can be compared. In this lesson, students will engage with real data presented as counts, ratios, and percentages to investigate issues of health inequity during the COVID-19 pandemic.

SOCIAL JUSTICE OUTCOMES

- I can recognize and describe unfairness and injustice in many forms including attitudes, speech, behaviors, practices, and laws. (Justice 12)

- I will speak up or take action when I see unfairness, even if those around me do not, and I will not let others convince me to go along with injustice. (Action 19)

ESSENTIAL MIDDLE GRADES CONCEPTS

- **Ratio and Proportion—** Grade 7: Analyze proportional relationships and use them to solve real-world and mathematical problems.

- **Statistics and Probability—** Grade 7: Draw informal comparative inferences about two populations.

MATHEMATICS PRACTICES

- Reason abstractly and quantitatively.

- Construct viable arguments and critique the reasoning of others.

- Attend to precision.

ABOUT THE LESSON

This lesson uses a launch–explore–summarize instructional model and is intended to take approximately 180 minutes to complete across three class periods.

Lesson 1: Students will consider when it is appropriate to use proportional reasoning in a real-world situation to reason about COVID-19 data in terms of comparisons of deaths to population rather than only considering a single quantity of counts of deaths.

Lesson 2: Students will explore injustice related to COVID-19 deaths by race and ethnicity by comparing percentages.

Lesson 3: Students will take action and engage in research to understand the problem and use what they learned from that research to brainstorm ways in which they can take action and then create a plan for action.

Resources and Materials

- PowerPoint resource to support lesson facilitation

- Excel graphs of data for PowerPoint resource

- Student Resource 1: *Taking Action Planning Sheet* (1 per group of students)

- Student Resource 2: *Hundreds Grids* (cut out the grids so that each student has at least 1 hundreds grid)

- Device with internet (1 per pair of students)

- Website: COVID Tracking Project "Data Summary" (https://bit.ly/3lwTpO3)

- Website: COVID Tracking Project "Racial Data Tracker" (https://covidtracking.com/race)

- Poster paper

- Markers, colored pencils, or crayons

- Video: "Kids Can Change the World," from Matt and Jake Webb, TED^X (https://bit.ly/31w1mfy)

- Video: "How to Change the World," from Kid President (https://bit.ly/3ly98wi)

- Copies of or digital access to the following news articles (1 article per student):

  + "Pandemic Report," by Brian S. McGrath, *Time for Kids*, June 23, 2020 (https://bit.ly/3xNsaUz)

  + "Cases on the Rise," by Ellen Nam, *Time for Kids*, July 7, 2020 (https://bit.ly/3lv2Iyc)

  + "Education Update," by Allison Singer, *Time for Kids*, August 5, 2020 (https://bit.ly/3ptEupk)

  + "Race to a Vaccine," by Allison Singer, *Time for Kids*, September 18, 2020 (https://bit.ly/2ZNA9nR)

Note: *You can adjust the data to reflect authentic COVID data in your state.*

LESSON 1 FACILITATION

Unpacking U.S. COVID-19 Data

The data for this lesson come from the COVID Tracking Project. Because of the dynamic nature of the data, consider updating the figures in the Teacher PowerPoint file to reflect the most recent data available (using the editable Excel spreadsheet also provided). The COVID Tracking Project "Data Summary" contains data through March 7, 2021 (https://bit.ly/3lwTpO3); consider accessing other COVID data sources for additional data.

Launch (10 minutes)

- Introduce the lesson by showing the bar graph, "COVID-19-Related Deaths by Race and Ethnicity" (PowerPoint slide 1). Say:

  + *COVID-19 has affected the lives of everyone. Data on cases and deaths flash across the news daily. These data can be presented in different ways, and how they are presented shapes how we are able to make sense of the situation. Here is a bar graph of the number of total COVID-19-related deaths in the United States by race and ethnicity.*

- Ask:

  + *What do you notice in the graph?*

  + *What do you wonder?*

  + *How do these data make you feel?*

  + *How does this impact your community?*

- **Special Note 1:** This initial notice/wonder is meant to give students a chance to begin to make sense of what the quantities are that they will be exploring during the lesson in a low-stakes environment. This will also give students a chance to relate their own personal experiences of the pandemic, as some students might know someone who has died and have a very emotional connection to the context of the lesson. You should give students the space to feel and share as appropriate. A common noticing will likely be that the largest number of deaths are of white people, followed by Black and Hispanic individuals.

- **Special Note 2:** If you have not previously discussed race and ethnicity in your class, this would also be an opportunity to ask students what they understand about race and ethnicity and begin to have a discussion about what it is and why it is important to discuss.

- To get students thinking about comparing quantities and to connect to what they have learned previously about fractions, begin to problematize looking at total counts alone. Ask probing questions such as these:

  + *Do you think there are the same number of white people in the United States as there are Black people?*

  + *Can we compare parts that come from different wholes?*

  + *Does it make sense to just compare totals in this situation?*

Explore (40 minutes)

- Show students the graph, "U.S. Population by Race and Ethnicity" (PowerPoint slide 2), and use the following questions to continue the discussion:

  + *What do you notice from looking at this graph?*

  + *How does this information relate to what you observed in the previous graph of COVID-19 deaths by race and ethnicity?*

The goal of these questions is to begin to prime students to reason multiplicatively instead of additively. It is assumed that students have seen the topic before. However, students often struggle with identifying situations appropriate for the use of proportions. If students do not begin to make the connections from the previous questions, ask more directed questions such as these:

  + *What do you notice about the size of these bars relative to each other as compared to the size of the bars in the previous graph?*

  + *Since white individuals make up such a large amount of the population compared to Black or Hispanic individuals, does it make sense to compare the number of deaths from those groups directly?*

- Once you have primed students to begin to think comparatively, have them work in groups to respond to the following prompt:

  + *How can you compare the number of white people who have died from COVID-19 to the number of Black people who have died from COVID-19, considering how much of the population each group makes up? You can use pictures, graphs, numbers, and words.*

- While students are working together, move around the room to respond to questions, ask probing questions of groups that seem to be struggling unproductively, and select student responses to present to the whole class.

  + *The goal is for students to create a ratio between the number of deaths and the totals for each population of white and Black people. Representations such as a tape diagram, pie chart, or side-by-side bar graphs might be used by students.*

- To begin to move students into reasoning proportionally, present the graph titled "COVID-19-Related Deaths per 100,000 by Race and Ethnicity" (PowerPoint slide 3).

  + This graph shows the number of COVID-19-related deaths per 100,000 by race and ethnicity relative to their population size in the United States.

  + It is best to start with ratios per 100,000 as it is easier to compare whole-number quantities, whereas the unit ratio would create a decimal quantity. Also, the decimal quantities may support beliefs that the COVID-19 pandemic is overrated and not many people are dying.

- Tell students to respond to the following questions in order in their groups:

  + *What does the y-axis represent in this graph?*

  + *What do the quantities above the bars represent in the graph?*

  + *How does this graph compare to the graph of the numbers of COVID-19-related deaths we saw at the beginning of class?*

  + *What conclusions do you draw from this graph?*

  + *Which graph do you think better represents what is happening?*

  + *What are you still wondering?*

The goal is for students to see the difference between a quantity and a comparison of quantities. One of the main takeaways from these questions should be that when considering the deaths relative to the population, there is a disturbing pattern that emerges in terms of Black and Hispanic people being disproportionately impacted by COVID-19.

  + It is important to highlight for students why it is important to reason proportionally in this situation.

+ It is also important to point out that the single quantity above each bar in the graph is the numerator of a ratio. For example, 92 per 100,000 Black or African Americans have died from COVID-19.

- Show slide 4 and say:

  + *We are going to dig deeper into the patterns we see in the graphs using data from the COVID Tracking Project (https://covidtracking.com/race).*

  When you go to the COVID Tracking Project webpage and scroll down the first figure, you will see one similar to that shown in the following graph ("Nationwide, Black people . . ."). Note that this graphic is updated daily as new data come in, so it may be slightly different than the image shown on the slide. Consider showing the actual website or updating the slide.

Nationwide, Black people are dying at 2.3 times the rate of white people.

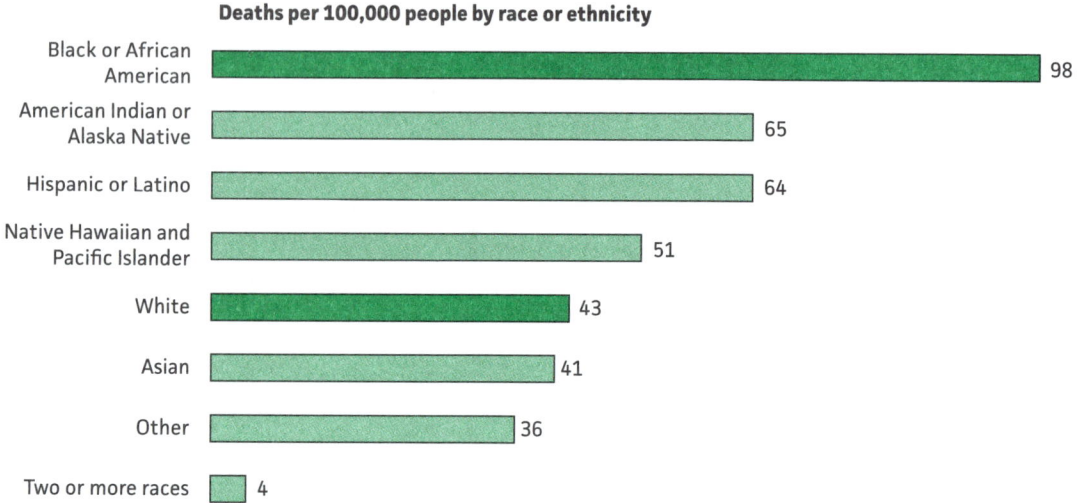

Deaths per 100,000 people by race or ethnicity

Black or African American	98
American Indian or Alaska Native	65
Hispanic or Latino	64
Native Hawaiian and Pacific Islander	51
White	43
Asian	41
Other	36
Two or more races	4

Source: COVID Tracking Project. (2022). The COVID Tracking Project. *The Atlantic* https://covidtracking.com. CC BY 4.0 (https://creativecommons.org/licenses/by/4.0/).

- Have students read the statement at the top of the graph, then in groups respond to the following questions:

  + *How does that statement make you feel?*

  + *What does that statement mean to you?*

  + *Where does the 2.3 come from in the statement?* [change 2.3 if using updated image]

  + *How could you make a similar statement about Hispanic or Latina/o people?*

  + *How could we compare Hispanic people's deaths to Black people's deaths?*

  + *What might be causing these differences?*

The goal is to get students using their proportional reasoning to think about the relationships between ratios and to see inequities in the COVID-19 deaths by race and ethnicity and how mathematics can be useful in reading the world.

Summarize (10 minutes)

- Use the following questions to facilitate a class discussion:

    + *When is it better to use ratios to compare quantities, rather than looking at a single quantity?*

    + *What patterns do ratios help you see that are not apparent from looking at individual quantities?*

    + *What might be causing the inequitable impact of COVID-19 on people of different races?*

- The purpose of the final discussion of the lesson is to come to some generalizable conclusions from the explorations of the day.

    + A big idea about ratios that should be emphasized is that ratios are a comparison between quantities.

    + Another big idea is that when comparing groups of different sizes, proportional reasoning and ratios are necessary.

    + The main social justice outcome is for students to see ratios help comparisons between social groups to interrogate inequities.

LESSON 2 FACILITATION

Comparing States

Preparation: Obtain a screenshot of the race and ethnicity data for your state to replace slide 6. See the COVID Tracking Project (https://covidtracking.com/race/dashboard).

Launch (15 minutes)

This Launch revisits the meaning of percentages. Even if your students are strong in operating with and computing percentages, it is suggested that you help students to create the visual in this section because it will help them visualize comparisons later in the Explore section of the lesson.

- Say: *Today you are going to be looking at data for our state in order to explore whether there are inequities related to race and ethnicity for deaths from COVID-19.*

- Show students just the race and ethnicity and percentage of population data for your state (PowerPoint slide 6) to get students thinking about the percentage of people in your state by race and ethnicity. Be sure to use the data from the COVID Tracker website to be consistent with the data you will use later.

  Note: *Texas is used as an example throughout this lesson.*

Race/ethnicity	Percentage of population
Black or African American alone	12%
Hispanic or Latino*	39%
Asian alone	5%
Native Hawaiian and Pacific Islander alone	<1%
American Indian or Alaska Native alone	<1%
Two or more races	2%
White alone	42%
Some other race alone	<1%

*Note. Hispanic or Latino ethnicity, any race. All other race categories in this table are defined as Not Hispanic or Latino.

Source: COVID Tracking Project. (2022). Racial Data Dashboard. *The Atlantic.* https://covidtracking.com. CC BY 4.0 (https://creativecommons.org/licenses/by/4.0/).

- Remind students that the word "percent" is composed of two roots: *per* meaning out of, and *cent* meaning one hundred. Say: *So 12% means that 12 out of every 100 people in Texas are Black or African American* (alone refers to the fact that they are not Black and Hispanic).

- Give each student a hundreds grid and have them shade in the percentage of the population represented by each race/ethnic group (see the following example).

+ **Note:** *All groups representing less than 1% of the population are left off. Prompt students to notice that the numbers add up to more than 100 squares. Feel free to discuss how rounding can lead to this discrepancy.*

- Say:

  + *If COVID-19 impacted all races and ethnicities equally, then what would a hundreds grid that depicted deaths by race and ethnicity for Texas look like?*

- Give students time to think and discuss why the grid should be identical (or approximately identical) if COVID-19 impacted everyone in the same way.

- Connect to the conversation from Lesson 1 about why it is important to examine these data as percentages, which are a special type of ratio, rather than as counts.

- During this discussion, connect back to the ratios from the previous lesson that were per 100,000 to help with student understanding of the context and mathematics.

Explore (30 minutes)

- Next, show students just the top part of the dashboard for your state (slide 7). Ask: *What does it mean that Texas has reported race and ethnicity data for 6% of cases and 98% of deaths?*

  + Help students make sense of this by comparing it to your classroom. For instance, say:

    $6\% = \frac{6}{100} = \frac{3}{50} = \frac{1.5}{25}$. *So for a class of 25 students, you would report the race and ethnicity of 1.5 students. Would this help anyone to understand what the racial or ethnic background of the class is?*

  + *In the case of Texas, the low reporting of race and ethnicity data for cases is highly problematic and means that we should not consider the case data to be representative of the cases across the state. The high reporting of race and ethnicity for COVID-19 deaths, however, means that these data should be fairly accurate and so we will focus on these.*

Texas

Texas has reported race and ethnicity data for:

The following tables reflect only those cases and deaths where race/ethnicity is known and reported by Texas. If this state's reporting percentages are low, interpret with caution.

6%
CASES

98%
DEATHS

Source: COVID Tracking Project. (2022). Racial Data Dashboard. *The Atlantic.* https://covidtracking.com. CC BY 4.0 (https://creativecommons.org/licenses/by/4.0/).

- Present the rest of the dashboard from your state (slide 8). Ask your students to respond to the following questions on paper and then in a pair–share chat with a neighbor:

  + *What do you notice about this table?*

  + *What do you wonder?*

  + *How does it make you feel?*

- Ask students to share their thoughts in a whole-group discussion. Make sure part of this conversation involves unpacking what conclusions cannot be made from these data due to missing data indicated in the notes in the graphs you have shared.

Cases and death by race/ethnicity

Race/ethnicity	Percentage of population	Percentage of cases	Percentage of deaths
Black or African American alone	12%	19%	11%
Hispanic or Latino*	39%	45%	56% ◆
Asian alone	5%	2%	2%
Native Hawaiian and Pacific Islander alone	<1%	- ①	- ①
American Indian or Alaska Native alone	<1%	- ①	- ①
Two or more races	2%	- ①	- ①
White alone	42%	34%	30%
Some other race alone	<1%	<1% ①	<1% ①

Notes

◆ Racial/ethnic disparity likely. <u>See why</u>.

⊙ Hispanic or Latino ethnicity, any race. All other race categories in this table are defined as Not Hispanic or Latino.

① This data should not be compared with percentage of the population. Texas does not specify if this race category is included under a grouping it labels "Other."

Source: COVID Tracking Project. (2022). Racial Data Dashboard. *The Atlantic.* https://covidtracking.com. CC BY 4.0 (https://creativecommons.org/licenses/by/4.0/).

- Now, invite students to work in partners or trios to investigate a state of their choosing. Tell them to prepare a summary of their findings to share at the end of the lesson.

  + You might organize these by creating a shared file for students to sign up for a state and enter their findings.

  + Remind them to attend to whether or not there are sufficient data reported to draw any conclusions about that state.

- Present the following focus questions:
    + *Are there health inequities in your state?*
    + *If so, which groups appear to be disproportionately affected?*

Summarize (15 minutes)

- Present students with the overarching question:
    + *Do health inequities exist in the United States with regard to the COVID-19 pandemic? If so, are there any patterns in the inequities?*

- Invite students to share what they find. You might do this by projecting the shared file that students entered their findings into and having them quickly share what they found.

- Ask the class:
    + *Do these data indicate that there are health inequities in the United States?*
    + *If so, what patterns have you noticed?*

  Remind students to justify all statements with data.

- Ask students to begin researching why these inequities might exist and to begin brainstorming ways they might take action to fight these inequities. Let them know you will begin the conversation about taking action in the next lesson.

TAKING ACTION

Lesson 3 Facilitation

Here, Taking Action is considered a third full lesson period. It is suggested that you collaborate with a teacher in another subject area, perhaps English Language Arts (ELA) or social studies, in order to coordinate efforts and integrate the action. For instance, students might decide to write a petition, which would fit well with an ELA lesson.

Launch (10 minutes)

- Begin by posting the following question on the board: *What actions can we take to address health inequities, especially inequities related to the COVID-19 pandemic?*

- Prime students' thinking and inspire them to take action by playing one of the following videos:
    + "Kids Can Change the World," from Matt and Jake Webb, TED^X (https://bit.ly/31w1mfy; watch 0:00–1:45)
    + "How to Change the World," from Kid President (https://bit.ly/3ly98wi)

- Ask students to turn and talk to a partner about their takeaways from the video. Then have a few students share out. Be sure to encourage the message that students can take action and make an impact.

Explore (45 minutes)

- The first step in taking action is to do research to understand the problem and to find possible solutions. Invite students to read one (or more) of the following articles. As they read each article, they should brainstorm actions they can take to address inequities related to the COVID-19 pandemic, adding to their responses from the Launch.

  + "Cases on the Rise," by Ellen Nam, *Time for Kids*, July 7, 2020 (https://bit.ly/3lv2Iyc). This article gives information on COVID-19 cases in July 2020, along with descriptions of policy changes that states with high rates of infection made right before their numbers skyrocketed. It also offers tips for prevention of COVID-19.

  + "Education Update," by Allison Singer, *Time for Kids*, August 5, 2020 (https://bit.ly/3ptEupk). This article describes how the COVID-19 pandemic has disrupted education for more than 1 billion children around the world and has increased inequities.

  + "Pandemic Report," by Brian S. McGrath, *Time for Kids*, June 23, 2020 (https://bit.ly/3xNsaUz). This article gives an overview of the pandemic at the time of reporting and suggests that what is needed is clear government leadership, including improved contact tracing.

  + "Race to a Vaccine," by Allison Singer, *Time for Kids*, September 18, 2020 (https://bit.ly/2ZNA9nR). This article describes the process for creating a vaccine and where a COVID-19 vaccine was in development at the time of writing. It also alludes to issues with equity in distributing the vaccine once it is approved.

- Once students have read and brainstormed on their own, move them into pairs or small groups with others who read the same articles to share their ideas and generate more ideas. Once you hear several good ideas floating around the room, pull the students into a whole-group discussion.

- Open the whole-group discussion by reorienting students to the social justice-related question for the day: *What actions can we take to address health inequities, especially inequities related to the COVID-19 pandemic?*

- Record all ideas, even if they initially seem unrealistic or too simplistic. Try to record all ideas quickly, allowing students to discuss ideas. The goal is to get ideas on the board and then have students either (a) vote on the ideas to choose one to do as a class or (b) choose actions that they want to take and form groups/committees to take those actions. Both options have merit, but the second one allows for a greater role for each student in the class.

- Once an action(s) has been chosen, place students into groups and have them use Student Resource 1 (*Taking Action Planning Sheet*) to plan their action. If there is time, they can begin working on their action steps. From here, the finalization of this action may take several forms. For example, you may have students complete this action at home, you may choose to give students 10 minutes each class period to work on it, or you may choose to give them more time.

  + **Note:** *It is recommended that students not be asked to work on this action entirely on their own time, as taking action should be considered an integral aspect of social justice education and should be given adequate in-class time so that students see it is valued.*

Summarize (5 minutes)

- Begin by asking students to reflect on the social justice question from Lesson 1: *How can mathematics help interrogate systemic injustice?*

- Ask students to share how they used mathematics over the last 3 days to learn about and uncover health inequities, how that knowledge may have guided the actions they designed, and whether they used any mathematics as they considered and planned for their action.

- Be sure to highlight the role that mathematics plays in understanding current events and in being able to see where inequities lie in order to take action and address them.

ONLINE RESOURCES

 These downloadable resources can be found online at
resources.corwin.com/TMSJ-MiddleSchool

▼ *Student Resource 1: Taking Action Planning Sheet*

Created by Melissa A. Gallagher

Names _____

_____

_____

_____

Date _____ Issue _____

Action Plan

What is the injustice that you are working to address? Write a coherent summary of what your problem is in 3–5 sentences below:

_____

_____

_____

What does your group want to do to be a "part of the solution"?

_____

_____

How will you "be a part of the solution"? Be specific.

_____

_____

When will this event/activity take place/be finished? (All projects **must** be finished by May 30)

_____

Where will this event/activity take place?

_____

▼ *Student Resource 2: Hundreds Grids*

▼ *PowerPoint resource to support lesson facilitation*

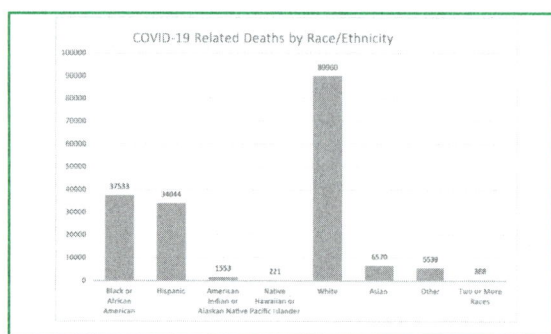

▼ *Excel graph resource (with editable graphs for PowerPoint)*

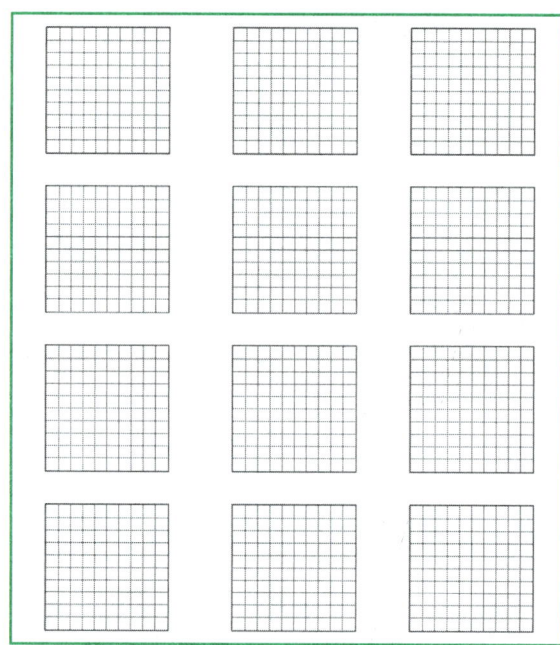

Black or African American	409,16,113	37,533	92
Hispanic	575,17,935	34,044	59
American Indian or Alaskan Native	26,99,073	1,553	58
Native Hawaiian or Pacific Islander	5,82,718	221	38
White	2349,04,818	89,960	38
Asian	175,74,550	6,570	37
Other	157,89,961	5,539	35
Two or More Races	104,35,797	388	4

LESSON 7.6 HEALTH INEQUALITIES: COVID-19 AND OTHER HEALTH CONDITIONS

Tashana Howse and Kendrick Savage

HEALTH INEQUALITY

Since March 2020, the COVID-19 pandemic has changed the way we live, work, and even socialize. Scientists and other professionals continue to work hard to improve vaccines and find treatments that will minimize the number of cases and deaths from the disease. Because COVID-19 is new, one would think that it would affect groups of people in the same way as well as at the same rate. However, the current health inequalities within certain U.S. communities indicate that this new disease will affect these individuals at a higher and/or faster rate. Many of the leading causes of death in our country are related to diseases that affect people of Color at higher rates than white people; this is now also true for COVID-19. Therefore, it is evident that action must be taken to combat this issue of health inequality in communities of Color, specifically Black communities.

CONTENT NOTE

The topic of COVID might be difficult and personal for many students. We encourage you to refer to Chapter 4 ("Dealing With Emotional or Sensitive Conversations") for guidance on how to approach the Launch of the lesson. If you would prefer to avoid discussing COVID at this time, an alternative activity for the Launch is provided in the Online Resources for this lesson.

DEEP AND RICH MATHEMATICS

The mathematics focus of this lesson is to understand ratios and rates as well as how to describe phenomena using ratios and rates. By examining quantities related to common health conditions within communities, students will discover how rates and ratios provide viable information to describe trends in data. They will analyze data and make conjectures about relationships among quantities. Trends related to health conditions will highlight health inequalities within certain communities and support students in understanding how mathematics helps to uncover issues within society. When issues are uncovered, they can be addressed by engaging in appropriate action.

ABOUT THE LESSON

This lesson uses a launch–explore–summarize instructional model and is intended to take approximately 200 minutes to complete across three class periods.

Lesson 1: Students recall prior knowledge of the relationship among decimals, fractions, and percentages by engaging in a data collection exercise. They will collect and organize data in a table and circle graph.

Lesson 2: Students are introduced to the definition of a ratio and rates and their relationship to fractions and percentages by working through contextualized ratio problems.

Lesson 3: Students use ratio and rate reasoning to analyze real-world problems. They use ratios to unpack data from the Centers for Disease Control and Prevention (CDC) related to various types of health cases.

Lesson 4: Students summarize their findings. Students will present their findings as a group using a modality of their choice.

Resources and Materials

- Teacher Resource 1: *Health Cases Activity Sheet* (1 copy per class, cut out)

- Teacher Resource 2: *Notes for Article Readings*

- Teacher Resource 3: *Excel File With Sample Solutions to Worksheet 4*

- Worksheet 1: *Health Cases* (1 per student)

- Worksheet 2: *Exploring Ratios in Context* (1 set per student)

- Worksheet 3: *Health Disparities Abstract Readings* (1 per student)

- Worksheet 4: *Examining the Number of Deaths in Georgia* (1 per student)

- Student Resource: *Excel File With Data for Worksheet 4* (optional)

- Computer with Excel or Google Slides (1 per group; optional)

- Copies of the following articles for the jigsaw activity may be searched by title (each student needs 1 article):

  + "COVID-19 Pandemic Highlights Racial Health Inequities," by Crystal Johnson-Mann et al., *The Lancet*, July 10, 2020 (https://bit.ly/3ohNIoY)

  + "COVID-19 and the Widening Gap in Health Inequity," by Helene J. Krouse, *Otolaryngology–Head and Neck Surgery*, May 5, 2020 (https://bit.ly/3xNlkya)

  + "Racial Disparity of Coronavirus Disease 2019 in African American Communities," by Ravina Kullar et al., *Journal of Infectious Diseases*, September 15, 2020 (https://bit.ly/3lybpHQ)

  + "COVID-19 and Health Disparities: The Reality of 'the Great Equalizer,'" by Stephen A. Mein, *Journal of General Internal Medicine*, May 14, 2020 (https://bit.ly/2ZNeg8a)

Note: The data for Lesson 2 are for the state of Georgia. You can adjust the data to reflect the values for your state.

LESSON 1 FACILITATION

Health Cases

Students recall prior knowledge of the relationship among decimals, fractions, and percentages by engaging in a data collection exercise. They will collect and organize data in a table and circle graph. Then, in groups, they will complete a task related to ratios and fractions.

Preparation: Prepare the introduction activity by cutting out the colored health cases squares in Teacher Resource 1 (*Health Cases Activity Sheet*). Place the colored squares in a basket for students to draw without looking.

Launch (10 minutes)

- As students enter the classroom, instruct them to draw a square from the basket and take it to their seat. Once all students have entered the classroom, say: *Today we are going to collect data on the colored squares that you selected. Those squares are representative of common health cases within our community, state, and country. Similar to what you may have done in your science class, this exercise is a simulation.*

- Distribute Worksheet 1 (*Health Cases*). As a whole class, complete the table for "Part 1: Health Cases in Our Community" by calling out the name of the health case. Tell students to raise their hands when their case is called and to record the count in the "number of cases" column in the table.

- Tell students to represent the data for each case as a fraction, decimal, and percentage.

- After completing the table, instruct students to represent the data in a circle graph. Review how to connect percentages to circle graphs as needed.

- Use the following questions to facilitate a class discussion:

    + *What do you notice about the data we collected and organized?*

    + *What do you wonder about the data we collected and organized?*

    + *What type of questions do you have from this exercise?*

Explore (30 minutes)
- Place students into groups of three to four and have them look at "Part 2: Health Cases in the Other Class" of Worksheet 1, which includes information from another class that simulated the same activity.

- Tell students to work with their group members to complete the task. Have students select a group recorder, as one completed paper will be collected from each group.

- As groups finish, collect data from the class based on their answers to the questions. This will enable groups to agree or disagree based on how they reasoned with the data.

    Note: *It is important that you do not provide corrections to group responses. This information will set the stage for Lesson 2 and provide the information needed to introduce ratios.*

Summarize (10 minutes)
- Use the following questions to facilitate a closing discussion about concerns related to these illnesses in our communities. Say: *Being aware of these cases is important, and it is worth studying to see if there is a pattern.*

- Ask the following questions:

    + *Does the amount of students in the class matter when determining who has "more" or "less"? Why or why not?*

    + *What do we mean when we ask if one class has "more" or "less" cases than the other?*

    + *What if we decide to look at this data in another way? How could we do that?*

+ *What if we break down the data in relation to races and ethnicities? What could this tell us?*

+ *How does mathematics help us understand the phenomena?*

LESSON 2 FACILITATION

Ratio and Rates

Students are introduced to the definition of ratio and rates and their relationship to fractions and percentages by unpacking data from the CDC. They will read the abstracts of two articles and reflect on the reading. In groups, students will then examine the number of deaths within their state related to cases explored in Lesson 1.

Launch (20 minutes)

- Revisit the work from Lesson 1, using differences among group responses for Worksheet 2 (*Exploring Ratios in Context*) to facilitate a brief discussion. Say to the class, *During this lesson, we will take a closer look at how we can use mathematics to gain an understanding of the health issues.*

- Introduce the term *ratio* and how it is related to but different from a *fraction*. Connect the term *ratio* to *rates* from the reading.

  Note: When reviewing data from both classes, fractions can be the same, but ratios may be different if the size of the class is different. When introducing ratio, it is important to indicate multiple ways to represent a ratio. Additionally, stress the part-to-part relationship for a ratio, which is different from the part-to-whole relationship of a fraction. Use information from Worksheet 2 to explain this difference.

- Introduce the students to more precise language for comparing ratios. A class could have the same number of coronavirus cases as another class, but they could have a "higher rate" of coronavirus cases if the total number of students is smaller.

- Distribute Worksheet 2. Tell students to complete the tasks individually first, then compare solutions with their group members.

  Note: The four questions on Worksheet 2 help students understand the concept of ratio in context.

- As students work in groups, monitor their progress and provide appropriate assistance as needed. Use purposeful selection, sequencing, and connecting techniques to call on groups to present their solutions to the class. Consider providing solutions to groups as they finish and then allowing them to begin the next task.

Explore (15 minutes)

- Provide each student with Worksheet 3 (*Health Disparities Abstract Readings*) and abstract readings of two articles to bridge the connection between Lessons 1 and 2. Tell them to complete the task individually, then have a group discussion.

- Facilitate a brief whole-class discussion.

Summarize (15 minutes)

- When students finish reading and answering questions, have them discuss their responses within their groups. When all groups are finished, use the following questions to facilitate a discussion about the social justice context:

  + *Based on your readings, which race and ethnicity is more impacted by COVID-19? How do we know?*

  + *How do you think we can use mathematics to unpack the truth about this situation?*

  + *What does this have to do with social justice?*

- For homework, assign each group member one of the following articles to read for the jigsaw activity (Teacher Resource 2, *Notes for Article Readings*). Also share with students the article reflection prompts listed on Worksheet 2.

  + "COVID-19 Pandemic Highlights Racial Health Inequities," by Crystal Johnson-Mann and colleagues, *The Lancet*, July 10, 2020 (https://bit.ly/3ohNIoY)

  + "COVID-19 and the Widening Gap in Health Inequity," by Helene J. Krouse, *Otolaryngology–Head and Neck Surgery*, May 5, 2020, (https://bit.ly/3xNlkya)

  + "Racial Disparity of Coronavirus Disease 2019 in African American Communities," by Ravina Kullar and colleagues, *Journal of Infectious Diseases*, September 15, 2020 (https://bit.ly/3lybpHQ)

  + "COVID-19 and Health Disparities: the Reality of 'the Great Equalizer,'" by Stephen A. Mein, *Journal of General Internal Medicine*, May 14, 2020 (https://bit.ly/2ZNeg8a)

LESSON 3 FACILITATION

Using Ratios to Understand Health Inequities

Students will use ratio and rate reasoning to analyze real-world problems. They will use ratios to unpack data from the CDC related to various types of health cases. They will make conjectures about U.S. COVID-19 causes and deaths and then devise a plan to tell the right story regarding the data.

Preparation: Two Excel files are available for this lesson. One is a sample solution for the tables in Worksheet 4. The other is a file with the data and blank table in Worksheet 4. If groups of students have access to a computer with Excel or Google Slides, you might choose to allow students to use these files instead of computing ratios and drawing graphs by hand.

Launch (15 minutes)

- Organize students into groups based on their assigned article for the jigsaw activity. Tell students to take notes as they discuss the reflection prompts so that they can effectively share this information during the class discussion. See Worksheet 2 (*Exploring Ratios in Context*).

- Circulate to each group to collect evidence from the discussions, then use the following questions to facilitate a class discussion about the articles and health disparities within the United States:

    + *How is COVID-19 affecting communities of Color?*

    + *Why is COVID-19 affecting communities of Color at a faster rate? How do you know?*

    + *Are there are other illnesses affecting communities of Color?*

Explore (20 minutes)

- Instruct students to examine data related to health cases using Worksheet 4 (*Examining the Number of Deaths in Georgia [Your State]*). Tell students they will unpack the data and discover ways of reasoning with the data to tell the true story. Encourage students to recall the ratio exercises completed for Lesson 2. Note that NH stands for "non-Hispanic."

- Use the following prompts as needed to assist students with getting started and to facilitate group discussion as they complete the task:

    + *Refer to the population total and population by race and ethnicity. What does this tell us about the population in Georgia [your state]?*

    + *Fill in the total column in Table 1 in Worksheet 4. Why might these totals be useful?*

    + *How can we determine the ratio of the number of deaths to the number of people by race and ethnicity?*

    + *What value should we use as the denominator for these ratios?*

    + *What can this ratio tell us? Let's check this out for all the races and ethnicities. Fill in the chart in Table 2 in Worksheet 4 with ratios instead of raw numbers.*

    + *Based on your ratios, which race and ethnicity has a higher rate of deaths?*

    + *Calculate the percentages of your ratios. What do you notice?*

+ *What conclusions can we make?*

+ *What actions can be taken to combat COVID-19?*

Summarize (15 minutes)

- Open the floor for comments and/or questions from the groups. Use chart paper to document points from the discussion.

- Use the following questions to facilitate a discussion and provide students with opportunities to share information from their home lives and/or surrounding communities.

    + *How does understanding ratios support us in understanding health inequalities in our home? Community? State? Nation?*

    + *How do we use the knowledge of health inequalities to inform and/or empower ourselves as well as our families, communities, and beyond?*

    Ask these social justice-related questions (if not answered already):

    + *What can we do to combat this issue?*

    + *How is this related to social justice?*

- Explain to students that their conceptions from this discussion will be used in the next lesson.

LESSON 4 FACILITATION

Lessons Learned

This lesson involves students making final summaries and presenting their findings.

Launch (5 minutes)

- Reestablish the groups. Reiterate to the students the great job that they have done over the past three lessons and how powerful it is to use mathematics as it relates to social justice issues.

- Speak to the students about how they can continue to help those in need because they took an interest in exploring the issue deeper with mathematics. Then recap the discussion points and social justice concerns from the previous lesson.

- Explain to students what is expected of them for the day and the process that they will follow for today's lesson, which is focused on the students' thoughts, experiences, findings, and ah-ha moments over the last three lessons. As a group, they will prepare a presentation to share what they have gained from the lesson on health inequalities.

Explore (30 minutes)

- Tell groups to prepare a presentation that unpacks their thoughts, experiences, findings, and ah-ha moments. The presentation must include the following:

  + *Describe the mathematics and/or mathematical tools used within the lesson. What mathematics did we engage in? Were these mathematical tools appropriate for the lesson? Explain why.*

  + *How were mathematics and/or mathematical tools used to unpack the social justice issue?*

  + *What aspects of the lesson definitely speak to social justice? Describe why this issue is a social justice issue.*

  + *Provide two extra points that stood out throughout the lesson that may or may not have been discussed. Explain why these points are important to you.*

  + *What actions do you think can be taken to ensure health equity?*

- Students can choose the modality for their presentation (e.g., PowerPoint, Flipgrid, Zoom, Padlet, Dotstorming, Jamboard, etc.).

Summarize (15 minutes)

- Open the floor to questions from classmates.

- Conclude the class discussion by reiterating how mathematics is a powerful tool that we can use to empower our communities, states, and world. Encourage students to always look to see how they can solve issues using mathematics if possible. Commend them on the great job they have done.

- Provide the students with next steps for sharing each group's planned actions with all stakeholders.

TAKING ACTION

Consider the following activities:

- Suggest that students share with their families what they have learned about health disparities within their racial and ethnic group, especially when referring to COVID-19. Encourage them to take action by continuing to follow CDC guidelines for specific health conditions. For example, recommendations for COVID-19 mitigation include wearing a mask, washing hands frequently, maintaining a social distance of 6 feet, and covering coughs and sneezes.

- Consult with your administrators to arrange opportunities for students to display their work on the school website and/or social media pages and encourage parents to watch their children's videos.

- Encourage students to contact local organizations and organize a health fair, workshops, or seminars for parents and local community members to provide health checkups to emphasize the importance of knowing your health status, providing advice for taking action, and offering ways to combat the disparities.

- Invite individuals from the health care profession and organizations to speak with your students about ways to continue their study of health issues within their own communities and to describe career opportunities.

ONLINE RESOURCES

online resources ↖ Available for download at **resources.corwin/TMSJ-MiddleSchool**

▼ *Teacher Resource 1: Health Cases Activity Sheet*

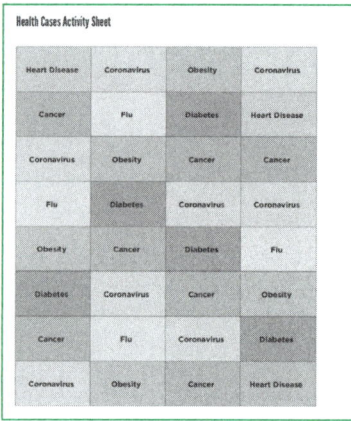

▼ *Teacher Resource 2: Notes for Article Readings*

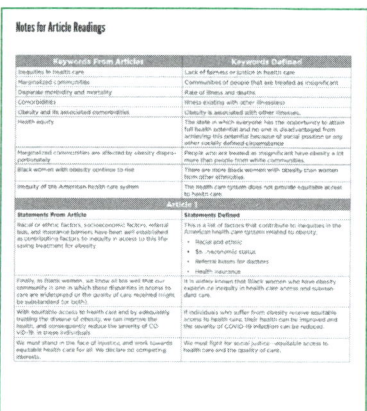

▼ *Teacher Resource 3: Excel File With Sample Solutions to Worksheet 4*

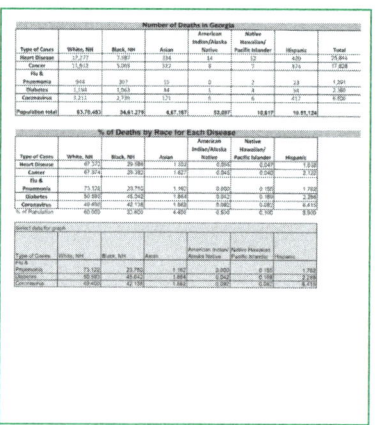

▼ Worksheet 1: Health Cases

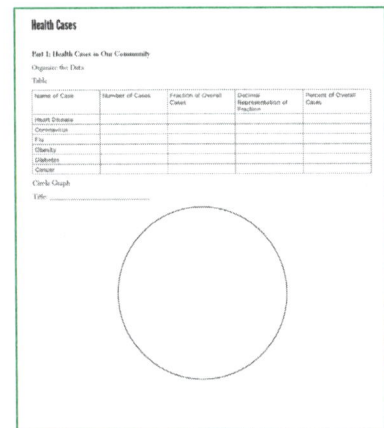

Health Cases

Part 1: Health Cases in Our Community

Organize the Data

Table.

Name of Case	Number of Cases	Fraction of Overall Cases	Decimal Representation of Fraction	Percent of Overall Cases
Heart Disease				
Coronavirus				
Flu				
Obesity				
Diabetes				
Cancer				

Circle Graph

Title: _____

▼ Worksheet 2: Exploring Ratios in Context

Exploring Ratios in Context

1. Which fraction is more $\frac{2}{5}$ or $\frac{1}{3}$? Explain your reasoning.

2. Mr. Williams has 33 students in his mathematics class. Sixteen of the students are boys. Write two ratios to describe Mr. Williams's class. Explain your reasoning.

3. Populations set up their restaurant to accommodate two different parties. In one section of the restaurant, there was a party of 18 people with enough seats for 54 people. At the same time, the other section of the restaurant had 14 people with seats for 30 people. Which section was more crowded? How do you know?

4. In a pet store, the ratio of cats to dogs is 4:5. There are a total of 28 dogs and cats at the pet store. How many cats are there?

▼ Worksheet 3: Health Disparities Abstract Readings

Health Disparities Abstract Readings

COVID-19 and the Widening Gap in Health Inequity

by Helene J. Krouse, PhD, RN

The coronavirus disease 2019 (COVID-19) pandemic has brought to light significant health inequities that have existed in our society for decades. Blacks, Hispanics, Native Americans, and immigrants are the populations most likely to experience disparities related to burden of disease, health care, and health outcomes. Increasingly, national and state statistics on COVID-19 report disproportionately higher mortality rates in blacks. There has never been a more pressing time for us to enact progressive and far-reaching changes in social, economic, and political policies that will shape programs aimed at improving the health of all people living in the United States.

Reflect on the reading:

What questions do you have?

What do you want to know?

COVID-19 Pandemic Highlights Racial Health Inequities

by Crystal Johnson Mann, Monique Hazzan, and Shaneeta Johnson

In 1966, Dr. Martin Luther King Jr. stated, "Of all the forms of inequality, injustice in health care is the most shocking and inhumane." These words remain ever true and relevant in our current climate of health care. The COVID-19 pandemic has unfortunately affected health care on a global scale and has magnified the inequities in access to health care that exist before. This pandemic has highlighted the equity gap in outcomes for marginalized communities, specifically the Black community, as starkly shown by the diagnosis morbidity and mortality from COVID-19 in individuals from these communities compared with the majority white population.

Reflect on the reading:

What questions do you have?

What do you want to know?

▼ Worksheet 4: Examining the Number of Deaths in Georgia

Examining the Number of Deaths in Georgia

Georgia's Total Population   10,617,423

Table 1.

	Number of Deaths in Georgia (~population of)					
Type of cases	White, NH (~6,370,453)	Black, NH (~3,461,279)	Asian (~462,767)	American Indian/Alaska Native (~53,087)	Native Hawaiian/Pacific Islander (~10,617)	Hispanic (~1,051,124)	Total (10,617,423)
Heart disease	17,277	7,587	334	14	12	420	
Cancer	11,912	5,005	322	8	7	374	
Flu and pneumonia	944	307	15	0	2	23	
Diabetes	1,194	1,063	44	1	4	54	
Coronavirus	3,211	2,739	121	6	6	417	
Total							

Data collected from the following websites: https://oasis.state.ga.us/oasis/webquery/qryMortality.aspx, https://www.census.gov/quickfacts/GA, https://covid.cdc.gov/covid-data-tracker/#cases_casesinlast7days, https://dph.georgia.gov/covid-19-daily-status-report

1. The data in the table provide the number of deaths due to different health cases in Georgia across different racial categories. What conclusions can you draw from these data? Does comparing the number of deaths give you a complete picture of how different diseases are affecting the population?

2. Fill in the total column in the table. Why might these totals be useful?

3. How can we determine the ratio of the number of deaths to the number of people by race?

4. Fill in the table below to represent the ratio of the number of deaths by race to the deaths by type of case. Then give the table a title.

▼ Student Resource 1: Excel File With Data for Worksheet 4

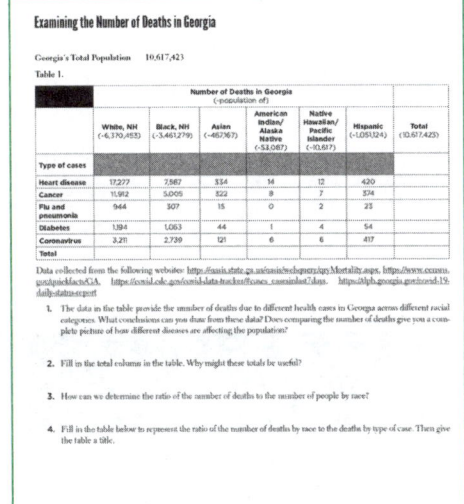

Number of Deaths in Georgia						
Type of Cases	White, NH	Black, NH	Asian	American Indian/Alaska Native	Native Hawaiian/Pacific Islander	Hispanic	Total
Heart Disease	17,277	7,587	334	14	12	420	25,644
Cancer	11,912	5,005	322	8	7	374	17,928
Flu & Pneumonia	944	307	15	0	2	23	1,291
Diabetes	1,194	1,063	44	1	4	54	2,360
Coronavirus	3,211	2,739	121	6	6	417	6,500
Population total	63,70,453	34,61,279	4,67,167	53,087	10,617	10,51,124	

% of Deaths by Race for Each Disease						
Type of Cases	White, NH	Black, NH	Asian	American Indian/Alaska Native	Native Hawaiian/Pacific Islander	Hispanic
Heart Disease						
Cancer						
Flu & Pneumonia						
Diabetes						
Coronavirus						
% of Population						

CHAPTER 8

ALGEBRA
Expressions, Equations, and Functions

In middle school, students apply and extend previous understandings of arithmetic to algebraic expressions from elementary school and include exponents as operators, offering increased opportunities to explore more complex situations mathematically. A focus is on reasoning about and solving one-variable equations and inequalities, generating expressions and equations from real-world contexts to solve real-life problems. As students progress through the middle grades, they begin to represent and analyze quantitative relationships between dependent and independent variables, making connections between proportional relationships, lines, and linear equations. Before they exit middle school, students should have experiences analyzing and solving linear equations and pairs of simultaneous linear equations that offer wonderful opportunities to determine when two contexts may be equivalent. In addition, students should explore linear functions and are introduced to nonlinear functions offering opportunities to explore how two real-life variables may relate to one another. Algebra provides a wonderful opportunity for teachers to include student exploration of mathematics and real-world situations by intertwining contexts of social injustice with these concepts.

LESSONS

Lesson No.	Lesson Title	Mathematics Focus Areas	Grade	Social Justice Topic	Authors
8.1	*Gerrymandering of Voting Districts*	• Expressions and Equations • Ratio and Proportions • Number	6 and 7	Civil Rights and Governmental Laws	Chuck Munter and Cara Haines
8.2	*National Team Pay Investigation*	• Algebra and Functions	8	Economic Inequality	Andrew Reardon
8.3	*The Black Vote in America: Impact of the 1965 Voting Rights Act*	• Algebra and Functions	8	Civil Rights and Governmental Laws	Chuck Munter and Cara Haines

LESSON 8.1 GERRYMANDERING OF VOTING DISTRICTS

Chuck Munter and Cara Haines

CIVIL RIGHTS AND GOVERNMENTAL LAWS

Ostensibly, the U.S. government operates like a democracy, which means that—through the election of government representatives—American voters have a voice in making decisions. But since at least 1812, representatives from both political parties have tried to undermine the power of particular voters by manipulating election results through the process of *gerrymandering*: strategically drawing boundaries of voting districts to ensure a party's advantage and/or suppress the vote of Black Americans. As an example of the latter, in 1957, the Alabama legislature tried to redraw the boundaries of the city of Tuskegee to exclude almost all of the majority of the Black population and to include all of the minority white population. But a 1960 Supreme Court ruling found that this case of gerrymandering denied citizens the right to vote based on race, thereby violating the 15th Amendment.

Today, there is still concern that in an attempt to stay in power, politicians are gerrymandering to disenfranchise particular groups of voters. In this lesson, students will have opportunities to (a) explore gerrymandering by using mathematics to model the fairness of district boundaries and (b) discuss potential solutions for addressing gerrymandering as a means of voter suppression. We hope that when implementing this lesson in isolation, teachers will honor its original intent by framing gerrymandering as a historical and modern-day tactic for suppressing the voices of Black voters in the United States.

DEEP AND RICH MATHEMATICS

This lesson provides an opportunity for students to investigate a real-world problem—gerrymandering—by mathematically modeling the fairness of voting district boundaries. Through this process, students will (1) clarify and make sense of the problem; (2) consider and define relevant variables and assumptions; (3) construct, test, and refine mathematical models; and (4) implement their final model to report results (i.e., whether a state's voting district boundaries are fair) (NCTM, 2020, p. 69).

- I can recognize and describe unfairness and injustice in many forms including attitudes, speech, behaviors, practices, and laws. (Justice 12)

- I am aware that biased words and behaviors and unjust practices, laws and institutions limit the rights and freedoms of people based on their identity groups. (Justice 13)

ESSENTIAL MIDDLE GRADES CONCEPTS

- **Expressions and Equations**—Grade 7: Solve real-life and mathematical problems using numerical and algebraic expressions and equations.

- **Ratio and Proportions**—Grade 6: Understand ratio concepts and use ratio reasoning to solve problems.

- **Number**—Grade 6: Apply and extend previous understandings of numbers to the system of rational numbers.

MATHEMATICS PRACTICES

- Reason abstractly and quantitatively.

- Construct viable arguments and critique the reasoning of others.

- Model with mathematics.

ABOUT THE LESSON

This lesson uses a launch–explore–summarize instructional model and is intended to take approximately 90 minutes to complete in one class period.

Resources and Materials

- Worksheet 1: *Gerrymandering of Voting Districts Task* (1 per student)

- Worksheet 2: *Considering the Fairness of Missouri's Voting District Boundaries*

- PowerPoint resource to support lesson facilitation

- Teacher Resource 1: *Gerrymandering of Voting Districts Task (Answer Key)*

- Calculator

- Video: "Explaining the Efficiency Gap," from WNYC (https://bit.ly/3rxYMRa)

LESSON FACILITATION

Launch Option 1 (20–30 minutes)

Prelaunch: To cultivate students' interest in investigating the fairness of voting district boundaries, you may consider acting out a scenario in which a simple majority does *not* determine the outcome of a vote. This might involve (1) inviting students to vote (in secret) on an issue that is of interest to them, (2) using the results to strategically assign students to groups—or "voting districts"—by **cracking** and **packing** the simple majority, and (3) declaring the winning outcome based on the voting district majority.

- For this to work, you will need to select an issue on which the class is likely to be roughly evenly split. Examples might include the following: (a) how the class would like to spend designated free time on a given day (e.g., going for a walk outside vs. playing an indoor game), (b) which snack the teacher will bring in the next day (e.g., cupcakes vs. donuts), or (c) which musical artist the class will listen to during groupwork. As an alternative, you could conduct a mock election in which students are prompted to vote for one of two candidates, each of whom is running on a platform that most students are likely to support (e.g., Candidate A: 4-day school week vs. Candidate B: Shorter school day). Whatever you choose, it is recommended that you survey the class about several things so that you can choose one for which there is just a slim majority.

- Begin the lesson by asking students to recall the vote from the previous day and to sit in the groups you created, which—if you cracked and packed the simple majority (dark green)—should look like this:

- Conduct the vote again, this time by eliciting and publicly recording votes, first from students in group 1, then from those in group 2, and so on. After the results have been recorded, determine the majority for each of the groups.

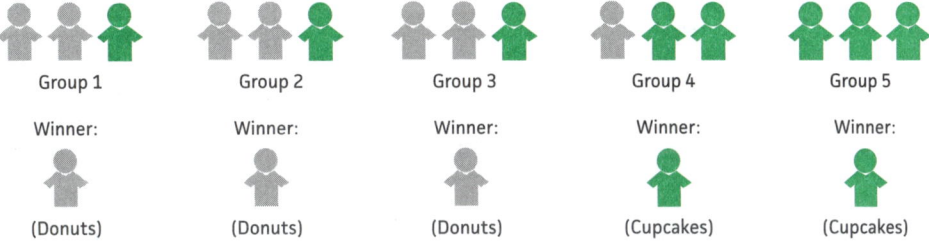

- Finally, determine the winner of the overall vote by finding the majority of the group outcomes, which in this case would be students who voted in favor of having donuts as a snack. Share the final results, and ask students to consider the following questions:

  + *Are the results fair? Why or why not?*

  + *What do you notice about the groups that allowed the minority to decide the outcome?*

- Use the scenario and prompts from the Launch wrap-up to discuss *gerrymandering* and introduce students to Missouri or your own state as a case for considering whether voting district boundary lines are fair.

Launch Option 2 (20–30 minutes)

- To ensure that students can access the context of the introductory task, initiate a discussion about *representation*. Consider eliciting students' ideas about scenarios in which a group of people might select representatives to act and/or make decisions on their behalf. Ask questions such as these:

  + *What does it mean to be a representative for a group of people?*

  + *Can someone offer an example of a scenario in which a group of people might select one or more individuals to represent them?*

- Highlight ideas that reflect a representative democracy: a government system in which groups of people elect individuals to represent them (perhaps the school has a student council, with representatives for each grade level).

- Distribute Worksheet 1 (*Gerrymandering of Voting Districts Task*). The teacher key is also available in the online resources (Teacher Resource 2).

- Prompt students to think about fairness in representation by describing the scenario depicted in the introductory task and allowing time for students to work through each of the questions. Consider structuring this time by allotting 3–5 minutes for questions 1 and 2 and 7–10 minutes for questions 3a–c.

Launch Wrap-Up for Option 1 or 2

- Spend the remaining 7–10 minutes briefly discussing gerrymandering as you use the following scenario to introduce Missouri or your own state as a case for considering whether voting district lines are fair. (**Note:** Consider contextual changes to your state and/or updates to the introduction and follow-up questions.)

  Since at least the 1812 Massachusetts state senate election and Governor Elbridge Gerry's redrawing of voting districts, including one in the Boston area that was said to look like a salamander (hence the word "Gerrymander"), people have tried to draw boundaries of voting districts to give their political party an advantage and/or racially discriminate. For example, in 1957, the Alabama legislature tried to redraw the boundaries of the city of Tuskegee to exclude almost all of the majority African American population and to include all of the minority white population. A 1960 Supreme Court ruling found this gerrymandering violated the 15th Amendment as it denied citizens the right to vote based on race. In 2018, two cases about gerrymandering made it to the U.S. Supreme Court—one from Wisconsin and one from North Carolina. The justices decided not to rule on either case, but concern still exists that politicians are cheating in order to stay in power.

In Missouri, based on the 2016 and 2020 presidential elections, about 59% of Missourians vote Republican.

Then ask the following questions.

1. *So, how many of the state's eight allotted U.S. representatives would you expect to be Republicans?*

Possible answer: "Four or five because 59% is a little more than half."

2. *What would the ratio need to be for you to suspect cheating (or "partisan gerrymandering") had taken place?*

Share that as of January 2021 (and the 117th Congress), six of the eight are Republican.

3. *Is this okay?*

Possible answer: The politicians who draw the voting districts would probably never admit it if they had cheated.

4. *So how could we tell?*

Explore (20–30 minutes)

- Distribute Worksheet 2 (*Considering the Fairness of Missouri's Voting District Boundaries*), as you invite students to work with a partner or in a small group to invent a rule (or mathematical model) that could be used to calculate the fairness of voting district boundary lines using the data from Missouri's (or your own state's) 2020 election.

- Visit with each group, ask questions about their processes, and take note of their rules. This will help with selecting and sequencing of rules later in the summary. If students are struggling to get started, encourage them to revisit ideas from the Launch (e.g., how students or representatives were grouped) and/or pose the following questions:

  + *What are some important variables to consider?*

  + *What might be some signs of cheating, or at least unexpected outcomes?*

  + *What did you notice about the groups in the drawing [from the Launch] that allowed the minority to decide the results?*

  + *If you were comparing two different sets of boundaries, could you determine which is more fair? How?*

  Here are some potential solutions for Worksheet 2 (that do not account for cracking and packing):

  + In solution 1, students do the following:

    1. Determine the proportion of the total number of votes that went to each party.

    2. Determine the proportion of representatives for each party.

    3. Find the difference between those numbers for each party.

    4. Decide that if either difference is more than $x\%$, then cheating occurred.

  + In solution 2, students do the following for each voting district:

    1. Measure the distance between the two furthest points on the boundary.

    2. Measure the distance between the two closest points on the boundary.

    3. Divide the larger distance by the smaller distance.

    4. Decide that if the quotient is greater than x, then cheating occurred in how the boundary was drawn.

Summarize (20–30 minutes)

- Use purposeful selection and sequencing to have students share their different rules for measuring the fairness of a state's drawing of voting districts and facilitate a discussion about the affordances and limitations of the rules. **Note:** If all students use the data (e.g., the Missouri data), then comparisons will all be grounded in the same election.

- Continue the discussion by inviting students' perspectives on whether the drawing of districts is fair. If they are unsure, ask *What additional information would be necessary to help make that determination?*

  + Students' ideas might resemble the efficiency gap model proposed by plaintiffs in the case from Wisconsin. In that model, votes cast in a district in which a party doesn't win and votes cast above the simple majority needed to win are considered "wasted votes." If the difference in wasted votes between parties (relative to the total number of votes) is large, that could be a sign that boundary lines have been gerrymandered. The "efficiency gap" is calculated by finding the difference between the number of wasted votes for each party and dividing it by the total number of votes. For example, in the Launch scenario of 50 people voting in Greenville, the green team wasted 5 votes and the gray team wasted 18 votes. You could calculate the efficiency gap as follows:

$$\frac{(5-18)}{50} = -26\% \text{ (in favor of the green team)}$$

- Make note of students' rules that resemble the efficiency gap, then show the video "Explaining the Efficiency Gap" (https://bit.ly/3rxYMRa). Have a brief discussion of the video and the equation: $E = (W_1 - W_2) \div N$, where E is the efficiency gap index, W_1 is the number of wasted votes for one party, W_2 is the number of wasted votes for the other party, and N is the total number of votes cast.

Larger differences between the numbers of wasted votes for two groups indicate that voting district boundary lines are unfair. It has been proposed that efficiency gaps of 7% and above should be considered cheating.

- After discussing the efficiency gap model, have students complete at least one of the following follow-up activities:

  + Calculate the fairness of Missouri's district boundaries and/or the groups/boundaries in the Launch scenario using the efficiency gap model.

  + Determine whether their state's district boundaries are fair. (**Note:** Your state data need to be provided.)

  + Visit the Gerrymandering Project website (https://fivethirtyeight.com/tag/the-gerrymandering-project/) to review different options for how districts could be drawn and the likely impact on election results.

"Wasted votes" are votes cast above the simple majority needed to win.

The "efficiency gap" is calculated by finding the difference between the number of wasted votes for each party and dividing it by the total number of votes.

The general equation for calculating an efficiency gap is: $E = (W_1 - W_2) \div N$, where E is the efficiency gap index, W_1 is the number of wasted votes for one party, W_2 is the number of wasted votes for the other party, and N is the total number of votes cast.

+ Have students share their ideas for how gerrymandering might be prevented at the state level (e.g., by ensuring that redistricting is a bipartisan—or *independent*—effort).

- Facilitate a brief discussion about what we, American voters, can do to address the issue of gerrymandering (e.g., contact state legislators, register to vote, participate in the census every 10 years).

TAKING ACTION

- After thinking through these issues, students or the school community may do the following:

    + Analyze the district in which the school or home is located, looking at how districting has changed or not changed over time and quantifying its fairness (or asking local politicians or public officials for their sense of the fairness of the district plan).

    + Track the ongoing Supreme Court consideration of gerrymandering issues and continue to spread awareness in the local community about what gerrymandering is and how it has historically disenfranchised certain groups.

- If students conclude that their state's voting district boundaries are not fair, teachers may consider supporting students in contacting state legislators to express concern.

- Gerrymandering happens in nonpolitical districts as well, such as school districts. Students might analyze how local school district boundaries are drawn as they assign children and households to elementary, middle, and high schools.

ONLINE RESOURCES

 **online resources** Available for download at **resources.corwin/TMSJ-MiddleSchool**

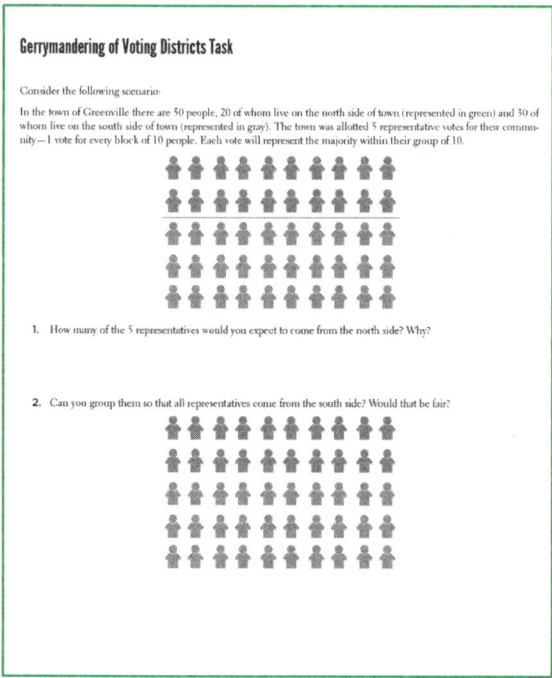

▲ *Worksheet 1: Gerrymandering of Voting Districts Task*

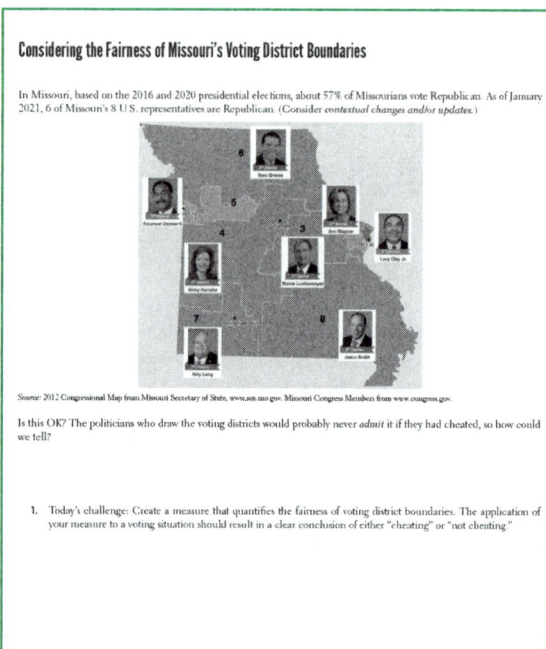

▲ *Worksheet 2: Considering the Fairness of Missouri's Voting District Boundaries*

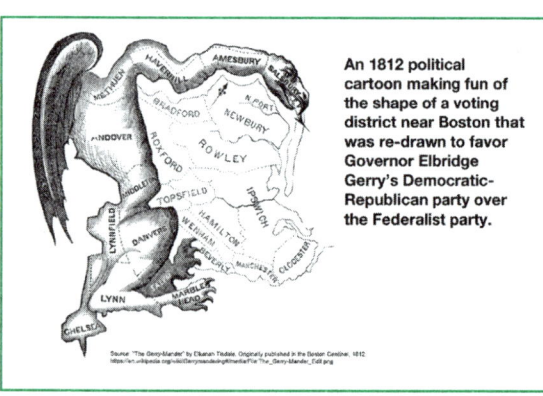

▲ *PowerPoint resource to support lesson facilitation*

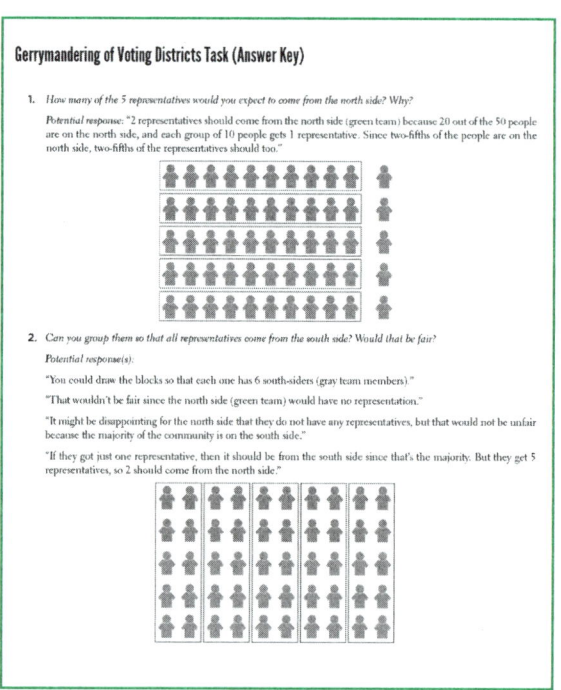

▲ *Teacher Resource 1: Gerrymandering of Voting Districts Task (Answer Key)*

LESSON 8.2 NATIONAL TEAM PAY INVESTIGATION

Andrew Reardon

ECONOMIC INEQUALITY

This lesson focuses on gender-based income inequality, using the U.S. Women's National Soccer team as a specific example for exploration. It provides an opportunity to educate students on the reality of the workplace and their positionality and power to create change. Students have an opportunity to investigate a social issue through a quantitative lens and to focus opinions and arguments on concrete data.

DEEP AND RICH MATHEMATICS

Students use systems of equations and the concept of having no solution over a given domain to understand the gender-based income inequality of the U.S. National Soccer teams. This project emphasizes the idea of what a solution to a system of equations truly means and gives an authentic context to anticipating where (if at all) lines intersect. The income structure of the soccer teams nicely sets up a linear system that students can then solve and interpret the meaning of their solution in real life.

ABOUT THE LESSON

This lesson uses a launch–explore–summarize instructional model and is intended to take approximately 90 minutes to complete across one class period.

Resources and Materials

- Worksheet 1: *Project: U.S. National Team Pay Investigation* (1 per student)

- Teacher Resource 1: *Answer Key for Worksheet 1*

- Video: "American Soccer's Gender Wage Gap," from *The Daily Show* (https://bit.ly/3xQCuLB)

SOCIAL JUSTICE OUTCOMES

- I am curious and want to know more about other people's histories and lived experiences, and I ask questions respectfully and listen carefully and non-judgmentally. (Diversity 8)

- I can recognize and describe unfairness and injustice in many forms including attitudes, speech, behaviors, practices, and laws. (Justice 12)

- I know that all people (including myself) have certain advantages and disadvantages in society based on who they are and where they were born. (Justice 14)

- I will work with friends, family and community members to make our world fairer for everyone, and we will plan and coordinate our actions in order to achieve our goals. (Action 20)

ESSENTIAL MIDDLE GRADES CONCEPTS

- **Expressions and Equations**—Grade 8: Analyze and solve linear equations and pairs of simultaneous linear equations.

This lesson is a unifying experience and really sets the tone for the class community early in the year for my classes. It is an opportunity for students to identify and begin to solve a problem as a cohesive group.

- Article: "Data: How Does the U.S. Women's Soccer Team Pay Compare to the Men?" by Laura Santhanam, *PBS NewsHour*, March 31, 2016 (https://to.pbs.org/3lrB6tG)

- Article: "Judge Says U.S. Women's Soccer Team Bound by No-Strike Clause," *ESPN*, June 3, 2016 (https://es.pn/3ppP0h4)

- Article: "100 Women: Is the Gender Pay Gap in Sport Really Closing?" by Valeria Perasso, *BBC News*, October 23, 2017 (https://bbc.in/3oj6lco)

LESSON FACILITATION

Launch (25 minutes)

- Distribute Worksheet 1 (*Project: U.S. National Team Pay Investigation*) and introduce students to the lesson by telling them that they should complete the "Video Reflection" as they view the video, "American Soccer's Gender Wage Gap," from *The Daily Show* (https://bit.ly/3xQCuLB).

- Facilitate a discussion of the video, emphasizing the importance of objective data that are presented in the video. Ask the following question as you encourage students to grapple with the implications of the data: *Should the women's team be paid equal to the men's team?*

- Organize students into three groups for a jigsaw activity and assign each group one of the following articles to read:

  + "Data: How Does the U.S. Women's Soccer Team Pay Compare to the Men" by Laura Santhanam, *PBS NewsHour*, March 31, 2016 (https://to.pbs.org/3lrB6tG)

+ "Judge Says U.S. Women's Soccer Team Bound by No-Strike Clause," *ESPN*, June 3, 2016 (https://es.pn/3ppP0h4)

+ "100 Women: Is the Gender Pay Gap in Sport Really Closing?" by Valeria Perasso, *BBC News*, October 23, 2017 (https://bbc.in/3oj6lco)

• Tell students that each group will read one article, answer summary questions about the article, and then share their responses with students from the other groups.

• After 10–15 minutes, have students form new groups that contain at least one member from each of the previous three groups. In this small group setting, a representative from each original group will present the findings from their original article. Instruct students that as each student presents their findings, everyone else should take notes in the space provided on the worksheet. This process repeats until representatives from each group have presented.

Explore (35 minutes)

• Tell students that they will work in small groups to investigate the pay for the men's and women's soccer teams as they complete the task. Tell them that they will also need to create a graph to share during the class discussion. (See Teacher Resource 1 for answers.)

• Use the following questions as you monitor student work and bring the whole class together to check for understanding:

+ *Is the women's team paid fairly?*

  - We define "fairly" as "equal pay for equal work."

  - In the context of the soccer teams, this means: *Given that the teams play the same number of games (complete the same amount of work), will there ever be a possible solution where they make the same amount of money?*

  - This is important because the answer is not clear—the women have a large base salary that the men don't receive, and the men get paid higher per-game bonuses. As a result, the answer is not straightforward and student analysis and interpretation is necessary to answer this question.

+ *What does it mean that the equations don't intersect in the real-world domain?*

  - In our analysis, we used an approximately average year with 22 games played, ending in a win or a loss. We then solve the system pictured below to get answers of $W = -0.89$ and $L = 22.89$. These answers are clearly out of the domain of real-world values—you cannot win -0.89 games, for example! What this tells us is that the equations do not intersect at a real-world value, which we can interpret as meaning the men's and women's teams will never make the same amount of money for the same amount of work—or that equal pay for equal work is not achievable given their current contracts.

+ *What would it mean if they did intersect?*

- If we reach a solution that is defined in the real-world context, then we can determine that there is a possible solution set of wins and losses that results in equal pay for equal work. This is not the case for this analysis, though.

Note: Encourage students to connect their responses to these questions with their solutions as they complete the worksheet.

Summarize (30 minutes)

- Facilitate a class discussion using the questions on Worksheet 1 as you have students share their solutions and graphs.

  Note: The solution is not included in the domain of playing 22 games. As a result, it is impossible for the teams to have the same income and there is no real solution!

- Provide students with the income values from the past 5 years and have them plot them on their graphs. Ask: *How does your graph represent the context of this problem?*

- Use the following questions to have students brainstorm in their groups actions they can take to share their findings and make a difference.

  + *What can I do about this problem?*

  + *Who should know about this problem?*

  + *What do you expect the graphs of the teams to look like?*

  + *In the future, what do you think the graphs will look like?*

TAKING ACTION

Have students compose a letter reflecting on what they learned about pay inequity and the need for change. They can write to an authentic audience of their choice, such as U.S. Men's Soccer players, the U.S. Soccer Federation, the president of the United States, or a local government official. A letter template is included in Worksheet 1.

ONLINE RESOURCES

 Available for download at **resources.corwin/TMSJ-MiddleSchool**

◄ *Worksheet 1: Project: U.S. National Team Pay Investigation*

Project: U.S. National Team Pay Investigation

Video Reflection
- After watching the video, "American Soccer's Gender Wage Gap" (https://bit.ly/3xQCuLB) for the first time:
1. One thing I wonder is

2. One thing I noticed is

Literature Investigation
3. What is the title of your article:

4. What does your article talk about:

5. My article talks about

6. Complete the following sentence:
   *Before reading my article I thought that _____, but now I know
   that _____*

7. When I share the information from my article with my peers, the three most important things are:
   First, _____
   Second, _____
   Third, _____

► *Teacher Resource 1: Answer Key for Worksheet 1*

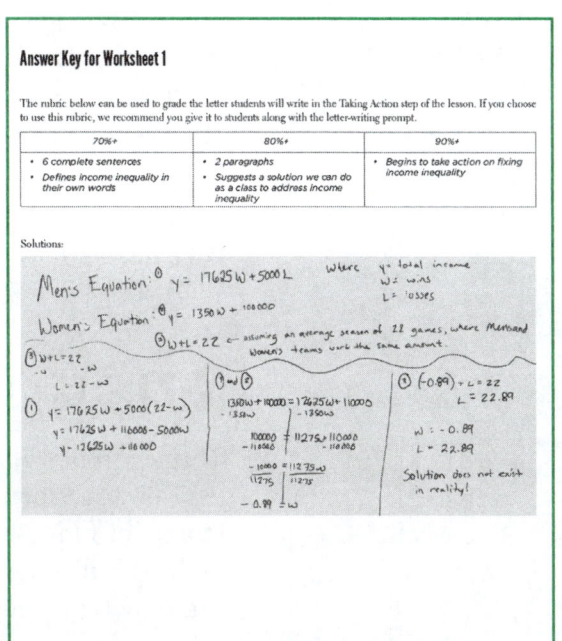

SOCIAL JUSTICE OUTCOMES

- I can recognize and describe unfairness and injustice in many forms including attitudes, speech, behaviors, practices, and laws. (Justice 12)

- I am aware that biased words and behaviors and unjust practices, laws and institutions limit the rights and freedoms of people based on their identity groups. (Justice 13)

ESSENTIAL MIDDLE GRADES CONCEPTS

- **Statistics and Probability—** Grade 8: Investigate patterns of association in bivariate data.

MATHEMATICS PRACTICES

- Reason abstractly and quantitatively.

- Construct viable arguments and critique the reasoning of others.

- Model with mathematics.

LESSON 8.3 THE BLACK VOTE IN AMERICA: IMPACT OF THE 1965 VOTING RIGHTS ACT

Chuck Munter and Cara Haines

CIVIL RIGHTS AND GOVERNMENTAL LAWS

An important story in U.S. history has been Black Americans' struggle for representation in governance—a struggle that began generations ago with obtaining (and now maintaining) the right and unencumbered access to vote. From determining how enslaved individuals would be counted toward Southern state representation in the writing of the Constitution to the nationwide voting rights protests—including the Selma to Montgomery march—of the 1960s and modern-day voter suppression tactics, Black citizens have consistently faced and worked to overcome obstacles to participating in so-called democratic processes. In this lesson, students will examine the impact of the Voting Rights Act of 1965, including mathematically modeling the increase over time in Black American representation in the U.S. Congress. One key aspect of this lesson is that it provides an opportunity to (1) highlight a story of (relative) success within the broader history of disenfranchisement and oppression of Black Americans and (2) connect the current state of congressional representation to the resilience and sacrifices made by Black Americans and others in the 1960s civil rights movement. However, it is also important to note that after more than a half-century, Black Americans' representation in Congress is still not yet to the minimum expectation of equaling their proportion of the population at large, thus emphasizing the long-lasting effects of previous (and current) injustice.

DEEP AND RICH MATHEMATICS

This lesson provides an opportunity for students to investigate patterns of association in bivariate quantitative data and explore the process of fitting a linear model to real-world data. In so doing, students will interpret the elements of linear functions (slope, *y*-intercept) in terms of the context and engage in inventing statistical means of assessing any given model's goodness of fit to the data.

ABOUT THE LESSON

This lesson uses a launch–explore–summarize instructional model and is intended to take approximately 90 minutes to complete across one period.

Resources and Materials

- Worksheet 1: *Impact of the 1965 Voting Rights Act* (1 per student)

- Teacher Resource 1: *Options for Launch Activity to Introduce the Lesson*

- PowerPoint resource to support lesson facilitation

- Clear rulers and/or (dry) spaghetti noodles

- Calculator

- Device with spreadsheet software (if that is a tool students are familiar with)

- Article: "The States Where Efforts to Restrict Voting Are Escalating," by Alex Samuels, Elena Mejía, and Nathaniel Rakich, *FiveThirtyEight*, March 29, 2021 (https://53eig .ht/31oMI9V)

Note: This lesson is currently written around the 117th Congress (2021–2023) and it is recommended that you update the data. For subsequent Congresses, teachers should update the number/percentage of Black members in the House of Representatives and the Senate.

LESSON FACILITATION

Launch (15 minutes)

- Begin by asking the following questions:

    + *Have you ever gone with your parents or other family members when they vote in an election?*

    + *If so, perhaps you've seen them get a sticker to wear?*

- Display slide 2 in the PowerPoint resource provided. Then ask:

    + *Where have you gone with them to vote?* (Possible responses may be schools, churches, community centers, libraries, etc.)

- + *Have they ever had any difficulty voting?* (Possible responses may be long lines, insufficient ID, getting to the polls, etc.)

- Explain the following information:

  Early on in the U.S., most states allowed only white male landowners to vote. Gradually, that changed.

  - + In 1865, the 13th Amendment to the U.S. Constitution abolished slavery.

  - + In 1868, the 14th Amendment guaranteed citizenship and equal protection under the law for the formerly enslaved and their descendants.

  - + And in 1870, the 15th Amendment gave all men, regardless of race, the right to vote.

  - + Finally, in 1920, the 19th Amendment guaranteed all women the right to vote in every state.

  Fast forward to 1964, almost a century after the 15th Amendment was ratified. In the election of Lyndon B. Johnson (D) over Barry Goldwater (R) for president of the U.S., turnout among Black voters outside the Southern states was 72%. But in the South, Black turnout was just 44% (slide 3 in the PowerPoint resource provided). (*Source:* "Black Turnout in 1964, and Beyond," by Alan Flippan, *New York Times*, October 16, 2014).

- Use the following question to facilitate a brief discussion: *What might have been the reason for the different Black voter turnout between the North and South?*

  Here are some potential responses:

  - + Students may think that there were fewer Black Americans living in the South (there weren't). If so, be sure they are reminded that these are percentages of the *Black American populations* in those regions: 72% of Northern Black Americans turned out to vote, while only 44% of those living in the South did so.

  - + Students may think there was a difference in interest between Northern and Southern Black Americans. If so, be sure the class establishes that there was actually considerable interest in voting among Black Southerners.

  - + Students may identify some form of voter suppression. If so, ask why white Southerners might have wanted to prevent their fellow Black citizens from voting.

- After students share their ideas, explain:

  - + Since the ratification of the 15th Amendment in 1870, Southern white officials had worked to prevent Black Americans from voting through a variety of tactics, including

- violence and intimidation,

- poll taxes (requiring a payment in order to register to vote), and

- literacy or comprehension tests,

all of which many white voters were exempt from through loopholes. For example, if they or a family member were allowed to vote prior to 1866 (something that applied to almost no Black voters), then they did not have to pay the tax, take the test, and so on.

- Present the questions from a 1964 Louisiana "literacy" test (slide 4 in the PowerPoint resource provided). Have a brief discussion and then share the following information:

  + A key focus of the U.S. civil rights movement of the 1950s and 1960s was ensuring Black Americans' ability to vote. Much of this culminated in the passage of the Voting Rights Act of 1965, which

    - prohibited racial discrimination in voting and

    - required some of the most offending states and counties to get federal approval before making any changes to their voting laws (Section 5, the "preclearance requirement").

- Display the graph, "Black Voter Turnout in South and elsewhere in U.S. Since 1965 Voting Right Act" (slide 5 in PowerPoint resource provided).

- Use the following question to facilitate a discussion of the shared information: *In the decades since, what impacts might the Voting Rights Act have had on U.S. politics and governance?*

  + Potential responses: more Black Americans voting, changes in laws, Barack Obama's presidency, and more elected Black American officials.

- Display the graph, "Number of Black Americans in the House and Senate, 1870–2007" (slide 6 in the PowerPoint resource provided).

- Ask: *What do you notice or wonder about the graph?*

  + Possible noticing: There has been an impact on representation. Many more Black Americans have served as state and national representatives since its passage.

- Ask: *What would adequate Black American representation in Congress look like?*

  + Potential responses:

    - All of them!

- More than white representation to make up for a history of not having a voice.

- At least equal to the proportion of Black Americans in the U.S. at large.

• Ask: *Using the graph, what would you do to predict what might happen in the next Congress or two?*

Explore (30 minutes)

• Let's examine the trend since the passing of the Voting Rights Act of 1965 (slide 7 in the PowerPoint resource provided) more closely. Say:

  + *In the current (117th) Congress, 60 of the 535 Congress members (11%) are Black American (57 of 435 in the House of Representatives and 3 of 100 in the Senate).*

  + *Currently, the U.S. population is about 13% Black American, which means the proportion of Congress that is Black American is getting very close to that of the nation at large.*

• Arrange students in small groups and provide a copy of Worksheet 1 (*Impact of the 1965 Voting Rights Act*), clear rulers or dry spaghetti noodles, a calculator, and a device with spreadsheet software (if that is a tool they are familiar with).

  Note: See Teacher Resource 1 for solutions to Worksheet 1.

• Prompt students to address the following scenario and questions:

  + *Given the trend since the Voting Rights Act of 1965, as shown in the graph and table, answer the following:*

    - *When do you expect Congress to be at least 13% Black American?*

    - *What equation would be a good fit for the data since 1965?*

    - *Show how you can use that equation to predict when Congress will become at least 13% Black American.*

• Use the following questions to monitor students' work and check their understanding:

  + *Do you expect the number of Black American members in the next Congress to increase, decrease, or stay the same? Why? What about the previous numbers makes you think that?*

  + *What are you looking for to determine whether a line is a good "fit" for these data?*

  + *How do you know your line is the best fit for these data? (How would you convince someone else yours is the best fit?)*

- Use purposeful selection, sequencing, and connecting techniques to call on groups to present their ideas as a class.

  Potential solutions and methods are as follows:

  + "Eyeballing" a line of best fit (pieces of spaghetti or clear rulers sometimes help with that process) and then (a) estimating the slope and y-intercept of the line or (b) choosing two points for the line to intersect and then writing the equation that includes those two points

  + Finding the line that goes through the most points

  + Finding a line that evenly splits the points above and below the line

  + Using the table of data to create a more systematic process for finding the best slope and y-intercept for the linear function (a less complex version of what linear regression accomplishes)

Summarize (45 minutes)

- Invite students to share (1) their different models for the data and (2) their predictions for when Congress will become at least 13% Black American based on those models.

  + If any groups used a systematic method to determine their line, you might consider ending with those strategies and inviting them to describe their process and what helped them decide what would be the "best" fitting function.

  + Be sure to support students in highlighting the meaning of the y-intercept and, in particular, the slope in their models.

- If time allows, introduce the idea of minimizing the sum of the distances between the point and the line by creating a spreadsheet that calculates those sums for each of the students' models and find out which of their models achieves the smallest sum.

- Use the following questions to facilitate a closing discussion:

  + *What do you think about the rate of change in Black American representation since 1965?*

  Possible responses: fast, slow, about what one would expect.

  + *Why might it be that the House of Representatives has greater Black American representation (13% in 2021) than does the Senate (3% in 2021)?*

  Note: This is related to the size of the voting blocks for those two chambers. Representatives in the House are elected by smaller, more locally focused voting districts, whereas senators are elected statewide. In Missouri in 2021, for example, two of eight representatives in the House were Black

American, elected by districts including St. Louis and Kansas City. Black Americans may be less likely to win a statewide election to the Senate. The Senate also upholds disproportionate state representation—for example, California has equal representation in the Senate to Wyoming. For many years, arguments have even been made to abolish the Senate, including from House members Victor Berger in 1911 and John Dingell in 2015.

> + *Where does that rate of change show up in your models today?*

Possible responses: the slope, the steepness/angle of our line.

> + *What does the slope mean in terms of representation in Congress?*

Possible response: a slope of 0.002 means an average of a 0.4% increase—or about two members—with every Congress.

Notes

- It has been several decades since the Voting Rights Act was passed, but it's possible that some students' older family members might still remember it. If so, you might encourage students to interview those family members about their experience and memory of it.

- In 2013, the Supreme Court ruled that Section 4(b) of the Voting Rights Act, the formula for determining which states should be subjected to Section 5 "preclearance" requirements, was outdated and therefore unconstitutional. Since then, many states have changed their laws to make voting more restrictive (e.g., voter ID laws, less opportunity for early voting), particularly since the 2020 national election.

- The 2022 U.S. midterm election is on November 8. It could be fun to use this lesson to spark interest in following news about (or even participating in!) that election, including examining (or even promoting) voter turnout and what the racial demographics of the new Congress will be.

TAKING ACTION

- Encourage students to investigate the modern tactics through which the voting of Black Americans (and other communities of Color) is still suppressed (e.g., gerrymandering of voting districts, ID laws, restricting opportunity for early voting). See the article, "The States Where Efforts to Restrict Voting Are Escalating," by Alex Samuels, Elena Mejía, and Nathaniel Rakich, *FiveThirtyEight*, March 29, 2021 (https://53eig.ht/3loMI9V).

- Have students brainstorm ways to support community members in navigating those tactics to ensure their votes are counted.

- Encourage students to join campaigns and other efforts striving to eliminate such tactics.

ONLINE RESOURCES

▼

Worksheet 1: Impact of the 1965 Voting Rights Act Task

Impact of the 1965 Voting Rights Act Task

In the 117th Congress, 60 of the 535 members (11%) in the House of Representatives and Senate are Black American. Currently, the U.S. population is about 13% Black American, which means that the proportion of Congress that is Black American is getting very close to that of the nation at large.

1) Given the trend since the Voting Rights Act of 1965, as shown in the graph and table below, when do you expect Congress to be at least 13% Black American? What equation would be a good fit for the data since 1965? Show how you can use that equation to predict when Congress will become at least 13% Black American.

Percent of U.S. Congress Who Were Black Since the Voting Rights Act of 1965

Year	% Black American	Year	% Black American	Year	% Black American	Year	% Black American
1965	1.1	1981	3.4	1997	7.5	2013	8.4
1967	1.3	1983	3.9	1999	6.9	2015	8.8
1969	2.1	1985	3.7	2001	6.9	2017	9.3
1971	2.4	1987	4.1	2003	7.1	2019	10.7
1973	3.0	1989	4.5	2005	7.7	2021	11.2
1975	3.2	1991	5.0	2007	8.4	2023	
1977	3.2	1993	7.5	2009	7.5	2025	
1979	3.0	1995	7.9	2011	8.0	2027	

▼

Teacher Resource 1: Options for Launch Activity to Introduce the Lesson

Options for Launch Activity to Introduce the Lesson

Option 1:
- Begin the lesson by giving portions (perhaps the math-related questions?) of one of the "literacy tests" that were given to Black Americans trying to register to vote.
- Lead a brief discussion to continue the Launch at "Explain the following information."

Option 2:
- Show the following clips from the 2014 film *Selma*:
  - 5:50 – 8:45: Annie Lee Cooper (Oprah Winfrey) attempting to register to vote in Dallas County, AL
  - 56:26 – 58:26 Scene depicting Dr. Martin Luther King, Jr., James Bevel, Hosea Williams, and other members of the Southern Christian Leadership Conference (SCLC) discussing a variety of voter suppression tactics that they want to see addressed in voting rights legislation
- Lead a brief discussion to continue the Launch at "Explain the following information."

Solutions and graphs for *Student Worksheet*

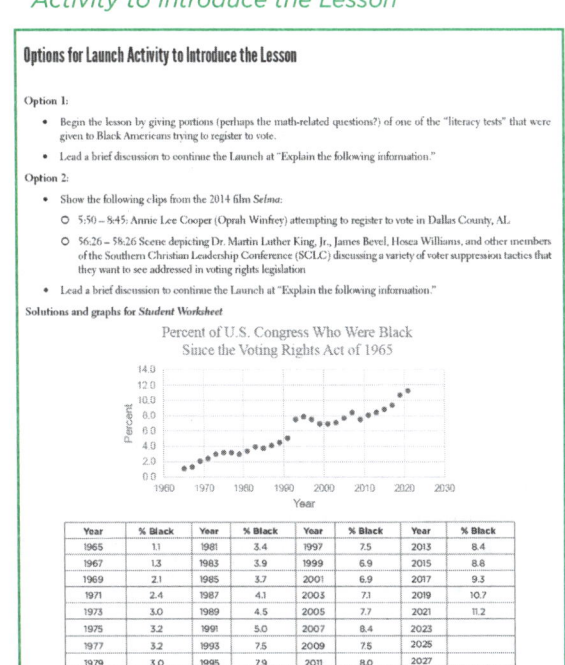

[NOTE: You might consider suggesting starting "year" at 0, or "years since 1965."]

▼

PowerPoint resource to support lesson facilitation

IMPACT OF THE VOTING RIGHTS ACT SLIDES

STATISTICS AND PROBABILITY

Early in middle school, students begin to develop an understanding of statistical variability. They begin with simple ideas like range and move toward more sophisticated measures like the mean absolute deviation. In addition to numerical summaries of variation, middle school students begin to summarize and describe distributions. They are encouraged to look at measures of center and how they relate to one another in different distributional shapes and other unusual data characteristics. Middle school students continue to develop these ideas by beginning to draw informal comparative inferences about two populations. Students use these understandings to make decisions about contextual situations using data. Students are also introduced to using random sampling to draw inferences about a population. Middle school students may be introduced for the first time formally to chance processes and asked to develop, use, and evaluate probability models. In the late middle grades, students begin to connect ideas of functions to data analysis by investigating patterns of association in bivariate data.

The conceptual category of statistics and probability exemplifies the use of context to make sense of mathematics. Franklin et al. (2007) emphasized the importance of context when engaged in statistical reasoning, including when developing questions, collecting data, analyzing data, and making inferences. With the underlying value of context, statistics and probability standards provide a natural opportunity for teachers to incorporate SJMLs while addressing their state's mathematical standards. Those who teach in a state in which statistics and probability standards are distributed across the middle school curriculum can create opportunities for students to examine instances of social injustice throughout their middle school pathway.

LESSONS

Lesson No.	Lesson Title	Mathematics Focus Areas	Grade	Social Justice Topic	Authors
9.1	*Playing With Data*	• Statistics and Probability	6	Economic Inequality	Odesma Dalrymple, Marissa Forbes, Celina Gonzalez, Kristin Komatsubara, Perla Lahana Myers, and Joi Spencer
9.2	*The Mathematics of Toxic Air Emissions*	• Statistics and Probability	6	Environmental Inequities	Queshonda Kudaisi and Oluwaseun Kudaisi
9.3	*Gender Pay Gap*	• Statistics and Probability • Ratio and Proportional Relationships	6 and 8 (Statistics and Probability) 7 (Ratio and Proportional Relationships)	Economic Inequality	Liza (Cope) Bondurant, Lee Inmon Dean, and Rebecca Hudson
9.4	*How Many Meals Can Minimum Wage Buy?*	• Statistics and Probability	8	Economic Inequality	Elizabeth O. Ayisi and Colleen Carman

SOCIAL JUSTICE OUTCOMES

- I can recognize and describe unfairness and injustice in many forms including attitudes, speech, behaviors, practices and laws. (Justice 12)

- I know that all people (including myself) have certain advantages and disadvantages in society based on who they are and where they were born. (Justice 14)

- I am concerned about how people (including myself) are treated and feel for people when they are excluded or mistreated because of their identities. (Action 16)

ESSENTIAL MIDDLE GRADES CONCEPTS

- **Statistics and Probability**— Grade 6: Develop understanding of statistical variability.

- **Statistics and Probability**— Grade 6: Summarize and describe distributions.

MATHEMATICS PRACTICES

- Reason abstractly and quantitatively.

- Construct viable arguments and critique the reasoning of others.

- Model with mathematics.

- Use appropriate tools strategically.

LESSON 9.1 PLAYING WITH DATA

Odesma Dalrymple, Marissa Forbes, Celina Gonzalez, Kristin Komatsubara, Perla Lahana Myers, and Joi Spencer

ECONOMIC INEQUALITY

Citizenship, democracy, and literacy are closely tied. One cannot reasonably hope to have a strong democracy without informed, thoughtful, and critical citizen-participants. The abilities of students to "read the world" are a portal into that citizenship and, ultimately, liberation. This lesson introduces students to a wide variety of data representations and leads them through a series of inquiries designed to help them read, reason about, and ultimately represent the world. As they move through the inquiries, students come to see that data can tell powerful stories, but that it is up to them to make sense of and decide what story the data tells. Likewise, students learn that data representations are susceptible to bias, and that they as readers and consumers of information can guard against that bias with specific tools. This lesson helps students consider questions such as these: *What are the values on the x-axis vs. the y-axis? Why? Are the different pictures presented to scale? What data might be missing? What data are highlighted?*

DEEP AND RICH MATHEMATICS

This lesson explores the power of understanding, analyzing, and creating different representations of data. The lesson also provides opportunities to explore several mathematical concepts including scale and ratio and proportion. For example, students consider what the expected outcomes might be if parks and green areas were distributed equally among communities and populations. They then compare an expected distribution to an actual distribution of green space. In this same vein, students consider the expected outcome if individuals were equally impacted by homelessness and then compare this to the realities in an actual community.

ABOUT THE LESSON

This lesson uses a launch–explore–summarize instructional model and is intended to take approximately 200 minutes to complete across four class periods.

Lesson 1: Students are introduced to data visualization and its power to increase the clarity and understanding of data and their patterns, trends, and relationships.

Lesson 2: Students work in groups to analyze different data representations to identify the intended story told by the visualization, its connection to the viewers, and questions that it sparks.

Lesson 3: Students work in groups to select data related to a social justice issue that is important to them and create their own data visualization.

Lesson 4: Students share their data visualizations and provide feedback to their peers.

Resources and Materials

- Student Resource 1: *Selected States' Data*

- Student Resource 2: *Data Representation Analysis, Creation, and Evaluation Tool*

- Teacher Resource 1: *Alternative Launch Activity for Lesson 2*

- Teacher Resource 2: *Additional Examples of Data Representations and Data Sets*

- PowerPoint resource to support lesson facilitation

- Sticky notes (optional for Lesson 4)

- Optional web-based resources for Taking Action:

  + Article: "Nature-Inspired Techniques Could Help Cities Become Greener," by Lucy Handley, *CNBC*, October 22, 2020 (https://cnb.cx/3DlTnPf)

  + Website: Tableau Public, "Viz of the Day" (https://tabsoft.co/3lwnmxU)

  + Website: Citizen Science (https://www.citizenscience.gov/#)

  + Website: Zooniverse (https://www.zooniverse.org/projects)

LESSON 1 FACILITATION

Introduction to Data Visualizations

Launch (10–15 minutes)

- Assign students to groups of three or four.

- Begin by posting the aerial image from Dar es Salaam, Tanzania (slide 2 in the Teacher PowerPoint Resource) and posing the following questions:

  + *What do you notice?*

  + *How does it make you feel?*

  + *What do you wonder?*

  Note: The aerial image of Dar es Salaam, Tanzania, is the simplest form of data representation—natural with no manipulation. Yet raw data can be compelling and tell a powerful story. The purpose of data visualization is to increase the clarity or understanding of data and their patterns, trends, and relationships. Students may believe that inequities between neighborhoods happen only in remote and faraway countries.

- Allow students 3 minutes to record their observations, feelings, and questions independently before sharing in groups and as a whole class.

Explore (20 minutes)

- Begin the discussion by sharing the following information about data visualizations (slide 4):

  + Data visualization is the communication of data in a visual manner.

  + The purpose of data visualization is to increase the clarity or understanding of data and its patterns, trends, and relationships.

  + Ultimately, it's an effective and efficient way to organize data and gain instant insights.

  + Visual representations are easier to understand than tables of raw data.

- Conclude by showing students the eight visual representations of the gender pay gap (slides 5–15). As each set of images is projected, use the following questions to facilitate group and class discussion:

  + *What is the story being told here?*

  + *What do the images lead you to notice?*

  + *What questions are you compelled to ask?*

  + *What emotions does this image make you feel? Why?*

Summarize (15 minutes)

- After viewing each of the images individually, display all the images on one screen (slide 16). Ask students to discuss the following in pairs before sharing with the class:

  + *Which representation is most powerful? Why?*

- Use the following questions to facilitate group and class discussion as you probe student understanding:

  + *What are the strengths/weaknesses of each representation?*

  + *How might you improve some of these representations?*

  + *What are some characteristics that some of these representations have in common?*

LESSON 2 FACILITATION

Analyzing Data Visualizations

> ### CONTENT NOTE
>
> The topic of COVID-19 might be difficult and personal for many students. We encourage you to refer to Chapter 4 (see "Dealing With Emotional or Sensitive Conversations") for guidance on how to approach the Launch of the lesson. If you would prefer to avoid discussing COVID at this time, an alternative activity for the Launch is provided in Teacher Resource 1 (*Alternative Launch Activity for Lesson 2*).

Launch (15 minutes)

- Begin by reading and discussing the following passage from the *National Geographic Kids* article, "What Is Coronavirus?" (https://bit.ly/3wPHnpD).

  Coronaviruses are a family of viruses that affect animals. Occasionally, coronaviruses have been known to move from animals to humans. The coronavirus we're talking about today is a new virus, which causes an illness called COVID-19.

  Coronavirus is mostly spread through the air, when people are in close contact with each other. This is why it's very important that we socially distance from other people, and wear a mask. Wearing a mask over your mouth and nose helps to stop your water droplets from reaching other people. If we all wear masks, we all keep our droplets to ourselves!

Scientists have created a vaccine for COVID-19, in record time! Vaccines give people protection from the virus. In the UK, they're being given to the most vulnerable people, like grandparents and hospital staff, first, because they are at the highest risk of catching coronavirus. Some people have already received the vaccine, which is brilliant news!

You might feel worried about coronavirus, about how it might affect you, your family and friends, and the changes that we are making to our daily lives at this time. The important thing to remember is that we are all in this together. Talk to your friends and family about your worries and work out how you will support each other during the coming weeks, ensuring you all stay happy and healthy.

Things You Can Do

+ Do things that make you happy like drawing, reading, and playing games.

+ Have a break from talking or thinking about coronavirus. Keep yourself busy and don't overcheck the news.

+ Don't believe everything that your friends tell you about coronavirus. Check the facts with a parent or trusted adult.

+ Your parents, family, and friends might be under more stress than usual, especially if they are working from home, or having to self-isolate, so think about things you can do to cheer them up. Perhaps you could help by tidying the house or writing them a note to brighten up their day?

● Next display COVID data from the Centers for Disease Control and Prevention (CDC) (slide 18) and ask students to discuss in their groups the following questions.

+ *What do you notice?*

+ *What do you wonder?*

+ *How does it make you feel?*

● Allow students 3 minutes to record their observations, feelings, and questions independently before sharing as a whole class/group.

● Have students share their observations, and use the following questions to extend the whole-class discussion:

+ *What story might this data tell?*

+ *How can data representation tell a compelling and powerful story?*

+ *How can data representations help us reveal, understand, and act upon bias and social inequities?*

- Students may notice that the representation displays the data but does not creatively tell a story with the data.

    + Follow up by asking, *What might make this representation more interesting or compelling?*

- Students may also misunderstand the meaning of 2.8× as the number of cases instead of 2.8 × the ratio (is it 2.8 times higher or 2.8 times as high?).

    + Check for understanding by listening to student interpretations and wonderings related to the representation. Additionally, ask: *What does 2.8× mean?*

It is important to consider the mathematics community that is established within the classroom when selecting representations to explore. Consider the resources in Chapter 4 for creating safe and brave spaces for fruitful discussions of sensitive topics.

Explore (20 minutes)

- Provide each group of students with Student Resource 2 (*Data Representation Analysis, Creation, and Evaluation Tool*) and either a printout of the data representation or a way to view it digitally.

- Select one of the following options for assigning data representations to the groups based on the level of understanding of data representations and students' readiness to explore complex/sensitive topics.

    1. Have all groups explore the same data representation. Use *Representations* A and A-2.

    2. Have each group explore one of two related data representations. Then have groups come together to form new groups with both representations. These groups present to each other and engage in discussion. Use *Representations* A and A-2 or *Representations* C and C-2.

    3. Have each group select a data representation that appeals to them, explore it, and then share an insight with the whole class.

- Share resources with the students so they can see other ways to creatively represent the data. Use the slides for Lesson 2 or provide students with examples 1–5 from Teacher Resource 2 (*Additional Examples of Data Representations and Data Sets*).

- Instruct students to respond to the following prompts as they explore the data representation:

  + *What story is being told through this visualization?*

    - *What patterns, relationships, and instant insights does it convey?*

    - *How does the representation aid our understanding of the data?*

  + *How does it connect to the viewer?*

    - *Does it align with who we are . . . our identities, emotions, interests, and values?*

    - *What bias(es) might this data and/or representation have?*

  + *What other questions does this lead you to ask?*

- Use the following questions as you monitor student progress and encourage small-group discussion:

  + *How can data representations tell a compelling and powerful story?*

  + *How can data representations help us reveal, understand, and act upon bias and social inequities?*

    - *What patterns, relationships, and instant insights are there in this data?*

    - *What story can be told from this data?*

    - *How would you visualize it?*

    - *How do you create a visualization to aid our cognition of the data?*

    - *How do you want it to connect to the viewer?*

    - *How will it align with who we are . . . our identities, emotions, interests, and values?*

    - *What other questions do you want to lead the viewer to ask?*

Summarize (15 minutes)

- Have each group present their representation by sharing highlights from their discussion. Encourage students to share their interpretation, insights, questions, and inequities observed.

- Tell students to consider the following questions as they listen to each presentation:

  + *Do you agree/disagree with the groups' interpretations?*

  + *How does the representation make you feel?*

  + *What questions does the data set spark for you?*

LESSON 3 FACILITATION

Creating Data Visualizations

Launch (5–10 minutes)

- Organize students into groups of two to four.

- Distribute Student Resource 1 (*Selected States' Data*) and instruct students to read and explore the data in Table 1: Selected State Data 2020.

- Have students discuss any patterns they are noticing in the data and various data visualizations. Pose the following questions:

  + *What do you notice?*

  + *What do you wonder?*

- This is an opportunity for students to make sense of the raw data and begin formulating ideas about how to represent the information.

Explore (20 minutes)

- Distribute and discuss Student Resource 2 (*Data Representation Analysis, Creation, and Evaluation Tool*). Tell students that they should use this tool to guide them as they create their data visualization.

- Have students recall from Lesson 2 the numerous ways that data can be represented, including music, video, drawings, sculpture, animations, and other artistic expressions.

- Instruct students to continue to explore Tables 1 and 2 (or select from data sets in Teacher Resource 2: *Additional Examples of Data Representations and Data Sets*) as they analyze the data to identify any trends, patterns, or important information.

- Provide students with the following prompts to facilitate small-group discussion as they analyze their data and create a data visualization:

  + *Observe the table of data.*

  + *What patterns, relationships, and instant insights are there in this data?*

  + *What story can be told from this data?*

  + *How would you visualize it?*

  + *How do you create a visualization to aid our cognition of the data?*

  + *How do you want it to connect to the viewer?*

  + *How will it align with who we are . . . our identities, emotions, interests, and values?*

+ *What other questions do you want to lead the viewer to ask?*

+ *Make a quick sketch.*

Summarize (20 minutes)

- Use the following questions to facilitate a whole-class discussion among students to reflect on their process, a-ha moments, challenges they are experiencing, and any lingering questions they have about their data sets or about data analysis.

  + *How do our identities, values, and interests influence our interpretations and representations of data?*

  + *What challenges did you face when selecting a visualization to aid your cognition of the data?*

  + *What other questions do you have about data analysis?*

- Have student groups share and review data visualizations within their group.

  + Each student should take 2–3 minutes to share about their interpretations, understanding, and reflections of the data.

  + Group members should use Student Resource 2 (*Data Representation Analysis, Creation, and Evaluation Tool*) to aid them in evaluating and providing feedback to their group members.

- Instruct students to use any takeaways from the whole-class discussion or feedback received from group members to prepare their data visualization to present during the next lesson.

LESSON 4 FACILITATION

Present and Evaluate

Launch/Explore (40 minutes)

Note on Critique: Critique can be a powerful tool to engage and build students' understanding of quality work. If this is the first time students engage in critique, model the process with them first by using an example that the class critiques together. Refer to Student Resource 2 in this lesson. Chapter 4 ("Models, Critique, and Descriptive Feedback") in the EL Education.org resource, *Leaders of Their Own Learning,* can also help you get started (https://bit.ly/3ruFA6y).

- Tell students that their task is to review the data visualization of each of their peers, excluding their group members, and provide constructive feedback. Written feedback should be provided using sticky notes or paper, or through an electronic tool such as Mural.

- Presentations can be facilitated in various ways. Two options are provided.

  ### Option 1: Gallery Review

  + Post each student's representation, virtually or physically around the classroom.

  + For the first gallery review, invite students to silently view each representation and consider the reflection questions.

  + For the second round of gallery reviews, provide the sentence starters below and invite students to post comments on their peers' data representations.

  ### Option 2: Whole-Class Presentations

  + Collect each student's data visualization in a digital format, such as PowerPoint or Google Slides, and provide access for students to view these on a device or present one at a time in a whole-class setting.

  + Tell students they should keep in mind the reflection questions and sentence starters when they provide feedback and comments.

  ### Reflection Questions

  + *What do you notice? Feel? Wonder?*

  + *What story is being told through this visualization?*

  + *What bias(es) might this data and/or representation have?*

  ### Sentence Starters

  + *This makes me feel . . . because . . .*

  + *I noticed . . . and it makes me wonder . . .*

  + *I think this representation is showing that . . .*

  + *This representation is powerful because . . . (or could be more powerful if . . .)*

- Finally, invite students to review the feedback from their peers and discuss in their groups to determine which feedback will guide their next steps for revision.

Summarize (10 minutes)

- Have students use the following prompts to reflect on the process of data analysis, creation and interpretation of visual representations, and evaluation.

  + *What feelings did the data evoke? Why?*

  + *What challenges did you face when analyzing the data? When evaluating the data? Why?*

In an effort to use more inclusive language, we use the term *gallery review* for what is commonly referred to as a "gallery walk."

+ *What considerations do you want to remember when interpreting and/or representing data?*

+ *What lingering questions do you have?*

+ *I think _____ and now I know_____.*

TAKING ACTION

Consider the following actions:

1. Invite students to search for and explore images of re-greening urban spaces, or read about it in the article "Nature-Inspired Techniques Could Help Cities Become Greener," by Lucy Handley, *CNBC*, October 22, 2020 (https://cnb.cx/3DlTnPf).

2. Invite an individual who does work to increase pay equity to speak with the class.

3. Encourage students to explore data each day using Tableau Public's "Viz of the Day" (https://tabsoft.co/3lwnmxU).

4. Introduce students to sites that provide opportunities for the public to volunteer.

   + Citizen Science (https://www.citizenscience.gov/#) is an official government website designed to accelerate the use of crowdsourcing and citizen science across the U.S. government. Some examples of projects they connect to social justice issues are as follows:

      - By the People (https://bit.ly/3rwsx4B)

      - U.S. Department of Agriculture People's Garden (https://bit.ly/3pns1mP)

      - Shared Air/Shared Action: Community Empowerment Through Low-Cost Air Pollution Monitoring (https://bit.ly/3G9oGOZ)

   + Zooniverse (https://www.zooniverse.org/projects) is the world's largest and most popular platform for people-powered research. This research is made possible by volunteers—more than a million people around the world who come together to assist professional researchers.

ONLINE RESOURCES

 Available for download at **resources.corwin/TMSJ-MiddleSchool**

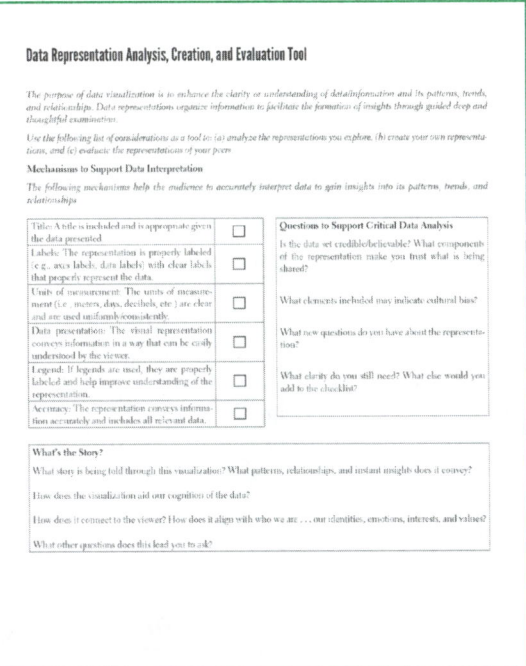

▲ *Student Resource 1: Selected States' Data*

▲ *Student Resource 2: Data Representation Analysis, Creation, and Evaluation Tool*

▲ *Teacher Resource 1: Alternative Launch Activity for Lesson 2*

 ▲ *Teacher Resource 2: Additional Examples of Data Representations and Data Sets*

 ▲ *PowerPoint resource to support lesson facilitation*

- I can recognize and describe unfairness and injustice in many forms including attitudes, speech, behaviors, practices, and laws. (Justice 12)

- I know how to stand up for myself and for others when faced with exclusion, prejudice, and injustice. (Action 17)

- I will speak up or take action when I see unfairness, even if those around me do not, and I will not let others convince me to go along with injustice. (Action 19)

ESSENTIAL MIDDLE GRADES CONCEPTS

- **Statistics and Probability—** Grade 6: Summarize and describe distributions.

MATHEMATICS PRACTICES

- Make sense of problems and persevere in solving them.

- Construct viable arguments and critique the reasoning of others.

- Use appropriate tools strategically.

- Attend to precision.

- Look for and make use of structure.

LESSON 9.2 THE MATHEMATICS OF TOXIC AIR EMISSIONS

Queshonda Kudaisi and Oluwaseun Kudaisi

ENVIRONMENTAL INEQUITIES

The lesson explores the impacts of toxic air emissions on an economically disadvantaged neighborhood in the midwestern United States. This lesson supports empowerment because it encourages students to examine injustice in the world and to make connections to what injustices may be occurring in their own environments. Students will recognize that power and privilege influence relationships on interpersonal, intergroup, and institutional levels and consider how they have been affected by those dynamics. This lesson also supports liberation because it encourages students to take action against injustice.

DEEP AND RICH MATHEMATICS

In this lesson, students are expected to go from one visual representation of numerical data to another representation of numerical data. Specifically, students are expected to use the information from a table containing toxic air emissions and average income by zip code; choose an appropriate graph to represent the data; graph the data; and finally, make observations and conclusions based on the data.

ABOUT THE LESSON

This lesson uses a launch–explore–summarize instructional model and is intended to take approximately 90 minutes to complete across one class period.

Resources and Materials

- Worksheet 1: *Emissions* (1 per student)

- Graph paper

- Colored pencils

- Video: "Landsman Requests EPA Air Quality Study of Winton Terrace," from WCPO 9, YouTube (https://bit.ly/3DgkBGV)

- Computer or device with internet access (1 per pair)

- Website: AirNow air quality data (https://www.airnow.gov/)

LESSON FACILITATION

Launch (15 minutes)

- Begin by distributing Worksheet 1 (*Emissions*) to students. Read the paragraph in question 1 together as a class. Then engage students in a whole-group discussion using the following questions:

  + *Are people entitled to clean air?*

  + *Whose interest should be considered more: companies or people who live in the community?*

- Share with students that mathematics can help us understand the impacts of toxic air emissions on different communities, and then use the following questions to engage students in a whole-group discussion:

  + *How do graphs help us to understand reality?*

  + *What are the various graphs that could be used to represent data?*

  + *How have graphs been used to promote an inaccurate view of reality?*

- Inform students that the lesson's focus is on toxic air emissions in the city of Cincinnati, Ohio.

Explore (60 minutes)

- Refer students to the table in question 2 in Worksheet 1, and ask them the following questions:

  + *How are the variables in the table related to each other?*

  + *In what ways can we include them in the same graph?*

- Next, instruct students to work individually as they use the information from the table to create a graph using "average household income" and "toxic air emissions" and then answer questions 3–5.

- As you monitor student progress, use purposeful selection and sequencing to have students share out when they have completed this portion.

- Show the video, "Landsman Requests EPA Air Quality Study of Winton Terrace," from a local news report on the air quality in the 45232 zip code (https://bit.ly/3DgkBGV).

- After showing the video clip, have students work with a partner and use the graphs they drew to discuss question 6.

- **Note:** Students will discuss what they noticed about the toxic air emissions and incomes in surrounding zip codes compared with Winton Terrace, their thoughts about why this may be, and what they think should be done about it. During the time that students are discussing, circulate around the room, attending to pairs and noting graphs and conversations to highlight in the whole-group discussion and lesson summary.

- After students have discussed question 6 in pairs, have students share out in a whole-group discussion. Then instruct students to complete question 7, where they will share their graphs with their partners, including how the graphs are similar and different.

- Conclude the exploration by instructing students to determine the air quality in their own context (i.e., city, town) by entering their zip code on the AirNow website (https://www.airnow.gov/). Students should then respond to question 8 on possible actions they can take toward better air quality in their community or in the community of others.

Summarize (15 minutes)

- Have students share out their responses and then use the following questions to facilitate a whole-group discussion:

    + *What did the graph help you to see about the air emissions by zip code?*

    + *Did the graph you chose help you to identify any patterns in average income and toxic air emissions by zip code? If so, what patterns did you identify?*

    + *After going through the lesson on toxic air emissions and learning about the potential impacts that it may have on communities, has your answer to these questions changed? "Are people entitled to clean air? Why or why not? Whose interest should be considered more: companies or people who live in the community? Why?"*

- Conclude the lesson by having students write a brief statement about the goals of the lesson. Have students share out.

    + Possible response: The focus of the lesson was to use mathematics to explore toxic air emissions in communities.

TAKING ACTION

Encourage students to take the following action:

- Write letters to their local and federal environmental protection agencies about areas that have bad air quality.

- Raise awareness of the issue to others in the school and in the community through local and social media campaigns about clean air.

ONLINE RESOURCE

 Available for download at **resources.corwin/TMSJ-MiddleSchool**

Emissions

1. Read the paragraph below.

   According to the Environmental Protection Agency [EPA] (2017), air pollution is "created by human activities such as vehicle use, industrial operations, and agriculture practices, or by natural events such as wildfires" (EPA, 2021a). The Clean Air Act (CAA) of 1970, a federal law that "regulates air emissions from stationary and mobile sources" (EPA, 2021b) was passed with the aim of protecting people and the environment. Despite such regulations, neighborhoods are still impacted by toxic air emissions. Some are very high in some areas and much lower in other areas. Below is a table that shows the number of toxic air emissions released in each zip code in Cincinnati as reported in 2016 (Knight, 2018). While there are more zip codes that exist in Cincinnati, no other zip codes reported emissions (Knight, 2018).

2. The table below displays data on the amount, in pounds, of toxic air emissions in 2016 as well as the average household income in certain zip codes in Cincinnati. Create a graph from the data in columns "Toxic Air Emissions (in pounds)" and "Average Household Income."

Zip Codes	Toxic Air Emissions (in pounds)	Average Household Income (in U.S. dollars)
45225	3584	$ 15,035
45231	15	$ 54,107
45215	4506	$ 45,021
45217	1123	$ 43,152
45227	52345	$ 47,450
45229	912	$ 20,618
45233	303	$ 74,819
45204	13501	$ 25,377
45232	75281	$ 12,554
45209	1509	$ 52,428
45216	1312	$ 30,413
45237	4384	$ 32,299
45212	11635	$ 38,190

   Source: The Cincinnati Enquirer and www.zipcodes.gov

3. What is the mean amount of toxic air emissions across all of the zip codes?

▲
Worksheet 1: Emissions

LESSON 9.3 GENDER PAY GAP

Liza (Cope) Bondurant, Lee Inmon Dean, and Rebecca Hudson

ECONOMIC INEQUALITY

In 2019, women made an average of $0.79 for every dollar made by men. In this lesson, students discuss differences between salaries of those who identify as males and females by examining the percent change of salaries across various careers and then doing a more in-depth study of a singular career to investigate the variables potentially responsible for the variations. These variables can include experience, postsecondary education, age, and so forth. Learning about gender pay gaps can provide opportunities for students to consider how this information could impact future salary negotiations.

DEEP AND RICH MATHEMATICS

The mathematical goals for this lesson are for students to be able to use their data from research to create a scatter plot and two-way table. Students also look at the specifics of their scatter plot and two-way table, such as clustering data, outliers, positive or negative association, and line of best fit. Along with creating the scatter plot and two-way table, students interpret their meaning in relation to the social justice topic, which helps them learn about the association between the variables. The interpretation of the scatter plot and two-way table should allow students to comprehend the difference in the pay scale and the different factors that affect the pay scale between females and males.

ABOUT THE LESSON

This lesson uses a launch–explore–summarize instructional model and is intended to take approximately 180 minutes to complete across three class periods.

Resources and Materials

- Student Resource 1: *Gender Pay Data Sheet* (1 per student; Excel version available)

- Student Resource 2: *Gender Pay Gap Extension Activity* (optional; 1 per student)

- Teacher Resource 1, Gender Pay Gap Desmos Exploration (https://bit.ly/31tqaV9)

  + Consider using the Teacher Guide (https://bit.ly/3rd4Z3U) to facilitate the Desmos Exploration.

- Teacher Resource: *Gender Pay Gap Extension Activity Teacher Key*

- Student Resource 3: *Gallery Review Student Prompts* (1 per student)

- Laptops (1 per group minimum, 1 per student ideally)

- Chart paper (1 piece per pair of students)

- Markers

- Graph paper

- Glue

- Tape

- Sticky notes

MATHEMATICS PRACTICES
- Model with mathematics.
- Use appropriate tools strategically.
- Construct viable arguments and critique the reasoning of others.

LESSON FACILITATION

Launch (25 minutes)

- Begin by arranging students in groups of three to four. Distribute Student Resource 1 (*Gender Pay Data Sheet*; printable and Excel versions are available) and ask students to determine the median salaries in 2019 for males and females in several careers that they are interested in.

- Review the different measures of the center of a data set (mean, median, and mode) as needed. Use the following question to probe students' understanding of the three measures:

  + *Why is the median a better estimate of a typical value than the mean for the salary data?*

- Steer the discussion to the conclusion that median is a better estimate of a typical value, because the distribution may not be symmetric (could be skewed or bimodal) and may contain values far from the center (outliers).

- Use the following example to model how to calculate the percent difference between the median salaries of males and females for one career.

  + General Formula:

  $$\% \text{ Salary Difference} = \left[\frac{(\text{Male Median Salary} - \text{Female Median Salary})}{\text{Male Median Salary}} \right] \times 100$$

  + Example:

  % Salary Difference (Male Median vs. Female Median for Secondary Teachers in 2018)

  $$= \left[\frac{(58,925 - 53,942)}{58,925} \right] \times 100$$

  = 8.46% (rounded to the nearest hundredth)

  On average, secondary male teachers make 8.46% more than secondary female teachers.

- Instruct students to determine the percent salary difference between the median salaries of males and females for each career that the group members are most interested in. After completing their calculations, students should explain their calculations to their peers. If there is a disagreement in their calculations, group members should resolve them through discussion.

- Use the following questions to lead a whole-class discussion. The purpose of the discussion is to answer any questions students have regarding calculating the percent difference, discuss interesting findings, and introduce students to the concept of a gender pay gap.

  + *How can you use percentages to describe increases or decreases between the salaries of males and females?*

  + *Were there any percent difference calculations that were difficult? What made them difficult to calculate?*

  + *For which professions did males have higher salaries than females?*

  + *What professions had the greatest percent difference between the salaries of males and females?*

  + *Which professions had the least percent difference between the salaries of males and females?*

- Share the definition of gender pay gap, according to the American Association of University Women (AAUW):

  + *The **pay gap** is the difference in men's and women's median earnings, usually reported as either the earnings ratio between men and women or as an actual pay gap.*

+ The pay gap is the % Salary Difference example we calculated.

+ In a 2021 report by AAUW, U.S. women working full time in 2020 were paid 83 cents for every dollar earned by men; thus, a gender pay gap of 17% exists between men and women.

- Engage students in a discussion regarding what variables may contribute to the pay disparities for the careers they are interested in. Pose questions such as these:

  + What factors may contribute to some professions having a greater gender pay gap than others?

  + What could help close the gender pay gap?

  Possible Responses

  + Students may point out that male athletes could be paid more than female athletes, because men's sports receive vastly more media coverage, television licenses, and sponsorship deals, which contribute to higher revenue.

  + They may also note that males may be better at salary negotiations than females.

  + If the female decides to have children and assumes the majority of the responsibility for raising the children, her male counterparts may be available to work more hours, hold more administrative duties, and travel more.

  + There are regional differences, and males may be more attracted to regions offering higher salaries.

  + Males may be more likely to pursue postsecondary education.

    - Finally, the salary differences may be based on the person's age and years of experience.

Explore (120 minutes)

- Introduce students to the *Gender Pay Gap Desmos Exploration* (https://bit .ly/31tqaV9) and say: *Working with a partner, you will create a scatter plot to explore changes in one salary over time and then create a line of best fit for the data.*

  + In the Desmos Exploration, students create two scatter plots (one for males and one for females) of the median salaries for their chosen career in 2015, 2016, 2017, 2018, and 2019. Consider using the Teacher Guide (https://bit.ly/3rd4Z3U) to facilitate the Desmos Exploration.

- Tell students that they will create a poster on chart paper that displays a summary of their exploration about the gender pay gap. Provide the following guidelines for the posters and share that technology tools such as Canva, Google Slides, or Microsoft PowerPoint can be used to create images for the poster.

+ Title (the career)

+ Data table

+ Graph (scatter plots and lines of best fit) on graph paper

+ Equations for the lines of best fit for males and females

+ Interpretations of what the slope and *y*-intercept mean for each line

Teacher Guide, Gender Pay Gap Desmos Exploration is a step-by-step planning resource available in the Online Resources for this lesson: https://bit.ly/31tqaV9

- Instruct students to work with a partner to explore changes in one salary over time using Student Resource 1 (*Gender Pay Data Sheet*). Tell students to consider the following questions:

  + *Which career are you most interested in researching more for the next phase of the lesson?*

  + *How did the salary data influence your decision?*

  + *Was there a small or large percent change between the salaries of males and females for the career you chose?*

- Use the following questions as a way to facilitate discussion among the students:

  + *Which quantity did you represent with the independent and dependent variables?*

  + *What do you notice about the data?*

  + *Are there any clusters of data?*

  + *Are there any outliers?*

  + *What is the direction of the associations?*

  + *Is there a strong or weak association for the data?*

- Remind students that they should create a linear model for males and females. Prompt students to use the Desmos settings to color code the lines so they can easily distinguish them (see page 5 of Desmos Basic Features for specific instructions for how to change the color).

- Use the following questions as you monitor student progress and assess their understanding:

  + *How did the technology (Desmos) help you represent, analyze, and interpret the data?*

  + *How close are the data points to each of the lines?*

  + *What are the slope and y-intercept for each equation?*

  + *What do the slope and y-intercept mean in this context?*

Summarize (35 minutes)

- After all groups have finished, display each group's poster around the room for students to do a gallery review.

- Provide students with Student Resource 3 (*Gallery Review Student Prompts*). Tell students that they should leave feedback for each display by placing sticky notes on each group's display.

- Let groups choose which poster to start with but tell students there can only be one group at each poster.

- As groups finish viewing the posters, allow students time to view the feedback left on their own display and, if necessary, to edit their posters based on the feedback.

- Use the following questions to facilitate a debrief discussion to consolidate student understanding:

  + *Which representation (data table, scatter plot, line of best fit) helped the most in determining the starting point and changes over time for the salaries of males and females?*

  + *How did you determine which gender had a lower starting salary in 2015?*

  + *How did you determine which gender had the greater rate of change in their salary from 2015 to 2019?*

In an effort to use more inclusive language, we use the term *gallery review* for what is commonly referred to as a "gallery walk."

TAKING ACTION

Consider the following actions.

Gender Pay Gap Extension Activity

- Encourage students to research additional data related to the gender pay gap and gender opportunity. This can be structured as an independent, small-group, or whole-class activity. Provide students with Student Resource 2 (*Gender Pay Gap Extension Activity*) to guide their research, and share with them that they may expand their research beyond the questions provided.

Reflection on the Life of Ruth Bader Ginsburg

- Encourage students to watch a movie or read a brief article and write a reflection on the life of Ruth Bader Ginsburg.

- Students can organize a movie viewing and discussion within their school or community:

  + *On the Basis of Sex* (2018; 2 hours)

  + *RBG* (2018; 1 hour and 39 minutes)

+ Article: "Ruth Bader Ginsburg Becomes First Woman to Lie in State: 8 Other Strides She Made for Women," by Sara M. Moniuszko, Maria Puente, and Veronica Bravo, *USA Today*, September 24, 2020 (https://bit.ly/3rvsS7H).

Research Equal Pay Legislation

- Students can research equal pay legislation to gather a historical view of the current issue.

  + The Equal Pay Act was passed in 1963 and prohibited employers from paying male and female workers different wages for "jobs the performance of which requires equal skill, effort, and responsibility, and which are performed under similar working conditions." However, it had many exceptions, which led to many people (including Ruth Bader Ginsburg) to advocate for pay equity in the 1970s and 1980s. The work continues today from the Paycheck Fairness Act of 2019 (which passed the House but failed the Senate in 2019) to Megan Rapinoe testifying before Congress on Equal Pay Day in 2021.

ONLINE RESOURCES

 Available for download at **resources.corwin/TMSJ-MiddleSchool**

Student Resource 1: Gender Pay Data Sheet

Gender Pay Gap Extension Activity

Use the following questions to research issues related to the gender pay gap and gender opportunity. The listed websites provide data sources to aid in constructing responses to the questions. The data sources may also be used to develop additional questions of interest that you can report on in your report.

1. How does the gender pay gap in the U.S. compare to the gender pay gap in other countries throughout the world? Which country has the greatest gender pay gap? Which country has the smallest?

   Source 1: Our World in Data, "Unadjusted Gender Gap in Average Hourly Wages, 2016" (https://bit.ly/3dd0USo)

2. What factors contribute to the gender pay gap?

   Source 1: Article, "Economic Inequality by Gender" by Esteban Ortiz-Ospina and Max Roser, Our World in Data, March 2018 (https://ourworldindata.org/economic-inequality-by-gender)

3. What is the difference between gender pay and gender opportunity?

   Source 2: Article, "30+ Gender Pay Gap Statistics You Should Know" by Baily Reiners, *Built in*, January 22, 2020 (https://builtin.com/recruiting/gender-pay-gap-statistics)

4. The gender pay gap can vary by region of the country. What does it look like where you live compared to other areas of the U.S.?

   Source 3: Interactive map, Gender Pay Gap by State, AAUW (https://bit.ly/3IgampR)

5. How has the gender pay gap changed in the past 50 years? Where do you see the future of the gender pay gap?

   Source 1: Our World in Data, "Unadjusted Gender Gap in Median Earnings, 1970 to 2016" (https://bit.ly/3ppUUyK)

6. Describe at least one woman from history who has helped pave the road to equal pay.

   Source 4: Article, "12 Surprising Women From History Who Paved the Road to Equal Pay" by Ellen Cranley, Business Insider, April 2, 2020 (https://bit.ly/3ryuhuZ)

7. What are some of the ways you can promote gender pay and gender opportunity equality?

   Source 1: See source from Question 2.

Student Resource 2: Gender Pay Gap Extension Activity

Gallery Review Student Prompts

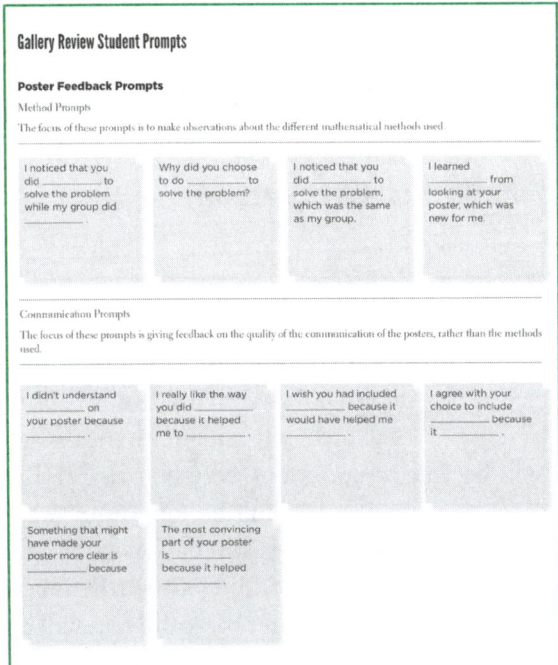

Student Resource 3: Gallery Review Student Prompts

Gender Pay Gap Extension Activity

Teacher Key

Note: The sources for each question are included in the Student Resource.

1. How does the gender pay gap in the U.S. compare to the gender pay gap in other countries throughout the world? Which country has the greatest gender pay gap? Which country has the smallest?

    The gender pay gap exists throughout the world. The greatest gender pay gap exists in South Korea. The smallest gender pay gap exists in Colombia.

2. What factors contribute to the gender pay gap?

    The largest factor impacting the gender pay gap is unexplained. This means it cannot be explained by a tangible variable.

    The factors that contribute to the gender pay gap (from most to least significant in 2010) include: occupation, industry, experience, race, and region. In 1980 education and unionization explained some of the differences in pay, but as of 2010 they had a negative impact on the gender pay gap.

3. What is the difference between gender pay and gender opportunity?

    A gender pay gap is one in which people are paid differently based on their gender, such as professional athletes. Megan Rapinoe's recent fight for equal pay for female professional soccer players is an example of a gender pay gap. In other careers, such as business managers, accountants, pharmacists, teachers, or lawyers, the gap is one of gender opportunity. Opportunities such as job title, experience level, industry, and location explain the differences in pay. Women may not be provided the same opportunities as men. For example, the majority of CEOs in the United States are males.

4. The gender pay gap can vary by region of the country. What does it look like where you live compared to other areas of the U.S.?

    My state is Mississippi, where women make 77% of what men make. Mississippi is ranked poorly (41st out of 50) for providing equal pay for males and females. I noticed Vermont is ranked best and Utah is ranked worst.

5. How has the gender pay gap changed in the past 50 years? Where do you see the future of the gender pay gap?

    The gender pay gap has decreased in the U.S. and worldwide over time. Hopefully, there will be no gender pay/opportunity gap in the future, but statistically it looks like it's going to take a very long time.

6. Describe at least one woman from history who has helped pave the road to equal pay?

    Beginning in 2013, Beyoncé was one of the first celebrities to use her star power to promote equal pay in interviews.

7. What are some of the ways you can promote gender pay and gender opportunity equality?

    Some ways include:

    (1) If you are responsible for hiring, don't hire based on gender. Rather, hire based on the candidates' qualifications and fit for the position. (2) Don't assume that certain careers are only for specific genders—be open-minded. (3) Support parents/caregivers in the workplace with family-friendly policies (i.e., paid leave, flexible schedules). (4) Vote for public leaders who support public policies that will help promote gender pay and gender opportunity equality (i.e., scholarships for higher education/training, child care, early childhood education).

Teacher Resource 1: Gender Pay Gap Extension Activity Teacher Key

SOCIAL JUSTICE OUTCOMES

- I can recognize and describe unfairness and injustice in many forms including attitudes, speech, behaviors, practices, and laws. (Justice 12)

- I know about some of the people, groups and events in social justice history and about the beliefs and ideas that influenced them. (Justice 15)

- I will speak up or take action when I see unfairness, even if those around me do not, and I will not let others convince me to go along with injustice. (Action 19)

ESSENTIAL MIDDLE GRADES CONCEPTS

- **Statistics and Probability—Grade 8: Investigate patterns of association in bivariate data.**

MATHEMATICS PRACTICES

- Reason abstractly and quantitatively.

- Construct viable arguments and critique the reasoning of others.

- Model with mathematics.

- Use appropriate tools strategically.

LESSON 9.4 HOW MANY MEALS CAN MINIMUM WAGE BUY?

Elizabeth O. Ayisi and Colleen Carman

ECONOMIC INEQUALITY

Young adults are often blamed for their debt, but is it really their fault? What about people living in poverty? This lesson examines the relationship between cost of living and minimum wage over time. Students will be able to determine whether their dollar amount really goes as far as it should. It won't be long until these young adults are applying for jobs, deciding whether to go to college and where, and/or determining how to budget their finances. This lesson may spark the debate over whether these choices will be enough for them to be financially comfortable, to reduce poverty in America, and to extend or develop how to empower themselves and other young adults.

DEEP AND RICH MATHEMATICS

Students will learn to build and analyze scatter plots while also estimating the line of best fit. They will conceptualize and analyze the slope, positive and negative association, and linear and nonlinear correlation. Additionally, students will calculate several slopes, compare slopes, and discuss the meaning of the slopes in a real-world context.

ABOUT THE LESSON

This lesson uses a launch–explore–summarize instructional model and is intended to take approximately 165 minutes to complete across three class periods.

Lesson 1: Students investigate the cost of living to answer the question: *Can you buy more Big Macs with an average salary in 2020 or with an average salary in 1968? Why?*

Lesson 2: Students compare the average rate of change of minimum wage and cost of living to understand what the slope of each line (minimum wage and cost of living) represents and how the steepness of each line affects affordability.

Lesson 3: Students will create a presentation of their argument about when they think Big Macs were more affordable.

Resources and Materials

- Teacher Resource 1: *Exit Slips* (1 per student)

- Worksheet 1: *Minimum Wage and Cost of Living* (1 per student)

- Worksheet 2: *Federal Minimum Wage* (1 per student)

- Worksheet 3: *Reflection* (multiple copies per student)

- Student access to computer with internet

- Website: Desmos.com graphing calculator

- Website: American Institute of Economic Research online cost of living calculator (https://bit.ly/31sLlXG)

- Website: FRED economic data, "Federal Minimum Wage Graph" (https://bit.ly/3I9ZDxf)

- YouTube video: "When Can You Afford More Big Macs? In 1968 or Now?" TikTok video by @genypastor (https://bit.ly/3rx4geE)

- Student access to Google Slides, PowerPoint, chart paper, or another tool to present their arguments

LESSON 1 FACILITATION

Minimum Wage and Cost of Living

Students will watch a TikTok video that introduces the question: *Can you buy more Big Macs with an average salary in 2020 or with an average salary in 1968? Why?* Students will then investigate the cost of living over this time frame through analysis of slope. Students will then make connections and prepare a presentation through research of historical events that may have impacted the cost of living.

Launch (10 minutes)

- Distribute Worksheet 1 (*Minimum Wage and Cost of Living*).

- Spark curiosity by showing the video, "When Can You Afford More Big Macs? In 1968 or Now?" (https://bit.ly/3rx4geE) and have students jot down some initial thoughts and observations on the worksheet (question 1.a).

- Present the Federal Minimum Wage graph on Worksheet 1 and have students label and analyze the graph.

  + Ask about units for *x*- and *y*-values and what these values represent.

  + Make a list of observations students can see. Focus on using precise vocabulary.

  + Notice that the data points on the graph are from the years 1968 to 2020.

- Tell students to make a hypothesis based on the video and the graph.

  + Encourage them to list what they notice and wonder.

  + Take a poll for each option and ask a few students to explain which option they chose and why.

  + Encourage students to think about whether wage is the only important factor in the affordability of Big Macs. Make a list of the factors students share.

Explore (35 minutes)

- Present the Cost of Living graph on Worksheet 1 and have students label and analyze the graph.

  + Ask students if they know what the cost of living means.

  + Ask about the units for *x*- and *y*-values and what these values represent. Have students describe what one of the (*x*,*y*) points represents.

  + Make a list of students' observations. Focus on using precise vocabulary.

  + Notice that the data points on the graph are from the years 1968 to 2020.

- Introduce the Online Cost of Living Calculator and discuss what information it gives and what each value represents.

- Tell students to work in their groups to complete Worksheet 1.

- While monitoring group work, ask students:

  + *What are the units of your rates of change?*

  + *What do your results mean?*

- Regroup when most groups have finished question 5 to highlight student work, including different representations and methods to find the slope between data points. Use purposeful selection, sequencing, and connecting techniques to call on groups to present their ideas as a class.

- Use the following questions to assess and facilitate student understanding of slope and context:

    + *When is it best to use each method?*

    + *Which slope was higher?*

    + *What does that mean in terms of buying power?*

    + *What would the graph look like if the slope was constant?*

    + *If the slope was constant at the same rate you found between 1968 and 1974, would the y-value for the year 2020 be higher or lower than the value shown on the graph in 2020?*

    + *Do you think that would be good or bad?*

- Tell students to continue working in their groups to complete question 6, where they will choose one line segment and investigate the historical impacts of that time period.

    + Encourage students to include words like *economy*, *major events*, *president*, and *minimum wage* in their search along with their years.

- While monitoring each group, do the following:

    + Ask students why they chose the specific time frame.

    + Encourage them to use precise vocabulary to relate to the graph, line segment, and slope when appropriate.

    + Ask: *Why do you think your event may have had an impact on cost of living?*

    + Discuss with groups who chose nonlinear intervals why the average rate of change of their interval may or may not be a good representation of their entire time frame.

    + Remind students that we are looking for possible causes and effects.

- Ask students to summarize their findings on the whiteboard, a single slide, or chart paper to present to the class. Have students do a gallery review of their peers' conclusions.

- Students can have solutions written on the board, on a sheet of paper, or on a Google slide. Give students the opportunity to see their peers' work. They may write questions/comments on their own sheet of paper or directly on their peers' work with a sticky note.

- Allow students to discuss the questions and comments from their peers as they prepare to present their findings to the whole class.

In an effort to use more inclusive language, we use the term *gallery review* for what is commonly referred to as a "gallery walk."

Summary (10 minutes)

- Each group will give a short presentation of their findings to the class. Presentations should include the following:

    + The time frame of the data their group chose

    + The slope of the line segment between the dates they chose

    + Their method for finding the slope

    + Any conclusions they came to

- As groups present, encourage students to record their observations on Worksheet 1. Encourage them to

    + record the slope and time frame of each group;

    + notice the difference in the representations and methods of finding slope; and

    + take notes on what is surprising from peers' presentations, and aspects they agree or disagree with.

- Pose the following questions to assess student understanding:

    + *What does a steeper slope represent?*

    + *What about a 0 slope?*

- Distribute or post the prompt for the Lesson 1 Exit Slip and have students record their answers for question 1. Use the following questions to have a brief discussion:

    + *Does a higher salary necessarily indicate more wealth?*

    + *How does cost of living affect a person/family?*

- Have students complete question 2 and then use the following questions to facilitate a discussion of the methods used to organize the slopes:

    + *What does this tell us about the cost of living over time?*

    + *How does a steeper slope or less steep slope affect daily living (if Big Macs are more/less affordable)?*

    + *What does that mean for available spending for other goods?*

    + *What can we conclude about the cost of living over time?*

LESSON 2 FACILITATION

Federal Minimum Wage

Students will use the Desmos graphing calculator to construct a scatter plot and line of best fit of federal minimum wage over the same time frame as Lesson 1.

Students will compare the average rate of change of minimum wage and cost of living to understand what the slope of each line (minimum wage and cost of living) represents and how the steepness of each line affects affordability.

Launch and Explore (50 minutes)

- Distribute Worksheet 2 (*Federal Minimum Wage*), and have students work in the same groups as Lesson 1 to complete the task. Remind them to use the same interval they studied for cost of living to investigate the federal minimum wage.

- Monitor students' progress throughout the activity, encouraging the appropriate use of vocabulary (linear/nonlinear association, outlier, positive/negative correlation) and connections to the context to analyze their graphs.

- Use the task questions and the following prompts to further probe students' thinking about minimum wage and cost of living and regroup as a class to review as needed.

Question 3

+ Encourage students to consider large jumps in the graph.

+ Ask: *If you draw a line segment between these two points, does this line represent all the points in the interval?*

Question 6

+ As students use Desmos to find the line of best fit for their scatter plot, provide support to find the slope and *y*-intercept. Ask these questions:

  - *Is the slope found using Desmos the same or different from the slope you found in Worksheet 1?*

  - *Why or why not?*

Question 10

+ As students compare the slope they found for cost of living in Lesson 1 and the slope they found for minimum wage, ask these questions:

  - *Do units matter?*

  - *Can you accurately compare the two slopes?*

Question 10e

+ As students use their values from minimum wage and with the cost of living calculator, ask these questions:

  - *Do you think the minimum wage values are fair?*

  - *Why or why not?*

- Encourage students to think about how even a few cents or dollars each month can accumulate over time.

+ Consider the budget of a typical family, and ask these questions:

  - *What are the necessities?*

  - *What should a family be able to afford on a full-time job?*

  - *What is fair?*

Summarize (5 minutes)

- Distribute or post the prompt for the Lesson 2 Exit Slip and have students record their answers.

LESSON 3 FACILITATION

Reflection

Students will create and present an argument about when they think Big Macs were more affordable using data from this unit to support their claim. The class will summarize how slope and lines of best fit can be used to understand the relationship between two variables. Students will consider how minimum wage can affect quality of life and whether the current federal minimum wage is fair.

Launch (5 minutes)

- Have students work in the same groups as Lessons 1 and 2 to prepare a presentation on when they think Big Macs are more affordable.

- Remind students that their presentation should include key vocabulary about their scatter plot, an analysis of slope, and historical relevance.

Explore (20 minutes)

- Monitor students as they work in their groups to complete their presentations.

- Use purposeful selection, sequencing, and connecting techniques to determine the order for groups to present.

Summarize (30 minutes)

- Students will need one copy of Worksheet 3 (*Reflection*) per group whose presentation they will observe. While each group presents their findings, prompt each student observing to use Worksheet 3 to record their notes on what they agree and disagree with, what surprised them, and whether the presentation made them reconsider their position.

- Ask students to brainstorm possible counterarguments to their presentations. How could they use mathematics to justify their position?

- Encourage students to ask their peers questions at the end of their presentations.

- If students use the same piece of information to support different claims, ask these questions:

  + *What additional information could be found to support one claim over the other?*

  + *What is the difference in arguments? What is in common?*

- Distribute or post the prompt for the Lesson 3 Exit Slip and have students record their answers.

- Have students briefly discuss in their groups and then have a class discussion. Use the following prompt to facilitate a closing discussion:

  + *Did this assignment change any perspectives or beliefs you held before we began this lesson?*

  + *How can you retain a sense of empathy?*

- Encourage students to do one of the options under Taking Action.

TAKING ACTION

Encourage students to take one of the following actions:

- Write a letter to their state representatives with their conclusions and justifications about affordability of goods and services over time. Persuade state representatives to support their claim about minimum wage in 2020.

- Make a poster with their findings and a call to action. Hang it up around their school or neighborhood.

- Educate others about cost of living and minimum wage.

- Speak at a city council meeting.

- Do a similar analysis for cost of their household cooking staples over time and share results with the community.

- Explore the Fight for $15 and worker strikes at fightfor15.org and around the web. Write an article for the school newspaper summarizing current events. Read some of the articles about McDonald's and make connections to what they read and the analysis they did in the previous lessons.

- Do an internet search of efforts (including collective actions and activism) to raise the minimum wage from the past and present. Ask students to consider whether there is any action in their community.

ONLINE RESOURCES

 Available for download at **resources.corwin/TMSJ-MiddleSchool**

Exit Slips

Day 1 Exit Slip:
1. Based on your exploration and your peers' findings, what is your answer to the question: Do you think you could buy more Big Macs with your salary in 1968 or this year? Why or why not? Did your answer change?

2. **Extension:** Organize the slopes of each group into categories or a list. Why did you choose this organization method? What does that tell us about cost of living over time?

Day 2 Exit Slip:
3. How could you justify your answer to the question, "Do you think you could buy more Big Macs with your salary in 1968 or this year?" What math can you use?

Day 3 Conclusion and Explanation:
4. When do you think people could buy the most Big Macs, in 1968 or now? Why? Did you change your mind since Day 1, when you saw the TikTok video?

5. Life is about more than just Big Macs. What about meals cooked at home? How does minimum wage affect quality of life?

▲
Teacher Resource 1: Exit Slips

Minimum Wage and Cost of Living

Part 1: Minimum Wage
1. A Look at Minimum Wage

   a. After watching the TikTok Video, list your observations:

   b. Look at the Federal Minimum Wage graph below and list your observations:

   c. Do you think you could buy more Big Macs with your salary in 1968 or this year? Why?

2. Federal Minimum Wage (y-axis is dollars per hour, x-axis is years since 1967)

▲
Worksheet 1: Minimum Wage and Cost of Living

Worksheet 2: Federal Minimum Wage

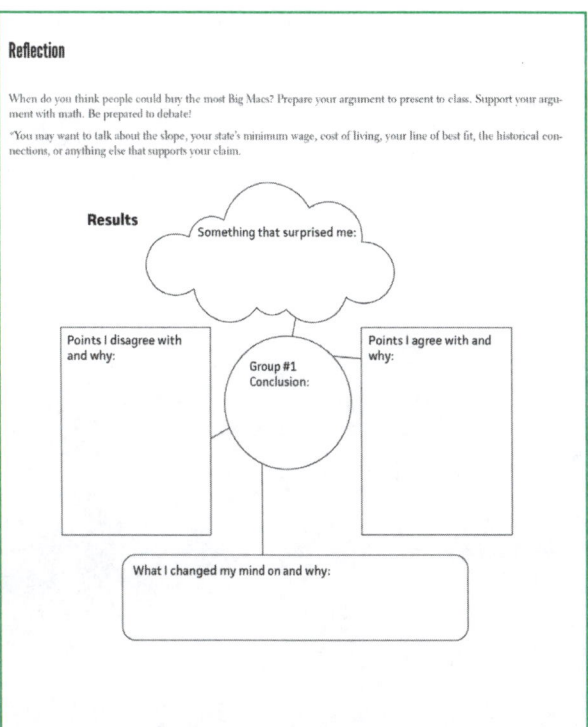

Worksheet 3: Reflection

CHAPTER 10

GEOMETRY

Middle school students develop a conceptual understanding of area, surface area, and volume. As teachers look at the standards, they will hopefully note consistency, which requires students to apply this understanding to real-world and real-life problems. Geometry as a spatial topic provides wonderful opportunities for students to explore different cultural artifacts and learn more about the larger world connecting these ideas with mathematical topics. A unique opportunity for later middle grade students is the opportunity to understand and apply the Pythagorean Theorem. Students could make connections of distances in their city or community that relate to voting booths, healthy food, and high-interest loan companies to government-assisted housing, their homes, and other areas to explore, understand, and respond to potential social injustices.

Using dynamically layered maps with dynamic geometry software and other technologies, students can dive deeply into modeling a variety of contexts. These software programs provide opportunities for teachers to connect students' geometric standards to real-life contexts that expand their connections to the community and empower them to use mathematics to understand and change the world. The following geometric lessons submitted by teachers from the field connect geometry and social justice.

LESSONS

Lesson No.	Lesson Title	Mathematics Focus Areas	Grade	Social Justice Topic	Authors
10.1	*Map Projections*	• Geometry and Measurement	6 and 7	Representation Justice	Lisa Skultety, Candace Joswick, Melissa Troudt, and Robin Keturah Anderson
10.2	*3D Modeling for Water*	• Geometry and Measurement	6 and 7	Human Rights	Mathew D. Felton-Koestler and Courtney Koestler
10.3	*Water Is Life— Our Collective Past, Present, and Future*	• Geometry and Measurement	6 and 7	Environmental Issues	Sara Rezvi and Tyrone Martinez-Black
10.4	*Accessible Playground*	• Geometry and Measurement	6 and 7	Ability	Maggie Lee McHugh and Jennifer Kosiak
10.5	*Investigating Areas to Determine Fairness*	• Geometry and Measurement • Ratio and Proportions • Statistics and Probability	8	Civil Rights and Governmental Laws	Jennifer Dao

- I can recognize and describe unfairness and injustice in many forms including attitudes, speech, behaviors, practices, and laws. (Justice 12)

- I will work with friends, family and community members to make our world fairer for everyone, and we will plan and coordinate our actions in order to achieve our goals. (Action 20)

ESSENTIAL MIDDLE GRADES CONCEPTS

- **Geometry—** Grade 6: Solve real-world and mathematical problems involving area, surface area, and volume.

- **Geometry—** Grade 7: Solve real-life and mathematical problems involving angle measure, area, surface area, and volume.

MATHEMATICS PRACTICES

- Construct viable arguments and critique the reasoning of others.

- Use appropriate tools strategically.

LESSON 10.1 MAP PROJECTIONS

Lisa Skultety, Candace Joswick, Melissa Troudt, and Robin Keturah Anderson

REPRESENTATION JUSTICE

Maps are "ubiquitous, authoritative, and taken for granted" (Krupar, 2015, p. 91). Consider the power of maps in everyday life—defining borders that may separate people and goods (sometimes causing contention), abstracting relationships between places and space, and documenting both topological changes in the earth and changes in political power and land control over time. Students often have significant misconceptions related to the relative size of various landmasses (e.g., belief that Greenland is larger than China) and have not questioned the "objectivity" or "truth" presented in maps.

DEEP AND RICH MATHEMATICS

In this lesson, students solve a real-world problem involving surface area and reasoning about the size of landmasses using mental models of countries and the Mercator projection. To interrogate the sizes of countries in relation to each other, students use concepts and reasoning related to geometric measurement. Comparing areas of countries, students encounter some conflict between what they thought they understood to be true and what they are currently learning.

ABOUT THE LESSON

This lesson uses a launch–explore–summarize instructional model and is intended to take approximately 270 minutes to complete across three class periods.

Lesson 1: Students use mathematical tools to interrogate the size of countries on the Mercator map.

Lesson 2: Students investigate different map projections and discuss how maps have embedded biases.

Lesson 3: Students continue to explore how map projections can be problematic and make recommendations to adopt a school map based on their investigations and discussions.

Resources and Materials

- Computers for pairs of students

- Teacher PowerPoint resource to support lesson facilitation: *Map Projection—Editable Teacher Template*

- Student PowerPoint resource to support lesson facilitation: *Map Projection—Interactive Student Slides* (1 per pair)

- Interactive Website: The True Size (http://thetruesize.com)

- Interactive Website: Metrocosm, "Eight Ways of Projecting the World" (https://bit.ly/3ojYUBs)

- Video: "Why All World Maps Are Wrong," from *Vox*, December 2, 2016 (https://bit.ly/3Dq69vX)

- Video: "Gall–Peters Projection" from "Somebody's Going to Emergency, Somebody's Going to Jail," *The West Wing*, season 2, episode 16, NBC, February 28, 2001 (https://bit.ly/3pHm8Bl)

This lesson is inspired by "Math, Maps, and Misrepresentation" in Gutstein (2006).

> Social Studies educators may offer possible extensions of this activity—discussions of disputed borders, how borders change over time and the varying reasons, the relationship between land sizes under political control, and other related topics like goods and services.

LESSON 1 FACILITATION

Students work in pairs or small groups to compare the size of two countries, first by appearance using a Mercator projection, then by estimating the area using tools in Student Resource 1 (*Map Projection—Interactive Student Slides*). As the class discusses their findings, it becomes clear that the predictions of country sizes based on the Mercator map projection are inaccurate, and their estimated areas reveal unanticipated findings.

Launch (40 minutes)

What Are the 10 Largest Countries? (10 minutes)

- Begin by asking students to close their eyes and envision the world. With a picture of the world in their mind, ask them to guess the 10 largest countries in the world and write them down.

  + Have students share their guesses with a partner.

- Project an image of a Mercator projection for all students to see.

  + Ask if anyone would like to revise their thinking about which are the 10 largest countries.

+ Discuss which countries are surprisingly larger or smaller than they recalled from memory.

- Project an image of the Mercator projection with the top 10 countries highlighted.

What Is the Purpose of a Map? (15 minutes)

- Ask students: *Why might it be important to have a true understanding of a country's size?*

- Allow students to discuss their ideas in groups and share with the whole class.

- Create a list of students' responses that can be referred back to later. This list will be revisited at the end of the lesson. Draw upon the following possible reasons during a class discussion if needed:

  + Perception of the population size, resources, or economy of a country

  + Perception of the heterogeneity of a country's citizens

  + How size can convey the importance of a country

How Can We Find the Size of a Country? (15 minutes)

- Discuss the following prompts in small groups:

  + *Aside from using the internet, how could we determine the size of the United States?*

  + *What information would we need?*

- Next, move to a whole-class discussion and select some groups to share their ideas.

- Project the image of the United States and ask students to provide other ideas after seeing the map.

- During this discussion, highlight a few key features to help support students during the Explore phase of the lesson:

  + In the United States, we generally consider the size of the country to be the square mileage.

  Identify the location of the map key.

  + Use the following sentence stem to support students' proportional reasoning by comparing the size of states: _____ *appears to be about ___ times larger/smaller than _____.*

Explore (40 minutes)

Exploring the Size of Countries (25 minutes)

- Organize students into pairs and assign each pair two countries to compare. Five suggested pairings are in Teacher Resource 1 (*Map Projection—Editable Teacher Template*).

    + Make sure that at least two pairings of students are assigned to the same set of countries. These two pairings will come together for a reflection activity later in the lesson sequence.

- Provide students with Student Resource 1 and have them look at the slide with the Mercator projection that magnifies their two countries of interest.

    + **Note:** If you select different countries for comparison, have students take screenshots of the countries on Google Maps and use the scale to create appropriately sized shapes for tiling to estimate the area.

    + Google Maps uses the Mercator projection, which distorts countries' areas. This distortion is relatively minimal when the map is zoomed in for countries that are "short" and centered around a few longitudinal lines. Avoid using "long" countries (like Chile) for comparison, which experience significant distortion in the Mercator projection due to its spanning of many latitude lines.

- Tell students to consider the relative size of the two countries and record their guess about how the countries compare, using the space provided in Student Resource 1.

- Have students estimate the size of each country using the provided tools and images. They can drag and drop the shapes to tile their countries.

- After calculating an estimate of the size of each country, the student partners will answer the final question in their section of the slide deck.

    + *How did the sizes of the countries compare? What did you learn?*

Discussion of Exploration With Peers (15 minutes)

- Once pairs arrive at a solution and final answer, form groups who investigated the same countries. Ask them to share their strategies and determine which estimate they feel is the most accurate of the size for each country. Discuss the following questions:

    + *What was your strategy to estimate the area?*

    + *Why did you select the shapes you used to estimate the area?*

    + *Which shapes did you start with and why?*

    + *Do you think that your estimate is an overestimate, an underestimate, or fairly accurate? Why?*

+ *Compare your results with your predictions. Why do you think your original predictions were correct or incorrect?*

+ *What did you notice as you worked? What do you wonder?*

Summarize (10 minutes)

• Provide time for students to individually summarize their findings from the pair exploration and small-group exploration. Tell them to record and retain their summary to refer to during the Lesson 2 Launch.

• Use the following prompts to guide student individual summary:

+ *Describe the strategy you and your partner used to estimate the size of the countries.*

+ *What are your calculated areas?*

+ *How was your group's strategy the same as or different from the other group you discussed your findings with?*

LESSON 2 FACILITATION

The Mercator Projection

Students review their findings from Lesson 1, that the Mercator projection does not accurately depict the size of countries. In pairs, students then explore through an online tool how the Mercator projection distorts the size of countries. Finally, students consider the social justice implications of the prevalence of the Mercator projection and the distortion it perpetuates.

Launch (15 minutes)

• Begin with a brief review of Lesson 1. Have students share findings and comments from the Lesson 1 exploration and the summaries created at the lesson.

+ Allow the country pair groups to share their solution strategies and help students identify themes from the discussion (e.g., countries that they thought were comparatively the same size but found them not to be, using large squares to estimate area is easy to calculate but not accurate for irregular borders).

+ Use the following prompts to guide the discussion:

  - *Share what your group of country pairings discovered and how you discovered it.*

  - *What are some common themes in what we learned?*

+ Consider the following strategies to facilitate students sharing:

- Students sketch their solution strategies on multiple whiteboards or model their solution strategies on a projected slide deck.

      - Facilitate a discussion to compare strategies by selecting and sequencing slides with different strategies.

    + During the discussion, use the following strategies to highlight common themes across the presentations:

      - Record phrases from students sharing about their strategies or discoveries in a space where the whole class can see the list.

- At the conclusion of sharing, asks students to notice similarities and differences across strategies and countries.

Explore (60 minutes)

Investigating the True Size Website (30 minutes)

- Begin the exploration with a whole-class demonstration to show students how to use the True Size website (https://thetruesize.com).

- Have students work in pairs and use the True Size website to investigate their initial list of the 10 biggest countries.

- Use the following questions as you monitor pairs' work and check for understanding:

    + *How many countries in the actual top 10 biggest countries were in the list you created?*

    + *What country in the actual top 10 biggest countries list is the most surprising to you? Why is that?*

    + *Choose one country. In what ways can you manipulate the country or other countries to investigate the true size? What do you notice as you perform those manipulations?*

    + *What are some ways to get a sense of the "true size" of the countries on the map?*

    + *What do you notice about the location of the country on the map and its size?*

    + *What do you wonder about seeing the top 10 countries and their location on the map?*

- Facilitate a brief whole-class discussion and have students share out findings and other "noticings and wonderings" from their exploration of the True Size website.

Why Is the Mercator Projection Problematic? (30 minutes)

- Use the following questions to continue whole-class or small-group discussions around why we need maps, what we use them for, and how the purposes of maps have changed over hundreds of years:

  + *What is the purpose of a map?*

  + *What information does a map need to provide?*

  + *Do you think that the purpose of a map has historically been the same? Why or why not?*

- Conclude the discussion by showing the *Vox* video, "Why All World Maps Are Wrong" (https://bit.ly/3Dq69vX).

  + The narrator demonstrates slicing a globe and trying to create a map on a flat plane and then describes various projections, the purpose for them, and some of the problems.

  + Stop the *Vox* video after the brief clip from *The West Wing* addressing how the Mercator projection is problematic.

  + **Note:** The full "Gall–Peters Projection" clip from *The West Wing* is available at https://bit.ly/3pHm8Bl.

Summarize (15 minutes)

- Use the following questions to facilitate whole-class or group discussion among the students:

  + Say to students: *In the video, the narrator comments, "One modern critique (of the Mercator projection) is that this distortion perpetuates imperialist attitudes of European domination over the Southern Hemisphere"* (timestamp 3:08). *Do you agree or disagree and why?*

  + Direct students to revisit the list of countries from the Launch portion of Lesson 1 as they consider why our understanding of countries' sizes might matter.

    - Students may resurface previous ideas about the perception of the relationships between the size of a country and power, population, and resources.

    - Encourage students to use specific examples to illustrate or refute their claims.

Traditionally, maps were necessary for sailing and navigation, and this is why the Mercator projection is a prominently used map. Yes, it obstructs the true size of countries, but it preserves the shape and direction, which would be necessary for sailing (see https://bit.ly/3ElWOXF).

LESSON 3 FACILITATION

Alternate Map Projections

Students work in pairs to consider other possible projections that have different distortions and demonstrate their understanding of both the mathematical and spatial justice focus of the lesson by writing a letter to argue for or against the use of the Mercator projection in schools.

Launch (10 minutes)

- Briefly revisit the top 10 largest countries by projecting the three map projections in the slides: the Mercator projection, the Winkle Triple projection, and the Behrman projection (the Behrman projection is the only projection of these three that preserves area, but it distorts shape).

- Ask students, *What differences and similarities do you notice between the three maps?* **Note:** It may be helpful for teachers to print out the three maps to allow students to compare the map projections side by side.

Explore (30 minutes)

- Introduce the activity by saying, *In the video, "Why All World Maps Are Wrong," the narrator described several alternative projections to the Mercator projection. Today you will investigate seven of these eight Metrocosm map projections* (https://bit.ly/3ojYUBs).

  + Show students that by clicking through the various projection names at the top of the website, they can explore what each projection does to the shape, size, and orientation of countries.

- Allow a few minutes for students to explore this website in pairs, then use the following prompts to facilitate a whole-class discussion:

  + Ask students to share what they noticed about the projections with the class.

  + If students have not already brought this up, it would be a good time to discuss how maps are often viewed as objective truth that simply portray the "facts." However, we saw in this lesson that this is not the case. By the nature of forcing a spherical object to be represented on a plane, some aspect has to be distorted. Each projection has to distort reality in some way. However, what is distorted has an impact on how we view our world.

Summarize (50 minutes)

- To summarize the lesson, share with students that they will write a letter demonstrating their understanding of the mathematical and social justice content.

  + Depending on each classroom's context, it may be appropriate to write a hypothetical letter (option 1) or a letter that will actually be delivered to educational stakeholders (option 2). While it provides students with a more meaningful and authentic experience to write letters that are being sent to appropriate leaders, every classroom may not be in a place where that would be appropriate. As such, you should select the option that works best for your context.

- **Option 1:** Pose the following hypothetical situation scenario to students.

  + *The local elementary school (insert the name of local elementary school) has decided to purchase new, large wall maps for every classroom. Officials figure they will simply replace every map with a new version of the same projection, the Mercator projection. Given what you learned in this lesson, do you think that this is the best option for the school?*

  + Instruct students to use the knowledge they gained about the area and size of countries, along with other concerns addressed in this lesson, to support their decision.

  + Write a letter to the administrators of (school name) addressing the following:

    - The school should purchase new Mercator projection maps. Why? What advantages does the Mercator projection have over the other projections? *or*

    - If you don't think that the school should purchase Mercator projections, which of the projections that you explored do you think the school should purchase? Why? What advantages does your chosen projection have over the Mercator projection?

- **Option 2:** Have students write a stakeholder letter.

  + Tell students that instead of crafting a letter to a hypothetical school board, students can write a similar letter to the stakeholders in their context, including their school and district leadership to affect change about map projections. Students can use similar questions to guide their thinking and support their written argument.

TAKING ACTION

- Have students compose letters advocating for an alternative to the traditional Mercator projection classroom map in local elementary or middle schools. These letters can be addressed to specific classroom teachers, grades, administrators, or school board members. Through changing the maps that support students' visualization of the world to reflect a more accurate representation of countries' sizes than the Mercator projection, younger students will be encouraged to have a better understanding of the true size of countries and continents in the world.

- The distortion of "objective truth" that most students believe is represented through a map is another key takeaway of this lesson. We often take "scientific representations" as objectively true and unable to be distorted. However, in map projections, the distortion is deliberate and unavoidable. School communities (teachers, administrators, and students) can build on this discussion intentionally in other courses to interrogate objective truths that are drawn upon in other disciplines.

ONLINE RESOURCES

 Available for download at **resources.corwin/TMSJ-MiddleSchool**

IMPORTANT NOTES
How Do We See Our World: Map Projection Investigation

- This slide deck is not intended for instruction with students.

- This slide deck allows you to make edits to the slide as necessary for you and your instructional setting. The Student Slides presentation has each slide imported as an image, so students cannot accidentally make changes. However, the various shapes (squares, triangles, and circles) that are available to tile over countries are movable.

- Some slides have additional notes in the "speaker notes" section to aid in facilitation.

- If you want to edit the slides, do so in this file. If you would like your edited slides to be unable to be manipulated by students, follow these steps:
  - With your edited slide selected, go to File > Download > PNG Image (this will download a high-quality image the size of the slide).
  - In this slide deck, insert a new blank slide.
  - Go to Slide > Change Background and a pop-up window should appear.
  - In that window, click "Choose Image" and locate your downloaded PNG file.
  - Copy the shape tiles from this file into the student slides so students can use them to tile their countries.

Teacher PowerPoint resource to support lesson facilitation: Map Projection—Editable Teacher Template

---

BEFORE YOU BEGIN
How Do We See Our World: Map Projection Investigation

- This slide deck is intended for student exploration.

- All images and text are imported as a background so they cannot be moved by students. However, the various shapes (squares, triangles, and circles) that are available to tile over countries are movable.

- Some slides have additional notes in the "speaker notes" section to aid in facilitation.

- If you want to edit the slides, do so in the Map Projection Editable Teacher Template. If you would like your edited slides to not be manipulated by students, follow these steps:
  - With your edited slide selected, go to File > Download > PNG Image (this will download a high-quality image the size of the slide).
  - In this slide deck, insert a new blank slide.
  - Go to Slide > Change Background and a pop-up window should appear.
  - In that window, click "Choose Image" and locate your downloaded PNG file.
  - Copy the shape tiles from the template into the student slides so students can use them to tile their countries.

Student PowerPoint resource to support lesson facilitation: Map Projection—Interactive Student Slides

SOCIAL JUSTICE OUTCOMES

- I can recognize and describe unfairness and injustice in many forms including attitudes, speech, behaviors, practices, and laws. (Justice 12)

- I will work with friends, family and community members to make our world fairer for everyone, and we will plan and coordinate our actions in order to achieve our goals. (Action 20)

ESSENTIAL MIDDLE GRADES CONCEPTS

- **Geometry—** Grade 6: Solve real-world and mathematical problems involving area, surface area, and volume.

- **Geometry—** Grade 7: Solve real-life and mathematical problems involving angle measure, area, surface area, and volume.

MATHEMATICS PRACTICES

- Make sense of problems and persevere in solving them.

- Construct viable arguments and critique the reasoning of others.

- Model with mathematics.

- Use appropriate tools strategically.

- Attend to precision.

LESSON 10.2 3D MODELING FOR WATER

Mathew D. Felton-Koestler and Courtney Koestler

HUMAN RIGHTS

The United Nations World Water Development Report provides insight on main trends concerning the state, use, and management of freshwater and sanitation in countries around the world. This global concern is one that we also see in the United States. This includes (1) examples of ongoing issues with access to safe drinking water such as the Flint water crisis in the United States and lack of access to water in other countries, (2) examples of "natural disasters" such as a tornado in Ohio that resulted in people needing emergency water until services had been restored, (3) a discussion about the socially constructed discourse of "natural disasters," and (4) information about the increasing frequency of "natural disasters" due to climate change. The lesson is contextualized with background information on water crises and access to fresh water and focuses on designing water containers for a hypothetical nonprofit that wants to bring water to people in need.

DEEP AND RICH MATHEMATICS

Throughout this lesson, students engage in mathematical modeling and design by considering a realistic context of bringing water to people in need and designing water containers to effectively do so. Students consider real-world variables, such as how many people they hope each container will serve, for how long, the amount of water they want to deliver, and the amount of material required to create the container. This process draws on the mathematical modeling process, visualizing 3D objects, using 3D modeling tools, calculating the volume and surface area of containers, and making conversions within units.

ABOUT THE LESSON

This lesson uses a launch–explore–summarize instructional model and is intended to take approximately 135–180 minutes to complete across three to four class periods.

Lesson 1: Students are introduced to the issue of water access.

Lessons 2–4: Students design water bottles.

Resources and Materials

- Worksheet 1: *Is It Natural?* (1 per student)

- Worksheet 2: *Info Sheet* (1 per student)

- Worksheet 3: *Water Containers: Individual Ideas* (1 per student)

- Worksheet 4: *Water Containers: Plan* (1 per pair of students)

- Base-10 blocks (at least one unit block and one mega-block)

- Various containers that hold liquid (reusable water bottle, 2-liter soda bottle, 1-gallon jug, water cooler jug, etc.)

- Teacher Resource 1: *Introduction to Tinkercad* (introductory lesson for software, with additional resources)

Videos

- "Why America's Drinking Water Crisis Goes Beyond Flint," from BBC News (https://bbc.in/3Dl0WWk)

- "The Sustainable Development Goals Explained: Clean Water and Sanitation," from the United Nations (https://bit.ly/3rz7rCQ)

- "Water Walk" featuring a poem by Martin Kiszko, from WaterAid (https://bit.ly/3DnUY7b)

- "Why Climate Change Makes Storms Stronger," from Yale Climate Connections (https://bit.ly/3GhAfUB)

Websites

- United Nations, "Water Scarcity" (https://bit.ly/3EmccmR)

- Centers for Disease Control and Prevention, "Global Water, Sanitation, & Hygiene (WASH) Fast Facts" (https://bit.ly/3xSiI25)

- Connecticut Democracy Center, "Writing to Your Elected Officials" (https://bit.ly/3xRfP1A)

- Charity Navigator (https://www.charitynavigator.org/)

Articles and Reports

- "When It Comes to Access to Clean Water, 'Race Is Still Strongest Determinant,' Report Says," by Nicole Acevedo, *NBC News*, November 27, 2019 (https://nbcnews.to/2ZU2TLS)

- "Watered Down Justice," by Kristi Pullen Fedinick et al., *NRDC*, March 27, 2020 (https://on.nrdc.org/3Dm2Upp)

LESSON 1 FACILITATION
Water Access

Launch (20 minutes)

- Introduce the lesson by sharing the following premise: *The nonprofit organization WATER³ wants students to help design water containers to bring water to people in need. During this series of lessons, you will work with a partner to design and then present a new water container.*

- Use the following questions to brainstorm ideas about why people may need water:

  + *Why might we need to deliver water to people?*

  + *Do you know of any examples of when this has happened?*

- Show the following videos to introduce students to the various challenges with accessing water. After each video, briefly pause and record what students notice, wonder, and feel in response.

  + *Access to clean water is a problem within the United States:* "Why America's Drinking Water Crisis Goes Beyond Flint," from BBC News (https://bbc.in/3Dl0WWk)

  + *A big-picture perspective:* "The Sustainable Development Goals Explained: Clean Water and Sanitation," from the United Nations (https://bit.ly/3rz7rCQ)

  + *A personalized perspective:* "Water Walk," from WaterAid, featuring a poem by Martin Kiszko (https://bit.ly/3DnUY7b)

- After watching the videos, ask students: *What do you notice about how water is transported to the homes?*

- Choose two or three facts from one of the following websites and do a simulation with the class where you figure out, if the class represented the world, how many classmates would not have access to clean drinking water or do not have a way to safely wash their hands at home.

  + United Nations, "Water Scarcity" (https://bit.ly/3EmccmR) (see the facts and figures to the right)

  + Centers for Disease Control and Prevention, "Global Water, Sanitation, & Hygiene (WASH) Fast Facts" (https://bit.ly/3xSiI25)

Explore (20 minutes)

- Explain to students that they will work in small groups to further investigate and discuss access to water and the human component to "natural" disasters. Distribute Worksheet 1 (*Is It Natural?*).

- **Note:** The excerpt from the article by Kendra Pierre-Louis ("There's Actually No Such Thing as a Natural Disaster," *Popular Science*, October 2, 2017, https://www.popsci.com/no-such-thing-as-natural-disaster) highlights the role of wealth and government policies but does not explicitly speak about race.

- As an extension activity, have students read and discuss the following two publications that highlight the role of race and racism when it comes to "natural" disasters and specifically access to safe drinking water:

  - "When It Comes to Access to Clean Water, 'Race Is Still Strongest Determinant,' Report Says," by Nicole Acevedo, *NBC News*, November 27, 2019 (https://nbcnews.to/2ZU2TLS)

  - *Watered Down Justice*, by Kristi Pullen Fedinick et al., *NRDC*, March 27, 2020 (https://on.nrdc.org/3Dm2Upp)

Summarize (20 minutes)

- Use the following prompts to lead a class discussion to summarize the main points from Worksheet 1 (*Is It Natural?*).

  - *According to the author, what is the difference between a hazard and a disaster?*

  - *Why does Kendra Pierre-Louis think natural disasters are not completely natural? What role does she think humans play in making them disasters?*

  - *Who does Kendra Pierre-Louis argue are most hurt by these disasters?*

  - *How do these ideas make you feel?*

  - *What do you agree with and disagree with in the article?*

  - *What could we do differently to lessen the impact of these disasters on the most vulnerable people?*

- **Optional** (connect to climate change): Watch the Yale Climate Connections video, "Why Climate Change Makes Storms Stronger" (https://bit.ly/3GhAfUB).

  - Ask students to share their feelings in reaction to the video and how this might increase the need for bringing water to people in need.

LESSONS 2–4 FACILITATION

Designing Water Bottles

Launch (20 minutes)

- Remind the class of the premise:

  - *The nonprofit organization WATER[3] wants their help in designing water containers to bring to people in need. While they could use exist-*

ing water containers, a manufacturing plant has recently reached out to WATER³ so they have an opportunity to design custom water containers that best fit WATER³'s needs. WATER³ has long-term efforts focused on improving water infrastructure (pipes, wells, sewer systems, etc.) for all people in the world. In the meantime, they also need effective containers to bring water to people in need.

- Have students recall some of the reasons people need access to water from the previous lesson's brainstorming activity. Remind students that people might need water in emergencies/disasters and due to chronic lack of access.

- Introduce the volume of water containers by showing students a base-10 block and explain that if it were empty it would hold 1 mL of water. Now show students a base-10 mega-block and ask: *How much water do you think it would hold?* (answer: 1 L of water).

- Say to students:

  For a warm up, we're going to look at some actual water containers and estimate how much each one holds. You probably used this before, it's a base-10 block. Each edge is 1 centimeter long, so it's a cubic centimeter. This means that if it were empty, it would hold 1 milliliter of water. And this is a base-10 mega-block. It's 10 blocks by 10 blocks by 10 blocks, so it's 1,000 blocks. That means it would hold 1,000 milliliters or 1 liter of water.

- Hold up other containers one at a time (cover or remove any volume information) and collect volume estimates for each container. Collect values that are too high, too low, and best estimates. Have students justify their reasoning as they share. Students should emphasize the reasoning behind their estimates, not getting the closest answer.

  Note: This works well in metric units (liters) as the two reference objects are in metric units. Here are some suggested objects:

    + A reusable 1-L water bottle

    + A 2-L soda bottle

    + A gallon jug (for milk or water, about 3.8 L)

    + A water cooler jug (usually 5 gallons or about 19 L)

- Use the following questions to have students brainstorm and discuss why humans need water:

    + *What do people need water for?*

    + *How much do you think they need every day?*

- Student responses will vary and can be categorized into the following three areas: People need water for basic survival, to help with cooking, and for hygiene.

- Now distribute Worksheet 2 (*Info Sheet*) and go over the information provided.

  + You may want to simplify the information and provide students with a single estimate for the amount of water each person needs per day.

  + Alternatively, consider leading a brief discussion and have the class decide as a whole before beginning to work on their own designs.

  Note: Simplifying the information may limit the number of decisions each pair of students has to consider in creating their own designs.

Explore (45–90 minutes)

- Distribute Worksheet 3 (*Water Containers: Individual Ideas*) and give students 10 minutes to work independently before sharing their ideas with a partner.

- Tell students that each person gets a chance to share uninterrupted, and then the other person can ask clarifying questions and discuss similarities and differences to their design. They will then have to decide on a design to build in 3D modeling software.

  + **Note:** If 3D modeling technology is not used, focus on the students' sketches. However, we felt that the use of 3D modeling software added a geometric visualization dimension to the task and helped students feel their designs were more "real" than they were during the initial planning stage.

- Distribute Worksheet 4 (*Water Containers: Plan*) to each pair of students and tell them that they must sketch out their plans ahead of time and discuss them with the teacher before they start to create the container. Use the following prompts when discussing design sketches with each pair of students:

  + Ask whether their values for questions 1–4 make sense (i.e., they have correctly computed the amount of water they need). This emphasizes aspects of mathematical modeling and design in which students must make real-world decisions and determine how they affect their mathematics (such as the amount of water needed).

  + Ask whether they have selected appropriate 3D shapes to approximate their design in the 3D modeling software. This emphasizes making mathematical models of the world in the area of geometry and knowledge of 3D shapes.

  + Remind students about how big the container will need to be. Remind them of the estimation activities from the warm up. If you want to emphasize computation of volume, then you may expand on the requirements here so that students must have specific dimensions and volume calculations to show their container will hold the amount of water needed.

- Remind students that they should take turns using the technology to build their design and have a brief discussion to determine which shape they need to create next and where to place it.

- As you circulate to monitor student progress and check for their understanding, remind students that they will need to present their container sketch to their classmates.

- Share the following with the whole class:

Presentations can take a number of formats, such as a poster with printouts for a gallery review or a slideshow with screenshots from the 3D modeling software used. The major focus should be on using your knowledge of the geometric concepts discussed in class and the context to provide an explanation for why you think this container will meet the needs of the people in need of water.

In an effort to use more inclusive language, we use the term *gallery review* for what is commonly referred to as a "gallery walk."

- Select several of the following guidelines for students to include in their presentations:

  + The number of people each container is for

  + How it will be used (survival, hygiene, and/or cooking)

  + How long they expect it to last

  + Dimensions and volume calculations for the amount of water held

  + How it will be used (e.g., Is it a backpack they wear? Do they set it up in their home/dwelling as a central water source? Is it delivered by truck?)

  + Any special design features

Summarize (45 minutes)

- Have students share their final products either as a gallery review or through short presentations. Tell students that as they listen to their classmates, they need to provide written or verbal feedback. Provide students with sentence starters such as these:

  + *What about . . . ?*

  + *Did you think about this situation: . . . ?*

  + *I really like this part of your design because . . .*

  + *That is similar to something we did because . . .*

- Lead a class discussion using the following prompts and then ask students to respond to each on an exit slip or in their mathematics journals:

  + *What math thinking did you do in this unit?*

  + *What did you learn about using math in the real world?*

  + *What did you learn about access to clean water?*

  + *What can we do to address this problem in the future?*

TAKING ACTION

- Once students complete this series of lessons, they can share knowledge gained about water access. This could take a variety of formats, including sending newsletters home to families and to the school community, and more broadly in the form of stories in local newspapers. Consider having students contact elected officials to encourage them to take action regarding access to safe water both domestically and internationally. The following guide provides a letter template to support students in their writing: Connecticut Democracy Center, "Writing to Your Elected Officials" (https://bit.ly/3xRfP1A).

- Students can also research nonprofit organizations related to clean drinking water and then have a class discussion about possible organizations to hold a fundraising event. The discussion should focus on the location of the group (local, national, and/or international) and any background information or values behind the group. The Charity Navigator website (https://www .charitynavigator.org/) provides some information about charitable groups and whether they focus on direct aid, policy change, or both.

ONLINE RESOURCES

 Available for download at **resources.corwin/TMSJ-MiddleSchool**

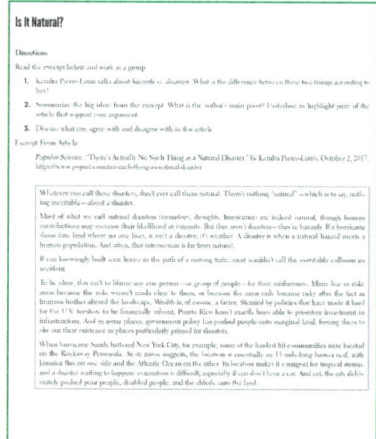

Worksheet 1: Is It Natural?

Worksheet 2: Info Sheet

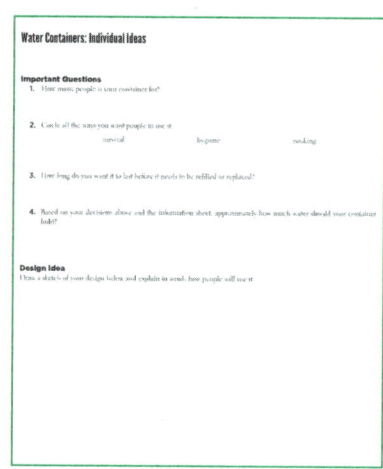

Worksheet 3: Water Containers: Individual Ideas

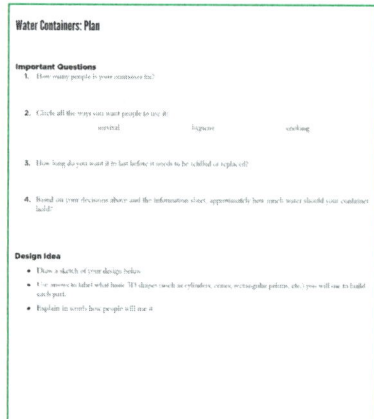

Worksheet 4: Water Containers: Plan

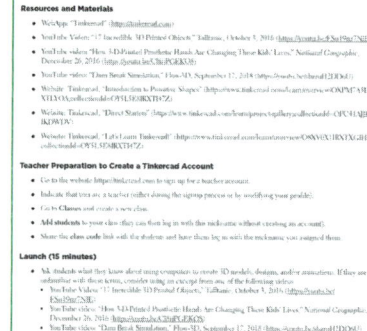

Teacher Resource 1: Introduction to Tinkercad

LESSON 10.3 WATER IS LIFE—OUR COLLECTIVE PAST, PRESENT, AND FUTURE

Sara Rezvi and Tyrone Martinez-Black

ENVIRONMENTAL ISSUES

We all rely on water to live. Water is life in many ways. But what happens when water is stolen? What happens when water is deliberately polluted by corporations? In this lesson, we highlight the importance of water on the individual level, on the community level, and on the actionable level as protectors of water.

A Note on a Healing-Centered Approach for Learners

It is important to recognize that conversations about land/territory and people's belonging may potentially trigger traumas or discomforts for immigrant and indigenous children. We encourage you to refer to Chapter 4 (see "Dealing With Emotional or Sensitive Conversations") for guidance on how to respond if children need social and emotional support during this lesson.

DEEP AND RICH MATHEMATICS

This lesson uses ratio and proportional reasoning and geometry to explore issues related to water usage. Students will use their mathematical reasoning and inference skills to calculate the rate of water in gallons per minute. They will estimate the amount of water usage per student in their respective homes. Students will also read map legends and make inferences about what map legends identify. Students will use estimation, inference, and analysis skills to determine the likelihoods of lead levels in homes in Flint, Michigan, based on adjacent models. In the final lesson, students will consider how we calculate map radii to determine, geographically, where and how we get our water.

- **Geometry— Grade 8:** Solve real-world and mathematical problems involving volume of cylinders, cones, and spheres.

- **Ratios and Proportions—** Grade 6: Understand ratio concepts and use ratio reasoning to solve problems.

- **Ratios and Proportions—** Grade 7: Analyze proportional relationships and use them to solve real-world and mathematical problems.

MATHEMATICS PRACTICES

- Make sense of problems and persevere in solving them.

- Reason abstractly and quantitatively.

- Construct viable arguments and critique the reasoning of others.

- Use appropriate tools strategically.

ABOUT THE LESSON

This lesson uses a launch–explore–summarize instructional model and is intended to take approximately 180 minutes to complete across three class periods.

Lesson 1: Students explore their own self-connection to water using estimation, reasoning, and inquiry to calculate water flow in gallons per minute and estimate how much water they use in their respective households.

Lesson 2: Students use the area (both rectilinear and circular) of residential regions to reason about the availability of clean/ potable water, make conjectures about the economic value of water, and consider what happens when communities are denied access to water.

Lesson 3: Students explore measurement of pollution to consider how communities are impacted by damage to an oil pipeline near their water sources.

Resources and Materials
- PowerPoint resource to support lesson facilitation

- Computers with internet access (1 for every 2 students)

- Worksheet 1: *My Water Usage* (1 sheet per student)

- Worksheet 2: *Exploring Maps and Water in Flint, Michigan* (1 sheet per student)

- Student Resource 1: *Exploring Maps* (1 for every 2 students)

Videos (for Teacher to Show)
- "Fan Video" from Kyle Alex Brett, featuring Beyoncé and Kendrick Lamar's song, "Nile" (https://bit.ly/31iFbty)

- "Water Is Life," music video from Will Evans (https://bit .ly/32YKWwT)

- "Water Tank," by Dan Meyer, WCYDWT (https://bit .ly/3ojpSJw)

- "5 Years Into Water Crisis, Little Miss Flint Hasn't Given Up," featuring Mari Copeny, *ABC News* (https://abcn .ws/3IsI6k2) (optional)

- "Flint's Water Crisis, Explained in 3 Minutes," explained by Joe Posner, *Vox* (https://bit.ly/3rzsVQ2)

Websites

- U.S. Geological Survey Online Calculator, "How Much Water Do You Use at Home?" (https://on.doi.gov/3IiDSLB) (1 link per student)

- Flint Service Line Map–Interactive (https://flintpipemap.org/map)

- Copies or digital access to these news articles (1 for each student):

  + "Treaties Still Matter—The Dakota Access Pipeline," *Native Knowledge 360°* (https://s.si.edu/3dnUOSx)

  + "Clean Water and Sanitation," United Nations (https://bit.ly/3xR8nmW)

Teacher Preparation/Background Articles

- "Flint Water Crisis: Everything You Need to Know," by Melissa Denchak, NRDC (https://on.nrdc.org/3ojtlHX)

- "Flint Water Crisis—Fast Facts," *CNN* (https://cnn.it/3IdY8hD)

- "Flint Map Shows Progress, Reveals Where Lead Likely Remains," by Stacy Woods, *NRDC* (https://on.nrdc.org/3on5eIo)

LESSON 1 FACILITATION

Self-Connection to Water

Students explore the themes of their own self-connection to water. Using estimation, reasoning, and inquiry, students calculate water flow in gallons per minute and estimate how much water they use in their respective households.

Set Up

- Tell students that in preparation for the lesson, they will need to gather information about the flow rate of water in their home (see slide 2 in Teacher Resource 1).

- Have students brainstorm ways they can gather information to determine the water **flow rate**. Ask: *How would I go about figuring this out?*

- Distribute Worksheet 1 (*My Water Usage, Home Assignment*) and conclude the discussion by suggesting that students take the following steps in preparation for the next class.

Flow rate: a measure of how many gallons of water per minute could potentially come out of a faucet.

+ Have students gather data for three faucets at home or in school (if necessary) by doing the following:

1. Select a container that has ounce measures (e.g., measuring cup, pint jar). Record the number of ounces for the container.

2. Turn on the water faucet (e.g., kitchen sink, bathtub, shower).

3. Time how long it takes to fill the container using a stopwatch or phone (in seconds).

4. Record the number of seconds it takes to fill the container.

5. Repeat for two water faucets.

Launch (15 minutes)

- Arrange students in groups of three to four.

- Have students complete a free write to respond to the following prompt: *What is your earliest memory of water? What is your connection to it?*

- Ask students: *What are some questions that surface as you're watching this video?*

- Show students "Fan Video" by Kyle Alex Brett featuring Beyoncé and Kendrick Lamar's song "Nile" (https://bit.ly/31iFbty).

  Here are some possible student questions:

  + *How do we use water?*

  + *What do we use it for?*

  + *Where does it go?*

  + *Who has access to water?*

  + *Who doesn't? Why?*

  + *Why are they attacking Black people in the video?*

  + *What is happening here?*

  + *Are they aiming the firehose at Black people?*

  + *When is that video from?*

- Use the following questions to facilitate small-group and class discussion:

  + *How is water described in your language or culture?*

  + *How is water used for hygiene and cleanliness in your community or family?*

  + *How is water used for spiritual or religious practices?* (e.g., Wudu or ablution for Muslims before ritual prayer daily, Baptism ceremonies, etc.)

+ *Why might someone need to have a sponge bath rather than an upright shower?* (attending to questions around ableism)

+ *What are some words for water you know?*

• Say: *Today, we are exploring our connections to water. I'm going to ask you to do some calculating and reflecting about water.*

• Use the following questions to begin a brief discussion:

+ *What is my personal connection to water?*

+ *Do I have to think about water as a social justice issue? Why or why not?*

Explore (20 minutes)

• Show students the first few minutes of the WCYDWT video, "Water Tank" (https://bit.ly/3ojpSJw), of a hose filling up a water tank (slide 7 in Teacher Resource 1).

• Use the following questions to facilitate a discussion:

+ *How do we calculate the flow of water?*

+ *What information do you need to calculate how fast (in gallons per minute) the water is flowing into the tank?*

Here are some potential student responses:

+ Volume of the tank because we need to know how many gallons fit in it

+ How long will it take for the tank to fill up?

• Gather estimates from students for the volume and time to fill the tank.

• Ask: *What variables do you have to account for? What variables can you ignore? How will you calculate this?*

• Highlight the following concepts during the discussion: *estimation, analysis, calculation, hypothesis,* and *correction.*

• Have students complete Part 1 of Worksheet 1 (*My Water Usage*) to calculate the water flow rate for filling up the tank (slide 10 in Teacher Resource 1).

+ The tank capacity is 6.5 gallons.

+ The time to fill the tank in the video is 8 minutes and 25 seconds.

• Have students complete Part 2 of Worksheet 1 to calculate their water flow rate for the faucets they gathered data from at home and compare in their groups (slide 11 in Teacher Resource 1). Tell students to use the following instructions for measuring and calculating the water flow rate at a faucet or shower.

Take the opportunity to help reinforce language around these concepts and allow for some sensemaking in home languages if these words are being introduced for the first time. Some questions you can ask include but are not limited to how students describe water in their home language in its various states (solid, liquid, gas) and how culturally or spiritually their communities might be connected to water (e.g., the Ganges River, the holy waters of Zamzam in Islamic history).

1. Select a container that has ounce measures (e.g., measuring cup, pint jar).

2. Turn on the water.

3. Time how long it takes to fill the container using your stopwatch or phone (in seconds).

4. Divide 60 by the number of seconds it took to fill your container (e.g., if it took 30 seconds to fill, then $\frac{60}{30} = 2$ gallons per minute).

5. The result is the flow rate of that application in gallons per minute.

- Ask: *How much water do you use in your own home? How would you go about figuring this out?*

- Tell students to complete Part 3 of Worksheet 1 to determine their water usage at home and for their family.

- Have students log on to the U.S. Geological Survey website, "How Much Water Do You Use at Home?" (https://on.doi.gov/3IiDSLB), to get a ballpark estimation of how much water they are using in their homes. Tell students to record the following:

  + Their total estimated water usage

  + Amount of water needed to take a bath

  + One "tip" they can try to save water

- Use the following questions to monitor students' understanding as they complete the task of determining their water usage:

  + *Are you surprised by the number you calculated? Why or why not?*

  + *What "tip" are you considering, and about how much water do you think it will save?*

- Use the following question to facilitate a brief class discussion:

  + *What does estimating help us figure out? What are the limitations of estimation?*

  + *What are some questions you could ask to make the estimation more precise?*

- As students are responding to the questions, capture responses on the board with student attribution next to their thoughts. The purpose of this is to continue facilitating the dialogue by giving students the ability to reference and build upon each other's thoughts.

Summarize (15 minutes)

- Ask students to write a summary of one connection to water they've explored for themselves during the lesson. Select a few students to share their summary.

- Use the following prompts to summarize and connect the lesson activities:

  + *Based on the video we watched at the beginning of this lesson, what is one way access to water is a social justice issue? What are some thoughts you have on ensuring water is accessible to all people?*

  + *What does Kendrick Lamar mean when he sings, "One time I took a swim in the Nile. I swam the whole way, I didn't turn around. Man, I swear. It made me relax when I came down. I felt liberated like free birds, I'm stimulated now." Have you ever felt free or liberated in water? If so, when was that? What was that like for you?*

LESSON 2 FACILITATION

Community Connection to Water

Students extend from their understanding of water use to consider what happens when water is denied, making a connection between community and water. Using the area (both rectilinear and circular) of residential regions, students reason about the availability of clean/potable water and make conjectures about the economic value of water.

Launch (10 minutes)

- Begin the lesson by showing a picture of protesters (slide 15 in Teacher Resource 1) and ask students: *Do you think water is a human right? Why or why not? What makes something a human right?*

  + Please note that students will look at the United Nations (UN) Sustainable Development goals in Lesson 3. These intro questions preview this in advance of the lesson.

- Tell students:

  Today we are exploring our connections to water and community. In particular, we'll use mathematics to explore the ongoing water crisis in two cities in Michigan. Now take out your work from the first lesson, where you estimated daily water usage and water usage for taking a bath. You will use this information as a comparison point during the lesson. First, I want you to compare it to the reality of this situation from Detroit in 2014.

- Show slide 16 as you share with students the following situation from Detroit in 2014:

"You're looking at mothers who are having to empty a case of water into a bathtub . . . They heat that water up, empty it into a bathtub. That means several trips from the kitchen to the bathroom. And if she's got three children, she's bathing the baby first, then the second child, then the oldest takes the bath in that same water."

(*Source:* "'It's Just Despair': Many Americans Face Coronavirus With No Water to Wash Their Hands," by Valaria Griffin, *NBC News*, March 26, 2020 (https://www.nbcnews.com/news/us-news/it-s-just-despair-many-americans-face-coronavirus-no-water-n1169351)

- Use the following questions to generate deeper understanding of the water issues in Detroit:

  + *How much water do you think this mother needed to bathe her children?*

  + *What do you think it would feel like to do this day after day?*

  + *How much water does this mother need in 1 week to bathe her children? One month? One year?*

Explore (40 minutes)

- Say, *Now we will explore the water issues in Flint, Michigan. As you watch this brief video, capture any numbers being reported in the story.*

- Show the video, "Flint's Water Crisis, Explained in 3 Minutes" (https://bit.ly/3rzsVQ2).

  + **Teacher Note:** More background on the Flint water crisis can be found in these articles:

    - "Flint Water Crisis: Everything You Need to Know," NRDC (https://bit.ly/3rzsVQ2)

    - "Flint Water Crisis—Fast Facts," *CNN* (https://cnn.it/3IdY8hD)

- Introduce students to the online Flint Service Line Map (https://flintpipe-map.org/map) and allow time for them to explore the tool. Ask students: *What are you noticing and wondering here? What are some questions you have?* (5–7 minutes)

- Record students' noticings and wonderings and encourage them to look back at the list as they are working on the next activity.

- Distribute Worksheet 2 (*Exploring Maps and Water in Flint, Michigan*) and Student Resource 1 (*Exploring Maps*) as you tell students that they will work with a small group to complete the task. (15–20 minutes)

- **Teacher Note:** You can find more background on this map in the NRDC article, "Flint Map Shows Progress, Reveals Where Lead Likely Remains," by Stacy Woods (https://on.nrdc.org/3on5eIo).

- Use the following questions to monitor and check students' understanding:

  - *What is happening in Flint, Michigan?*

  - *How long has the Flint crisis been occurring?*

  - *How do we calculate map radii to determine geographically where and how we get our water?*

  - *What is our community connection to water?*

  - *How does that connect back to me and the way I use and think about water?*

- Show PowerPoint slides 17–19 from Teacher Resource 1 as you use purposeful selection, sequencing, and connecting techniques to call on groups to share their responses as students are completing the worksheet. Use the following prompts to guide whole-group discussion of students' responses to the worksheet questions. (10–15 minutes)

  - Questions 2 and 3: What different ways did students approach the problem for question 2? What measurement tools did students use (rectilinear or circular) to address the second question? Why?

    Question 3: Have students share out what is missing/obscured in the graph presented by Governor Rick Snyder. What motivations would he have for presenting this graph as truth? What additional questions do you have now that we're thinking about this graph in more detail?

    - **Teacher Note:** From Chojnacki et al. (2017): "Governor Rick Snyder announcing improved Flint water testing results on April 22, 2016. The tweet was quickly deleted, and results then worsened over the summer. Note that the plotted points of the line do not correspond to the y-axis labels, and x-axis is not linear in time" (p. 5).

- Before transitioning to the summary, pose the following questions:

  - *What is the cost of human greed?*

  - *How has human greed impacted the most vulnerable in both Detroit and Flint, Michigan?*

Summarize (5 minutes)

- **Option 1:** Ask students to write a summary of one connection to water they've explored for themselves during the lesson. Select a few students to share their summary.

- **Option 2:** Show the video on Mari Copeny, "5 Years Into Water Crisis, Little Miss Flint Hasn't Given Up" (https://abcn.ws/3IsI6k2) (slide 20).

- Then ask students: *What actions did she take? Why?*

- Assign students the following independent activity/assignment (slide 21).

  + Family Water Interview Assignment: What happens when community water is polluted? Interview another teacher, family member, or community member about their connections to water, if they remember being around polluted water, and what was done about it in their own respective lifetimes.

- Tell students that during the next lesson, they will explore ways they can begin to apply what they are learning.

LESSON 3 FACILITATION

Protecting Our Connections to Water

Students build on notions of water availability to the cultural/biological health associated with water. Starting with distances from certain regions to affected populations, students explore measurement of pollution to consider how communities would be impacted by damage to an oil pipeline near their water sources.

Launch (10 minutes)

- Tell students that during today's lesson, they will explore how governmental decisions can impact a community's access to water and how those in the community responded.

- Begin the lesson by showing students the "Water Is Life" images (slide 23). Ask: *What do you notice about the image? What connections is the image making about water and living beings?* (3 minutes)

- Select several students to share their reflections. Capture their thoughts on a board with student attribution so that students can build, connect, and expand on each other's commentary. (5 minutes)

Explore (35 minutes)

- Have students read page 1 of "Clean Water and Sanitation: Why It Matters" (https://bit.ly/3xR8nmW), a UN Sustainable Development data/ infographic to reinforce the global perspective on the importance of water.

- Ask: *What would you do if your water were jeopardized?*

- Provide students digital access or print copies of "Treaties Still Matter" (https://s.si.edu/3dnUOSx) and tell them to read the introduction and "Opposing Perspectives."

- Have students compare the two maps under "Opposing Perspectives" and discuss in their groups the following questions:

  + *How do the ratios in the "Why It Matters" document influence how you read those maps?*

  + *What other ratios would be significant in making a case for the placement of the pipeline?*

  Here are some possible student answers:

  + People per square mile

  + Number of people per distance from pipeline

  + Area of potential spills versus distance to nearest population

  + White people per Sioux People

  + Sioux land per colonized land

- Ask:

  + *How do we calculate map radii to determine geographically where and how we get our water? How does money factor into this?*

  + *What is the distance between the projected pipeline and the water supply?*

  + *What happens when the pollution is intentional?*

Summarize (10 minutes)

- Tell students that during the lessons, we have seen how communities responded to water issues. Now we will take a look at how the community responded to the Dakota Access Pipeline. Show the video "Water Is Life" by Will Evans (https://bit.ly/32YKWwT) (slide 24).

- Use the following questions to facilitate discussion among students:

  + *What is one protection you can offer to future generations affected by water damage?*

  + *How might personal connections to place, family, and community inspire an individual to take informed action?*

  + *What does water access look like in your local communities?*

  + *Who has access to clean water? How do you know?*

TAKING ACTION

Encourage students to consider the following activities:

- **Information Campaign:** Prepare an information campaign for community members about local water issues impacting the area. Consider the following questions: Who has clean water? Who can pay for clean water? Is that guaranteed for the future? Why or why not?

- **Journalism:** Connect with a local environmental scientist (such as a biologist, environmentalist, naturalist, etc.) and interview them about the histories of the local water issues within the community. Publish this story in the local newspaper.

- **Stakeholder Interview:** What happens when community water is polluted? Interview another teacher, family member, or community member about their connections to water, if they remember being around polluted water, and what was done about it in their own respective lifetimes.

- **Political Activism:** Like Mari Copeny (Little Miss Flint), what are some actionable steps students can take within their communities to ensure that water is a human right for all (legislative letter writing, community organizing, teach-ins, etc.)?

ONLINE RESOURCES

 Available for download at **resources.corwin/TMSJ-MiddleSchool**

How much water do I use in my own home?

How do I figure out the flow rate of water?

Instructions for measuring the water flow rate at a faucet or shower:
1. Select a container that has ounce measures (e.g., measuring cup, pint jar). Record the number of ounces for the container.
2. Turn on the water faucet (e.g., kitchen sink, bath tub, shower).
3. Time how long it takes to fill the container using a stopwatch or phone (in seconds).

- Record the number of seconds it takes to fill the container.
- Repeat for two other water faucets.

https://www.e-tankless.com/measuring-flow-rate.php

PowerPoint resource to support lesson facilitation

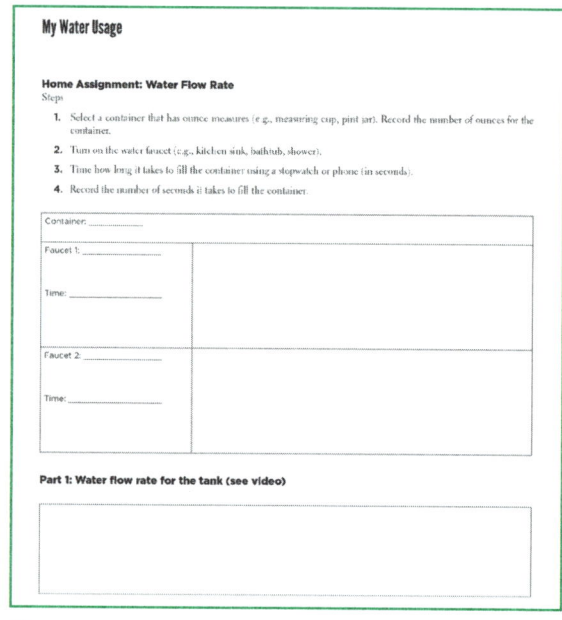

Worksheet 1: My Water Usage

Exploring Maps and Water in Flint, Michigan

1. What can you infer about the map legend here? In particular, what can you tell about the difference between homes in Flint having copper versus lead piping?

2. You are an environmental scientist trying to determine, with **mathematical modeling**, where homes in Flint, Michigan might have elevated lead in their water. You are looking at this section of a Flint map attempting to predict where you could color in homes with either high or known quantities of lead. Based on the image above, where would you begin? Can you explain your reasoning?

3. On April 22, 2016, the then-governor of Michigan, Rick Snyder, tweeted out this graph on Twitter. The tweet was quickly deleted. Can you come up with some mathematical issues with why this tweet may have been deleted so quickly?

Part 4: Family Water Interview (use notebook paper if additional space is needed)

▲

Worksheet 2: Exploring Maps and Water in Flint, Michigan

Exploring Maps

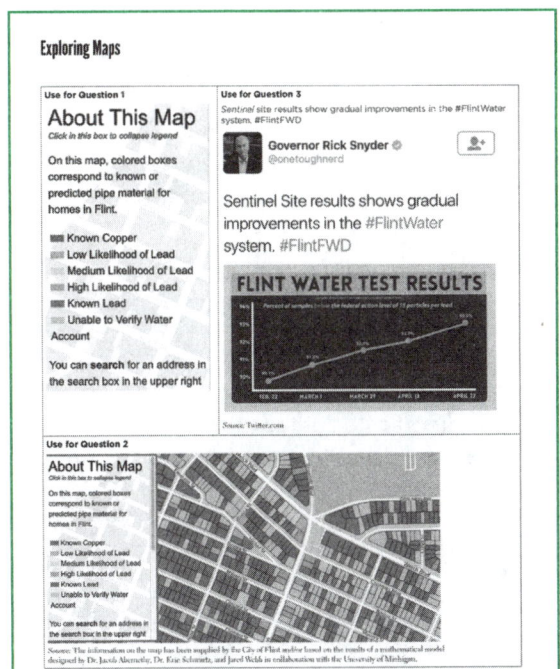

▲

Student Resource 1: Exploring Maps

- I am curious and want to know more about other people's histories and lived experiences, and I ask questions respectfully and listen carefully and non-judgmentally. (Diversity 8)

- I can recognize and describe unfairness and injustice in many forms including attitudes, speech, behaviors, practices, and laws. (Justice 12)

- I will work with friends, family and community members to make our world fairer for everyone, and we will plan and coordinate our actions in order to achieve our goals. (Action 20)

ESSENTIAL MIDDLE GRADES CONCEPTS

- **Geometry—** Grade 6: Solve real-world and mathematical problems involving area, surface area, and volume.

- **Geometry—** Grade 7: Draw, construct, and describe geometrical figures, and describe the relationships between them.

MATHEMATICS PRACTICES

- Construct viable arguments and critique the reasoning of others.

- Model with mathematics.

- Use appropriate tools strategically.

- Attend to precision.

LESSON 10.4 ACCESSIBLE PLAYGROUND

Maggie Lee McHugh and Jennifer Kosiak

ABILITY

How can we use mathematics to develop solutions to the lack of playgrounds for children with disabilities? According to the U.S. Census Bureau (2020), approximately 3 million children in the United States have a disability, yet most parks and playgrounds are not designed with these children in mind. During this lesson, students will develop empathy as they examine the concept of ableism, including the stereotypes that children with disabilities (physical, visual, and/or cognitive) encounter in their lived experiences. Through the lens of designing an accessible playground, students will be empowered to take action at the local or state level.

DEEP AND RICH MATHEMATICS

Students research accessible playground equipment before designing blueprints for an accessible playground by plotting geometric figures on the coordinate plane. As they design their blueprints, students will extend their understanding of elementary geometry and measurement concepts by applying multiplicative reasoning to determine how scale factor impacts the lengths and area of geometric figures. Students will also use their mathematical voice to make a recommendation to local or state officials regarding building an accessible playground.

ABOUT THE LESSON

This lesson uses a launch–explore–summarize instructional model and is intended to take 180 minutes to complete across three class periods.

Lesson 1: Students explore the idea of accessibility by connecting their playground experiences with the recognition that not all children have the opportunity to play. Students research items for their accessible playground, focusing on estimating dimensions in comparison to their classroom.

Lesson 2: Students learn about Americans with Disabilities Act (ADA) compliance and practice advocating for accessible playgrounds, and therefore children with disabilities, increasing student capacity to be change agents. Students apply their understanding of scale to determine the area of a composite shape that represents an accessible playground with equipment shown in a scaled drawing.

Lesson 3: Students formalize a definition of accessibility. Students apply their knowledge of plotting geometric figures in the coordinate plane to create a blueprint of their accessible playground, comparing scale area to actual area.

Resources and Materials

- Worksheet 1: *Research Template* (shared Google Doc or 1 set per Playground Team)

- Worksheet 2: *Math for All Playground* (shared Google Doc or 1 copy per student)

- Posterboard (1 per Playground Team)

- Grid paper, assorted sizes (1 cm, $\frac{1}{2}$ inch, $\frac{1}{4}$ inch)

- Markers, sharpies, or colored pencils

- Rulers

- Scissors

- Tape or glue

Videos

- "Inclusive Playground for Children With Disabilities Opens in Oakland County," by Priya Mann and Dane Kelly, *ClickOn Detroit*, June 5, 2020 (https://bit.ly/3pj4PWI)

- "A Playground for Everyone, No Matter Your Age or Ability," by Megan Thompson and Melanie Saltzman, *PBS NewsHour*, October 6, 2019 (https://to.pbs.org/3rzcwuP)

Commercial Websites

- AAA State of Play, "Single Seat Swing Platform With Frame" (https://bit.ly/3lzVEAq)

- AAA State of Play, "Playground Equipment, Handicap Accessible" (https://bit.ly/3dmEt0o)

- Playground Outfitters, "Wheelchair Accessible" (https://bit.ly/3pisxTa)

LESSON 1 FACILITATION

Accessible Playgrounds and Area

Launch (20 minutes)

- **Option 1:** Take students outside to a nearby playground if possible.

- **Option 2:** Project pictures of a local playground from multiple viewpoints. Have students think about other playgrounds they have played on.

- Ask students to respond verbally or in writing to this question: *What do you notice?*

- Make this statement: *In 1959, the United Nations declared that all children have the right to play. Given this information, what do you wonder about as you look at this playground?*

- After looking at the playground or photos of the playground, use the following questions to promote classroom discussion:

    + *Who has access to play on this playground? Who does not?*

    + *What does accessibility mean?* (broaden the definition beyond a person in a wheelchair)

    + *Where do you see accessible features in our society?* (ramps, doors, braille, large print, audio recordings, etc.)

    + *Do you see accessible items in this playground? What items could be included to be more accessible?*

- Highlight the importance of using people-first language when talking about children with disabilities (versus a disabled or handicapped child). Ask students to listen for that language in the video.

- Show a video about accessible playgrounds in the news:

    + "Inclusive Playground for Children With Disabilities Opens in Oakland County" by Priya Mann and Dane Kelly, *ClickOn Detroit,* June 5, 2020 (https://bit.ly/3pj4PWI)

    + "A Playground for Everyone, No Matter Your Age or Ability," by Megan Thompson and Melanie Saltzman, *PBS NewsHour,* October 6, 2019 (https://to.pbs.org/3rzcwuP)

- Ask students to reflect on the video using the following sentence stems:

    + *When watching the video, I was intrigued by . . .*

    + *One thing I noticed was . . .*

    + *Before today, I never thought about . . .*

    + *I wonder . . .*

- Ask students: *What might be some barriers to creating more accessible playgrounds?*

Explore (30 minutes)

- Select an item that could be part of an accessible playground. For example, show the picture of the Single Seat Swing Platform With Frame (https://bit.ly/3lzVEAq).

- Ask students to estimate the dimensions (length and width) of the swing. Use the website to discuss the words *dimensions*, *footprint*, and *use zone* in relation to playground equipment.

- Have students turn and talk.

    + *What's the difference between the footprint and use zone?*

        - *Footprint:* the dimensions of a playground item *if* it were rectangular

        - *Use Zone:* the dimensions of the footprint plus additional length on all four sides to ensure proper space between playground items in case a child falls from or jumps off of the equipment

    + *Why would it be important to know the use zone?*

- Form Playground Teams of three to four students each. Each individual student will research and find at least one commercial playground item to include in the team's accessible playground.

    Here are some potential websites for students to research:

    + AAA State of Play, "Playground Equipment, Handicap Accessible" (https://bit.ly/3dmEt0o)

    + Playground Outfitters, "Wheelchair Accessible" (https://bit.ly/3pisxTa)

    + Playground Outfitters, "Special-Needs Playground Equipment" (https://bit.ly/3djidog)

    Note: Many sites do not include the footprint or cost, so students would need to contact a vendor. Consider contacting vendors and requesting digital catalogs with the appropriate information prior to these lessons.

- Students include their items on the shared *Research Template* in Worksheet 1.

Summarize (15 minutes)

- After researching their items, student teams share their findings with the entire class. Use the following questions to promote discussion:

    + *Why did you choose this item?*

    + *Who has access to play or use this item? Is anyone not able to use the item?*

    + *Using our classroom, can you estimate the dimensions or footprint of this item? What about the use zone?* (e.g., about $\frac{1}{2}$ of the classroom, two classrooms, just a corner of the classroom)

LESSON 2 FACILITATION

ADA and Scale

Launch (15 minutes)

- Use the Inside-Outside Circles strategy for this activity.

    + Create two concentric circles, one "inside" the other.

    + As you pose each statement, determine if the "inside" or "outside" partner will go first.

    + After each round, move the "inside" or "outside" circle by one or two steps to the right or left.

- Help students develop advocacy skills by asking partners to respond to each of the following statements. Say: *Some people might argue against building an accessible playground. How might you respond to each of these statements?*

    + Accessible playgrounds may not appeal to all children.

    + Accessible playgrounds might separate children with disabilities from other children.

    + Accessible playgrounds do not provide enough challenge for older children who may get bored.

    + Accessible playgrounds may not reach the needs of all children with disabilities, such as children with autism spectrum disorder or children with visual impairment.

    + Accessible playgrounds are much more expensive and take up more space than regular playgrounds.

- Provide students with brief information about the Americans with Disabilities Act (ADA) of 1990.

- Discuss the following ideas:

    + *What is the Americans with Disabilities Act (ADA) of 1990?*

    + *What does it mean to be ADA compliant?*

    + *Why do you think ADA compliance would relate to playgrounds?*

    + *Who benefits from ADA compliance? Who is not included in ADA compliance?*

> **Brief Info About the ADA**
>
> The Americans with Disabilities Act (ADA) of 1990 is a civil rights law that made it illegal to discriminate against people with disabilities. The ADA requires that newly constructed or renovated government or public buildings and public spaces be designed with accessibility in mind. In 2012, the U.S. Department of Justice put new standards forward, including creating accessible playgrounds and parks for both children with disabilities and caregivers with disabilities. Some of the new guidelines include the following:
>
> - Sandboxes must have an accessible route or ramp to the border of the box.
>
> - Ramps to higher levels must use accessible routes.
>
> - Play areas must include accessible ground-level play pieces such as swings, slides, and spring rockers.

Explore (30 minutes)

- Activate prior knowledge by asking students the following questions:

    + *What do you know about scale?*

    + *Have you ever seen a scale on a map or blueprint?*

    + *What is the function or role of a scale?*

- Provide students with a copy of Worksheet 2 (*Math for All Playground*). Prompt students to think about the following:

    + *What do you notice about the playground item?*

    + *What accessible features are apparent to you?*

    + *Who has access? Is anyone left out?*

- Ask students to predict what would be a good scale for the Math for All playground equipment. Discuss ideas; for example, a wheelchair ramp should be at least 3 feet in width.

- Tell students that the scale used for this blueprint was 1 scaled unit, which is the same as $2\frac{1}{2}$ feet (or 3 feet or $\frac{3}{4}$ meter; choose a scale that will challenge students appropriately).

- Have students work in mixed groups of three to determine the following:

    + Side lengths for horizontal and vertical pieces

    + Total length and width of the playground if it were a rectangle (footprint)

    + Total area of the playground equipment

- Also have students estimate the dimensions of a rectangular use zone (at least 4 feet longer on all sides) and the area of the use zone.

- Monitor student progress throughout the activity to select and sequence the presentations of the Math for All playground task.

Summarize (15 minutes)

- In Playground Teams, have students look at the items selected for their playground. Have students consider their largest items as well as the smallest items. Have students consider space needed between items.

- Student teams will determine a scale for their playground blueprint given that the entire blueprint will need to fit on a 17×22-inch sheet of paper (i.e., four sheets of 8.5×11-inch paper that form a rectangle with centimeter grid paper, $\frac{1}{2}$-inch grid paper, or $\frac{1}{4}$-inch grid paper). Teams must justify their reasoning.

- Consider asking questions such as these:

  + *What influenced you to choose your scale?*

  + *If you were to change your scale to ___, how would that impact your blueprint?*

LESSON 3 FACILITATION

Playground Design and Coordinate Plane

Launch (5 minutes)

- Guide students to formalize a definition of *accessibility*. What does accessibility mean in relation to children and playgrounds?

- Lead students to discuss other features that might be included to increase accessibility at their playground. This may include accessible picnic tables, accessible benches, walkways with ramps, rubberized turf (versus mulch), and so on.

Explore (45 minutes)

- Recalling the mathematical concepts of area and scale, have students estimate the size of objects for their playground by cutting paper to the approximate size and shape of the playground equipment. Tell Playground Teams to begin brainstorming the layout for their blueprint on paper. Playground Teams may consider adding some of the other features discussed in the Launch.

- Partner Playground Teams together to allow for feedback on their overall design and accessibility features. Provide sentence stems to facilitate feedback:

    + *What would happen if you . . .*

    + *One suggestion would be . . .*

    + *Perhaps you could redo this part in order to . . .*

    + *Something that might make this stronger is to . . .*

    + *It might be useful to consider . . .*

- Provide Playground Teams the items to create their blueprint, including grid paper (centimeters, $\frac{1}{2}$ inch, or $\frac{1}{4}$ inch); poster board; tape; rulers; sharpies; markers; and so on.

- Playground Teams assemble their items to create a coordinate plane (using the four pieces of grid paper taped onto the poster board as each of the four quadrants). Using their brainstorm as a guide, students draw the items for their playground onto the coordinate plane, placing at least one item in each quadrant.

- Students list the following information on the poster board:

    + Title of the playground

    + Three to four sentences explaining how the playground meets the needs of all children using the definition of accessibility created during the Launch

    + Coordinates of the vertices for each playground item

    + Chosen scale

    + Scaled dimensions and area of each item

    + Short reflection on how their scaled dimensions and scaled area relate to the actual dimensions and area of each item

Summarize (10 minutes)

- Playground Teams will share their blueprints for other students to explore via a gallery review.

- During the gallery review, peers will complete feedback via the Glows and Grows strategy, writing one glow and one grow on a sticky note.

    + Share one "glow" (or strength) of the Playground Team's blueprint.

    + Share one "grow" (or area of improvement) for the Playground Team's blueprint.

In an effort to use more inclusive language, we use the term *gallery review* for what is commonly referred to as a "gallery walk."

TAKING ACTION

- Before embarking on this series of lessons, examine the parks and playgrounds in your local area, including playgrounds at local schools. Consider having students create a blueprint for a local playground that is not accessible.

- Students can use their blueprints to write advocacy letters to the local parks and recreation department or a principal of a local elementary school about (re)designing a playground to be accessible and compliant with ADA recommendations. Students could include blueprints of their design and possibly a comprehensive budget (including accessible turf) to enhance their position.

- Consider inviting adults and/or children (if appropriate) with various disabilities to share their lived experiences. Challenge students to design a playground specifically for the various guest speakers to enhance their lived experiences. Guide students to move from having empathy for the other to being advocates for change with these guest speakers. (Be careful not to instill a "savior" mentality in which people who are able-bodied must solely advocate for people with disabilities.)

ONLINE RESOURCES

 Available for download at **resources.corwin/TMSJ-MiddleSchool**

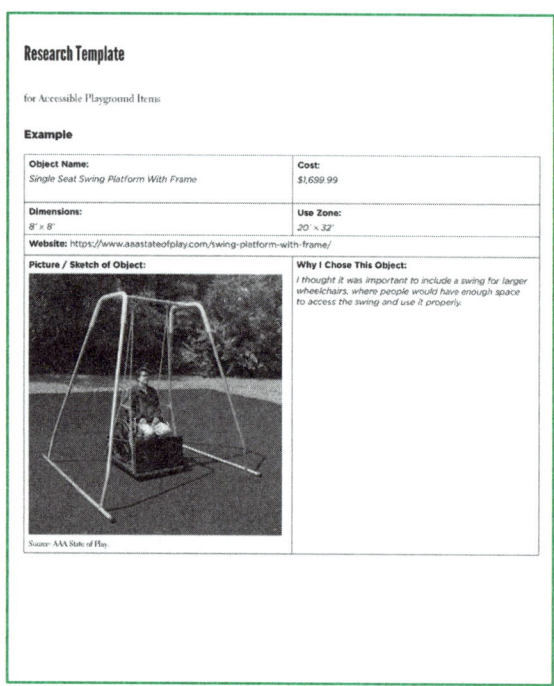

▲
Worksheet 1: Research Template

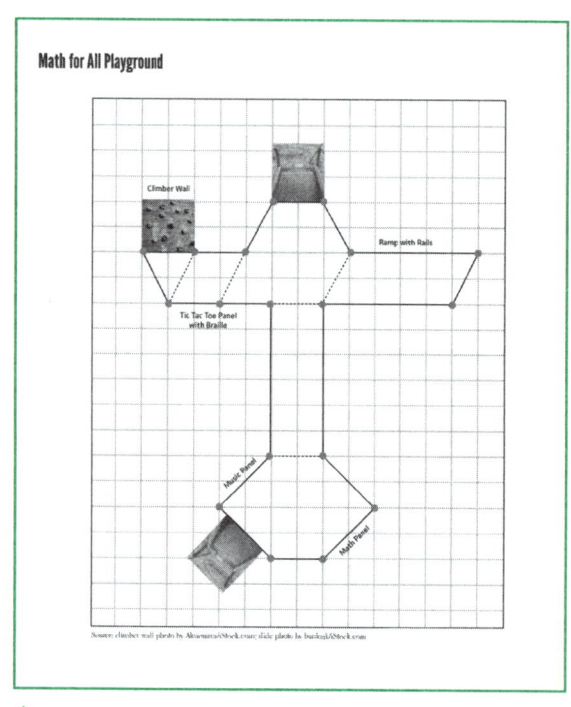

▲
Worksheet 2: Math for All Playground

LESSON 10.5 INVESTIGATING AREAS TO DETERMINE FAIRNESS

Jennifer Dao

CIVIL RIGHTS AND GOVERNMENTAL LAWS

There are several laws and practices that impact an individual's voting rights and "fair outcomes" of an election. This lesson allows students to share and develop an understanding of the inequity of voting and **gerrymandering** through the exploration of equal vs. equivalent areas using mathematics. This knowledge will empower them to share their voices on how gerrymandering negatively impacts and favors certain political parties depending on where the lines are drawn as they better understand **cracking** and **packing** in the system of gerrymandering. Teachers should research and understand the issue of gerrymandering in detail and how it impacts voting outcomes in their local communities.

DEEP AND RICH MATHEMATICS

Students apply their knowledge of finding the area of triangles and rectangles to calculate the areas of irregular polygons. This skill is used to solve the task of allocating land as students gain an understanding of the following: *What does equal mean? What does equivalent mean? Does equal always imply fairness?* Students integrate their knowledge of fractions, ratios, and calculating areas to support their sensemaking and reasoning in regard to gerrymandering.

ABOUT THE LESSON

This lesson uses a launch–explore–summarize instructional model and is intended to take approximately 250 minutes to complete across five class periods.

Lesson 1: Students explore the idea that "equal" does not always determine fairness by showing claim, mathematical evidence, and reasoning of their findings.

Lesson 2: Students find the areas of irregular shapes on a dot grid via the Surround and Subtract method or the Decompose and Rearrange method.

SOCIAL JUSTICE OUTCOMES

- I can recognize and describe unfairness and injustice in many forms including attitudes, speech, behaviors, practices, and laws. (Justice 12)

ESSENTIAL MIDDLE GRADES CONCEPTS

- **Geometry**—Grade 7: Solve real-life and mathematical problems involving angle measure, area, surface area, and volume.

- **Ratio and Proportion**—Grade 7: Analyze proportional relationships and use them to solve real-world and mathematical problems.

- **Statistics and Probability**—Grade 7: Investigate chance processes and develop, use, and evaluate probability models.

Gerrymandering: manipulation of voting boundaries to favor a political party.

Cracking: diluting the voting power of a party across many districts.

Packing: concentrating a party's voting power in one district to reduce their voting power.

Lesson 3: Students examine blue (diamond) and red (circle) squares in the Equal vs. Equivalent Areas task to determine whether equal always means equivalent and to see if the majority or popular vote always wins.

Lesson 4: Students gain an understanding of gerrymandering by exploring three different levels of games.

Lesson 5: Students research mathematical and social studies definitions related to gerrymandering and create a presentation to summarize their understandings from the four lessons.

Note: Taking Action is included as Lesson 5.

Resources and Materials

- Computer with internet access (1 per pair of students)

- Worksheet 1: *Island Inheritance Task* (1 per student)

- Worksheet 2: *Irregular Areas*

- Worksheet 3: *Equal vs. Equivalent Areas Group Task* (1 per group of 3–4 students)

- One of the following to accompany Lesson 3:

    + Dry-erase markers and whiteboard or plastic sheet protector (1 per group of 3–4 students)

    + Day After Election Digital Desmos Activity (for Worksheet 3) (https://bit.ly/3lAlJz4)

    + Digital copy of the figure from Worksheet 3 on Jamboard

- Worksheet 4: *Gerrymandering Games*

- Worksheet 5: *Equal vs. Equivalent Areas Project*

- Worksheet 6: *Gerrymandering Extension Questions*

Online Games

- Polytrope, "District" (http://polytrope.com/district/)

- Game Theory Test, "GerryMander: A Voting District Puzzle Game" (http://gametheorytest.com/gerry/)

- USC Annenberg Center, "The ReDISTRICTING Game" (http://www.redistrictinggame.org/)

Video

- "Gerrymandering: How Drawing Jagged Lines Can Impact an Election," from Christina Greer, *TED-Ed* (https://bit .ly/3xZKREE)

Copies or Digital Access to These News Articles

- "The Math Behind Gerrymandering and Wasted Votes," Patrick Honner, *Quanta*, October 12, 2017 (https://bit .ly/3GgxVx3)

- "Investigating Gerrymandering and the Math Behind Partisan Maps," by Patrick Honner and Michael Gonchar, *New York Times*, November 30, 2017 (https://nyti .ms/3EtUspK)

- "Gerrymandering Background," NCTM (https://bit .ly/3Eqo9bc)

- "The Geeks Who Put a Stop to Pennsylvania Partisan Gerrymandering," Issie Lapowsky, *Wired*, February 20, 2018 (https://bit.ly/32Z9xBF)

- "The Mathematicians Who Want to Save Democracy," by Carrie Arnold, *Nature*, June 7, 2017 (https://bit.ly/3omiLA6)

- "Rig the Election With Math!" *FiveThirtyEight*, October 28, 2016 (https://53eig.ht/3Ga5ctE)

This lesson was adapted from Lesson 8.2 (*Gerrymandering*) by Sven A. Carlsson in *High School Mathematics Lessons to Explore, Understand, and Respond to Social Injustice* by Berry et al. (2020).

LESSON 1 FACILITATION

Island Inheritance Task

Launch (5 minutes)

- Set the stage and present to students Worksheet 1 (*Island Inheritance*). Start with the focus question by saying: *Today in class we will explore the question: "Are equal and fair the same?"* Have students record this focus question in their notebooks.

- Tell students that they will also explore this question: *How can I use my knowledge of equal vs. equivalent areas to determine a fair way to split the island for me and my 8 family members in the Island Inheritance task?*

- Say to students: *Imagine you are on an island, and you have 8 family members. You and your family just inherited this island and want to share it equally among one another. Draw how you would fairly split up the island and provide your claim, mathematical evidence, and reasoning on the worksheet.*

Explore (20 minutes)

- Have students complete the task within groups. Encourage them to discuss their solutions with each other.

- As you visit groups and listen to their conversations and ideas, think of how you would like to summarize students' findings. Use purposeful selection, sequencing, and connecting techniques to call on groups to share their ideas as a class.

Summarize (20 minutes)

- Pose the following question to have groups begin sharing: *Does equal area always mean it's fair? Why or why not?*

- Create a class anchor chart or Google slide to document students' thinking process as they share to the class what they think so far when asked the question, using their knowledge of equal area to determine fairness.

- During group sharing, students should recognize that there are many ways to divide the island, and even if there may be equal area, they may argue that it isn't fair because one family member may be landlocked and the other gets to be surrounded by water and so on.

- Highlight the key idea of the task—that having an equal number of squares does not mean that each member is getting an equivalent section of the island.

LESSON 2 FACILITATION

Finding Area of Irregular Shapes on a Dot Grid

Launch (5 minutes)

- Hand out Worksheet 2 (*Irregular Areas*) and give students time to work on problems 1, 2, and 3 to find the area of the shapes on the dot grid.

- Discuss together what the dot grid means, and what one unit means. Ask: *Does it mean number of dots or number of lines when you connect?* Also be sure to discuss how to label the area as units squared.

- Explain to students that the task is to find the areas of the other irregular shapes using what they know about how four connected dots make one unit squared.

An anchor chart is an artifact of classroom learning. Like an anchor, it holds students' and teachers' thoughts, ideas, and processes in place. Anchor charts can be displayed as reminders of prior learning and built upon over multiple lessons (Learning for Justice, https://bit.ly/3DkmPFq).

- Discuss problem 4 in detail and use student strategies and thinking to highlight two methods to find irregular areas: (1) the Surround and Subtract method and (2) the Decompose and Rearrange method.

Explore (30 minutes)

- Instruct students to work with their groups to complete the task and check in with one another.

- As you monitor individual and group progress, use purposeful selection, sequencing, and connecting techniques to determine groups to present their ideas.

- As the shapes become more complex, encourage students to use strategies from some of the simpler shapes to help solve the more complex shapes.

- Some students may develop an easier strategy instead of counting the inside area of the irregular shape; they may draw a rectangle or square that frames the shape, determine the area of the rectangle or square, and subtract the "white or negative" space to find the exact area.

- Other students may choose to divide the inside area of the irregular shape into known areas and get the correct answer.

- Identify some incorrect solutions to share during the discussion. Discussing incorrect areas highlights the value of students' thinking and gives students opportunities to provide constructive feedback and respectfully disagree.

- Question 9 prompts them to design their own irregular shape with an answer key. You might choose to have them make their own shapes for review later or have them fold them so only their shape shows (not the answer) and switch with a classmate to find the areas of each other's shapes. Highlight pairs who found the area using different strategies.

Summarize (15 minutes)

- Facilitate a class discussion by having individual students or groups present their work to the class, and discuss strategies on how to find areas of irregular shapes on a dot grid. Make a class poster of the strategies.

- Use the following questions to connect the lesson to the Island Inheritance task.

  + *When do you think people would use knowledge to find irregular areas in the real world?* Have students generate a class list of running ideas. (Examples: finding the area of an irregular space to have enough carpeting or hardwood flooring, finding the area of an irregular region on a map, etc.)

  + *How can you apply your knowledge of finding areas of irregular shapes to the Island Inheritance task?*

  + *What role do you think areas of irregular shapes will play in the real world to determine fairness as related to gerrymandering?*

LESSON 3 FACILITATION
Equal vs. Equivalent Area

Launch (5 minutes)

- Share with students how what we saw in the Island Inheritance task in the previous lesson will help us better understand voting in the United States, the Electoral College, and how gerrymandering plays a role in the process.

- Use the following questions to facilitate a brief discussion:

  + *How does the Island Inheritance task relate to our society?*

  + *Is there anything in your life or anything you've heard about that requires things to be split up?*

  + *Are they always equal? Are they always fair?*

  + *What does "Which color (or shape) won?" mean?*

  + *Which color (or shape) is there more of in a single group? In the whole group? The majority? Clarify.*

- Share with students that today they will continue to explore the question from the previous lesson: *Does equivalent always mean equal? And does equal always mean equivalent? Why or why not?*

- Note that students will also explore this question: *How can I apply my knowledge of definitions of equal and equivalent to areas in the real world to determine fairness as related to gerrymandering?*

Explore (35 minutes)

- Pass out two copies of Worksheet 3 (*Equal vs. Equivalent Areas Group Task*) to each group, and have students work in their groups with a partner to complete the task.

- Provide students with dry-erase markers and a whiteboard or plastic sheet protector. Also consider providing a digital image of the figure from Worksheet 3 using an online tool such as Jamboard or direct them to use the Desmos version of the worksheet.

- As you monitor group progress, pay attention to how students are approaching the problem. Use the worksheet questions to further probe students' thinking about the areas.

  + Question 6 ("Divide the squares into 5 equal groups where the other team can win.") will spark a lot of conversation and will be a challenge for students to figure out. Give time for students to productively struggle, and encourage them to try different ways.

- As an optional challenge to students who might finish early, give them a printout of dot paper and say: *Make as many other irregular shapes as you can that have the same areas you found today.* Ask them to label and show their work.

Summarize (10 minutes)

- Use the following questions to facilitate a closing discussion:

  + *Does equivalent always mean equal? And does equal always mean equivalent? Why or why not?*

  + *How can you apply your knowledge of definitions of equal and equivalent to areas in the real world to determine fairness as related to gerrymandering?*

- Once students are convinced it is possible for red (circles) to win depending on how the lines are drawn, share that this is what politicians have decided to do by redistricting and drawing lines on a map, and that they will explore how this works in the next lesson.

LESSON 4 FACILITATION

Gerrymandering Games

Launch (5 minutes)

- Distribute Worksheet 4 (*Gerrymandering Games*) and ask students: *What is gerrymandering?*

- Tell students that they will play several games during this lesson to help them gain an understanding of the concept of gerrymandering.

- Introduce students to three online gerrymandering games:

  + Level 1: "District" from Polytrope is a good introduction (http://polytrope.com/district/).

  + Level 2: "GerryMander: A Voting District Puzzle Game" from Game Theory Test offers good visuals and practice (http://gametheorytest.com/gerry/).

  + Level 3 (optional): "The ReDISTRICTING Game" from USC Annenberg Center offers more in-depth practice if you're interested (http://www.redistrictinggame.org/).

Explore (30 minutes)

- Have students explore and play games of each level depending on their choice to answer the questions: *What is gerrymandering? Does equal area always mean it's fair? Why or why not?*

- Tell them to discuss the outcomes in their groups as they play the games and to take notes on any key terms and definitions. Encourage students to create a visual for key terms and definitions.

- As you monitor groups, remind students to take notes and to discuss the outcomes and strategies they are using for each game.

Summarize (10 minutes)

- Facilitate a class discussion by having students share the key terms, definitions, and strategies they used. Ask: *What role does gerrymandering play when it comes to* equal *and* fair?

- Share with students that during the next lesson they will create a presentation to synthesize their understanding of gerrymandering.

LESSON 5 FACILITATION

TAKING ACTION

Note: This is considered as a fifth full lesson period.

Student Synthesis Research Project and Presentations

Launch (5 minutes)
- Distribute Worksheet 5 (*Equal vs. Equivalent Areas Project*) and discuss the expectations. Tell students they can work individually, with a partner, or in a group of three to four students. The worksheet includes the following parts:

  + Part 1: Research. Students take notes to include in the presentation.

  + Part 2: Create a Presentation. The product that students create should be a choice of options such as a Google Slides presentation, poster, or infographic presentation.

  + Project Checklist. Students complete this checklist before they submit their work.

Explore (40 minutes)
- As you monitor student groups, remind them that their research should encompass what they have learned throughout the four lessons.

- Use the Worksheet 5 questions to assess student understanding of the concepts from the lessons.

Summarize (5 minutes)

- Share with students that throughout the rest of the year, groups/individuals will have the opportunity to share their presentations.

- Encourage students to do the following:

  + Complete Worksheet 6 (*Gerrymandering Extension Questions*) and use their state's map on a dot plot to try and determine if the voting districts are fair.

  + Create a school newsletter to share with others in their school, home, and community.

  + Write a letter to a local government official to share their findings individually, in a group, or as a whole class.

ONLINE RESOURCES

online resources Available for download at **resources.corwin/TMSJ-MiddleSchool**

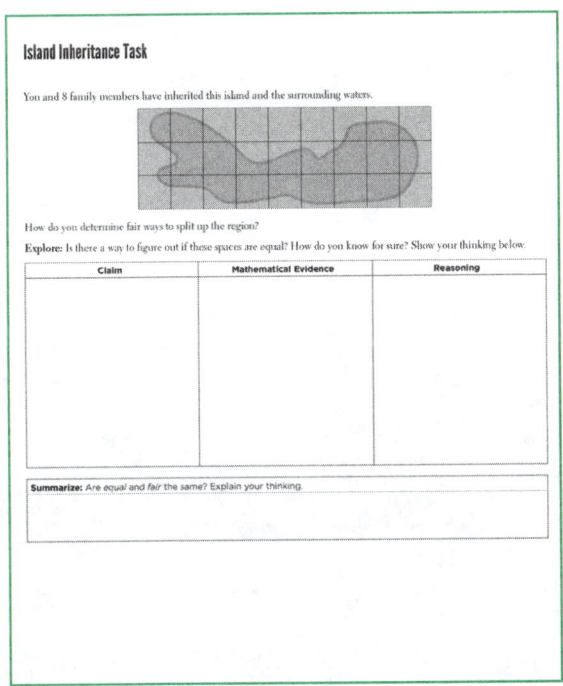

▲ *Worksheet 1: Island Inheritance Task*

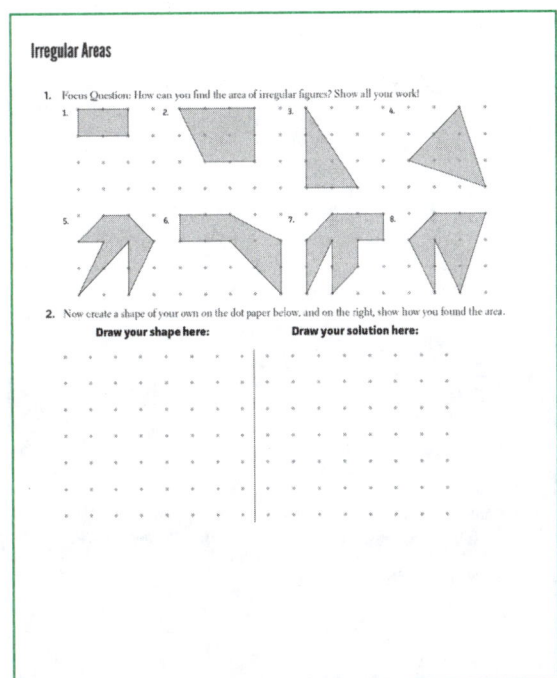

▲ *Worksheet 2: Irregular Areas*

Equal vs. Equivalent Areas Group Task

Focus Question: Does equivalent always mean equal? Why or why not?

Group accountability: Complete the tasks and questions below together with your group.

Individual accountability: Write your answers to the questions in your notebook.

Math Questions:

1. How many total squares are there?
2. What color (shape) represents a bigger area?
3. What is the percentage of red (circles)?
4. What is the percentage of blue (diamonds)?
5. Divide the squares into 5 equal groups. Which color (or shape) won? How do you know?
6. Divide the squares into 5 equal groups where the other team can win. How did you create those regions?

Equal vs. Equivalent Areas Group Task

Focus Question: Does equivalent always mean equal? Why or why not?

Group accountability: Complete the tasks and questions below together with your group.

Individual accountability: Write your answers to the questions in your notebook.

Math Questions:

1. How many total squares are there?
2. What color (or shape) represents a bigger area?
3. What is the percentage of red (circles)?
4. What is the percentage of blue (diamonds)?
5. Divide the squares into 5 equal groups. Which color (or shape) won? How do you know?
6. Divide the squares into 5 equal groups where the other team can win. How did you create those regions?

Worksheet 3: Equal vs. Equivalent Areas Group Task

Gerrymandering Games

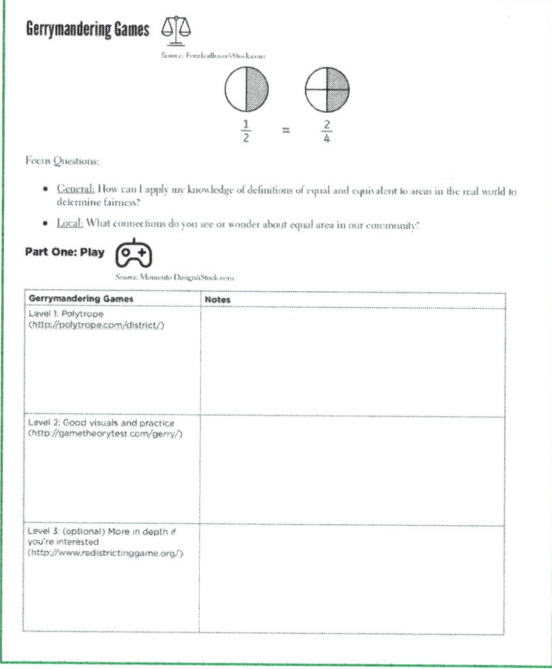

Source: Fenykheshey/iStock.com

$$\frac{1}{2} = \frac{2}{4}$$

Focus Questions:

- <u>General:</u> How can I apply my knowledge of definitions of equal and equivalent to areas in the real world to determine fairness?
- <u>Local:</u> What connections do you see or wonder about equal area in our community?

Part One: Play

Source: Momento Design/iStock.com

Gerrymandering Games	Notes
Level 1: Polytrope (http://polytrope.com/district/)	
Level 2: Good visuals and practice (http://gametheorytest.com/gerry/)	
Level 3: (optional) More in depth if you're interested (http://www.redistrictinggame.org/)	

Worksheet 4: Gerrymandering Games

Equal vs. Equivalent Areas Project

Source: Fenykheshey/iStock.com

$$\frac{1}{2} = \frac{2}{4}$$

Part One: Research

Source: Momento Design/iStock.com

Answer these questions in your notebook.

Mathematical Definitions:

1. What does **equal** mean? What does **equivalent** mean?
2. Explain the difference between equal and equivalent. Does equivalent always mean equal? Why or why not?
3. Support your reasoning with mathematical evidence (example: fractions, area of shapes, pictures, tables, graphs, etc.).

Social Studies Definitions:

4. What is gerrymandering? (Watch the video and select *at least 2* articles)
   - Watch TED-Ed Video: "Gerrymandering: How Drawing Jagged Lines Can Impact an Election" (https://bit.ly/3xZKREE)
   - Article: "The Math Behind Gerrymandering and Wasted Votes" (https://bit.ly/3GgxVx3)
   - Article: "Investigating Gerrymandering and the Math Behind Partisan Maps" (https://nyti.ms/3EtUspK)
   - Article: "Gerrymandering Background" (https://bit.ly/3Eqo9be)
   - Article: "The Geeks Who Put a Stop to Pennsylvania Partisan Gerrymandering" (https://bit.ly/3ZZ9xBF)
   - Article: "The Mathematicians Who Want to Save Democracy" (https://bit.ly/3omiLA6)
   - Article: "Rig the Election With Math!" (https://5eig.ht/3Ga5ctE)

   As you explore and research, answer the following:
   - What did you learn?
   - What surprised you?
   - What do you still wonder about?

5. What does "redistricting" mean?
   - Analyze the Atlas of Redistricting (https://5eig.ht/3Ga6Rzo)

Worksheet 5: Equal vs. Equivalent Areas Project

Gerrymandering Extension Questions

In your Inverse Variation Project, you determined the area of 1 through 5 equally divided sections of your country's area—pretending that it was in the shape of a rectangle. Although the notion of rectangular-shaped countries is not realistic, sectioning land for political reasons is a real occurrence. Please feel free to research online to help you answer the questions that follow. All responses should be written in 2 or more complete sentences.

1. If you could divide Evanston into 5 to 6 areas with approximately the same number of people living in each area, what would be the best way to do it? Justify.

2. Define the term *redistricting* in your own words.

3. Define the term *gerrymandering* in your own words.

4. How do redistricting and gerrymandering work in Evanston?

5. Who do you think should decide how legislative districts are drawn? Justify.

6. Why did President Obama, among others, want to reform (change) the redistricting process?

Worksheet 6: Gerrymandering Extension Questions

PART
III

NEXT STEPS

ADVICE FROM THE FIELD

We believe that the voices of the lesson authors and reviewers, those engaged in this work from day to day, can be meaningful to you and others interested in teaching mathematics for social justice (TMSJ). Hearing from other teachers who are dedicated to TMSJ can offer insights and extra enthusiasm for implementing the lessons in this book (Chapters 6–10) and for creating your own social justice mathematics lessons (SJMLs; Chapter 12).

The lesson authors and reviewers all have experience with teaching mathematics for social justice in their classrooms. All have also used one or more of the lessons in this book. We asked them to share some of their experiences, successes, challenges, advice, and feedback from students and stakeholders. Specifically, we asked these questions:

- What social and mathematical student experiences should be considered for teachers implementing this lesson? How did you or may you navigate this during implementation?

- If applicable, what has been most valuable about your (and your students') experiences when implementing a TMSJ lesson or lessons?

- What additional bits of advice would you offer anyone implementing this or other TMSJ lessons?

- As a mathematics educator committed to social justice, how do you take care of or support yourself and your community to sustain and persist?

In this chapter, we report their responses, using their words as much as possible. We hope their experiences will provide some inspiration and insight as you set forth to implement the SJMLs in this book or create your own and as you refine your identity while teaching mathematics for social justice. We position advice from the field in successes, challenges, self-care, and additional advice. We summarize this chapter with the value of TMSJ.

SUCCESSES IMPLEMENTING SJMLs

The lesson authors, reviewers, and testers of lessons in the book see middle school students as insightful. Bethany Chan says middle school students are "much more insightful than we think, so we need to give students more credit when discussing these important and serious social justice topics." Those active in this work believe that TMSJ effectively engages students, demonstrates the usefulness of mathematics, provides autonomy and choice, and builds stronger communities.

Engaging Students

TMSJ can help you engage students during the middle grades. This is a period of human development when students are beginning to understand more deeply the world around them and their communities.

> *While implementing TMSJ lessons, my students have been more engaged and I have been able to learn so much more about my students. As a white female, I have also been able to learn so much more about the injustices in our society through my students' lived experiences.*
>
> —Nichole Campbell

Many of the lessons include tasks, worksheets, activities, and online materials that deeply engage students. Liza (Cope) Bondurant found "the interactive Desmos components very engaging" for students. Gavi Gelbart found that "students were interested in learning about an issue that they had heard of outside of the math classroom" while testing lessons. Gavi also said that "several parents from one of my classes were pleased to hear we had studied [this] topic." Generally, as Jennifer Dao notes regarding the work of TMSJ, "giving students time to think and collaborate and discuss—sharing their thoughts and opinions—is valuable in my experience."

Using Mathematics to Shed Light

K. Elizabeth Hammonds notes that middle school students enjoy having a "connection to [the] real world. Kids like learning in the context of which issues can be solved with math." Students participating in TMSJ lessons can examine the world around them and their communities critically. Dr. Natalie Odom Pough described how middle school students "were able to see how powerful big businesses and their endless reach are when difficult decisions are being made." You can support middle school students through using mathematics to explore, understand, and respond to social injustices.

> *The results of the activity left the students surprised, and most of them expressed concerns regarding the inadequate payment received by the workers. They were also appalled by the difference [in the] percentage of money received by the retailers and the workers. Students engaged in conversations*

around how clothing industries and fast fashion keep people working in garment factories.

—Bethany Chan, Debasmita Basu, Rebecca Ellis, Frances K. Harper, and Jennifer Ruef

Many students shared they saw the relevance of math and its connection to the real world from this week-long series of lessons that connected their learning of irregular areas and how it connects with determining whether areas are equal and/or equivalent . . . my student teacher shared he was impressed how students never asked "what is the purpose of math" because they did see the purpose of math in class and its connections to the real world.

—Jennifer Dao

Providing Student Autonomy

As described in the framework for this book and the next chapter, you are encouraged to allow students to generate questions from their own experiences and community that are meaningful to them to discuss in the mathematics classroom and use mathematics to understand more deeply. Colleen Carman said, "My students feel more connected to a topic when they have a choice. They feel like detectives who become experts on what they're studying, and the social justice aspect helps them feel connected to learning more." When you use TMSJ lessons, you are empowering students to be agents of change.

Neutral facilitation is powerful—if your students can reach an outcome independently, they are far more likely to engage and be inspired. It can feel disingenuous if students feel like their facilitator leads them to a pre-determined outcome (even if there is only one way to interpret the data). Facilitating from a student-centered and neutral stance allows students to make their own decisions and implement their analysis. The projects and their associated requirements are designed to set appropriate expectations— trust your students to get there independently!

—Andrew Reardon

Each year, students thoroughly enjoy working on the blueprints to their playground. Students have expanded the idea of a playground to water parks, adventure parks, roller coaster/amusement parks, zoos, fishing ponds, carnivals, etc. Creativity, student voice, choice, and a growing empathy for children with disabilities have led to students being internally motivated to successfully create blueprints of playgrounds (etc.) that highlight not only their mathematical skills but also their understanding of inclusivity.

—Maggie Lee McHugh and Jennifer Kosiak

Building Community

As Peggy Nayar noted, TMSJ lessons have the opportunity to "buil[d] stronger and stronger relationships throughout the year." This work takes place among middle school students, teachers, and the larger community. Nicole Kelley described this impact on students in Lesson 7.1 (*Hey Google, Who's a Mathematician?*), saying that "the strengths are in the opportunities for students to see themselves reflected as mathematicians . . . [the] focus on mathematicians, specifically those of under-represented populations, will go a long way to help expose the bias towards white male mathematicians in our current standards." Maggie Lee McHugh described this impact at all three levels: "[A]s I move deeper into my journey as a social justice educator, I recognize the vital role reflection plays in the classroom. The more time students reflect on their work as mathematicians and as young adults who can impact their community, the deeper impact the experience has on students."

This work is reshaping communities in many positive ways. Kelly, a lesson reviewer, described this impact by saying, "I really enjoyed learning and thinking about new ways to present mathematics that would bring new perspectives and awareness to students." In addition to just changing perspectives and awareness, those involved in this work have seen larger community improvements.

The lesson worked well in the summer camp we ran. Students were particularly engaged in using the 3D modeling software and came up with several creative ways of thinking about the problem. Designs included wearable backpacks, stackable containers with collapsible handles, a large truck-based water delivery system, and containers with built-in food [and] water filters and food storage. Students considered various issues in their designs, including how easy they would be for individuals to use and carry, refilling options, and stacking for transportation.

—Mathew D. Felton-Koestler and Courtney Koestler

We were shocked to learn how many visits to food shelves were increasing in our community. Many of our students expressed relief at knowing this was not just happening to their families. A sense of shame, that we did not want our students to have, seemed to have been somewhat lifted. We had great student engagement and conversation and better homework completion and assessment scores than we had ever had before. It was humbling for us to realize how our students were directly impacted by hunger. Being food insecure in a cold Minnesota winter, with limited access to transportation, is not easy to endure. Watching our students make announcements for our food drive, carrying our canned goods to the food shelf, packaging meals, being kind to one another—regardless of culture or gender or economic backgrounds—filled us with pride and excitement to teach this again.

. . . having spent this entire school year to date with a focus on social justice, we knew that using culturally relevant topics and themes of social justice [was] more important than ever. We were shocked to see how tobacco and vaping companies targeted not only teens and preteens, but all people of Color and those in the LGBTQ communities as well. It was humbling for us to realize how our students were directly targeted by tobacco and vaping companies. Watching our students create movie trailers, commercials, and other forms of public service announcements empowered us as teachers and our students as learners. Parents commented how happy they were to see their students using technology to do more than play games.

—Nichole Campbell and Peggy Nayar

PLAN FOR AND RESPOND TO CHALLENGES

Our lesson authors, reviewers, and testers of lessons reported a few challenges that you may already feel or predict. One challenge speaks to the ways stakeholders (e.g., parents/caregivers and the school community) may misunderstand your intentions when bringing in a social issue to your mathematics classroom. Another challenge surrounded the importance of making plans to support students when discussing heavy topics that might overwhelm them.

Involve Stakeholders

One lesson author pointed to the concern stakeholders have for middle school students engaging in controversial issues. This insight seems to suggest productive ways to inform them about how you plan to engage students thoughtfully.

The teacher should consider that some stakeholders may feel this lesson should not be taught. Stakeholders may feel the lesson is too political. The teacher should highlight the fact that students use math to arrive at their own conclusions and are supplementing the resources the teacher provides with their own research.

—Elizabeth O. Ayisi and Colleen Carman

A lesson author noted that it is crucial for you to communicate with stakeholders that the mathematics centers the conversation and your perspective remains objective:

The main pushback I receive is about whether the mathematics will be too easy or too hard for students. So, when I start with implementing or writing lessons, I usually start with the standards I want to teach and introduce that through the lens of social justice. Explaining this to stakeholders and show-ing them lessons usually clears up concerns.

—Colleen Carman

Connect to Students' Lived Experiences

One lesson author discussed the challenges of facilitating heavy topics that connect to middle school students' lives. Specifically, how might you overcome negative information about a student's community, and how can you ensure that students thoughtfully consider the ideas and information they learn? Here are two responses to these issues.

Teachers are communicating the social and mathematical goals of the lessons to stakeholders by inviting community reflections on their connections to water. Students will be invited to interview a stakeholder in their life about their respective memories of water, pollution, and any actionable steps that were taken.

—Sara Rezvi and Tyrone Martinez-Black

Many teachers face the challenge of students asking whether the mathematics they are learning relates to their lives. One lesson author wrote:

Most often, teachers who teach directly from the curriculum, and focus on skill and drill, [and use the] direct method of instruction get responses from students like "When will we ever use math in real life?" Teachers may choose to share the importance of connecting mathematics to the real world to help students see its relevance, because using and applying mathematics to help better understand and make sense of the world allows students to be critical thinkers.

—Jennifer Dao

ADDITIONAL ADVICE TO COLLEAGUES IMPLEMENTING SJMLs

Our lesson authors also provided some other words of advice based on their experiences in their classrooms. They suggest that you broaden the scope of social justice work, prioritize students' needs, work with others, and build a strong classroom community. As you plan your SJML, give careful attention to ensuring the quality of and strong focus on mathematics.

Prioritize Students' Learning Experiences

The lesson authors, reviewers, and testers of lessons described how you should prioritize middle school students' learning experiences in your mathematics classrooms by centering them during lessons.

I would say first people/students need to understand that TMSJ lessons are not always about race. Students have to be educated about the broad scope of the subject and why it is necessary to be implemented in the classroom. Once those conversations are had, then it gets a lot better, especially showing students and young teachers that there is another area where mathematics can be used inside the classroom.

—Kendrick Savage

Do your best to center student voices and allow them to come to their own conclusions.

—Gavi Gelbart

Be a Lifelong Learner

More than one lesson author pointed to the importance of continuously learning about students' lives to be informed about issues that relate to them. Being a lifelong learner can enhance an understanding of your students' world and provide you with more opportunities to improve their lives through highlighting teaching mathematics for social justice.

I think it is important for educators to be lifelong learners just as it is for students when they are learning. To always collaborate with colleagues, discuss important events, read the news, and be up to date with current events.

—Jennifer Dao

Learn about the context yourself. These tasks require delving into complex social issues and they are opportunities for both the teacher and the students to learn together in dialogue.

—Travis Weiland

Engage in the lesson as a student first; this will allow you to organize your thoughts and be knowledgeable of the experiences that the students must have while they engage in the lesson.

—Tashana Howse

SUPPORT YOURSELF TO SUSTAIN AND PERSIST

Engaging in social justice work is often arduous and taxing on teachers. To sustain the work, you need to practice effective self-care. Next, the lesson authors give some advice to help others take care of themselves while doing this work.

Self-Care

The lesson authors share that self-care is an integral component of all social justice and provide some advice on ways to take care of yourself.

[I] try to know when to step back and take breaks when I can. I also try to remember that I cannot nor is it my responsibility to fix everything; in other words, I try to give myself some grace. I find it important to make time for myself and my mental health. Taking walks and time away from work and technology.

—Travis Weiland

I ground myself in asking: "What's best for students?" It can be easy to get frustrated with the many flaws in the education system; however, grounding myself in the work I do with students and seeing the positive impact on their outcomes over time keeps me motivated.

—Andrew Reardon

Don't Do This Work Alone

The lesson authors explain the importance of finding a community for support while engaging in this work.

Find someone to do the work with and lean into the discomfort of the work.

—Nichole Campbell

Build up your tribe. Find people (colleagues, community members, administrators, parents/guardians, Twitter PLN, etc.) who share your commitment. Purposefully create time to interact with SJ educators who share your vision. Read. Reflect. Accept.

—Maggie Lee McHugh

CONCLUSION

We hope that the words of a few experienced educators who teach mathematics for social justice help connect some of the ideas of this book, provide some insight and confidence to implement a SJML, and inspire you to ensure your students have a meaningful experience with mathematics. To support this work, NCTM (2020) noted

> *The topics explored during middle school have the potential to, and should, move beyond a utilitarian emphasis to one where students develop a positive mathematical identity, have a voice to understand and critique the world through mathematics and spark curiosity to experience the wonder, joy, and beauty of mathematics. (p. 16)*

We close this chapter of voices from the field with comments about how important TMSJ can be and share some closing thoughts.

The Value of Teaching Mathematics for Social Justice

The lesson authors, reviewers, and lesson testers shared many statements describing the value of TMSJ and its relationship to students and other stakeholders. Those in this work see value in TMSJ to build larger and deeper positive social perspectives, empower students to be positive agents of change, and engage in deep mathematics. We hope these comments are encouraging for others as they pursue or continue this work.

Deep and Broad Social Understanding

Hochieh Lin and Joanne Baltazar Vakil saw this work as moving students "to consider not only the environmental issues, but also the social issues involved." In fact, engaging with this book and the TMSJ community itself can deepen and broaden perspectives. Dr. Natalie Odom Pough stated, "Although we are on the East Coast, connecting with what is happening in California allowed participants to see how political decisions adversely impact individuals." Others also share their insights on lessons from the book and TMSJ from the field.

> *Student experiences should include engagement in discussions related to healthy living, American health care, and health care access. These discussions may happen in or outside the classroom. It's important for them to converse with their families and others in their communities to understand each topic above. Regarding mathematics, students should engage in real-world experiences, which require them to represent, work with, and make sense of ratios, fractions, and percents. They need experience with data analysis.*
>
> —Tashana Howse

Most items we interact with on a daily basis have a long supply line and require many different people to be involved before the item gets to us, no matter the cost of the item. Helping students to first recognize the supply line before jumping into the lesson may help to make the material more real and relatable.

—Rebecca Ellis

[Our lessons are] an opportunity for students to: (1) learn more about—and perhaps become more interested in—voting and the democratic process; (2) recognize the persistence of calculated voter suppression, especially for Black Americans; and (3) use mathematics as a tool for mathematizing and critiquing the world.

—Chuck Munter and Cara Haines

Deep and Engaging Mathematics

We *teach mathematics* for social justice, so TMSJ done effectively makes mathematics a central, critical, and essential component. TMSJ is valuable because

TMSJ becomes just one more natural application for describing and measuring the world, rather than a particular message that students must internalize.

—Rebecca Ellis

Thus, the lesson authors and others doing work in TMSJ make mathematics a focus area:

TMSJ lessons highlight the critical need to understand mathematical concepts in order to make change in our society. By implementing lessons centered on social justice, students see the authentic reasons for learning topics and apply their understanding of mathematics, leading to deeper, more meaningful learning.

—Maggie Lee McHugh

As a result of this lesson, they will be able to use their experience with graphing points on a coordinate plane, reading maps, and distance to educate others regarding the lengths some must travel to access a full-service grocery store. Advocacy for increased access will be a natural next step following this lesson.

—Becky Evans, Emmalee Bielenberg, Julia Novosad, and
Cassie Ruettiger

Students will be able to justify and critique the reasoning of others through analyzing mathematical evidence. Moreover, by understanding areas and the definitions of equal versus equivalent, students will be able to determine whether or not decisions and actions reflect fairness based on their deep understanding of mathematics of equal or equivalent areas.

—Jennifer Dao

Despite the fact that some students may struggle to make the connection between cost of living and minimum wage, through investigations of slopes, they will be able to build their mathematical and contextual comprehension . . . We also believe that through scaffolding, a thorough analysis of slope, conversations, immediate feedback, purposeful reflection, and appropriate use of tools, this lesson promotes critical thinking and problem solving while remaining accessible to all students.

—Elizabeth O. Ayisi and Colleen Carman

Empowerment of Students

The long-lasting impact of TMSJ is to empower students to be authors and doers of mathematics and to empower them to understand and respond to injustices by using mathematics. Becky Evans, Emmalee Bielenberg, Julia Novosad, and Cassie Ruettiger shared how students are "empowered to communicate their learning through engaging with their local government and/or community grocery stores" in their lesson. Others invested in this work also focus on the need to empower students:

Social justice lessons need an action component. Because the content is sometimes heavy, it is essential that students are empowered to act on what they learn and co-create in the lesson. . . . Mathematical arguments can change our communities and the world. There is hope in action.

—Jennifer Ruef

By studying the social as well as the mathematical aspects of the catastrophe, students can feel empowered by realizing the numbers will tell them a lot of what they need to know. In turn, students can take this same empowerment into the world to make sure no other group of people/city will suffer such a horrible situation. Lastly, teachers should know this lesson seeks to address the three social justice pedagogical goals (Gutstein, 2006): (1) reading the world with mathematics, (2) writing the world with mathematics, and (3) developing positive cultural and social identities.

—Kendrick Savage and Tashana Howse

Closing Thoughts From Our Lesson Authors

It is our experience that beginning to implement SJMLs is a bit of a frightening proposition for mathematics teachers, perhaps yourself. There isn't supposed to be controversy in the mathematics classroom, and mathematics is often beloved because there is no gray, just black and white. These mathematics classroom norms make it difficult to feel comfortable shifting those expectations.

As mathematics teachers, we may feel comfortable supporting classroom dialogue that compares different approaches to solving a mathematics problem. While we can sometimes help students identify one approach as incorrect or less generalizable, it can be challenging to imagine having a similar conversation about social issues. As authors of this book, each of us has felt the same trepidations. Two strategies have been beneficial for us. The first is to recognize that we will learn and improve on how well we support student-led discussions of complex issues, and at the same time, we already have a lot of tools as experienced teachers to manage these well. And second—as mentioned before—is to work with others. Collaboration leads to much greater insight, resources, and solutions and provides the support and confidence to take on something new!

CREATING SOCIAL JUSTICE MATHEMATICS LESSONS FOR YOUR OWN CLASSROOM

You may enter this chapter from different backgrounds with regard to your experience enacting social justice mathematics lessons (SJMLs) as well as with regard to your goal for enacting social justice lessons. This chapter provides some suggestions to get you started with planning your own lessons and/or units. Grab a copy of the SJML Planner template (Appendix F) and you are on your way.

This chapter also provides additional details about the SJML framework used to design the lessons in this book. This chapter will first revisit the seven elements of the SJML design framework we found helpful to prepare the lessons in this book. Next, we suggest steps to develop SJMLs of your own.

SETTING A FRAMEWORK FOR AN EFFECTIVE SJML

In Chapter 5, we identified seven design principles building on our earlier discussions of what made for an effective SJML. Those seven elements of our SJML framework are as follows:

1. Building and Sustaining Beloved Community

2. Equitable Mathematics Teaching Practices

3. Authentic, Challenging Question or Concern

4. Social and Mathematical Understanding

5. Social and Mathematical Investigation

6. Social and Mathematical Reflection

7. Action and Public Product

These principles emphasize creating opportunities for students to achieve both mathematical and social justice goals (Gutstein, 2006). An investigation to achieve these goals should emerge from an authentic and challenging question or concern of the students in a Beloved Community; their interests and curiosities should drive their learning. And students should, in some way, act on what they learn. Finally, the lessons should be designed in such a way to ensure the opportunity to implement equitable mathematics teaching practices, such as those discussed in Chapter 4 (Figure 4.3) and revisited in Figure 12.1.

Figure 12.1. Five Equity-Based Teaching Practices (Aguirre et al., 2013)

Equity-Based Practices

1. Go deep with mathematics.
2. Leverage multiple mathematical competencies.
3. Affirm mathematics learners' identities.
4. Challenge spaces of marginality.
5. Draw on multiple resources of knowledge (mathematics, culture, language, family, community).

Let's revisit the seven design elements of the SJML framework we use in this book. Our aim is to establish similar guiding principles for you before you start designing your own SJML and provide a few additional resources to draw upon.

Element 1: Building and Sustaining Beloved Community

Earlier in this book, we encouraged you to learn your students' social, cultural, and academic funds of knowledge so that you can build on student assets and establish a Beloved Community in your classroom—essential work to engage in before and alongside envisioning and implementation of SJMLs (see Chapter 2). You can do this with a focus toward building students' self-awareness and social awareness. Setting time aside in your instruction to build this practice in students is not natural and will not take place in the limited time you have them in your mathematics class unless you deliberately focus on developing these attributes.

We have argued that building and sustaining a Beloved Community requires guiding students toward becoming responsible decision makers, focusing on self-management, and building relationship skills. You must be careful to assume that SJMLs or other tasks in themselves promote these attributes. You will need to purposefully provide time for deep examination of your positionality and teaching context; establish and maintain mathematics classroom community commitments; enact mathematics community builders and icebreakers; learn about your students as people who do mathematics; use mathematics autobiographies and storytelling; and implement regular temperature checks, in relation to and beyond mathematics.

Even these deliberate efforts to build a Beloved Community can fall short if deeper biases (conscious or subconscious) are felt by your students. Are you sending micro-messages, small or subtle messages that are communicated through your teaching practices without saying a word? This may include positioning, looks, gestures, grouping, tone of voice, question difficulty, or the framing of feedback.

PAUSE AND REFLECT

Take a moment to reflect on your Beloved Community and how your teaching practices affect it before moving on. How have your teaching practices and tasks conveyed a message of belonging?

Element 2: Equitable Mathematics Teaching Practices

The Equitable Mathematics Teaching Practices introduced in Chapter 4 form the foundation for SJMLs and help you establish a classroom culture and environment that is ready to take on the challenge of integrating social justice issues with mathematics. Here we briefly review the five equity-based teaching practices for the mathematics classroom identified by Aguirre et al. (2013), restated in Figure 12.1. We feel these five teaching practices go a long way toward achieving Gutstein's (2006) mathematics and social justice goals: to read the world with mathematics; to write the world with mathematics; and to develop positive social, cultural, *and* mathematical identities (revisit these aims in Figure 1.4).

We have argued in this book that to effectively TMSJ, you must ground the lesson in rich and meaningful mathematics, attending to mathematical goals. Going deep with mathematics means that we engage students in high-cognitive-demand tasks; engage students in a discourse-rich classroom; and support students in analyzing, comparing, justifying, and proving their solutions.

Further, we've encouraged you to develop a classroom culture in which there are many ways of being mathematically smart. When you know these varied assets students bring, you can leverage varied mathematical competencies to create the need for student collaboration on complex problems in which they rely on the knowledge and skills each brings. To achieve this, tasks in a SJML should offer multiple entry points, allowing students with varying skills, knowledge, and levels of confidence to engage with the problem and make valuable contributions. In this way, you can use a student's previous mathematics knowledge as a bridge to promote new mathematics understanding.

You can also build on the social, cultural, family, and community knowledge your students bring to your class. Tap into the mathematics knowledge and experiences related to students' culture, community, family, and history as resources for building further understanding. Invite students to teach the class more about a context in which they are experts. Ask students to draw connections to similar experiences or contexts they know. Recognize and strengthen multiple language forms, including connections between mathematics language and everyday language, to include the affirmation of and support for multilingualism. In doing so, you position your students as competent learners and doers of mathematics.

Element 3: Authentic, Challenging Question or Concern

A SJML must be grounded in your students' questions or concerns, creating opportunities for authentic and challenging engagement with both social and mathematical issues. We've identified topics of injustice to help us sort through possible issues, and try to open our minds and hold us accountable to consider a wide variety of issues, including ability, civil rights and governmental laws, economic inequality, education, environment, gender and sexual identity, health, human rights, race and ethnicity, faith and religion, rights and activism, and representation (see Appendix D). Contexts such as these can encourage students to observe patterns, critique information, learn to ask questions, and reflect. Sometimes a shift in mindset can be helpful, as Sara Rezvi and Tyrone Martinez-Black note in Lesson 10.3 (*Water Is Life—Our Collective Past, Present, and Future*):

> *When we hear the phrase "water is life" from Indigenous activists in Standing Rock and beyond, what questions are brought up for those of us who are settlers in the United States? In American society, water is assumed to be drinkable, potable, hygienic, and perpetually flowing. Yet, as in the cases of Flint, Michigan and Standing Rock, access to water for Black, Indigenous, and other people of Color continues to be severely limited due to human greed, capitalism, legacies of colonization, and white supremacy. The purpose of this lesson series is to take a deep look from a social justice and mathematical perspective that allows students to interrogate their relationship to water. We explicitly use First Nations' perspectives on the relationship of people to water in past, present, and future forms. We refer to this way of knowing as an alternative to Western conceptions of water as property, expendable, and commodifiable. Instead, we direct students to interrogate their relationship to future generations and the Blackfoot nation epistemology of cultural perpetuity.*

Asking students what they notice; what they wonder; what they feel; and how they may want to act upon their noticings, wonders, and feelings (based on @Laurie_Rubel) can be an excellent way to structure the initial stages of problem-posing. Expanding on noticing and wondering with feeling and acting makes the mathematics more relevant to the student and also focuses on the importance of taking action.

The Question Formulation Technique (QFT) developed by The Right Question Institute (https://rightquestion.org; Rothstein & Santana, 2011) expands on the notice-and-wonder strategy in a different way by offering a simple, yet rigorous, step-by-step process to help students produce, improve, and strategize on how to ask and follow through on *their own* questions (Figure 12.2). You may find this a useful resource to design early stages and structure your overall lesson.

Figure 12.2. Steps of the Question Formulation Technique

Step	Explanation
1. Design a Question Focus	Identify the focus of question formulation.
2. Introduce the Rules	Establish four rules for producing questions: • Ask as many questions as you can. • Do not stop to discuss, judge, or answer the questions. • Write down every question exactly as it is stated. • Change any statement into a question.
3. Introduce the Question Focus and Produce Questions	Present the Question Focus, keeping explanation to a minimum. Students generate a list of questions, following the rules.
4. Improve Questions	Categorize questions as open or closed. Discuss the value of each type of question. Change questions from one type to the other.
5. Prioritize Questions	Prioritization instructions should bring participants back to the central objective. Students share why they selected their priority questions.
6. Discuss Next Steps	How will questions be used? Decisions should align with priority instructions.
7. Reflect	Students can reflect: What did you learn? How can you use what you learned? This step helps students think metacognitively about how they used questioning to learn.

Element 4: Social and Mathematical Understanding

Tasks in this book are grounded in mathematically driven investigations of social contexts. By pursuing questions or concerns of your students (Element 3) and implementing tasks that build upon their own reasoning and problem solving (Element 2), a SJML has strong potential to positively impact your students' learning of mathematics and about the world as well as their social, cultural, and mathematical identities.

An effective SJML must explicitly identify both mathematical and social justice goals (NCTM, 2014). Consider two questions to help establish these goals and assess if they've been achieved (DuFour et al., 2016): "What do you want all students to know and be able to do?" and "How will you know if students learn it?"

These questions emphasize the need for you to critically think about your lesson goals and design of assessment tools. We recommend that you identify goals for your lesson of three distinct types:

1. **Mathematics content**—what we want students to know and be able to do (e.g., see Mathematical Essential Concepts in Appendix C)

2. **Mathematics proficiencies, practices, and processes**—how we want students to show what they know and can do (see Figure 4.1)

3. **Social justice outcomes**—how we want students to demonstrate their understanding of and response to an issue (see Figure 4.2)

To get a handle on your students' understanding of the mathematics content, process, and social justice goals you established for the lesson, we invite you to consider two additional questions from DuFour et al. (2016) in your planning:

1. What will you do if they don't get it? That is, how will you respond when some students do not learn?

2. What will you do if they do? That is, how will you extend the learning for students who are already proficient?

Both questions ask you to consider how you will respond as a result of your formative assessment during the SJML. While many of us have expertise in responding to these questions for mathematics, it is a new landscape for many of us to think about these questions in the context of our social justice lesson goal(s). We offer a couple of recommendations. First, allow yourself to be a learner as you implement SJMLs. In your initial design and implementation of a SJML, focus on the assessment of the social justice goal; don't feel like you have to know how to respond to support students who are not there yet or those who are excelling.

> **Allow yourself to be a learner as you implement these SJMLs.**

Second, we do encourage you to not remain in the mode of only assessing and not responding or adjusting instruction. Therefore, we offer one initial strategy for each of the two questions posed by DuFour et al. (2016). For students who may not yet achieve your social justice learning goal, we encourage you to ask them to tell you more about their reasoning or rationale behind a stance they have taken or a recommendation or decision they have made. Pressing students to articulate their reasoning often allows them to catch contradictions in their own thinking. Many of you are familiar with this same strategy for mathematical learning. Lastly, consider asking students who may be ready for an extension on your social justice goal questions like these: "Might there be a context where you would feel differently?" or "Why might a certain person disagree with your conclusion?" Using questions like these helps build and sustain a Beloved Community, increasing students' self- and social awareness.

> **Ask students to tell you more about their reasoning or rationale behind a stance they have taken or a recommendation or decision they have made.**

Elements 5: Social and Mathematical Investigation

In designing a lesson, establishing the goals and determining how to assess student achievement of them is an important first step. Next is to design the learning experience. The problems posed by students to drive these lessons lend themselves to effectively launch an investigation that promotes reasoning and problem solving (NCTM, 2014). Several instructional frameworks align well with the design of such tasks, such as inquiry-based learning (University of Texas at Arlington, 2017), problem-based learning (Schettino, 2016), or Project-Based Learning (PBL; Buck Institute for Education, n.d.). The resources identified in Figure 12.3 can offer additional information to support your understanding and design of PBL-style investigations for your SJML.

Figure 12.3. Project-Based Learning Resources: Design Elements and Teaching Practices

Resource	Key Features
PBL Works (Buck Institute for Education, n.d.)	**Design Elements** • Challenging problem or question • Sustained inquiry • Authenticity • Student voice and choice • Reflection • Critique and revision • Public product
Project-Based Teaching: How to Create Rigorous and Engaging Learning Experiences (Boss & Larmer, 2018)	**Teaching Practices** • Build the culture • Design and plan • Align to standards • Manage activities • Assess student learning • Scaffold student learning • Engage and coach
Rigorous PBL by Design: Three Shifts for Developing Confident and Competent Learners (McDowell, 2017)	**Designing for Student Confidence and Competence** • Clarity—understanding expectations of learning upfront • Challenge—structuring and sequencing learning through projects • Culture—knowing and acting collectively

An effective investigation driven by both mathematical and social justice goals begins with a problem posed by your students. Often, an effective next step is to structure an opportunity for students to learn more about the problem and the social contexts. Students are encouraged to wonder and ask questions, providing an opportunity for them to create a meaningful and personal connection to the issue.

An important next step is to frame a mathematical inquiry that will begin to answer questions. The QFT helps you structure the initial student questions into this mathematical inquiry. The investigation should conclude with students taking action, a form of a public product (Element 6), in which they determine what should be done with their new knowledge, the results of their learning.

Element 6: Social and Mathematical Reflection

People need opportunities to reflect on experiences in order to solidify insights and create lasting understanding (National Research Council, 2000). So you should use intentional instructional practices that foster forethought as well as self-reflection, or metacognitive tasks. An effective SJML includes opportunities for reflection, which can be prompted by questions such as those in Figure 12.4.

Figure 12.4. Questions That Promote Reflection and Metacognition

Questions That Promote Reflection and Metacognition

1. What are the main ideas of today's lesson?

2. Was anything confusing or difficult?

3. If something isn't making sense, what question might we ask?

4. What is most important to include in the notes I am keeping?

5. What were some of the most interesting discoveries I made while working on this project? About the problem? About myself? About others?

6. Now that it's over, what are my first thoughts about this overall project? Are they mostly positive or negative? If positive, what comes to mind specifically? Negative?

7. What were some of my most challenging moments and what made them so?

8. What were some of my most powerful learning moments and what made them so?

9. How well did my team and I communicate overall?

10. What were some things my teammates did that helped me to learn or overcome obstacles? How did I help others during this process? How do I feel I may have hindered others?

11. What would I do differently if I were to approach a similar problem again?

12. In what moments was I most proud of my efforts?

Characteristics of a task can also promote reflection about mathematics, social issues, and how the two inform one another. These characteristics include ensuring a high level of cognitive demand, the questions posed by peers, and discussion about solution methods or ideas with peers, as orchestrated by you, the teacher (Smith & Stein, 2018). Understated to this point in the design of a SJML, investigation is the intentionality to ensure a discourse-rich classroom (NCTM, 2014). Through that discourse, students reflect on their learning. Because SJML investigations emerge from student questions, they lend themselves to many opportunities for discussion of ideas; plan for how you will facilitate meaningful mathematical discourse (Figures 4.4 and 4.5). A well-designed SJML supports the teacher to consider

- the frequency and type of discourse opportunities,

- the types of questions you will use as you monitor and assess student understanding, and

- the possibilities for emotional or divisive responses about the social justice issue (Figure 4.6).

The strategies you choose to use to support equitable student discourse (Cohen & Lotan, 2015) are critical to how you will position students as competent learners and doers of mathematics and build a Beloved Community.

Here is a final note regarding the importance of reflection. As you listen to how your students respond to the social justice learning goal, you likely will better understand the goal yourself. We encourage you to return to the Social Justice Standards and Social Justice Grades 6–8 outcomes (see Figure 4.2) developed by Learning for Justice (2016) to consider how you might refine SJMLs and specifically the learning goals you've set for your students.

Element 7: Action and Public Product

Gaining a deeper understanding of and awareness about an issue is a personal growth outcome that might be expressed in the way an individual goes about living their lives and interacting with those in the world. An end result of investigating an issue of social injustice is often a deeper understanding and awareness that connects to

- identity—how we view ourselves;

- diversity—how we view others and their perspectives; and

- justice—how we view fairness and unfairness, unequal power relations, and the impact of bias.

Sometimes that understanding and awareness leads to a response or actions that may involve others. It is our position that unless some form of action is included, a SJML misses a key quality of social justice—for students to see themselves as able to have an impact on their world. The Social Justice Standards identified by Learning for Justice (2016; see Appendix D) identify five Action outcomes appropriate for the middle grades (Figure 12.5).

Learning for Justice Social Justice Action Standards, Middle School Outcomes	
Anchor Standard	**Middle School Level Outcome**
Action 16	I am concerned about how people (including myself) are treated and feel for people when they are excluded or mistreated because of their identities.
Action 17	I know how to stand up for myself and for others when faced with exclusion, prejudice, and injustice.
Action 18	I can respectfully tell someone when his or her words or actions are biased or hurtful.
Action 19	I will speak up or take action when I see unfairness, even if those around me do not, and I will not let others convince me to go along with injustice.
Action 20	I will work with friends, family, and community members to make our world fairer for everyone, and we will plan and coordinate our actions in order to achieve our goals.

Source: Reprinted with permission of Learning for Justice, a project of the Southern Poverty Law Center. www.learningforjustice.org

These outcomes ask that you provide opportunities for students to express empathy, recognize their own responsibility, speak up with courage and respect, make principled decisions about when to take a stand against bias, or plan and carry out collective action. A well-designed SJML creates the opportunity for students to use mathematics to take these actions or present their public product. Students can be asked to share how mathematics has helped shape their understanding of the social justice issue or consider how they may need to teach others to better understand the issue through mathematics. The action element, necessary for an effective SJML, will be most relevant and meaningful if it is developed by the students, in collaboration with you. This action step should consider the level of involvement with other students, teachers, administrators, community members, and other stakeholders. Andrew Reardon shared how the use of SJMLs "is a unifying experience and really sets the tone for the class community early in the year for my classes. It is an opportunity for students to identify and begin to solve a problem as a cohesive group."

PAUSE AND REFLECT

Reflect on the seven elements summarized in Figure 5.1. In which elements should you place more deliberate effort to effectively teach a SJML?

Summary

An effective SJML rests on building and sustaining a Beloved Community. We view the question of learning about both mathematics and social issues to be structured as an interactive pair of cycles (Figure 5.1), grounded in equitable mathematics teaching practices in a Beloved Community. The outer cycle emphasizes our role as actors in the world. Our actions often originate from the posing of an authentic problem or question about the world. As we learn about that problem, we then take action on that problem. Taking action and reflecting upon that action often result in a more refined understanding of the problem and sometimes raise new questions. Within this action cycle is a learning cycle. Prompted by the authentic problem, investigation develops understanding, which is enhanced through reflection and further investigation—of a social issue and mathematical topic.

The teacher can utilize students' interests and concerns about their school, community, nation, or world to draw upon this action cycle to engage the learning cycle, one that develops both mathematical learning and better understanding of the social context in which we live. Next we discuss recommendations for *how* you, the teacher, might design your own SJML in such a way that it achieves the cycles of learning and action present in this SJML framework. We begin with, if possible, collaborate!

> **Recommendation: Collaborate.** *Work with colleagues. It is so much easier to develop ideas when you are able to bounce them back and forth. This colleague could be a fellow teacher (at the same school site or a different one), a mathematics coach, a university faculty member, a community member, and so on. Consider who at your school or in your community may be a resource for insight into the social issue, a thought partner on the mathematics, or someone who can give feedback.*

GETTING STARTED

Specific challenge: Writing your own SJML seems daunting in your first approach. In addition to the many instructional concerns addressed earlier in the book, writing your own lesson adds the specific challenge of doing well to support your students to achieve **both mathematical and social justice goals**. In this section, we provide an outline and suggestions from our experience writing lessons of our own. Our aim in students' experience of the lesson is that their generative theme is foregrounded, yet they experience, learn, and value mathematics as crucial to understanding and acting upon that generative theme. Here is a list of the seven steps we recommend you follow in developing a SJML, followed by elaboration of each step:

1. Learn about relevant social injustices.

2. Identify the mathematics.

3. Establish your goals.

4. Determine how you will assess your goals.

5. Create a social justice question for the lesson.

6. Design the student resources for the investigation.

7. Plan for reflection and action.

Step 1: Learn About Relevant Social Injustices

We suggest that your first step in designing a SJML is to seek out potential social justice issues (Element 3 of our SJML framework). Listen to and talk with your students, informally or formally, as a part of instruction.

> **Recommendation: Build on your students' interests.** *We have suggested that effective TMSJ builds from concerns or questions students have about their school, community, nation, or world, what Freire (1970/2000) refers to as generative themes. Get to know your students and their community. What are relevant issues in the community? What specifically are your students interested in? You can find out in various ways, including quick surveys, class discussions, parent connections, and more (see Chapter 2).*

Conversations with your students will teach you their ideas, hopes, doubts, fears, and questions they have about their family, friends, school, community, nation, or world. These conversations help you to identify the social justice topic and problem that can serve as a generative theme for your lesson. Debasmita Basu, who co-developed Lesson 7.2 (*The True Cost of That $29 T-Shirt in the Store Window*), stated:

> *When my team and I started writing our lesson on sweatshops, we started with some knowledge and basic understanding of the issue. As we explored*

more, our perspectives modified. We listened to each other's thoughts on sweatshops, the working conditions there, and our ethical responsibilities towards buying products made in sweatshops. It was indeed rewarding to experience how our thoughts found a common ground and helped us gain an in-depth understanding of the issue.

Your next step will be to learn all you can about the issue. Many issues will have local or community meaning or impact. Your students will identify social issues and potential injustices right in your area that might directly involve your students and community members. Consider reaching out to community organizations that focus on responding to the topic or issue. Attend a meeting and scour their website. Find a member of the organization to speak with. We have rarely found it helpful to ask what mathematics they use to understand or respond to the injustice; however, your deeper understanding of the issue will help you to make that connection. Consider the following from Peggy Nayar, who co-developed Lesson 7.4 (*Smoking and Vaping: Targeting of Marginalized Communities by the Tobacco Industry*):

Sending letters home to families at the start of every unit, informing families not only of the mathematics content, but the social justice topic as well, promoted strong family engagement. It encouraged more family input, [provided] new ways for me as the teacher to look at aspects of topics, [and] gave support for any teachers fearful of trying to teach with a social justice lens. Having a supportive administrator was helpful. Analyzing results and data allowed us to show improvement that could not be negated.

In addition, we recommend reading the news (television, internet, papers, YouTube, etc.) with an eye toward the mathematics in the headlines. For example, the 2021 heatwave on the West Coast makes an excellent context for understanding outliers (Bhatia et al., 2021). Many news outlets post stories to their internet sites, which allows you to share articles and videos with your students.

However, this also reminds us that all sources of information come with bias, whether it be news or other forms of media or information—even the community group. We've encouraged you to collaborate with school colleagues to develop allies in your work. This is an excellent opportunity to involve the school librarian or other teacher to (1) learn more about bias in information sources and (2) explore how to counteract and balance the information you receive as well as present to students. Thanheiser and Koestler (2021) have found that one helpful way to question any source is to take a critical literacy stance and to ask the following questions:

- Who writes the stories?

- Who benefits from the stories?

- Who is missing from the stories?

Our first step still asks you to learn about the questions and concerns of your students. However, rather than immediately pursuing them in conjunction with your students, we encourage you to learn more about the topic on your own first. Our intent is that you've identified a few potential social justice issues from which to build a SJML aligned with the mathematics content you must teach. Sometimes lessons exist that can be modified to adapt to your students' interests. Lesson 8.1 (*Gerrymandering of Voting Districts*) focuses on Missouri but can easily be adapted to any state.

Step 2: Identify the Mathematics

The lesson you design must allow you to go deep with mathematics aligned to the course that you teach, SJML Element 2. You may be able to make the argument that you are focused on the mathematics teaching practices, but in many cases you will be best served to ground your lesson in the content you are responsible to teach.

But what mathematics is involved in the potential issues you've learned from your students in Step 1? We have a few strategies that have worked for us. A first strategy is to begin with the content; as you listen to your students, read the news, and collect potential contexts to explore in the classroom, categorize them by content areas that can be addressed. Often the same context allows for multiple different mathematics topics to be explored (see Lesson 6.2, *Cor(o)ner Stores and Food Apartheid*, which includes addition and subtraction of signed numbers; absolute value; and rate, ratio, and proportion). Collecting and categorizing contexts is one way to build a collection of contexts for various content areas (see Chapters 6–10 in this book). Often, however, it seems harder to think with a particular standard—sometimes the mathematics is so specific, the connection is not apparent. We recommend using the *Cluster* name—a descriptor for a group of related standards, used in the Common Core State Standards as well as many state standards. We have found it easier to ask ourselves, "How might the context of this social issue allow us to 'build new functions from existing functions'?" (NGCBP & CCSSO, 2010) or "How might the context of this social issue allow us to use a similarity transformation to determine a length, area, or volume by using proportional relationships?" (NCTM, 2018).

A second approach that has worked well for us is to look at a few of our favorite, well-designed activities. Often the context of the activity provides insight into how it could be modified or replaced with the context of the generative theme. In this way, much of the rich mathematics is already in place, and we are modifying the context to match students' questions or concerns about a social issue or injustice.

Step 3: Establish Your Goals

Having married the potential for the mathematics you must teach with a context of social injustice that is of interest or concern for your students, you are now ready to begin to identify what you want students to learn. We encourage you to set both mathematics and social justice goals for your lesson, Element 3 of the SJML framework.

Recommendation: Develop the mathematics and the social justice goals to support each other. *This is not easy. Especially for those who teach middle school mathematics courses, you are often tightly constrained to ensure students learn specific district or state content standards. At the same time, you want to allow your students to explore social justice topics that are of interest to them. Thus, you need to look for contexts that (a) are of interest to your students and (b) lend themselves to explore a specific mathematics topic.*

You might choose to write these in a more formal manner, as learning outcomes. The mathematics goals should focus on the mathematics content standards as well as possibly the mathematics teaching practices. Here you should draw upon your local expectations and resources. Plan your social justice goals to build student understanding with regard to the Identity, Diversity, Justice, and/or Action domains. Here, we strongly encourage you to utilize the Learning for Justice Social Justice Standards identified for students in middle grades (see Appendix D).

We encourage you to only write a small number of learning outcomes. While it can be easy to see many connections mathematically and to the Social Justice Standards, when you articulate too many goals it can be difficult to maintain focus on student learning of any of them. For example, Lesson 6.3 (*Billionaire Power*) could be used at many levels for many mathematical purposes and many social justice purposes. However, having a clear focus for the current implementation helps guide the discussions when implementing it. Consider setting only one mathematical and one social justice learning outcome. This will allow you to maintain focus on what students should be learning. We agree with Debasmita Basu that "we should redefine the 'learning of mathematics.' Instead of treating the subject as a gatekeeper of future opportunities, we should help the learners to see the true potential of mathematics, its presence in our lives, and its immense power to understand the world around us."

Recommendation: Balance the mathematics and social justice goals. *This is also not easy. After establishing both the mathematics and the social justice goals, they will need to be balanced during enactment so as to not shortchange either of them.*

Step 4: Determine How You Will Assess Your Goals

To help maintain your focus on student learning of both mathematics and social justice outcomes, next identify how you will summatively and formatively assess student achievement of those goals. While summative sometimes is only thought of for the end-of-unit tests, we suggest that at the end of a lesson segment, a summative assessment is appropriate to the degree to which that lesson segment has come to a close. That is not to say the assessment must be a high-stakes test, or even counted for points. You might find a way to identify the extent to which students achieve the mathematical and social justice learning goals in their final product and the action step they take.

Of course, during instruction you can be intentional to formatively assess what students are learning as well. Consider including a brief check in during a lesson to check on one or more of your learning goals. Plan to pose social justice and mathematics questions; prepare the questions you will use to assess and to advance students' thinking. An exit ticket or similar reflection activity at the end of a class period can also provide good insight for you on the extent you are reaching your goals for the whole class and each student. For example, see Lesson 9.4 (*How Many Meals Can Minimum Wage Buy?*) and Lesson 10.2 (*3D Modeling for Water*), which both include several exit slips.

Step 5: Create a Social Justice Question for the Lesson

Following on the insight provided by analyzing your students' interests through a mathematical lens, you are likely able to shape a question that can drive the investigation, SJML Element 4. And while it seems ideal for the students to identify the question and then you follow their investigation, that is likely to make it very hard for you to achieve the goals you've established for the lesson. Here is where we have found the QFT (Rothstein & Santana, 2011) to be powerful. The method provides an opportunity for your students to develop the questions for the exploration.

In short, the QFT method is useful for you to engage students in developing multiple questions about an issue, and then narrowing them to what seems most interesting and feasible. We have found it to be frequently the case that some of the questions identified by students are very close to the question you have considered in advance, and sometimes the student question is better.

At that point, in your role guiding the discussion, you can identify one of your students' questions as something you think will be both interesting and powerful to better understand, and improve, the issue. The previous student conversation has created an interest and commitment to the many different questions; we have found this technique to be effective in corralling the ideas and interests of many students into one area for further study. Using the QFT can help you build and sustain a Beloved Community and SJMLs that directly relate to your students.

Remember, one aim for your SJML is for your students to see how mathematics can be used to help them understand the issue. The question or questions used to drive the lesson might be explicit or grounded in a mathematical approach to better understanding social injustice. For example, in Lesson 7.2 (*The True Cost of That $29 T-Shirt in the Store Window*), students are prompted to ask additional questions of the context to learn more about the companies producing those shirts.

Step 6: Design the Student Resources for the Investigation

This next step can feel like a daunting task, and we don't mean to minimize it. Developing the resources to support student learning of the mathematical and social justice learning outcomes—through exploration driven by the social justice question—forms the heart of the student investigation, SJML Element 5.

You might begin trying to identify the print and digital resources your students might draw upon to learn more about the social issue, likely a few of the same that you used to better understand the issue. It is our experience that materials with the following qualities are most effective. The resources should

- require no more than 15 minutes of reading time,

- use brief segments of video,

- be directed at an adolescent audience,

- identify common assumptions and misunderstandings about the context, and

- be paired to offer opposing perspectives on a controversial issue.

You will also need to structure the mathematical investigation of the lesson. In our work, we have accomplished this in one of two ways. Sometimes we just rely on our experience with many different curricula and similar mathematical tasks to identify a series of prompts or tasks that allow students to better understand the context through deepening their understanding of mathematics. The immediate challenge we recognize in this approach is how difficult it can be to ensure a high level of cognitive demand. You should structure the lesson for students to explore, reason, create, justify, generalize, and/or apply, rather than having students recall, compute, or calculate before interpreting the meaning in the context of the lesson. The latter approach not only creates more shallow mathematical understanding, but also causes students to lose ownership in the investigation process. For example, in Lesson 6.4 (*Middle School Mathematics to Explore People Represented in Our World and Community*), students could identify decimals, percentages, and fractional parts for different characteristics. Once they do that for the world and their home town, a fraction comparison problem could be situated in comparing that characteristic across the world and the home town. This could lead to including different ways to compare fractions as same-size pieces, same number of pieces, benchmarking, greater number of larger pieces, and equivalent fractions (Thanheiser et al., 2016).

The most success we've had designing the investigation portion of the SJML is to gain inspiration from or modify strong lessons we've used in the past. Using the structure and sequence of questions from an established, quality lesson is often the most reliable way to ensure a lesson that promotes reasoning, problem solving, and quality engagement of your students.

Ensure that when the tasks are implemented there is plenty of opportunity for student-to-student discourse. Often it is most effective to write this into the student resources, or at the very least into the teacher notes. You can provide opportunities for discourse by assigning pairs or small groups to debrief a print or digital resource, collaborate on problem solving, plan for a presentation and discussion of ideas with classmates, and more. Revisit the discourse section of Chapter 4 (especially Figures 4.4, 4.5, and 4.6) to help you plan these opportunities into your investigation. An example of the use of "Dealing With Emotional or Sensitive Conversations" can be seen in Lesson 7.6 (*Health Inequalities: COVID-19 and Other Health Conditions*).

As a final bit of advice, we return to our initial recommendation: if you haven't already been developing the lessons with others, consider asking a colleague for feedback on the first draft of the student resources. You might even try the activity with a small group of students and solicit their feedback.

Step 7: Plan for Reflection and Action

The final element of an effective SJML is to ensure opportunity for reflection and action, SJML Elements 6 and 7. Consider the following questions: What individual, small-group, and whole-class opportunities will you provide for your students to reflect on the lesson and consider possible actions they can take to address an injustice? How will your students share what they have learned about the social justice issue? How might they put to use their mathematics knowledge and skills developed in this lesson?

We find ourselves often getting stuck on the idea of writing a letter or presenting at the school board, city council, community center, and so on. While these are often good options, we have found the five Social Justice Standards in the Action domain (Appendix D) prompt us to consider different ideas for taking action; collective action is not the only option. Students might commit to stand up to the exclusion, prejudice, or bias surrounding the lesson topic. You might support them in deciding what an appropriate and effective response might be.

Better yet, ask your students how they want to take action.

FINAL WORDS

If you have read to this point in the book, you have already begun your journey to integrate social justice issues into your classroom. We encourage you to use and share with others *Middle School Mathematics Lessons to Explore, Understand, and Respond to Social Injustice* as a resource to help as you engage in all phases of preparing to teach a SJML. Our hope is that this book will inspire others to empower their students as learners and doers of mathematics who can use mathematics as a tool to address issues of social injustice. Each of us has a role to play in shaping the future of the mathematics education community, so connect with others who embrace TMSJ, who are using SJMLs in their classroom, who are continuing to learn about the related topics discussed in this book, who are taking action to make a difference, and who are members of a TMSJ community willing to share and learn. We hope this book will begin to enable groups of educators to evaluate and recognize the needs of local areas, create ideas for dismantling injustices, and become more inclusive and accountable for teaching the whole student. We invite you to connect with us through Facebook at Mathematics Lessons to Explore Injustice (https://www.facebook.com/groups/178098736933840) and Twitter @SJMathematics.

APPENDIX A: ADDITIONAL RESOURCES

Additional Examples of Social Justice Mathematics Lessons

Classroom Fruition: http://classroomfruition.blogspot.com/p/social-justice-problems.html

Dingle, M., & Yeh, C. (2021). Toolkit for "Mathematics in context: The pedagogy of liberation." *Learning for Justice.* https://www.learningforjustice.org/magazine/spring-2021/toolkit-for-mathematics-in-context-the-pedagogy-of-liberation

Gutstein, E., & Peterson, B. (Eds.). (2013). *Rethinking mathematics: Teaching social justice by the numbers* (2nd ed.). Rethinking Schools.

Felton-Koestler, M. D., Simic-Muller, K., & Menéndez, J. M. (2017). *Reflecting the world: A guide to incorporating equity in mathematics teacher education.* Information Age.

Harper, F. (n.d.). *Solving world problems: Equity & social justice in mathematics.* https://francesharper.com

Karaali, G., & Khdjavi, L. S. (2019). *Mathematics for social justice: Resources for the college classroom.* MAA Press.

Math and Social Justice: A Collaborative MTBoS Site: https://sites.google.com/site/mathandsocialjustice/curriculum-resources

Ontario Ministry of Education. (n.d.). *Teaching mathematics through a social justice lens.* https://www.dcp.edu.gov.on.ca/en/curriculum/elementary-mathematics/grades/g8-math/strands#strand-a

Osler, J. (2007). *RadicalMath.* http://www.radicalmath.org

Raygoza, M. C. (2016). Striving toward transformational resistance: Youth participatory action research in the mathematics classroom. *Journal of Urban Mathematics Education, 9*(2), 122–152.

Stocker, D. (2006). *Maththatmatters: A teacher resource linking math and social justice.* CCPA Education Project. Canadian Centre for Policy Alternatives.

Stocker, D. (2017). *Maththatmatters 2: A teacher resource linking math and social justice.* CCPA Education Project. Canadian Centre for Policy Alternatives.

Wright, P. (n.d.). Teaching maths for social justice. https://maths-socialjustice.weebly.com

Wright, P. (2016). *Teaching mathematics for social justice: Meaningful projects for the secondary mathematics classroom.* Association of Teachers of Mathematics.

Learn More About Elements of TMSJ (Figure 1.1)

Aguirre, J., Mayfield-Ingram, K., & Martin, D. B. (2013). *The impact of identity in K–8 mathematics: Rethinking equity-based practices.* National Council of Teachers of Mathematics.

Bartell, T. G. (2018). *Toward equity and social justice in mathematics education.* Springer.

Benjamin Banneker Association, Inc. (2017). *Implementing a social justice curriculum: Practices to support the participation and success of African-American students in mathematics.* http://bbamath.org/index.php/2017/11/19/the-benjamin-banneker-social-justice-position-statement/

Boston, M., Dillon, F., Smith, M., & Miller, S. (2017). *Taking action: Implementing effective mathematics teaching practices in Grades 9–12.* National Council of Teachers of Mathematics.

Cohen, E. G., & Lotan, R. A. (2014). *Designing groupwork: Strategies for the heterogeneous classroom* (3rd ed.). Teachers College Press.

Drake, C., Aguirre, J. M., Bartell, T. G., Foote, M. Q., Roth McDuffie, A., & Turner, E. E. (2015). *TeachMath learning modules for K–8 mathematics methods courses.* Teachers Empowered to Advance Change in Mathematics Project. http://www.teachmath.info/

Flores, A., & Malloy, C. (2008). *Mathematics for every student, responding to diversity, Grades 9–12.* National Council of Teachers of Mathematics.

Greer, B., Mukhopadhyay, S., Powell, A. B., & Nelson-Barber, S. (Eds.). (2009). *Culturally responsive mathematics education.* Routledge.

Gutstein, E. (2006). *Reading and writing the world with mathematics.* Routledge.

Horn, I. (2012). *Strength in numbers: Collaborative learning in secondary mathematics.* National Council of Teachers of Mathematics.

NCSM & TODOS: Mathematics for ALL. (2016). *Mathematics education through the lens of social justice: Acknowledgment, actions, and accountability.* https://www.todos-math.org/socialjustice

Smith, M. S., & Stein, M. K. (2011). *5 practices for orchestrating productive mathematics discussions.* National Council of Teachers of Mathematics.

Stinson, D. W., & Wager, A. A. (2012). *Teaching mathematics for social justice: Conversations with educators.* National Council of Teachers of Mathematics.

Su, F., & Jackson, C. (2020). *Mathematics for human flourishing.* Yale University Press.

White, D. Y., Fernandes, A., & Civil, M. (Eds.). (2018). *Access and equity: Promoting high quality mathematics: Grades 9–12*. National Council of Teachers of Mathematics.

Teaching for Social Justice

Cult of Pedagogy. *A collection of resources for teaching social justice.* https://www.cultofpedagogy.com/social-justice-resources

Facing History and Ourselves: https://www.facinghistory.org

Learning for Justice: https://www.learningforjustice.org

New York Collective of Radical Educators (NYCoRE). *Revolutionizing the classroom: Transforming mainstream curriculum into social justice teaching.* http://www.nycore.org/newsite/wp-content/uploads/revolutionizingtheclassroom.pdf

Southern Illinois University, Edwardsville. *Resource list for including social justice issues in curricula.* https://www.siue.edu/education/diversity/resources.shtml

Online Communities Interested in Teaching (Mathematics) for Social Justice

Antiracist Education Now: https://www.facebook.com/ARENWashington

Equity and Social Justice in Mathematics Education—Facebook: https://www.facebook.com/groups/178344199241717/

Graphs in the World

Instagram: @graphsintheworld

Facebook: https://www.facebook.com/graphsintheworld

Teaching on Days After: Dialogue & Resources for Educating Toward Justice—Facebook: https://m.facebook.com/groups/teachingondaysafter

Teaching Social Justice Resource Exchange—Facebook: https://www.facebook.com/groups/831353196972485

Resources for Building Your Own Social Justice Mathematics Lessons

DATAJUSTICE Project: https://datajusticeproject.net

EdGap: http://edgap.org

Gallup: https://www.gallup.com

Gapminder: https://www.gapminder.org

GLSEN: https://www.glsen.org

The Institute for Justice: https://ij.org

New York City Department of Education. (n.d.). *Statistics and social justice.* Retrieved from https://www.weteachnyc.org/resources/resource/high-school-math-statistics-and-social-justice/

O'Neil, C. (2017). *Weapons of math destruction: How big data increases inequality and threatens democracy.* Broadway Books.

Smith, D. J. (2009). *If America were a village: A book about the people of the United States.* Kids Can Press.

Smith, D. J. (2011). *If the world were a village: A book about the world's people.* Kids Can Press Ltd.

Teaching Mathematics for Spatial Justice: http://www.mathforspatialjustice.org

U.S. Census Bureau: https://www.census.gov/en.html

What's Going on in This Graph? (*New York Times*): https://www.nytimes.com/column/whats-going-on-in-this-graph

Available for download at
resources.corwin/TMSJ-MiddleSchool

APPENDIX B: LESSON RESOURCES

Lesson 6.1: Food Apartheid: Graphing and Understanding Access to Healthy Food

- Teacher Resource 1: *Image of Complete Grocery Store Map* (Day 1)
- Teacher Resource 2: *Grocery Store Coordinates*
- Desmos activity: *Map of Lincoln*, to be shared with students (https://bit.ly/3luWAWD)
- Video: "Food Deserts in DC: Let's Talk," from NPR (https://bit.ly/31l2gM1)
- Blog post: "Food Apartheid: Racialized Access to Healthy Affordable Food" by Nina Sevilla, *NRDC Expert Blog*, April 2, 2021 (https://on.nrdc.org/3rteFYP)
- Website: U.S. Department of Agriculture (USDA) Food Access Map (https://bit.ly/32WfX4z)

Lesson 6.2: Cor(o)ner Stores and Food Apartheid

- Worksheet 1: *Cor(o)ner Store: Where Healthy Food Comes to Die*
- Worksheet 2: *Grocery Trek*
- Worksheet 3: *A Call to Action*
- Student Resource 1: *Task Card*
- Teacher Resource 1: *Grocery Trek Optional Scaffolds and Extensions*
- Teacher Resource 2: *Additional Lesson Resources*
- Video: "Verify: Living in a Food Desert" (https://bit.ly/3dcIRzh)
- Video: "Food Deserts in DC," from NPR (https://bit.ly/31l2gM1)
- Article: "'Food Apartheid,' Not 'Desert,'" University of Texas at Austin Campus Environmental Center (https://bit.ly/3G947SL)
- Website: Worldometer, "United States Population" (https://bit.ly/3xNEvIc)
- Website: U.S. Department of Agriculture Economic Research Service, "Food Access Research Atlas" (https://bit.ly/3lr4wbm)

Optional

- Video: "Why Is the 1% So White?" from Francesca Ramsey of MTV Decoded (https://bit.ly/3ld0ixP)
- Documentary series: "Race: The Power of an Illusion" (https://www.racepowerofanillusion.org)
- Article: "The Future of Healing: Shifting From Trauma Informed Care to Healing Centered Engagement" by Shawn Ginwright, May 31, 2018 (https://bit.ly/3Dh3Vze)
- Article: "Why Do Corner Stores Struggle to Sell Fresh Produce?" by Sam Bloch, *The Counter*, February 21, 2019 (https://bit.ly/3xOJNTJ)
- Article and video: "West Oakland Food Desert Blooms With a Single Produce Market," from KTVU (https://bit.ly/3dcOldl)
- Video: "Grown in Oakland," from *The Atlantic* (https://bit.ly/3xL5nlY)

(Continued)

Lesson 6.3: Billionaire Power

- Book: *How Much Is a Million?* by David M. Schwartz (1985)
- Worksheet 1: *Proportional Reasoning*
- Worksheet 2: *Top 10 Billionaires*
- Worksheet 3: *The Billionaire Complex*
- Article: "Uber, Other Gig Companies Spend Nearly $200 Million to Knock Down an Employment Law They Don't Like – and It Might Work," by Faiz Siddiqui, *Washington Post*, October 26, 2020 (https://wapo.st/3EvyhiA)
- Article: "Uber and Lyft Used Sneaky Tactics to Avoid Making Drivers Employees in California, Voters Say. Now, They're Going National," by Faiz Siddiqui and Nitasha Tiku, *Washington Post*, November 17, 2020 (https://wapo.st/3plyEGm)

Lesson 6.4: Middle School Mathematics to Explore People Represented in Our World and Community

- Teacher Resource 1: *Additional Resources*
- Book: *If the World Were a Village: A Book About the World's People* (2nd ed.) by David J. Smith (2011)
- Website: "100 People: A World Portrait" statistics (https://bit.ly/3G48v5w)
- Video: "100 People: A World Portrait," from 100 People (https://www.100people.org/the-100-people-project-an-introduction/)
- Website: Ed-Data (https://www.ed-data.org/)
- Website: CIA "The World Factbook" (https://www.cia.gov/the-world-factbook/)

Lesson 7.1: Hey Google, Who's a Mathematician?

- Worksheet 1: *Who's a Mathematician?*

Websites

- Google Image search (https://images.google.com)
- Not Just White Dude Mathematicians, by Dr. Annie Perkins (https://bit.ly/3lxaDuN)
- Mathematically Gifted and Black, by Dr. Erica Graham, Dr. Raegan Higgins, Dr. Candice Price, and Dr. Shelby Wilson (https://mathematicallygiftedandblack.com)
- Lathisms, by Dr. Pamela E. Harris, Dr. Alicia Prieto-Langarica, and Dr. Luis Sordo Vieira (https://www.lathisms.org)
- Mathematicians of the African Diaspora, originated by Scott Williams and supported by the National Association of Mathematicians and the Educational Advancement Foundation (https://www.mathad.com/home)
- Indigenous Mathematicians, founded by Kyle Dahlin, Rebecca Garcia, Ashlee Kalauli, Marissa Loving, and Kamuela Yong (https://www.indigenousmathematicians.org/)

Lesson 7.2: The True Cost of That $29 T-Shirt in the Store Window

- Worksheet 1: *How Should Money Be Allocated?*
- Worksheet 2: *How Is Money Allocated?*
- Worksheet 3: *What Is Fair?*
- Video: "True Cost Clothing Industry" (https://francesharper.com/clothing-industry-video/)
- Website: Educational Video Center (https://evc.org/)
- Website: Two Dollar Challenge (http://twodollarchallenge.org/our-story/)

Lesson 7.3: Majority and Power

- PowerPoint resource to support lesson facilitation
- Video: "All About the Supreme Court," from *History Channel* (https://bit.ly/32L70La)
- Video: "What Is the Legislative Branch of the U.S. Government?" from *History Channel* (https://bit.ly/3rx103a)
- Website, Supreme Court, "About the Court" (https://bit.ly/3rBYSqZ)
- Website, U.S. Senate, "Party Division" (https://bit.ly/3ElHeLo)
- Online collaboration resource: Mentimeter (https://www.mentimeter.com/) or Google Slides

Lesson 7.4: Smoking and Vaping: Targeting of Marginalized Communities by the Tobacco Industry

- Teacher PowerPoint resource to support lesson facilitation
- Teacher Resource 1: *Vaping Facts With Numbers From TheTruth.com*
- Teacher Resource 2: *Proportional Relationships Project*
- Student PowerPoint: *Truth Fact Activity*
- Worksheet 1: *Constant of Proportionality Guided Notes*
- Worksheet 2: *Constant of Proportionality*
- Video: "Read Between the Lies: Black Lives/Black Lungs," from Lincoln Mondy (https://bit.ly/3DjfAxr) and others from The Truth.com (https://www.thetruth.com/)

Lesson 7.5: Health, Race, and Ratios

- PowerPoint resource to support lesson facilitation
- Excel graphs
- Student Resource 1: *Taking Action Planning Sheet* (1 per group of students)
- Student Resource: *Hundreds Grids* (cut the grids out so that each student has at least 1 hundreds grid)
- Website: COVID Tracking Project, "Data Summary" (https://bit.ly/3lwTpO3)
- Website: COVID Tracking Project, "Racial Data Tracker" (https://covidtracking.com/race)
- Video: "Kids Can Change the World," from Matt and Jake Webb, TEDX (https://bit.ly/31w1mfy)
- Video: "How to Change the World," from Kid President (https://bit.ly/3ly98wi)
- Article: "Pandemic Report," by Brian S. McGrath, *Time for Kids*, June 23, 2020 (https://bit.ly/3xNsaUz)
- Article: "Cases on the Rise," by Ellen Nam, *Time for Kids*, July 7, 2020 (https://bit.ly/3lv2lyc)
- Article: "Education Update," by Allison Singer, *Time for Kids*, August 5, 2020 (https://bit.ly/3ptEupk)
- Article: "Race to a Vaccine," by Allison Singer, *Time for Kids*, September 18, 2020 (https://bit.ly/2ZNA9nR)

Lesson 7.6: Health Inequalities: COVID-19 and Other Health Conditions

- Teacher Resource 1: *Health Cases Activity Sheet*
- Teacher Resource 2: *Notes for Article Readings*
- Teacher Resource 3: *Excel File With Sample Solutions to Worksheet 4*
- Worksheet 1: *Health Cases*
- Worksheet 2: *Exploring Ratios in Context*
- Worksheet 3: *Health Disparities Abstract Readings*
- Worksheet 4: *Examining the Number of Deaths in Georgia*
- Student Resource: *Excel File With Data for Worksheet 4* (optional)

(Continued)

(Continued)

Articles

- "COVID-19 Pandemic Highlights Racial Health Inequities," by Crystal Johnson-Mann et al., *The Lancet,* July 10, 2020 (https://bit.ly/3ohNIoY)

- "COVID-19 and the Widening Gap in Health Inequity," by Helene J. Krouse, *Otolaryngology–Head and Neck Surgery*, May 5, 2020 (https://bit.ly/3xNIkya)

- "Racial Disparity of Coronavirus Disease 2019 in African American Communities," by Ravina Kullar et al., *Journal of Infectious Diseases*, September 15, 2020 (https://bit.ly/3lybpHQ)

- "COVID-19 and Health Disparities: The Reality of 'the Great Equalizer,'" by Stephen A. Mein, *Journal of General Internal Medicine*, May 14, 2020 (https://bit.ly/2ZNeg8a)

Lesson 8.1: Gerrymandering of Voting Districts

- Worksheet 1: *Gerrymandering of Voting Districts Task*

- Worksheet 2: *Considering the Fairness of Missouri's Voting District Boundaries*

- PowerPoint resource to support lesson facilitation

- Teacher Resource 1: *Gerrymandering of Voting Districts Task (Answer Key)*

- Video: "Explaining the Efficiency Gap," from WNYC (https://bit.ly/3rxYMRa)

Lesson 8.2: National Team Pay Investigation

- Worksheet 1: *Project: U.S. National Team Pay Investigation*

- Teacher Resource 1: *Answer Key for Worksheet 1*

- Video: "American Soccer's Gender Wage Gap," from *The Daily Show* (https://bit.ly/3xQCuLB)

- Article: "Data: How Does the U.S. Women's Soccer Team Pay Compare to the Men?" by Laura Santhanam, *PBS NewsHour*, March 31, 2016 (https://to.pbs.org/3lrB6tG)

- Article: "Judge Says U.S. Women's Soccer Team Bound by No-Strike Clause," *ESPN*, June 3, 2016 (https://es.pn/3ppP0h4)

- Article: "100 Women: Is the Gender Pay Gap in Sport Really Closing?" by Valeria Perasso, *BBC News*, October 23, 2017 (https://bbc.in/3oj6lco)

Lesson 8.3: The Black Vote in America: Impact of the 1965 Voting Rights Act

- Worksheet 1: *Impact of the 1965 Voting Rights Act: Student Worksheet*

- Teacher Resource 1: *Options for Launch Activity to Introduce the Lesson*

- PowerPoint resource to support lesson facilitation

- Article: "The States Where Efforts to Restrict Voting Are Escalating," by Alex Samuels, Elena Mejía, and Nathaniel Rakich, *FiveThirtyEight*, March 29, 2021 (https://53eig.ht/31oMI9V)

Lesson 9.1: Playing With Data

- PowerPoint resource to support lesson facilitation

- Student Resource 1: *Selected States' Data*

- Student Resource 2: *Data Representation Analysis, Creation, and Evaluation Tool*

- Teacher Resource 1: *Alternative Launch Activity for Lesson 2*

- Teacher Resource 2: *Additional Examples of Data Representations and Data Sets*

Optional Web-Based Resources for Taking Action

- Article: "Nature-Inspired Techniques Could Help Cities Become Greener," by Lucy Handley, *CNBC*, October 22, 2020 (https://cnb.cx/3DITnPf)

- Website: Tableau Public, "Viz of the Day" (https://tabsoft.co/3lwnmxU)

- Website: Citizen Science (https://www.citizenscience.gov/#)

- Website: Zooniverse (https://www.zooniverse.org/projects)

Lesson 9.2: The Mathematics of Toxic Air Emissions

- Worksheet 1: *Emissions*
- Video: "Landsman Requests EPA Air Quality Study of Winton Terrace," from WCPO 9, YouTube (https://bit.ly/3DgkBGV)
- Website: AirNow air quality data (https://www.airnow.gov/)

Lesson 9.3: Gender Pay Gap

- Student Resource 1: *Gender Pay Data Sheet*
- Student Resource 2: *Gender Pay Gap Extension Activity*
- Student Resource 3: *Gallery Review Student Prompts*
- Teacher Resource 1: *Gender Pay Gap Extension Activity Teacher Key*
- Teacher Guide: Gender Pay Gap Desmos Exploration (https://bit.ly/31tqaV9)

Lesson 9.4: How Many Meals Can Minimum Wage Buy?

- Teacher Resource 1: *Exit Slips*
- Worksheet 1: *Minimum Wage and Cost of Living*
- Worksheet 2: *Federal Minimum Wage*
- Worksheet 3: *Reflection*
- Website: Desmos.com graphing calculator
- Website: American Institute of Economic Research online cost of living calculator (https://bit.ly/31sLIXG)
- Website: FRED economic data, "Federal Minimum Wage Graph" (https://bit.ly/3I9ZDxf)
- Video: "When Can You Afford More Big Macs? In 1968 or Now?" TikTok video by @genypastor shown on YouTube (https://bit.ly/3rx4geE)

Lesson 10.1: Map Projections

- Teacher PowerPoint resource to support lesson facilitation
- Student PowerPoint resource to support student exploration
- Interactive Website: The True Size (http://thetruesize.com)
- Interactive Website: Metrocosm, "Eight Ways of Projecting the World" (https://bit.ly/3ojYUBs)
- Video: "Why All World Maps Are Wrong," from *Vox*, December 2, 2016 (https://bit.ly/3Dq69vX)
- Video: "Gall–Peters Projection" from "Somebody's Going to Emergency, Somebody's Going to Jail," *The West Wing*, season 2, episode 16, NBC, February 28, 2001 (https://bit.ly/3pHm8BI)

Lesson 10.2: 3D Modeling for Water

- Worksheet 1: *Is It Natural?*
- Worksheet 2: *Info Sheet*
- Worksheet 3: *Water Containers: Individual Ideas*
- Worksheet 4: *Water Containers: Plan*
- Teacher Resource 1: *Introduction to Tinkercad* (introductory lesson for software, with additional resources)

(Continued)

(Continued)

Videos

- "Why America's Drinking Water Crisis Goes Beyond Flint," from BBC News (https://bbc.in/3DlOWWk)
- "The Sustainable Development Goals Explained: Clean Water and Sanitation," from the United Nations (https://bit.ly/3rz7rCQ)
- "Water Walk" featuring a poem by Martin Kiszko, from WaterAid (https://bit.ly/3DnUY7b)
- "Why Climate Change Makes Storms Stronger," from Yale Climate Connections (https://bit.ly/3GhAfUB)

Websites

- United Nations, "Water Scarcity" (https://bit.ly/3EmccmR)
- Centers for Disease Control and Prevention, "Global Water, Sanitation, & Hygiene (WASH) Fast Facts" (https://bit.ly/3xSiI25)
- Connecticut Democracy Center, "Writing to Your Elected Officials" (https://bit.ly/3xRfP1A)
- Charity Navigator (https://www.charitynavigator.org/)

Articles and Reports

- "When It Comes To Access to Clean Water, 'Race Is Still Strongest Determinant,' Report Says," by Nicole Acevedo, *NBC News*, November 27, 2019 (https://nbcnews.to/2ZU2TLS)
- "Watered Down Justice," by Kristi Pullen Fedinick et al., *NRDC*, March 27, 2020 (https://on.nrdc.org/3Dm2Upp)

Lesson 10.3: Water Is Life—Our Collective Past, Present, and Future

- PowerPoint resource to support lesson facilitation
- Worksheet 1: *My Water Usage*
- Worksheet 2: *Exploring Maps and Water in Flint, Michigan*
- Student Resource 1: *Exploring Maps*

Videos (for Teacher to Show)

- "Fan Video," from Kyle Alex Brett, featuring Beyoncé and Kendrick Lamar's song "Nile" (https://bit.ly/31iFbty)
- "Water Is Life," music video from Will Evans (https://bit.ly/32YKWwT)
- "Water Tank," by Dan Meyer, WCYDWT (https://bit.ly/3ojpSJw)
- "5 Years Into Water Crisis, Little Miss Flint Hasn't Given Up," featuring Mari Copeny, *ABC News* (https://abcn.ws/3lsl6k2) (optional)
- "Flint's Water Crisis, Explained in 3 Minutes," explained by Joe Posner, *Vox* (https://bit.ly/3rzsVQ2)

Websites

- U.S. Geological Survey Online Calculator, "How Much Water Do You Use at Home?" (https://on.doi.gov/3IiDSLB) (1 link per student)
- Flint Service Line Map-Interactive (https://flintpipemap.org/map)
- "Treaties Still Matter—The Dakota Access Pipeline," *Native Knowledge 360°* (https://s.si.edu/3dnUOSx)
- "Clean Water and Sanitation," United Nations (https://bit.ly/3xR8nmW)

- "Flint Water Crisis: Everything You Need To Know," by Melissa Denchak, NRDC (https://on.nrdc.org/3ojtlHX)
- "Flint Water Crisis—Fast Facts," *CNN* (https://cnn.it/3ldY8hD)
- "Flint Map Shows Progress, Reveals Where Lead Likely Remains," by Stacy Woods, *NRDC* (https://on.nrdc.org/3on5elo)

Lesson 10.4: Accessible Playground

- Worksheet 1: *Research Template*
- Worksheet 2: *Math for All Playground*
- Video: "Inclusive Playground for Children With Disabilities Opens in Oakland County," from Priya Mann and Dane Kelly, *ClickOn Detroit*, June 5, 2020 (https://bit.ly/3pj4PWI)
- Video: "A Playground for Everyone, No Matter Your Age or Ability," from Megan Thompson and Melanie Saltzman, *PBS NewsHour*, October 6, 2019 (https://to.pbs.org/3rzcwuP)
- Website: AAA State of Play, "Single Seat Swing Platform With Frame" (https://bit.ly/3lzVEAq)
- Website: AAA State of Play, "Playground Equipment, Handicap Accessible" (https://bit.ly/3dmEtOo)
- Website: Playground Outfitters, "Wheelchair Accessible" (https://bit.ly/3pisxTa)

Lesson 10.5: Investigating Areas to Determine Fairness

- Worksheet 1: *Island Inheritance Task*
- Worksheet 2: *Irregular Areas*
- Worksheet 3: *Equal vs. Equivalent Areas Group Task*
- Worksheet 4: *Gerrymandering Games*
- Worksheet 5: *Equal vs. Equivalent Areas Project*
- Worksheet 6: *Gerrymandering Extension Questions*
- Day After Election Digital Desmos Activity (for Worksheet 3) (https://bit.ly/3lAlJz4)

Online Games

- Polytrope, "District" (http://polytrope.com/district/)
- Game Theory Test, "GerryMander: A Voting District Puzzle Game" (http://gametheorytest.com/gerry/)
- USC Annenberg Center, "The ReDISTRICTING Game" (http://www.redistrictinggame.org/)

Video

- "Gerrymandering: How Drawing Jagged Lines Can Impact an Election," from Christina Greer, *TED-Ed* (https://bit.ly/3xZKREE)

Copies or Digital Access to These News Articles

- "The Math Behind Gerrymandering and Wasted Votes," by Patrick Honner, *Quanta*, October 12, 2017 (https://bit.ly/3GgxVx3)
- "Investigating Gerrymandering and the Math Behind Partisan Maps," by Patrick Honner and Michael Gonchar, *New York Times*, November 30, 2017 (https://nyti.ms/3EtUspK)
- "Gerrymandering Background," NCTM (https://bit.ly/3Eqo9bc)
- "The geeks Who Put A Stop to Pennsylvania Partisan Gerrymandering," by Issie Lapowsky, *Wired*, February 20, 2018 (https://bit.ly/32Z9xBF)
- "The Mathematicians Who Want to Save Democracy," by Carrie Arnold, *Nature*, June 7, 2017 (https://bit.ly/3omiLA6)
- "Rig the Election With Math!" *FiveThirtyEight*, October 28, 2016 (https://53eig.ht/3Ga5ctE)

APPENDIX C: ESSENTIAL MIDDLE GRADES CONCEPTS

CONTENTS

Ratios and Proportional Relationships
- Grade 6: Understand ratio concepts and use ratio reasoning to solve problems.

- Grade 7: Analyze proportional relationships and use them to solve real-world and mathematical problems.

The Number System
- Grade 6: Apply and extend previous understandings of multiplication and division to divide fractions by fractions.

- Grade 6: Compute fluently with multidigit numbers and find common factors and multiples.

- Grade 6: Apply and extend previous understandings of numbers to the system of rational numbers.

- Grade 7: Apply and extend previous understandings of operations with fractions.

- Grade 8: Know that there are numbers that are not rational, and approximate them by rational numbers.

Expressions and Equations
- Grade 6: Apply and extend previous understandings of arithmetic to algebraic expressions.

- Grade 6: Reason about and solve one-variable equations and inequalities.

- Grade 6: Represent and analyze quantitative relationships between dependent and independent variables.

- Grade 7: Use properties of operations to generate equivalent expressions.

- Grade 7: Solve real-life and mathematical problems using numerical and algebraic expressions and equations.

- Grade 8: Work with radicals and integer exponents.

- Grade 8: Understand the connections between proportional relationships, lines, and linear equations.

- Grade 8: Analyze and solve linear equations and pairs of simultaneous linear equations.

Functions
- Grade 8: Define, evaluate, and compare functions.

- Grade 8: Use functions to model relationships between quantities.

Geometry
- Grade 6: Solve real-world and mathematical problems involving area, surface area, and volume.

- Grade 7: Draw, construct, and describe geometrical figures, and describe the relationships between them.

- Grade 7: Solve real-life and mathematical problems involving angle measure, area, surface area, and volume.

- Grade 8: Understand congruence and similarity using physical models, transparencies, or geometry software.

- Grade 8: Understand and apply the Pythagorean Theorem.

- Grade 8: Solve real-world and mathematical problems involving volume of cylinders, cones, and spheres.

Statistics and Probability
- Grade 6: Develop understanding of statistical variability.

- Grade 6: Summarize and describe distributions.

- Grade 7: Use random sampling to draw inferences about a population.

- Grade 7: Draw informal comparative inferences about two populations.

- Grade 7: Investigate chance processes and develop, use, and evaluate probability models.

- Grade 8: Investigate patterns of association in bivariate data.

 Available for download at **resources.corwin/TMSJ-MiddleSchool**

APPENDIX D: SOCIAL JUSTICE TOPICS, STANDARDS, AND GRADE LEVEL OUTCOMES

Social Justice Topic
Ability
Civil Rights and Governmental Laws
Class
Economic Inequality
Environmental Issues
Gender and Sexual Identity
Health Inequality
Immigration
Opportunity
Race and Ethnicity
Resistance and Oppression
Rights and Activism
Faith and Religion
Representation
World Diversity

Social Justice Standards (Learning for Justice, 2016)	
Identity	
1	Students will develop positive social identities based on their membership in multiple groups in society.
2	Students will develop language and historical and cultural knowledge that affirm and accurately describe their membership in multiple identity groups.
3	Students will recognize that people's multiple identities interact and create unique and complex individuals.
4	Students will express pride, confidence, and healthy self-esteem without denying the value and dignity of other people.
5	Students will recognize traits of the dominant culture, their home culture, and other cultures and understand how they negotiate their own identity in multiple spaces.

Social Justice Standards (Learning for Justice, 2016)

Diversity

6	Students will express comfort with people who are both similar to and different from them and engage respectfully with all people.
7	Students will develop language and knowledge to accurately and respectfully describe how people (including themselves) are both similar to and different from each other and others in their identity groups.
8	Students will respectfully express curiosity about the history and lived experiences of others and will exchange ideas and beliefs in an open-minded way.
9	Students will respond to diversity by building empathy, respect, understanding, and connection.
10	Students will examine diversity in social, cultural, political, and historical contexts rather than in ways that are superficial or oversimplified.

Justice

11	Students will recognize stereotypes and relate to people as individuals rather than representatives of groups.
12	Students will recognize unfairness on the individual level (e.g., biased speech) and injustice at the institutional or systemic level (e.g., discrimination).
13	Students will analyze the harmful impact of bias and injustice on the world, historically and today.
14	Students will recognize that power and privilege influence relationships on interpersonal, intergroup, and institutional levels and consider how they have been affected by those dynamics.
15	Students will identify figures, groups, events, and a variety of strategies and philosophies relevant to the history of social justice around the world.

Action

16	Students will express empathy when people are excluded or mistreated because of their identities and concern when they themselves experience bias.
17	Students will recognize their own responsibility to stand up to exclusion, prejudice, and injustice.
18	Students will speak up with courage and respect when they or someone else has been hurt or wronged by bias.
19	Students will make principled decisions about when and how to take a stand against bias and injustice in their everyday lives and will do so despite negative peer or group pressure.
20	Students will plan and carry out collective action against bias and injustice in the world and will evaluate what strategies are most effective.

APPENDIX E: LESSONS BY ESSENTIAL MIDDLE GRADES CONCEPTS, SOCIAL JUSTICE GRADES 6-8 OUTCOMES, AND SOCIAL JUSTICE TOPICS

Lessons by Essential Middle Grades Concepts

Grade	Essential Middle Grades Concepts	Lesson Title
	Number	
6	Apply and extend previous understandings of numbers to the system of rational numbers.	6.2 Cor(o)ner Stores and Food Apartheid 8.1 Gerrymandering of Voting Districts
7	Apply and extend previous understandings of operations with fractions.	6.4 Middle School Mathematics to Explore People Represented in Our World and Community
7	Solve real-world and mathematical problems involving the four operations with rational numbers.	6.3 Billionaire Power
	Ratios and Proportions	
6	Understand ratio concepts and use ratio reasoning to solve problems.	6.2 Cor(o)ner Stores and Food Apartheid 7.1 Hey Google, Who's a Mathematician? 7.2 The True Cost of That $29 T-Shirt in the Store Window 8.1 Gerrymandering of Voting Districts
7	Analyze proportional relationships and use them to solve real-world and mathematical problems.	6.4 Middle School Mathematics to Explore People Represented in Our World and Community 7.1 Hey Google, Who's a Mathematician? 7.2 The True Cost of That $29 T-Shirt in the Store Window 7.4 Smoking and Vaping: Targeting of Marginalized Communities by the Tobacco Industry 9.3 Gender Pay Gap

Grade	Essential Middle Grades Concepts	Lesson Title
	Algebra: Expressions, Equations, and Functions	
6	Apply and extend previous understandings of arithmetic to algebraic expressions.	6.1 Food Apartheid: Graphing and Understanding Access to Healthy Food
7	Solve real-life and mathematical problems using numerical and algebraic expressions and equations.	8.1 Gerrymandering of Voting Districts
8	Analyze and solve linear equations and pairs of simultaneous linear equations.	8.2 National Team Pay Investigation
	Statistics and Probability	
6	Summarize and describe distributions.	9.2 The Mathematics of Toxic Air Emissions 9.3 Gender Pay Gap
7	Draw informal comparative inferences about two populations.	6.4 Middle School Mathematics to Explore People Represented in Our World and Community
8	Investigate patterns of association in bivariate data.	8.3 The Black Vote in America: Impact of the 1965 Voting Rights Act 9.3 Gender Pay Gap
	Geometry	
6	Solve real-world and mathematical problems involving area, surface area, and volume.	10.1 Map Projections 10.2 3D Modeling for Water 10.4 Accessible Playground
7	Draw, construct, and describe geometrical figures, and describe the relationships between them.	10.4 Accessible Playground
7	Solve real-life and mathematical problems involving angle measure, area, surface area, and volume.	10.1 Map Projections 10.2 3D Modeling for Water

Lessons by Social Justice Standards: Grades 6–8 Outcomes

	Social Justice Outcomes, Grades 6–8 (Learning for Justice, 2016)	Lesson Title
	Identity	
3	I know that overlapping identities combine to make me who I am and that none of my group identities on their own fully defines me or any other person.	7.6 Health Inequalities: COVID and Other Health Conditions

(Continued)

(Continued)

	Social Justice Outcomes, Grades 6–8 (Learning for Justice, 2016)	Lesson Title
5	I know there are similarities and differences between my home culture and the other environments and cultures I encounter, and I can be myself in a diversity of settings.	10.3 Water Is Life—Our Collective Past, Present, and Future
	Diversity	
7	I can accurately and respectfully describe ways that people (including myself) are similar to and different from each other and others in their identity groups.	6.4 Middle School Math to Explore People Represented in Our World and Community
8	I am curious and want to know more about other people's histories and lived experiences, and I ask questions respectfully and listen carefully and non-judgmentally.	6.4 Middle School Math to Explore People Represented in Our World and Community 8.2 National Team Pay Investigation 10.4 Accessible Playground
10	I can explain how the way groups of people are treated today, and the way they have been treated in the past, shapes their group identity and culture.	7.1 Hey Google, Who's a Mathematician? 10.3 Water Is Life—Our Collective Past, Present, and Future
	Justice	
12	I can recognize and describe unfairness and injustice in many forms including attitudes, speech, behaviors, practices, and laws.	6.1 Food Apartheid: Graphing and Understanding Access to Healthy Food 7.2 The True Cost of that $29 T-Shirt in the Store Window 7.3 Majority and Power 7.4 Smoking and Vaping: Targeting of Marginalized Communities by the Tobacco Industry 7.5 Health, Race, and Ratios 8.1 Gerrymandering of Voting Districts 8.2 National Team Pay Investigation 8.3 The Black Vote in America: Impact of the 1965 Voting Rights Act 9.1 Playing With Data 9.2 The Mathematics of Toxic Air 9.3 Gender Pay Gap 9.4 How Many Meals Can Minimum Wage Buy?

Social Justice Outcomes, Grades 6–8 (Learning for Justice, 2016)		Lesson Title
		10.1 Map Projections
		10.2 3D Modeling for Water
		10.4 Accessible Playground
		10.5 Investigating Areas to Determine Fairness
13	I am aware that biased words and behaviors and unjust practices, laws and institutions limit the rights and freedoms of people based on their identity groups.	8.1 Gerrymandering of Voting Districts
		8.3 The Black Vote in America: Impact of the 1965 Voting Rights Act
		9.3 Gender Pay Gap
14	I know that all people (including myself) have certain advantages and disadvantages in society based on who they are and where they were born.	6.2 Cor(o)ner Stores and Food Apartheid
		6.3 Billionaire Power
		6.4 Middle School Math to Explore People Represented in Our World and Community
		7.3 Majority and Power
		7.6 Health Inequalities: COVID and Other Health Conditions
		8.2 National Team Pay Investigation
		9.1 Playing With Data
		10.3 Water Is Life—Our Collective Past, Present, and Future
15	I know about some of the people, groups and events in social justice history and about the beliefs and ideas that influenced them.	9.3 Gender Pay Gap
		9.4 How Many Meals Can Minimum Wage Buy?
	Action	
16	I am concerned about how people (including myself) are treated and feel for people when they are excluded or mistreated because of their identities.	7.4 Smoking and Vaping: Targeting of Marginalized Communities by the Tobacco Industry
		9.1 Playing With Data
17	I know how to stand up for myself and for others when faced with exclusion, prejudice, and injustice.	9.2 The Mathematics of Toxic Air
19	I will speak up or take action when I see unfairness, even if those around me do not, and I will not let others convince me to go along with injustice.	7.4 Smoking and Vaping: Targeting of Marginalized Communities by the Tobacco Industry
		7.5 Health, Race, and Ratios
		9.2 The Mathematics of Toxic Air
		9.4 How Many Meals Can Minimum Wage Buy?

(Continued)

(Continued)

	Social Justice Outcomes, Grades 6–8 (Learning for Justice, 2016)	Lesson Title
20	I will work with friends, family and community members to make our world fairer for everyone, and we will plan and coordinate our actions in order to achieve our goals.	6.1 Food Apartheid: Graphing and Understanding Access to Healthy Food 6.2 Cor(o)ner Stores and Food Apartheid 6.3 Billionaire Power 7.2 The True Cost of that $29 T-Shirt in the Store Window 7.3 Majority and Power 8.2 National Team Pay Investigation 10.1 Map Projections 10.2 3D Modeling for Water 10.4 Accessible Playground

Lessons by Social Justice Topic

Social Justice Topic	Lesson Title
Ability	10.4 Accessible Playground
Civil Rights and Governmental Laws	7.3 Majority and Power 8.1 Gerrymandering of Voting Districts 8.2 National Team Pay Investigation 10.5 Investigating Areas to Determine Fairness
Economic Inequality	6.3 Billionaire Power 8.3 The Black Vote in America: Impact of the 1965 Voting Rights Act 9.3 Gender Pay Gap 9.4 How Many Meals Can Minimum Wage Buy?
Environmental Issues	9.2 The Mathematics of Toxic Air 10.3 Water Is Life—Our Collective Past, Present, and Future
Health Inequality	6.1 Food Apartheid: Graphing and Understanding Access to Healthy Food 6.2 Cor(o)ner Stores and Food Apartheid 7.4 Smoking and Vaping: Targeting of Marginalized Communities by the Tobacco Industry 7.5 Health, Race, and Ratios 7.6 Health Inequalities: COVID and Other Health Conditions

Social Justice Topic	Lesson Title
Human Rights	7.2 The True Cost of that $29 T-Shirt in the Store Window 10.2 3D Modeling for Water
Representation	7.1 Hey Google, Who's a Mathematician? 10.1 Map Projections
World Diversity	6.4 Middle School Math to Explore People Represented in Our World and Community

 Available for download at **resources.corwin/TMSJ-MiddleSchool**

APPENDIX F: SOCIAL JUSTICE MATHEMATICS LESSON PLANNER

PART I

CONTEXT

Essential Mathematics Concepts	Mathematics Practices

Social Justice Issue or Standards (Middle School Outcomes)

CONTEXT

Purpose

Audience

Allies

Timing

WHEN (in unit)

Circle one: Beginning Middle End Special Lesson	Number of periods: _____

Benefits

HOW

Circle one: Mathematical Tasks   Three-Act Tasks   Project-Based Learning   Other

OUTCOME/ACTIONS

PART II

Introduction/Engagement:

What will the teacher do?	What will students do?

Investigation/Exploration:

What will the teacher do?	What will students do?

Share and Discuss:

What will the teacher do?	What will students do?

Taking Action:

REFERENCES

Aguirre, J., Mayfield-Ingram, K., & Martin, D. B. (2013). *The impact of identity in K–8 mathematics: Rethinking equity-based practices.* National Council of Teachers of Mathematics.

Anderman, L. H. (2003). Academic and social perceptions as predictors of change in middle school students' sense of school belonging. *Journal of Experimental Education*, 72(1), 5–22. https://doi.org/10.1080/00220970309600877

Bartell, T. G., Wager, A., Edwards, A., Battery, D., Foote, M., & Spencer, J. (2017). Toward a framework for research linking equitable teaching with the standards for mathematical practice. *Journal for Research in Mathematics Education*, 48(1), 7–21.

Benjamin Banneker Association. (2017). *Implementing a social justice curriculum: Practices to support the participation and success of African-American students in mathematics.* http://bbamath.org/index.php/2017/11/19/the-benjamin-banneker-social-justice-position-statement/

Berry, R. Q., III, Conway, B. M., IV, Lawler, B. R., & Staley, J. W. (2020). *High school mathematics lessons to explore, understand, and respond to social injustice.* Corwin.

Bhatia, A., Fountain, H., & Quealy, K. (2021, June 29). How weird is that heatwave in Portland, Seattle, and Vancouver? *New York Times*. https://nyti.ms/3rv2qet

Boaler, J. (2006). Urban success: A multidimensional mathematics approach with equitable outcomes. *Phi Delta Kappan*, 87(5), 364–369.

Boss, S., & Larmer, J. (2018). *Project based teaching: How to create rigorous and engaging learning experiences.* Association for Supervision and Curriculum Development.

Buck Institute for Education. (n.d.). *PBLWorks: Project-based learning for all.* https://www.pblworks.org

Chojnacki, A., Dai, C., Farahi, A., Shi, G., Webb, J., Zhang, D. T., Abernathy, J., & Schwartz, E. (2017). A data science approach to understanding residential water contamination in Flint. Association for Computing Machinery. https://dx.doi.org/10.1145/3097983.3098078

Cochran-Smith, M. (2004). *Walking the road: Race, diversity, and social justice in teacher education.* Teachers College Press.

Cohen, E. G., & Lotan, R. A. (2014). *Designing groupwork: Strategies for the heterogeneous classroom* (3rd ed.). Teachers College Press.

Delgado, R. (1990). When a story is just a story: Does voice really matter? *Virginia Law Review*, 76(1), 95–111. https://doi.org/10.2307/1073104

Deutsch, A. (1944). The first US census of the insane (1840) and its use as pro-slavery propaganda. *Bulletin of the History of Medicine*, 15(5), 469–482. https://www.jstor.org/stable/44446305

Doxtdator, B. (2018). *The humanity of our students isn't up for debate.* https://longviewoneducation.org/humanity-students-isnt-debate/

DuFour, R., DuFour, R., Eaker, R., Many, T. W., & Mattos, M. (2016). *Learning by doing: A handbook for professional learning communities at work* (3rd ed.). Solution Tree.

Emdin, C. (2016). *For white folks who teach in the hood. . . and the rest of y'all too: Reality pedagogy and urban education.* Beacon Press.

Facing History and Ourselves. (n.d.). *Teaching current events educator guide*. https://www
.facinghistory.org/educator-resources/current-events/teaching-current-events-
educator-guide

Featherstone, H., Crespo, S., Jilk, L. M., Oslund, J. A., Parks, A. N., & Wood, M. B.
(2011). *Smarter together! Collaboration and equity in the elementary math classroom*.
National Council of Teachers of Mathematics.

Fennell, F. M., Kobett, B. M., & Wray, J. A. (2017). *The formative 5: Everyday assessment
techniques for every math classroom*. Corwin.

Flippan, A. (2014, October 16). Black Turnout in 1964, and Beyond. *New York Times*.
https://www.nytimes.com/2014/10/17/upshot/black-turnout-in-1964-and-beyond
.html

Frankenstein, M. (1983). Critical mathematics education: An application of Paulo
Freire's epistemology. *Journal of Education, 165*(4), 315–339. https://www.jstor.org/
stable/42772808

Franklin, C., Kader, G., Mewborn, D., Moreno, J., Peck, R., Perry, M., & Scheaffer, R.
(2007). *Guidelines for assessment and instruction in statistics education (GAISE)
report: A preK–12 curriculum framework*. American Statistical Association.

Freire, P. (2000). *Pedagogy of the oppressed* (M. B. Ramos, Trans.). Continuum. (Original
work published 1970)

Garcia, A. A. S. J. (1974). *Generative themes: A critical examination of their nature and
function in Paulo Freire's educational model* (Master's thesis, Loyola University).
https://ecommons.luc.edu/luc_theses/2683

Gholson, M., Buenrostro, P., Mann, L., Gutstein, E., & Hoover, M. (2017). Inside criti-
cal/radical mathematics education: A video exploration. In A. Chronaki (Ed.),
Proceedings of the Ninth International Mathematics Education and Society Conference
(pp. 154–158). University of Thessaly Press.

Ginwright, S. (2016). *Hope and healing in urban education: How urban activists and
teachers are reclaiming matters of the heart*. Routledge.

Ginwright, S. (2018). The future of healing: Shifting from trauma informed care to healing centered engagement. *Medium*. https://ginwright.medium.com/the-future-of-
healing-shifting-from-trauma-informed-care-to-healing-centered-engagement-
634f557ce69c

González, N., Andrade, R., Civil, M., & Moll, L. (2001). Bridging funds of distributed
knowledge: Creating zones of practices in mathematics. *Journal of Education for
Students Placed at Risk, 6*(1–2), 115–132. https://doi.org/10.1207/S15327671ES
PR0601-2_7

Gutstein, E. (2006). *Reading and writing the world with mathematics: Toward a pedagogy
of social justice*. Routledge.

Guzmán, L. D., & Craig, J. (2019). The world in your pocket: Digital media as invitations
for transdisciplinary inquiry in mathematics classrooms. *Occasional Paper Series,
2019*(41), 6. https://educate.bankstreet.edu/occasional-paper-series/vol2019/iss41/6

Hackett, M., Wood, M. B., Picazo, C., Stautman, S., Valentine, J., & Wheeler, G. (2020,
January 25). *Developing a collaborative learning environment for your mathematicians*. Mathematics Educator Appreciation Day Conference, Tucson, AZ.

Hardré, P. L., Ling, C., Shehab, R. L., Nanny, M. A., Nollert, M. U., Refai, H., . . . &
Wollega, E. D. (2013). Teachers in an interdisciplinary learning community:
Engaging, integrating, and strengthening K–12 education. *Journal of Teacher
Education, 64*(5), 409–425. https://doi.org/10.1177/0022487113496640

Harper, F. (2019). A qualitative metasynthesis of teaching mathematics for social justice in action: Pitfalls and promises of practice. *Journal for Research in Mathematics Education, 50*(3), 268–310.

Harvey, M. R. (1996). An ecological view of psychological trauma and trauma recovery. *Journal of Traumatic Stress, 9*(1), 3–23. https://doi.org/10.1002/jts.2490090103

Hoffer, W. W. (2016). *Developing literate mathematicians: A guide for integrating language and literacy instruction into secondary mathematics.* National Council of Teachers of Mathematics.

Horn, I. (2012). *Strength in numbers: Collaborative learning in secondary mathematics.* National Council of Teachers of Mathematics.

Horn, I. S. (2017). *Motivated.* Heinemann.

Huinker, D., & Bill, V. (2017). *Taking action: Implementing effective mathematics teaching practices, Grades K–5.* National Council of Teachers of Mathematics.

Jansen, A. (2020). *Rough draft math: Revising to learn.* Stenhouse Publishers.

Jansen, A., Cooper, B., Vascellaro, S., & Wandless, P. (2017). Rough-draft talk in mathematics classrooms. *Mathematics Teaching in the Middle School, 22*(5), 304–307. https://doi.org/10.5951/mathteacmiddscho.22.5.0304

Johnson-Mann, C., Hassan, M., & Johnson, S. (2020, August). COVID-19 pandemic highlights racial health inequities. *Lancet Diabetes and Endocrinology, 8*(8), 663–664. https://doi.org/10.1016/S2213-8587(20)30225–4

Kahne, J., Middaugh, E., & Allen, D. (2015). Youth, new media, and the rise of participatory politics. In D. Allen & J. S. Light (Eds.), *From voice to influence: Understanding citizenship in a digital age* (pp. 35–58). University of Chicago Press.

The King Center. (n.d.). *The Beloved Community.* https://thekingcenter.org/about-tkc/the-king-philosophy/

Kokka, K. (2019). Healing-informed social justice mathematics: Promoting students' sociopolitical consciousness and well-being in mathematics class. *Urban Education, 54*(9), 1179–1209. https://doi.org/10.1177/0042085918806947

Kokka, K. (2020). Social justice pedagogy for whom? Developing privileged students' critical mathematics consciousness. *The Urban Review, 52,* 778–803. https://doi.org/10.1007/s11256-020–00578-8

Krouse, H. J. (2020, July). COVID-19 and the widening gap in health inequity. *Otolaryngology–Head and Neck Surgery, 163*(1), 65–66. https://doi.org/10.1177/0194599820926463

Krupar, S. (2015). Map power and map methodologies for social justice. *Georgetown Journal of International Affairs, 16,* 91. https://www.jstor.org/stable/43773699

Kullar, R., Marcelin, J. R., Swartz, T. H., Piggott, D. A., Macias Gil, R., Mathew, T. A., & Tan, T. (2020). Racial disparity of coronavirus disease 2019 in African American communities. *Journal of Infectious Diseases, 222*(6), 890–893. https://doi.org/10.1093/infdis/jiaa372

Ladson-Billings, G. (1995). But that's just good teaching! The case for culturally relevant pedagogy. *Theory Into Practice, 34*(3), 159–165. https://doi.org/10.1080/00405849509543675

Ladson-Billings, G. (2009). *The dreamkeepers: Successful teachers of African American children.* Jossey-Bass. (Original work published 1994)

Ladson-Billings, G. (2019, April). *Are we still solving for X? The pedagogical practices limiting student success in mathematics.* Opening keynote presentation at the National Council of Teachers of Mathematics 2019 Annual Meeting & Exposition, San Diego, CA.

Ladson-Billings, G. (2021). I'm here for the hard re-set: Post pandemic pedagogy to preserve our culture. *Equity & Excellence in Education, 54*(1), 68–78. https://doi.org/10.1080/10665684.2020.1863883

Ladson-Billings, G., & Tate, W. (1995). Toward a critical race theory of education. *Teachers College Record, 97*(1), 47–68. https://doi.org/10.1177/016146819509700104

Larnell, G. V., Bullock, E. C., & Jett, C. C. (2016). Rethinking teaching and learning mathematics for social justice from a critical race perspective. *Journal of Education, 196*(1), 19–29. https://doi.org/10.1177/002205741619600104

Lawler, B. R. (2012). The fabrication of knowledge in mathematics education: A postmodern ethic toward social justice. In A. Cotton (Ed.), *Towards an education for social justice: Ethics applied to education.* Peter Lang.

Learning for Justice. (2016). Social justice standards: The anti-bias teaching tolerance framework. https://www.learningforjustice.org/

Levaasseur, K., & Cuoco, A. (2003). Mathematical habits of mind. In H. L. Schoen (Ed.), *Teaching mathematics through problem solving: Grades 6–12* (pp.27–37). Reston, VA: NCTM.

Lounsbury, J. H. (2010). This we believe: Keys to educating young adolescents. *Middle School Journal, 41*(3), 52–53.

McDowell, M. (2017). *Rigorous PBL by design: Three shifts for developing confident and competent learners.* Corwin.

McGee, E. O. (2020). *Black, brown, bruised: How racialized STEM education stifles innovation.* Harvard Education Press.

Mein, S. A. (2020). COVID-19 and health disparities: The reality of "the Great Equalizer." *Journal of General Internal Medicine, 35*(8), 2439–2440. https://doi.org/10.1007/s11606-020–05880-5

Menakem, R. (2014). *My grandmother's hands: Racialized trauma and the pathway to mending our hearts and bodies.* Central Recovery Press.

Michaels, S., O'Connor, C., & Resnick, L. B. (2008). Deliberative discourse idealized and realized: Accountable talk in the classroom and in civic life. *Studies in Philosophy and Education, 27*(4), 283–297. http://doi.org/10.1007/s11217-007–9071-1

Moses, R. P., & Cobb, C. E. (2001). *Radical equations: Civil rights from Mississippi to the Algebra Project.* Beacon Press.

NCSM & TODOS: Mathematics for ALL. (2016). *Mathematics education through the lens of social justice: Acknowledgment, actions, and accountability.* https://www.todos-math.org/socialjustice

National Council of Teachers of Mathematics. (1989). *Curriculum and evaluation standards for school mathematics.* Author.

National Council of Teachers of Mathematics (2000). *Principles and standards for school mathematics.* NCTM.

National Council of Teachers of Mathematics. (2014). *Principles to actions: Ensuring mathematical success for all.* Author.

National Council of Teachers of Mathematics. (2018). *Catalyzing change in high school mathematics: Initiating critical conversations.* Author.

National Council of Teachers of Mathematics. (2020). *Catalyzing change in middle school mathematics: Initiating critical conversations.* Author.

National Governors Association Center for Best Practices, Council of Chief State School Officers. (2010). *Common core state standards: Mathematics.* Author.

National Research Council. (2000). *How people learn: Brain, mind, experience, and school.* National Academies Press.

National Research Council. (2001). *Adding it up: Helping children learn mathematics.* J Kilpatrick, J. Swafford, and B. Findell (Eds.). Mathematics Learning Study Committee, Center for Education, Division of Behavioral and Social Sciences and Education. Washington, DC: National Academy Press.

A Pathway to Equitable Math Instruction. (2021). *SEADTheme guidebook: Belonging.* https://equitablemath.org/

Powell, A. B. (1995). Critical mathematics: Observations on its origins and pedagogical purposes. In Y. M. Pothier (Ed.), *Proceedings of 1995 annual meeting of the Canadian Mathematics Education Study Group* (pp. 103–116). Mount Saint Vincent University Press.

Pulido, I. B., Miglietta, A., Cortez, G. A., Stovall, D., & Aviles de Bradley, A. (2013). Re-framing, re-imagining, and re-tooling curricula from the grassroots: The Chicago Grassroots Curriculum Taskforce. *Current Issues in Comparative Education, 15*(2), 84–95. https://files.eric.ed.gov/fulltext/EJ1016202.pdf

Raygoza, M. C. (2016). Striving toward transformational resistance: Youth participatory action research in the mathematics classroom. *Journal of Urban Mathematics Education, 9*(2), 122–152.

Rothstein, D., & Santana, L. (2011). *Make just one change: Teach students to ask their own questions.* Harvard Education Press.

Rubel, L. (2017). Equity-directed instructional practices. *Journal of Urban Mathematics Education, 10*(2), 66–105. https://doi.org/10.21423/jume-v10i2a324

Schettino, C. (2016, September 26). Aspects of problem-based teaching. *Mathematics Teacher.* https://www.nctm.org/Publications/MT-Blog/Blog/Aspects-of-Problem-Based-Teaching/

Seda, P., & Brown, K. (2021). *Choosing to see: A framework for equity in the math classroom.* Dave Burgess Consulting Inc.

Seider, S., Clark, S., & Graves, D. (2020). The development of critical consciousness and its relation to academic achievement in adolescents of color. *Child Development, 91*(2), e451–e474. https://doi.org/10.1111/cdev.13262

Shirude, S. (2019, October 21). Math is life. Life is a story. So why aren't we telling stories in math class? *Woke Math.* https://esmathteacher.wordpress.com/blog-feed/

Silver, E. A., & Mills, V. L. (2018). *A fresh look at formative assessment in teaching mathematics.* National Council of Teachers of Mathematics.

Skovsmose, O. (1994). *Towards a philosophy of critical mathematical education.* Kluwer Academic.

Smith, D. J. (2011). *If the world were a village: A book about the world's people* (2nd ed.). Kids Can Press.

Smith, M. S., & Stein, M. K. (2018). *5 practices for orchestrating productive mathematics discussions* (2nd ed.). National Council of Teachers of Mathematics.

Steele, D. M., & Cohn-Vargas, B. (2013). *Identity safe classrooms, Grades K–5: Places to belong and learn.* Corwin.

Tate, W. F. (2013). Race, retrenchment, and the reform of school mathematics. In E. Gutstein & B. Peterson (Eds.), *Rethinking mathematics: Teaching social justice by the numbers* (2nd ed., pp. 42–51). Rethinking Schools.

Thanheiser, E., & Koestler, C. (2021). If the world were a village: Learning mathematics while learning about the world. *Mathematics Teacher Educator, 9*(3), 202–228. https://doi.org/10.5951/MTE.2020.0021

Thanheiser, E., Olanoff, D., Hillen, A., Feldman, Z., Tobias, J. M., & Welder, R. M. (2016). Reflective analysis as a tool for task redesign: The case of prospective elementary teachers solving and posing fraction comparison problems. *Journal of Mathematics Teacher Education, 19*(2–3), 123–148. https://doi.org/10.1007/s10857-015–9334-7

TODOS: Mathematics for ALL. (2021). *The mo(ve)ment to prioritize antiracist mathematics: Planning for this and every school year.* https://www.todos-math.org/assets/images/The%20Movement%20to%20Prioritize%20Antiracist%20Mathematics%20final%203.0_v6.pdf

Turner, E., Aguirre, J., Drake, C., Bartell, T. G., Roth McDuffie, A., & Foote, M. Q. (2015). Community mathematics exploration module. In C. Drake et al. (Eds.), *TeachMath learning modules for K–8 mathematics methods courses.* Teachers Empowered to Advance Change in Mathematics Project. http://www.teachmath.info

University of Texas at Arlington. (2017). *How inquiry-based learning can work in a math classroom.* http://bit.ly/2mtkma1 (Updated 2021)

U.S. Census Bureau. (2020). *American Community Survey: Children characteristics.* https://data.census.gov/cedsci/table?q=children%20with%20disabilities&tid=ACSST1Y2019.S0901&hidePreview=false

Vasquez, V. M. (2016). *Critical literacy across the K–6 curriculum.* Taylor & Francis.

Vasquez, V. M., Janks, H., & Comber, B. (2019). Critical literacy as a way of being and doing. *Language Arts, 96*(5),300–311.

Wigfield, A., & Eccles, J. (2002). The development of competence beliefs, expectancies for success, and achievement values from childhood through adolescence. In A. Wigfield & J. Eccles (Eds.), *The development of achievement motivation* (pp. 91–120). John Wiley & Sons.

Wolpow, R., Johnson, M. M., Hertel, R., & Kincaid, S. O. (2009). *The heart of learning and teaching: Compassion, resiliency, and academic success.* Office of the Superintendent of Public Instruction, Compassionate Schools.

INDEX

**JENNIFER M. BAY-WILLIAMS,
JOHN J. SANGIOVANNI, ROSALBA SERRANO,
SHERRI MARTINIE, JENNIFER SUH**

Because fluency is so much more
than basic facts and algorithms

Grades K–8

**KAREN S. KARP,
BARBARA J. DOUGHERTY,
SARAH B. BUSH**

A schoolwide solution for students'
mathematics success

Elementary, Middle School, High School

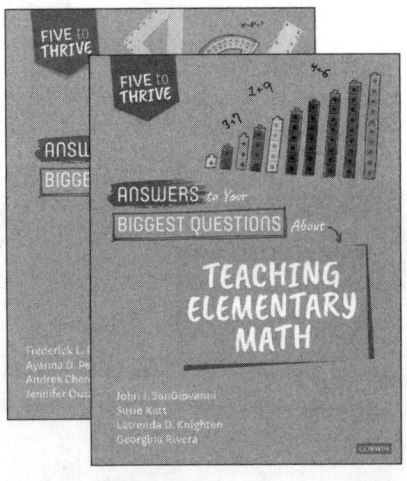

**JOHN J. SANGIOVANNI, SUSIE KATT,
LATRENDA D. KNIGHTEN, GEORGINA RIVERA,
FREDERICK L. DILLON, AYANNA D. PERRY,
ANDREA CHENG, JENNIFER OUTZS**

Actionable answers to your most pressing questions
about teaching elementary and secondary math

Elementary, Secondary

**SARA DELANO MOORE,
KIMBERLY RIMBEY**

A journey toward making
manipulatives meaningful

Grades K–3, 4–8

CORWIN

A SAGE Publishing Company

Helping educators make the greatest impact

CORWIN HAS ONE MISSION: to enhance education through intentional professional learning.

We build long-term relationships with our authors, educators, clients, and associations who partner with us to develop and continuously improve the best evidence-based practices that establish and support lifelong learning.

NATIONAL COUNCIL OF
TEACHERS OF MATHEMATICS

The National Council of Teachers of Mathematics supports and advocates for the highest-quality mathematics teaching and learning for each and every student.